P9-CMV-068

1890
1939

From Challenger to Joint Industry Leader, 1890-1939

A History of Royal Dutch Shell

Joost Jonker & Jan Luiten van Zanden

Published under licence from Boom Publishers, Amsterdam, initiating publishers and publishers of the Dutch edition.

OXFORD
UNIVERSITY PRESS

2007

[1]

From Challenger to Joint Industry Leader, 1890-1939 is one of four volumes of the work entitled *A History of Royal Dutch Shell*, written by a team of four authors associated with Utrecht University, Jan Luiten van Zanden, Stephen Howarth, Joost Jonker and Keetie Sluyterman. It is the result of a research project which was supervised by the Research Institute for History and Culture and coordinated by Joost Dankers.

The other volumes are:

Stephen Howarth and Joost Jonker
Powering the Hydrocarbon Revolution, 1939-1973

Keetie Sluyterman
Keeping Competitive in Turbulent Markets, 1973-2007

Jan Luiten van Zanden
Appendices. Figures and Explanations, Collective Bibliography, and Index, including three DVDs

Contents

Chapter 1
From opposite ends towards a common purpose, 1890-1907

Chapter 2
An enterprise waiting to be made, 1907-1914

Chapter 3
Surviving the crucible, 1914-1919

Chapter 4
At the peak of its power

Chapter 5
The contours of a mature corporation: management and staff relations

Chapter 6
The evolution of technology, research, and marketing

Chapter 7
Taxing times, 1929-1939

OXFORD
UNIVERSITY PRESS

Great Clarendon Street, Oxford OX2 6DP

Oxford University Press is a department of the University of Oxford.
It furthers the University's objective of excellence in research, scholarship,
and education by publishing worldwide in

Oxford New York

Auckland Cape Town Dar es Salaam Hong Kong Karachi
Kuala Lumpur Madrid Melbourne Mexico City Nairobi
New Delhi Shanghai Taipei Toronto

With offices in

Argentina Austria Brazil Chile Czech Republic France Greece
Guatemala Hungary Italy Japan Poland Portugal Singapore
South Korea Switzerland Thailand Turkey Ukraine Vietnam

Oxford is a registered trade mark of Oxford University Press
in the UK and in certain other countries

British Library Cataloguing in Publication Data
Data available

ISBN: 978-0-19-929878-5

General Introduction

The twentieth century has been called the century of oil, and rightly so, because during those hundred years the modern oil industry – which only began in 1859 – grew into a business of great sophistication that underpinned much of the global economy. Its raw material, hydrocarbons in liquid or gaseous form, provided an unrivalled range of products that transformed civilization with astonishing rapidity. In terms of social influence, not least among these products was kerosene or paraffin, the first widely available source of artificial light. Thereafter oil's use and influence became ever more widespread, enabling the revolutionary transport systems of automobile and aeroplane, and massively enhancing the capabilities of sea transport. Oil and gas also gave rise to a chemical industry that radically changed our material culture and environment, both for better and for worse, via plastics, detergents, pesticides, and innumerable other new products. Humankind had never known source materials as flexible and versatile as petroleum (crude unrefined oil) and natural gas; and, quickly and inevitably, military applications were found for their myriad products. In many a war of the twentieth century access to oil was a key to victory, and many a war was fought in order to gain or protect access to oil. Understanding the history of oil is therefore essential for understanding the history of the past century, and the early years of the twenty-first century offer no indication that oil and gas will do anything but continue to increase in strategic importance for the foreseeable future.

The economies of scale inherent to this very capital-intensive industry meant that, almost from the start, it was dominated by big companies. From the 1880s it was controlled by the Standard Oil Trust, created by John D. Rockefeller. From 1902 this enterprise was increasingly challenged by a combination of other interests – Dutch, British, and for a time also French – which came together in 1907 to form the Royal Dutch/Shell Group of Companies (Shell, or the Group), Rockefeller's only major rival at the time. Moreover, in 1911, after a series of new anti-trust laws, the Standard Oil Trust was broken up into its constituent parts. By then Shell's operations were worldwide, whether in exploration, production, transport, refining, or marketing, with companies wholly or partly owned by the Group working in every inhabited continent, and the Group's history forms the subject of these three volumes.

The Royal Dutch/Shell Group may reasonably be called a unique enterprise. Of its two parents, one was Dutch – the Royal Dutch Company for the Exploitation of Petroleum Wells in the Dutch East Indies, in 1949 more simply renamed the Royal Dutch Petroleum Company – and one was British, The 'Shell' Transport & Trading Company. The merger of their business interests in 1907 created a Group with a dual nationality, 60 per cent being owned by Royal Dutch and 40 per cent by Shell Transport. This hybrid structure proved to be surprisingly resilient, surviving intact for almost a century until July 2005, when the interests of the two parents were transferred to one new company, Royal Dutch Shell plc.

The Royal Dutch/Shell Group was one of the first of all cross-border mergers in the corporate age. It was also one of the most successful and enduring, a model which few have succeeded in matching, and a key objective of our study is to explain the long-term success of this intricate organization. We explore the Group's evolution against the background of the general social, economic, and political changes of the twentieth and early twenty-first centuries, focusing on three features that made Shell unique – its dual nationality, its global character, and its organizational structure – and in so doing we offer an analysis of the competitive advantages and disadvantages of these features in the international oil industry.

To examine the Group's role in a full hundred years (and more) of the oil industry in almost every part of the world, and to achieve that in a relatively limited space and time, has been a highly stimulating challenge. It will be helpful to indicate at once what this resultant work is, and what it is not. The authors recognized early on that it would be quite impossible for them, and probably tedious for the readers, to give a detailed reconstruction of oil exploration, refining, transport, marketing, and branding in the more than 150 countries in which the Group has been active, and to explain the decisions that lay behind all those actions. Local decisions sometimes influence the top, but this is not a history seen from the refinery floor. Nor is it a history of each of the many dozens of operating companies and joint ventures that have made up the Group. Instead, given that top-level decisions always influence the local, we have focused on decision-making at the top; on strategy, and the structure that followed from it; and on long-term performance. More specifically, our research concerns were articulated into five specific areas, as follows.

Firstly we sought to establish the basic facts of Shell's *operational spread*: what did it do, where, when, and why? Though basic, these facts were not simple. Over time, the Group entered many new regions, while old ones had to be abandoned. In the 1910s, for example, it entered the United States, a move of great strategic significance, and from the 1920s it diversified from its upstream and downstream activities in oil into chemicals, gas, and other activities. Some of these still remain, others have been sold or abandoned; some were undertaken alone, and others were shared with other companies. Similarly, the contribution of different regions and activities to Group sales and profits changed over time. Indonesia, for example, was highly profitable before 1929, but became a liability after 1945.

Secondly, we investigated the Group's *internal organization*. With dual nationality and a complex and often changing division of labour between two head offices on opposite sides of the North Sea, how were the different parts of this worldwide Group kept together? How did the whole global enterprise function? How did managers strike a balance between centralization and decentralization, between the shared goals and strategies of a truly integrated oil venture, and the importance of taking decisions at the lowest possible level? The most prominent theme to emerge from this investigation was that, in response to its dual nature, the Group very quickly developed its own special philosophy and strategy—initially keeping the business together via a strong team of managers whose frequent long journeys maintained close ties between the parents and the operating companies, and later by developing a matrix structure which tried to balance the forces of region and function. At the same time, the enterprise also had to find a balance between the interests of shareholders and other stakeholders, including the staff.

Our third area of research covered *competition and performance*. In its early years the Group fought an intense and often hostile competition with the Standard Oil Trust and, after 1911, with the companies that developed from the broken-up Trust – especially with the one that became the largest and most powerful of them all, Standard Oil of New Jersey (later Exxon, and subsequently Exxon Mobil). Though not so powerful at the time, the Group's other main rival was the British-based Anglo-Persian Oil Company (later Anglo-Iranian, and subsequently BP), and we present the often rocky relations between it and the Group. Slowly more cooperative attitudes developed, resulting in the famous Achnacarry cartel of 1928, to which Shell attached more importance than did most companies in the industry. After the Second World War the relationships between the oil companies again changed, and the rise of the countervailing power of OPEC in the 1960s and 1970s drastically altered the competitive structure of the industry. Thus, at different periods, different combinations of competition and cooperation were crucial ingredients of the Group's strategy, and we assess its success on the basis of comparisons with the market share, profitability and size of its main competitors.

Our fourth focus was on *innovation* – a vital element in creating competitive advantages, and a factor that we soon saw came under several different headings where the Group was concerned. Its attitude towards technical research and development, and its relations with other research bodies, were essential strands concerning all fields, from exploration, production and refining to transport, oil products and chemicals. Technological innovations were found to be closely connected to markets; for instance, developments in the cracking process gave the Group great flexibility in choosing which oil products were to be manufactured in response to changes in market demand. Innovation in Shell also extended to methods of selling, which contributed to its industry-wide reputation as a successful marketer. These interconnected topics are dealt with in some detail.

Finally, we examined the *role of politics*. The point has already been made that during the twentieth century oil and politics became closely interconnected. The decision of Royal Dutch, in 1907, to merge its business with Shell Transport, and not to take over the company, was justified by Deterding, the founder of the Group, by the fact that this would give access to the British Empire, an access that would be lost if the Group were to become a Dutch company only. Since 1907, links between politics and oil have

[2]

remained very close; in order to survive Royal Dutch/Shell became involved in the great conflicts of the twentieth century, including the two World Wars, and the Russian, Mexican, Indonesian, and Iranian revolutions. In the 1960s and 1970s the rise of OPEC fundamentally changed the power balances within the oil industry. How the Group dealt with these changes and conflicts will therefore also be covered by this study.

Moving now to the structure of the work, we have divided this history into three chronological volumes. The first, by Joost Jonker and Jan Luiten van Zanden, starts with the essential pre-history of the Group – that is, the separate foundation and development of its parent companies – and then covers the development of the Group itself up to the outbreak of the Second World War. The second volume, by Stephen Howarth and Joost Jonker, moves the story on from that point to the eve of the 'First Oil Shock' in the early 1970s and demonstrates how during that period and despite a number of handicaps (such as the loss of Indonesia as a major producing area, a weak position in the Middle East, and a strong orientation to the fragmented European market) the Group nonetheless managed to keep up with its main competitors. The third volume, by Keetie Sluyterman, shows how the Group responded to OPEC's disruption of the industry's established order and the globalization of markets in the 1990s. It brings our analysis up to and beyond the problems which led to the creation of Royal Dutch Shell plc.

We should point out that in general we have preferred to use the contemporary names of the different companies, of the Group itself and of its competitors, in the ways they were used in any given period and as they appeared in our sources. We decided likewise to stay close to our sources when using certain measures and units, especially those of crude oil production.

Finally in this General Introduction we wish to make clear the relationship between the Group and ourselves as researchers and authors. None of us is an employee of any of the Group's companies. Published to mark the Group's centenary in 2007, the work we present here is the fruit of our independent research on the subject resulting from a commission. We have had the full support and cooperation of Group management, with unrestricted access to the Group archives and every assistance in gaining interviews with key persons from the Group's history. Moreover, we were free to discuss and research the topics we considered most relevant for understanding the Group's development.

The progress of the project was monitored by an editorial committee, with an equal number of economic historians and company representatives. They read the initial outline that formed the basis for the whole project, as well as all the chapters. The authors are greatly indebted to the members of the committee for their enthusiasm, unwavering support, and constructive criticism. The academic side was represented throughout by Professor Karel Davids (Free University) and Professor Geoffrey Jones (Harvard Business School). For Shell, Jeroen van der Veer took part in the whole process; Jyoti Munsiff, Adrian Loader, and Sir Philip Watts were members of the committee for a shorter or longer period of time. Some chapters were also read by experts on specific topics. Pieter Folmer acted as the indispensable link between the authors and Shell. This history could not have been written the way it is without the care, courtesy, and attention which he lavished on it.

We trust that our approach will present readers with an informative and thought-provoking picture of the Group's strengths and weaknesses during its first century of existence.

Jan Luiten van Zanden
Stephen Howarth
Joost Jonker
Keetie Sluyterman

Introduction

During the half century preceding the Second World War, the oil industry developed at break-neck speed. World crude output rose from 12 million tons or 90 million barrels in 1890 to 280 million tons in 1938. Product range also widened spectacularly, from the lamp oil with which it all started to a huge variety of oil-derived products, ranging from fuels and lubricants to explosives, synthetic chemicals, and pesticides.

The first three chapters of this volume are organized chronologically, commencing with the pre-history of the Group and ending with the aftermath of the First World War. During these years, the Group expanded very rapidly, successfully fought off challenges from the Standard Oil Trust, and survived major setbacks such as the nationalization of the oil industry in the Soviet Union. By 1920 the Group had overtaken Standard Oil of New Jersey (Jersey, Jersey Standard), the biggest of the survivors from the break-up of the Trust in 1911, and was the world's leading crude producer and joint leader of the industry. The focus of these three chapters therefore lies in explaining how two fairly small companies, Royal Dutch and Shell Transport, from countries that had no indigenous crude oil production, managed to achieve this astonishingly powerful expansion.

For this explanation, the historiography of oil has always laid great emphasis on the role of Henri Deterding, the visionary founder of the Group who dominated its management until his retirement in 1936. Deterding was one of the world's most successful businessmen of his generation, but as often as not the descriptions of his role rely on F. C. Gerretson's history of the Group's formative years. Between 1917 and 1922 Gerretson had been secretary of Bataafsche Petroleum Maatschappij, handling the correspondence of both Colijn and Deterding in that role, an experience that made a deep impression. He greatly admired both Royal Dutch and Deterding, and although highly detailed and often beautifully written (as befitted a published poet) his history is a rather biased account of both the man and the company. Other accounts of Deterding's career have similarly been either for or against the man, so to redress the balance, this first volume aims to give a proper and even-handed assessment of Deterding, putting his achievements in perspective and analysing his contribution in relation to those of his fellow directors.

After the first three chronological chapters, chapters 4-6 are organized thematically and cover various aspects of developments during the interwar period. Chapter 4 focuses on access to oil, which was relatively easy before the First World War but increasingly fraught with difficulties after it. The war had also increased the existing managerial strains within the Group, leading to a profound power struggle at the top which (as seen in chapter 5) was only partly resolved during the later 1920s. A profound transformation in research and manufacturing, and in marketing and branding (analysed in chapter 6) shows the emergence of the Group as a mature and cohesive corporation. Finally, chapter 7 deals with the challenges arising during the 1930s: overproduction, extreme economic nationalism, and totalitarianism. These circumstances showed that Deterding had lost his touch. His resignation in 1936 exposed the chauvinist forces that were still at work in the Group's top management.

Acknowledgements

The authors owe a great debt to the two Royal Dutch Shell archivists in London and in The Hague, Veronica Davies and Rob Lawa, without whose assistance we could not have handled the huge diversity of material within the short time available to us. The Rothschild Archive, London, kindly gave us permission to consult the records from the French Rothschild bank now deposited at the Centre des Archives du Monde du Travail in Roubaix, France. At Shell Oil, Houston, Jack Doherty's hospitality and deep knowledge of the company greatly facilitated our research. Kanada Hardy was untiring in her efforts to bring up archival material; Laura Linda and Hector Pineda helped us to understand the intricacies of the board records. In The Hague, Greg Lewin and Jan Verloop materially assisted us with their knowledge of oil technology; any remaining mistakes on that subject are entirely due to us. The start of the project benefitted materially from the support of Rob van der Vlist, until his retirement General Attorney of Royal Dutch. We are much obliged to Maurits van Os for the zest with which he tackled mountains of archival material, and to Martijn Eickhoff for research in German archives. The student assistants Peter Koudijs, Suzanne Lommers, Christiaan van der Spek and Robbert van den Bergh did a great job in collecting the statistical data from company records and reports and from many other different sources. Beppe Kessler gave us free access to the Kessler papers stretching back into the late 19th century. Zoe Deterding sent us interesting material from the Deterding family. Nico van Horn ferreted out books and other useful material. Finally, J. B. A. Kessler shared with us memories about his father and about Deterding, bringing that distant past so much closer to us.

Joost Jonker
Jan Luiten van Zanden

[1]

Chapter 1

From opposite ends towards a common purpose, 1890-1907

For a thirty-year-old industry, oil was still a remarkably immature business in 1890. The main product flows originated in only two countries, the US and Russia. Production concentrated on a single item, lamp oil or kerosene. Demand was nowhere near being saturated, creating a seller's market for established concerns and wide opportunities for new entrants, provided they could muster sufficient capital and expertise. During the early 1890s, two new contestants entered the fray. Both targeted the huge and rapidly growing Asian market as their objective, but they did so with radically different approaches. The outcome was a surprising one, with the bigger and ostensibly stronger of the two succumbing to its rival and becoming the junior partner in a binational group.

A drilling site at Besitang, Sumatra, shows the immediately recognizable form of a derrick in the background. The rough building in the foreground is the boiler house and forge.

[2]

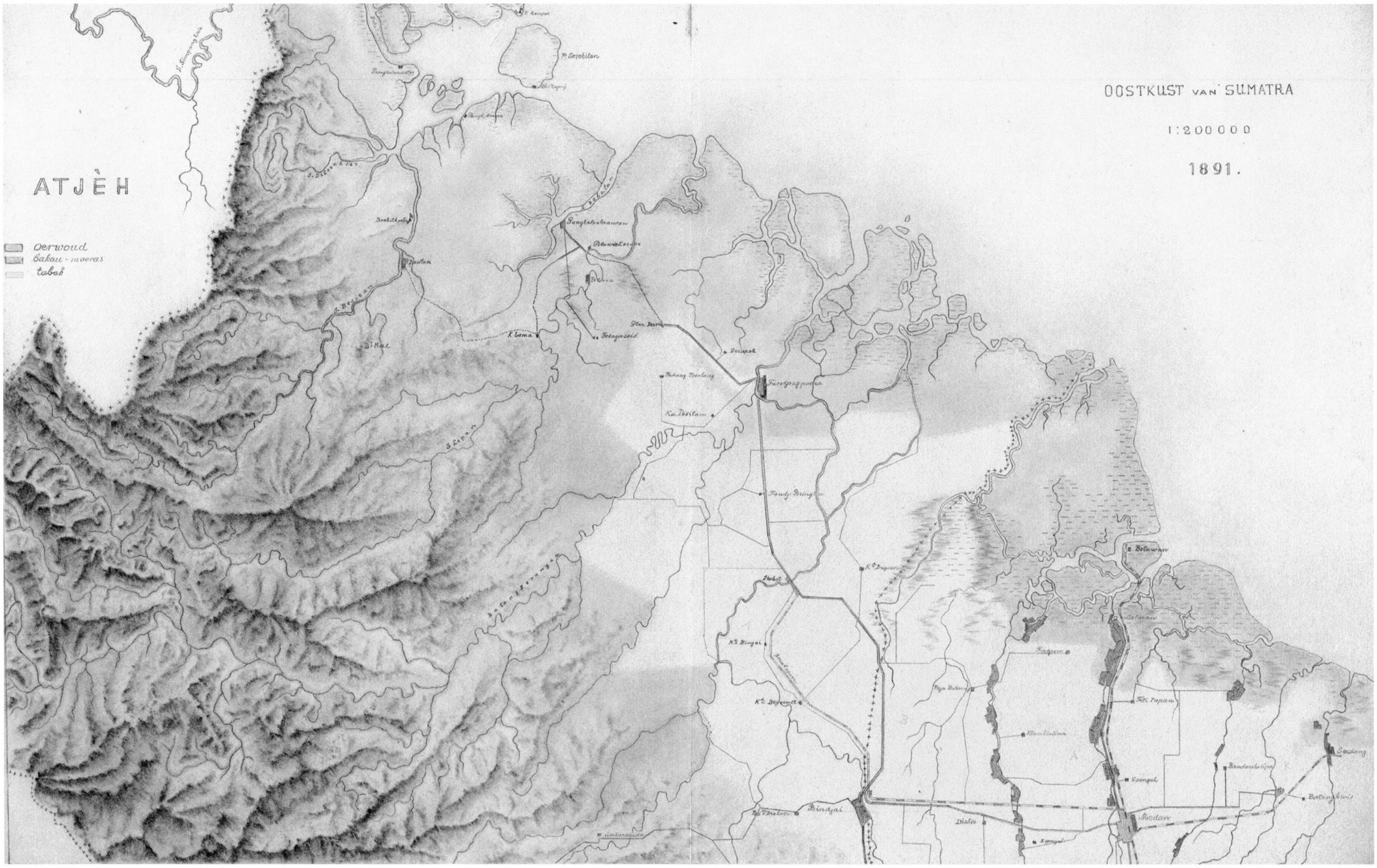

Two celebrations The venues could not have been more dissimilar. Here – a recent settlement in the tropical jungle cut into three sections by the tracks of a light railway, sleeping barracks made from wood and long grass, some Chinese shops, factory buildings, a saw mill, oil refining stills, a hospital, a fort, and a makeshift prison.[1] There – a festively decorated British shipyard, flags and bunting everywhere, the ground for once swept clean; materials, tools, and equipment stored out of sight, on the quayside a freshly painted platform for speeches and the inevitable bottle of champagne for crashing against the hull. No surprise then that the crowds taking part looked so very different. On the north-east coast of Sumatra along the Babalan River a motley crew of perhaps 400 men, made up of a handful of American drillers and refinery workers, European engineers, builders, and administrators, Chinese tinkers and labourers, Sikhs, and a colourful variety of Indonesian labourers, all dressed in their usual work gear, even though it was a Sunday, and tensely waiting for the great moment: a rough pioneers' society and thus no women to be seen. Allowing them onto the site was against official policy, though from time to time the site manager had groups of women shipped in to boost morale.[2] By contrast, at West Hartlepool a festive and well-dressed company, all top hats, frock coats, summer dresses, ribbons, and hats, umbrellas ready for the overcast and rainy sky, the yard's owners and managers, businessmen from London, an Alderman of the City of London with his wife and daughter, perhaps an MP, no doubt local dignitaries, presumably few workers that Saturday.

And yet, for all their outward differences, the occasions were similar in that both marked a watershed, a long-awaited day when planning and preparation at last produced tangible results. Both were to have momentous consequences in the world of oil. And in the end, though the two business concerns emerging from these celebrations followed very different trajectories, both occasions

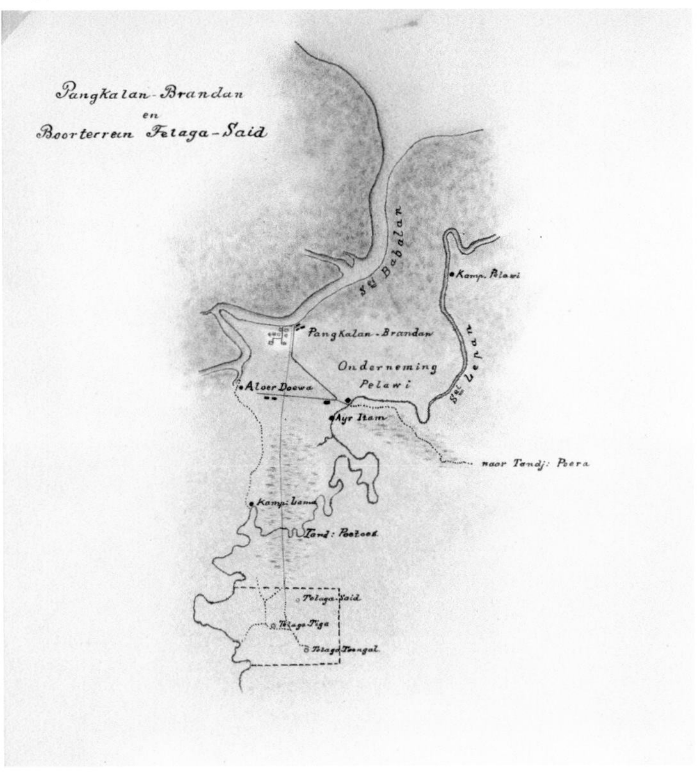

Royal Dutch's first area of operations: the east coast of Sumatra, 1891 (left), and Pangkalan Brandan and the drilling field at Telaga Said, 1892 (above).

would come to belong to the history of a single business concern. On 28 February 1892, after almost ten years of exploration and two more of arduous work to get production under way, crude oil finally flowed from the Royal Dutch wells at Telaga Said through the ten kilometres of pipeline into the refinery stills on the Babalan riverside, to loud cheers and toasting.[3] Exactly three months later, on 28 May, Mrs Fanny Samuel christened the SS *Murex*, the first of six oil tankers ordered by her husband's London trading firm of M. Samuel & Co. as part of a plan to break into the oil trade by shipping bulk kerosene to Asia. Those beginnings mirrored the different starting positions of the two concerns, Royal Dutch beginning operations upstream as producer and refiner, Samuel & Co. entering midstream as transporter of and wholesaler in oil. Both focused on the Asian market, however, and that factor would prove to be decisive in bringing them together.[4]

An emerging world market Both ventures aimed to capture their share of a booming trade. After the invention of the kerosene burning lamp and the subsequent discovery of huge crude deposits in the state of Pennsylvania during the 1850s, the American oil industry had taken off in an unprecedented way. From 1861, US traders started exporting kerosene, first to Europe and then around the world, meeting a great and surging demand for the cheap illuminant. Both industry and exports were increasingly dominated by John. D. Rockefeller's Standard Oil Company, established in 1870. Reorganized into a trust in 1882, Standard Oil became the byword for monopoly power. Around 1885 the company controlled some 80 per cent of marketing in the US, with a hold on export markets to match.

The rapid development of kerosene lighting and of the American oil industry stimulated crude production elsewhere in the world, notably around Baku on the Caspian Sea, then a recently annexed outpost of the Russian Empire. In 1873 Robert and Ludwig Nobel, two Swedish businessmen active in Russia, bought a refinery there, the basis for a rapidly expanding, integrated oil company which soon sold its product around the empire, denting American imports. Seeking a way to export their kerosene to markets in Europe and Asia, other producers started building a railway across the Caucasus to Batum on the Black Sea, which the Ottoman Empire had surrendered to Russia in 1878 at the Peace of San Stefano. When they ran into financial difficulties, the Paris Rothschild bank, which wanted Russian oil for its Fiume refinery on the Adriatic coast, lent the money to finish the line, which opened in 1883. Both the Nobels and the Rothschilds immediately began to export Russian oil to Europe and then to Asia, in fierce competition with Standard Oil. In 1886, the Rothschilds reorganized their Russian oil interests into an integrated company which became known by the acronym Bnito, the initials of its Russian name, translated into English as the Caspian and Black Sea Petroleum Company.

Meanwhile the oil fever had spread to other parts of the world, one of them also an outpost of empire, the Dutch East Indies.

Het Loo 10 April 1890.

Ingevolge de bevelen van Zijne Majesteit den Koning heb ik de eer ter kennis van Uw Hoog Edel Gestrenge te brengen, dat Zijne Majesteit heeft goedgevonden gunstig te beschikken op Uw verzoek van 4 April ll., en Uw Hoog Edel Gestrenge mitsdien vergunt, de Maatschappij bedoeld in Uw request den naam te doen voeren van: Koninklijke Nederlandsche Maatschappij tot exploitatie van Petroleumbronnen in Nederlandsch Indië, onder voorwaarde evenwel dat hierin, ook voor de toekomst, slechts zal worden gezien een bewijs van Zijner Majesteits zedelijken steun.

Uw Hoog Edel Gestrenge gelieve de verzekering te aanvaarden mijner gevoelens van bizondere hoogachting

de Adjudant
Particulier Secretaris des Konings

Hoog Edel Gestr. Heer J. B. A. de Gelder
Hoofd ingenieur Chef v. d. dienst der Bataviasche havenwerken Amb. lid v. d. Raad van Ned. Indië.

[4]

Gathering the fruits of empire It was an odd slip of the pen for a man so well versed in the affairs of state to make. Perhaps the King's secretary was absent-minded that day, burdened by his responsibilities in a royal household headed by an ailing, capricious monarch. Or perhaps, having bypassed the usual channels to obtain the King's approval of the request for a Royal warrant, Captain SMS. de Ranitz made the mistake in his urgent desire to notify the applicants.

Whatever the reason for his slip, in writing the official reply, De Ranitz added the adjective *Nederlandsche* or Dutch to the name of the prospective company, turning what was already a long name into an impossible tongue-twister. On opening his letter, the board discovered that their venture had been rechristened into Koninklijke Nederlandsche Maatschappij tot Exploitatie van Petroleumbronnen in Nederlandsch-Indië. Thus the name by which the company was to become famous, Royal Dutch, was created by a Court intrigue compounded by a professional error.[5] Even so the fact that the board had sought, and obtained, this peculiar favour highlights a key component of the new company's business strategy. From its inception, the Royal Dutch demonstrated a marked talent for networking, for getting things moving with a quiet word or through pulling strings, ensuring that its arguments were heard in the right places, even if the company did not necessarily get its way. Indeed, networking had been a crucial factor in developing the first oil concession, second in importance only to the character of Aeilko Jans Zijlker (1840-90) in nursing his discovery of oil into a proven concession.

Zijlker got to know the area as manager of a Sumatra tobacco company.[6] In 1880, he realized the potential for producing kerosene from the ubiquitous oil seepages there. Having secured a provisional concession, Zijlker obtained financial backing from a consortium of businessmen in Batavia (now Djakarta). Through his brother, a member of the Dutch Parliament, he obtained a personal introduction to the Governor-General of the Dutch East Indies. Zijlker's plans suited overall Government policy very well, for several reasons. Developing an oil industry would increase Dutch presence and power on Sumatra, still only nominally under Dutch control. Second, it would help to reduce the dependence on US kerosene. Finally, with the presence of oil a well-known fact, the arrival of prospectors in droves was simply a matter of time, raising the spectre of foreign companies such as the feared Standard Oil establishing themselves. This would endanger the two first objectives, so a Dutch initiative was very welcome indeed.

And Zijlker, a dynamic man blessed with a single-minded determination to achieve his self-imposed goal, but somewhat gauche and bereft of any engineering experience or deeper knowledge about oil, needed all the help he could get. The colonial administration began by helping him obtain a proper concession. Then, in the spring of 1884, the Governor-General agreed to let the

The hand of history: an unexplained slip of the pen brought an unexpected name, when the King's secretary wrote 'Royal Dutch Company' instead of 'Royal Company'.

Aeilko Jans Zijlker (1840-1890), the first man to recognize the commercial potential of Sumatran oil.

[6]

The final page of Zijlker's concession. Sold to Royal Dutch, this concession became the foundation stone of the company's success.

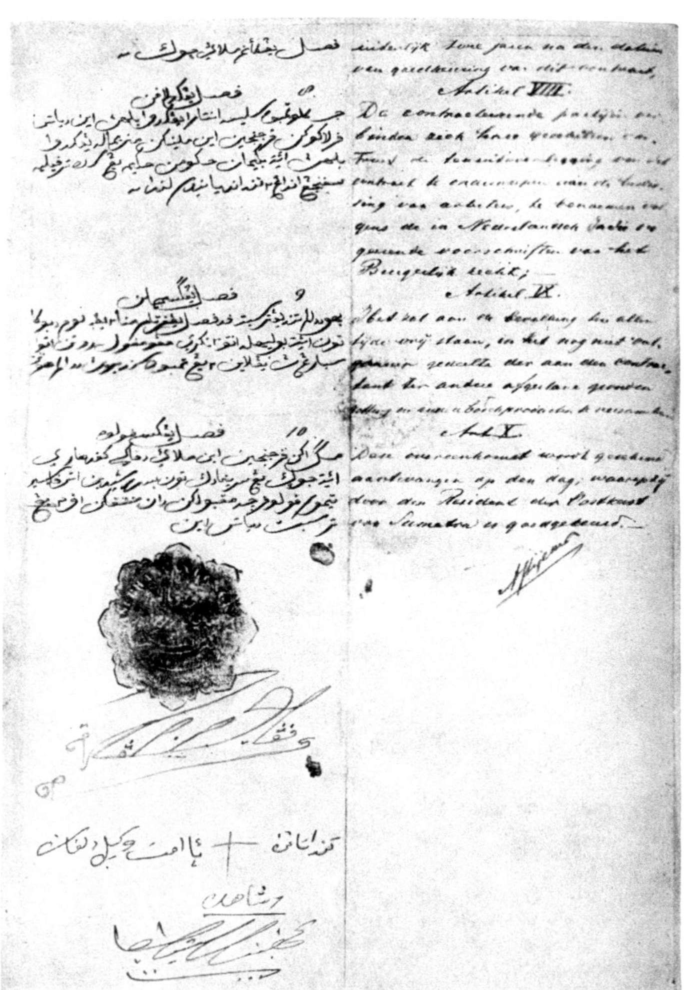

[5]

Mining Department do the exploratory drilling, thus saving Zijlker the trouble and expense of getting in a foreign crew. The goodwill gesture turned out to be rather a mixed blessing, for the project now became tied to the rack of bureaucratic infighting. Even so oil was struck within two months, and in June 1885 the Telaga Tunggal well came in, destined to become famous for its long productivity.[7] However, it took four years to bring the prospecting to a satisfactory conclusion, at a cost ten times the original estimate of 20-30,000 guilders.[8]

Late in 1889 Zijlker sailed for the Netherlands to set up an exploitation company. After almost ten years, he had realized his ambition, but he would not have achieved his goal without the Mining Department engineers, struggling in the dense and marshy Sumatran jungle, fending off occasional marauding bands of Acheh warriors to complete a drilling programme for which no one had given them clear instructions, under the critical gaze of someone whose practical experience in the field fell quite short of his vision. Meanwhile another Mining Department engineer, Adriaan Stoop (1856-1935), showed how to do it. Stoop obtained backing from a group of family members and acquaintances to prospect for oil, and acquired a concession on Java, near the port city of Surabaya. He had meanwhile also obtained permission to go on a Government-sponsored field trip to study the US oil industry at close quarters. On his return, he took temporary leave from his post to work his

[7]

In June 1885 the first commercially successful discovery of oil in eastern Sumatra was made using an Aalborg drill (diagram, centre). This rotary tool (left), hand-driven with a small steam engine, was awkward, slow and labour-intensive and was redesigned by J. A. Huguenin into 'an elegant, light rig, able to drill holes up to 300 metres'(diagram, right). The Huguenin design was later overtaken by the larger and more powerful Pennsylvanian drilling system, also known as the Californian system, seen here (right) in action in South Palembang in 1915.

[8]

GRONDPEILWEZEN IN NEDERLANDSCH-INDIË.
Vooraanzicht.
Schaal 1:10.
INSTALLATIE
VAN DEN BOORBOK
VOOR KLEINE BORINGEN
volgens de methode van
J.A.HUGUENIN.

[9]

[10]

A gusher at Telaga Said, Royal Dutch's first Indonesian concession, in 1890. The human figures show the small scale of the derrick and the gusher.

[11]

concession. Drilling began in January 1888, the first well came in three months later, and in April 1889 the Dordtsche Petroleum Maatschappij, as the venture came to be called, sold its first lamp oil, having spent only a fraction of the money that had been needed in Langkat.[9] Stoop did not have to struggle in a remote and hostile environment, he owed his rapid success more to his technical know-how in guiding explorations and setting up production.

Zijlker arrived in the Netherlands with a complete blueprint for the new company in his pocket. The Mining Department engineer who had led the second drilling campaign to its successful conclusion, Reindert Fennema, had written a very thorough and comprehensive report about the prospects for setting up an oil company on the concession, complete with detailed estimates of costs and revenues, plus designs for the refinery, a port facility on the Babalan river to ship products, and other necessary amenities such as roads and railroad tracks to connect the main establishment with the drilling site.[10] Moreover, Zijlker had shared the passage to Europe with N. P. van den Berg (1831-1917), arguably the colony's most prominent businessman, who threw his full weight behind Zijlker's venture, helping to translate Fennema's report into a business plan and a prospectus to float the business.[11] Van den Berg's wide social network supplied key board members, H. D. Levyssohn Norman as chairman and J. A. de Gelder as general manager, and also facilitated the acquisition of the royal warrant.[12] Zijlker sold his concession to the company and wisely chose to continue prospecting in the Dutch East Indies rather than assume a managerial position. Within six months the articles of association had been approved and the share capital of 1.3 million guilders successfully floated on the Amsterdam stock exchange.[13] The business was up and running from a room in De Gelder's home on the Celebesstraat in The Hague, the natural choice as a company seat for colonial enterprises like Royal Dutch since access to Parliament and the colonial administration weighed heavier with them than relations with Amsterdam finance. Consequently, the company did not develop close ties with banking and the securities trade, an atmosphere of deep mutual suspicion remaining the norm.[14]

These early months in the company's history are remarkable for showing the contours of a business policy which was to become the backbone of Royal Dutch's later strength. The vertebrae are best summarized as a clear sense of objectives, a tacit consensus about technology as instrumental to the business, close financial controls, and, after initial mishaps, a conspicuous ability to foster management talent. The company's objectives largely derived from its remote location. The business had to be fully integrated, combining production, refining, packaging, transport to the point

[12]

The first reservoir at Telaga Said, an open brick tank, was built in 1898 at a time when the connection between the well and the pipeline to the Pangkalan Brandan refinery was not working well. Oil was pumped from here to the refinery.

of sale, and a sales network, at a time when specialization on one or perhaps two links in the supply chain was very much the industry norm.[15] A Sumatran market did not exist, so sales would necessarily be international. Langkat's proximity to all main Asian ports was to prove a distinct advantage here, provided the company shipped the kerosene itself.[16] By contrast, the Dordtsche sold directly from the refinery to wholesalers, and thus Stoop hardly invested in transport or sales, departments which were to prove decisive in the later struggle for market control.[17] The central position of technology manifested itself in De Gelder's orders for equipment, which included a telephone system, still quite a novelty, for connecting the drilling sites with Pangkalan Brandan, as the main site on the Babalan river came to be called.[18] De Gelder, a very experienced engineer, designed the necessary installations

The case factory in Pangkalan Brandan, 1893, with cases under construction and piling up ready for use. The cases were made by hand to carry two tin oil cans holding four Imperial gallons each, at a cost of 1.5 guilders per case. The wood was difficult to work with, so this was cheap at the price; but before long tens of thousands of cases were being made every month.

In front of the tinshop at Pangkalan Brandan, 1892, the first shipment of cases is ready to depart in the company's diminutive narrow-gauge train. To the left of the picture next to the engine stands Kessler, with felt hat and walking stick, while other people pose by the solitary carriage with its benches and curtains – a scene described by W. P. H. du Pon (trainee and later works manager) as 'a still life that filled us with pride.'

[13]

[14]

and ordered the equipment.[19] In 1892, two young engineers were sent to the US to study the market for drilling equipment and refinery installations and find the right suppliers.[20] Both the designs and the budget were based on Fennema's data and Stoop's report from his US field trip. The board does not appear to have subjected them to much scrutiny. Nor did it waste much time discussing the hiring of a chemical engineer in 1895, acknowledging the need for technical expertise in maximizing revenues.[21] By contrast, expenditure and the cash balance figured at nearly every meeting. The perceived importance of a sound administration was also apparent in the appointment of a bookkeeper to accompany the site manager to Langkat.[22]

In the end paper controls and debates did little to relieve the board's utter dependency on its overseas manager. H. Stutterheim, the Langkat manager, faced a task of mindboggling magnitude and complexity: to clear a huge area of virgin jungle and erect the refinery, the brickworks, the sawmill, a water purification system, other workshops, stores, office, and residential accommodation, and then install the equipment; to build a quay strong enough to unload the expected heavy materials, such as railway locomotives; to lay ten kilometres of railway track on an embankment through the jungle to the drilling site, which included crossing a river with a bridge 44 metres long; to lay the pipeline to the refinery; to build the derricks and begin drilling; and to organize the whole administrative process needed to control progress against plans, and to keep track of cash, costs, materials, wages, workers. When the final plans arrived in December 1890, the site was not yet ready, but ships had already begun unloading building materials and equipment, creating a mounting pile of stores on the quayside, half-sorted, unchecked for damage or completeness, and exposed to the full rigours of the climate.[23]

Nor could the project have progressed much faster for a lack of hands. The works required an estimated 400 labourers, but few were to be found anywhere in the remote and thinly populated area. They had to be recruited from Java, Singapore, Penang, or elsewhere, through labour touts, and brought in. This slow process yielded a poor labour force. The conditions were not very attractive: rudimentary accommodation, poor food, no amenities, and hard

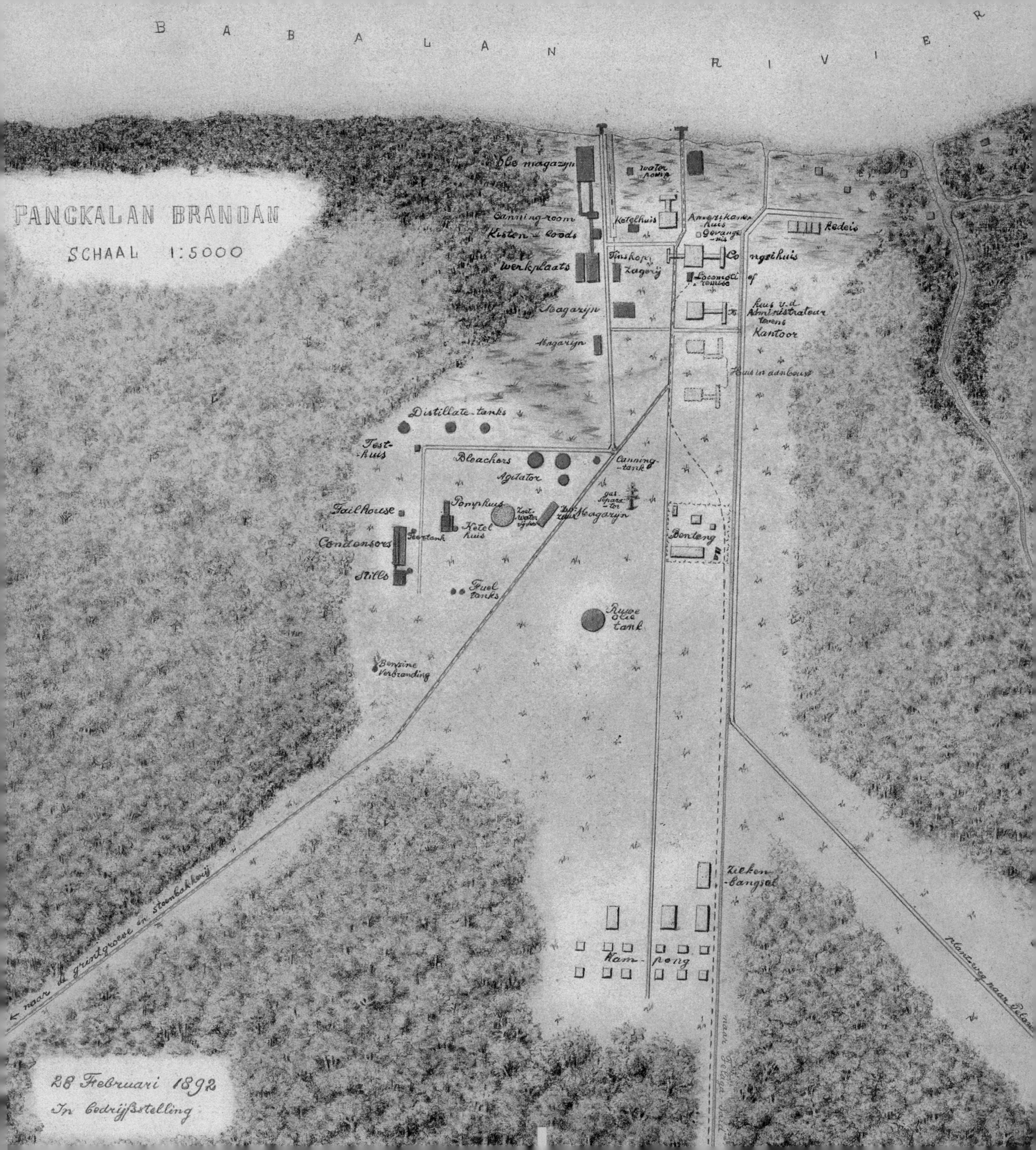

B A B A L A N R I V I E R

PANGKALAN BRANDAN

SCHAAL 1:5000

6be magazyn

water pomp

Canning-room

Kisten & Coods

Ketelhuis

Amerikaner-huis

Gevange-nis

Redeis

Tinskop & Zagerij

Werkplaats

Congsihuis

Locomotief remise

Magazyn

Huis v.d Administrateur tevens Kantoor

Magazyn

Huis in aanbouw

Distillate-tanks

Test-huis

Bleachers

Canning-tank

Agitator

gas separa-tor

Failhouse

Pomphuis

Zoet-water vijver

Magazyn

Ketel huis

Teertank

Condensors

Benteng

Stills

Fuel tanks

Ruwe Olie tank

Benzine Verbranding

Zieken-Bangsal

Kam-pong

← naar de grintgroeve en steenbakkerij

plantweg naar Pelaw

naar Telaga Said

28 Februari 1892

In bedrijfsstelling

A map of Pangkalan Brandan, 1892. At the bottom is the 'native village' and hospital; in the centre, the circular crude oil storage tank, and above it to the right, the fort. Left of centre are the primitive refinery buildings, while the buildings at the top include among others the tinshop, case factory, canning room, oil store, American quarters, European quarters, the Manager's house and office – and the prison. The diagonal track on the right comes in from Telaga Said, and the one on the left goes out to a gravel pit where waste products – everything except kerosene – were thrown away and burnt. 'The gigantic blaze that could be seen at night was for years a well-known landmark for shipping in the Strait of Malacca.'

work with ten-hour days and seven-day working weeks, punctured only by one *hari besar* or holiday every fortnight following pay day.[24] Consequently, nearly all labour was indentured, forced to work during a certain period in order to clear off debts to the tout, and thus not very motivated to give their best, or even to stay: there was a constant drain of workers slipping away into the jungle or onto ships leaving the quay. To stem the flow, the colonial government had passed special legislation allowing the employers to punish labourers in breach of their contract. The resulting repression rendered the difference between indentured labour and slavery an academic one, without really solving the original problem of labour shortage. Circumstances only began to change when a Medan lawyer published an indictment of the prevailing labour conditions in 1902.[25]

Finally, on top of these challenges, Stutterheim was expected to secure further concessions. The news of oil in the Indonesian archipelago had started a veritable prospecting fever, forcing Royal Dutch to safeguard its strategic interests. Zijlker had suddenly died in December 1890 on his way back to the Dutch East Indies, so that task also came down to Stutterheim, who successfully acquitted his multifarious responsibilities, but really had too little time to concentrate on Langkat itself.[26] In March 1891, the board noted that he had spent only 24 out of 139 days on the site, so the works had hardly been supervised at all.[27] The project seemed to spin out of control: the date for the start of production slipped, progress reports became fewer, money ran short without revenues being in sight. De Gelder declined to take charge on the spot, and was lucky enough to find an extraordinarily gifted manager to head a commission of enquiry: J.B. August Kessler (1853-1900).

Emigratie-, Verkoop- en Commissiekantoor.
Telegramadres ESAS, Soerabaia.
LEVEREN:
Flink, jong en gezond Werkvolk,
zoowel MADUREEZEN, JAVANEN en SOENDANEEZEN als CHINEEZEN
voor Land- en Mijnbouw-Ondernemingen.
RISICO van desertie van boord voor onze rekening.
Wij hebben met onze Koeliezendingen het meeste succes en zijn bereid copie tevredenheidsbetuigingen ter inzage op te zenden.
Leveren ook:
... en Javaansche Ambachtslieden.

[17]

H.H. Hoofd-Administrateurs en Administrateurs, zoomede Slagers en Ondernemers van Mijnbouw!
LEVERING tegen de LAAGSTE PRIJZEN:
Gecastreerd, degelijk Madureesch of Oost-Java Trek-vee
van 300—375 K.G., voorzien van certificaat van den Veearts.
Prachtige Madureesche Slachtstieren.
Flink, jong en gezond Oost-Java Werkvolk,
MANNEN of VROUWEN voor LAND- en MIJNBOUW, tegen **f 60.**— per volwassen persoon, franco BELAWAN.
Belast zich met het KOOPEN en VERZENDEN van:
Savoeneesche en Rottineesche Rij- en Wagenpaarden,
uitstekend geschikt voor BERGTERREIN.
Minzaam aanbevelend,
H. LEEKSMA Kzn., Soerabaia.

[16]

Advertisements from the newspapers *Deli-Courant*, (above) and *Sumatra Post* (below), 1902. 'Robust, young and healthy' men are offered as workers, and on the same page, castrated oxen are offered as draught animals – men and oxen alike 'at the lowest prices'.

[18]

[19]

PRAUW & KOELIELOON

SOLDEERLOON & SURVEILLANCE A/ BOORD

zooals die voor petroleum uit Langkat zullen worden berekend.

PLAATS	PRAUW-LOON p.¢.	KOELIE-LOON p.¢.	SOLDEERLOON p.¢.	SURVEILLANCE A/ BOORD p. DAG.
Batavia By lossing t/m 2000 ¢ p.dag.	5 cts.	3½ cts.	7 cts.	F.3.- tot F.5.- plus tambanga huur.
2000-6000" " "	7 "	3½ "	7 "	
6000-10000 ¢ p"	10 "	4 "	8 "	
Cheribon	4½/5 cts.	3½ "	5 "	F. 7,50
Soerabaya naar gelang van kwantiteiten	7½ tot 10 cts.	3 tot 4 cts.	5 "	F. 5.- p.dag

Batavia 28 December 1894.

Vignettes of the imperial age – led by local guides and guards with three rifle-carrying Europeans as overseers (above), Chinese coolies carry the cash for workers' payment through the jungle; and (below) a record of rates of pay, contrasting those for European surveyors with the rates for coolies and others.

Rebuilding operations

When Royal Dutch approached him, Kessler had been at a loose end for a few years after the disappointing end to his career in a prominent Java merchant firm. Some of his character traits showed a marked resemblance to Zijlker: the stamina in pursuing long-term objectives; the irascibility when thwarted, born from an inability to understand opposition to self-evident goals; an unbelievable capacity for sheer hard work. Zijlker had the angular character of a lone ranger, however, whereas Kessler's sometimes strained behaviour emanated from the inner tension between his formal bearing as a gentleman and his leadership drive, from the necessity to sustain polite discussion whilst craving to go into action. In the well-known photograph of Kessler at his desk, throwing a combative glare to someone outside the picture, he resembles nothing so much as a tiger caged in a suit.[28] Though a novice in oil and without technical training, he quickly developed a remarkable, intuitive grasp of the industry in all its aspects, founded on his keen sense for figures. With these qualities, Kessler nursed Royal Dutch through its teething troubles and great adolescent crisis.[29]

Kessler knew that funds would last until February 1892, giving him just over three months' time to start production.[30] The works were still only half-finished, and now bad weather began to take its toll. Torrential rain washed away part of the railway track, so the steam boilers could not be transported to the well heads. A new route was cleared through the jungle laying tracks on tree trunks felled over the marshy underground. Just when it was ready the Lepan river flooded again, destroying the work done, and again delaying the drilling of the vital new wells, needed to attain the projected production of 500 cases of two four-gallon (imperial) tins a day.[31] Heavy seas interrupted the food supply, forcing Kessler to send 80 people on an emergency trip to get rice from the nearest settlement, 20 kilometres across swamps and jungle. Labourers began to desert the site in droves, demoralized and fearful of rumours about marauding Achinese warriors. Kessler drove them hard, and would have wanted to drive them even harder, demanding from everyone the energy and commitment which he showed himself, omnipresent during the day, giving instructions

[20]

J. B. A. Kessler (1853-1900), Royal Dutch's insightful and determined leader in its earliest days, is seen here above as a young man, and below, in his office just a couple of months before his untimely death.

[21]

[22]

In this image from 1896, Kessler stands beside the 30-year-old Deterding, with An, one of Kessler's daughters, close by.

Figure 1.1
Royal Dutch, number of cases produced (left scale) and average cost and revenue per case, 1892-1900.

here, cajoling workers there, inspecting progress on the pipeline from the drilling site where people worked waist-high in the mud for days on end; and then working deep into the night to check the administration and write reports. For all the exertions and sacrifices of the multinational crew, the triumph of that Sunday morning when production started was truly his.

With production under way, only the market needed developing. In April 1892 Kessler set out for Penang with a few cases of Langkat kerosene, now dubbed Crown brand. A press presentation in the form of a comparative test between Crown brand, Russian, and Standard Oil's Devoes kerosene, created favourable publicity.[32] Kessler appointed H.J. Martijn as sales agent in Penang; within months Martijn had enquiries from Japan about sales of Crown kerosene.[33] His mission completed, Kessler returned to the Netherlands, succeeding De Gelder as general manager in November 1892.[34] Unfortunately the start of production and sales failed to lift the company's fortunes. The Asian market was being flooded with kerosene, as exporters scrambled for market share in anticipation of Samuel & Co.'s bulk shipments, the first of which arrived at Singapore with the *Murex* in August 1892.[35] A revolving credit arranged by the imaginative and dynamic Penang agent of the Nederlandsche Handel-Maatschappij (NHM), H.W.A. Deterding (1866-1939), helped to keep the Royal Dutch afloat. The new manager, James Waddell, showed initiative, such as the launch of a cheaper brand and attempts to sell refinery residue as ship boiler fuel, but he failed to raise production and cut costs fast enough to turn the tide.[36] A cash crisis loomed, and then a series of misfortunes appeared to signal the beginning of the end, for production dropped. The Royal Dutch's share price plummeted, and attempts to get emergency funding from Amsterdam bankers foundered on rumours that Standard Oil had managed to obtain a huge concession in Langkat, effectively encircling Royal Dutch.[37] Having convinced himself of the company's essential viability, Kessler once again set out for Sumatra.[38]

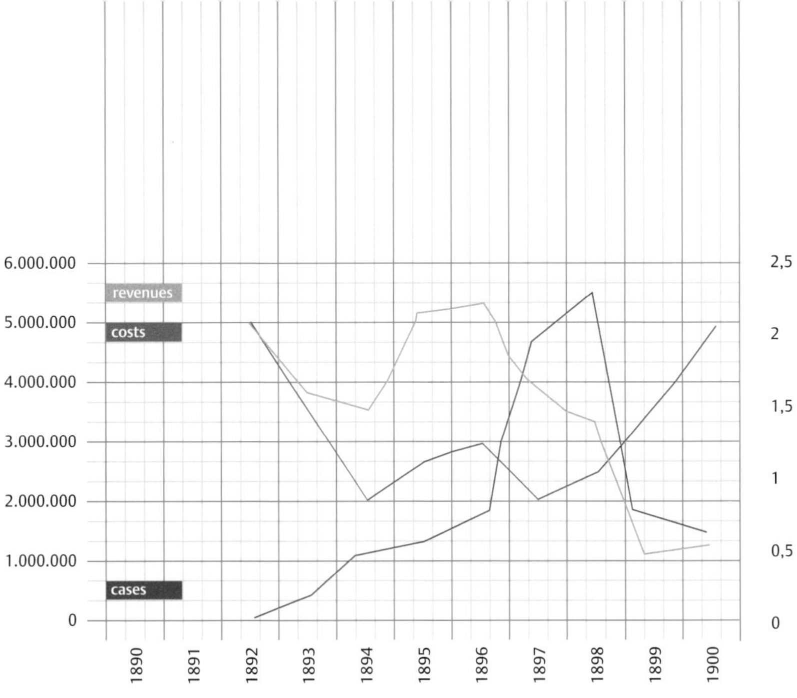

Two images from Pangkalan Brandan, 1900. The top one shows part of the wharf operated by the business, with the steam launch *De Disselwerf* visible in the upper right corner. The bottom one shows the eastern entrance to the installation, seen from the loading and unloading jetty. The materials lying around were a 'chronic phenomenon', according to Du Pon, because of the lack of storage space.

During his second sojourn, Kessler set himself four goals: raise output, cut production costs, expand the sales network, and obtain further firm concessions. The first two objectives had to be achieved largely through increasing productivity and retaining cash coming in, for there were no more funds for buying new equipment.[39] Within a few weeks of his arrival, Kessler had succeeded in raising production to 30,000 cases a month, or double Fennema's original target, partly by drilling new wells, and partly through the introduction of better refining methods which boosted the kerosene yield from 35 to 48 per cent of the crude oil, whilst also allowing a better separation of fractions.[40] As production rose, unit costs of course fell (Figure 1.1).[41] Packaging remained a crucial bottleneck, however. The customary kerosene sales unit of two four-gallon tins in a wooden crate represented almost 80 per cent of unit cost, without much scope for productivity gains.[42] The growing network served by Samuel & Co.'s tankers gave further urgency to switching to bulk, but adopting bulk transport required substantial investment, first of all in port installations at a new location.[43] During the autumn of 1893, Kessler decided to build a loading port about twelve kilometres north-west of Pangkalan Brandan on the Aru Bay, with storage tanks fed by a pipeline from the refinery. To relieve the mounting

[23]

[24]

[25]

[26]

[27]

[28]

[29]

Inside and outside a drilling operation: the boiler house (top and second), the drilling floor, and the derrick structure.

A detailed map of the oilfields surrounding Pangkalan Brandan, 1898.

Helping to overcome the serious difficulties of moving timber through the dense jungle to the sawmills, two elephants were procured as hauliers, coming into company service on 10 April 1895.

stocks of kerosene, temporary loading facilities were built on the chosen site. The reduction in freight cost achieved in this way immediately opened up new markets for Crown Oil, until then limited to the Straits. From November 1894, Royal Dutch began sending its products to China, Java, India, and Thailand.[44] On top of all this work, Kessler still found time to lay down procedures for all operations and to document progress by taking photographs, which he himself processed at night in an improvised darkroom.[45] Selected pictures were sent over to convince The Hague of money well spent, swelling Kessler's already voluminous correspondence. He kept the board up to date with monthly telegrams stating production and available cash, supported by a full written report every four to five weeks, which included copies of correspondence with the sales agents, and extensive administrative and technical data. In addition, Kessler sent a copy of his diary to The Hague, wrote confidentially to individual directors, and still found time for letters to his family.[46] Security and secrecy were prime concerns for Kessler, who had developed a siege mentality following the company's earlier predicaments, and was quick to see enemies.[47] Time and again he emphasized the importance of keeping technical data confidential, of not boasting about the company's performance, and above all of not admitting outsiders onto the sites, which were fenced and patrolled by a fifty-man security force. Visitors such as a Standard Oil delegation in 1897 were treated to selected sights, after Stoop had managed two years earlier to trespass and gain a good impression of production capacity.[48] Even so news and rumours travelled quickly, and Standard's agents throughout Asia had a notable knack for milking crew members on leave for information which was assiduously collected and sent to New York.[49] Though still keen to hold all the strings himself, Kessler came to rely more and more on his right-hand man Hugo Loudon (1860-1941), a young engineer who, sent out as technical manager, soon proved equally valuable for his tact and negotiating skills.[50]

Rising revenues finally lifted the company out of the danger zone. The year 1893 already showed a small profit, which was put to the reserves, but from 1894 profits flowed generously. Kessler, having returned to The Hague in the spring of 1895, kept a firm hand on the purse strings through cautious valuations and a policy of continuous investment tied to rigorous depreciation of assets. During 1892-94, with the company close to bankruptcy, he had managed to finance 600,000 guilders' worth of investment through retained earnings.[51] Shareholders were finally rewarded for their patience, receiving a decent 8 per cent dividend for 1894, but 44 per cent for 1895, 46.5 per cent for 1896, and 52 per cent for 1897. Investors rapidly jumped on the bandwagon, pushing share prices to par by mid-1894, and to 900 per cent of par in 1896, enabling the company to issue new shares.[52]

[31]

A drilling location in the jungle of Sumatra, early 20th century. In the distance a small derrick built on the Californian pattern stands among the trees. Nearer stand storage tanks, with a few workers visible on the path; and in the foreground, a European strikes a casual pose for the photographer high

[32]

A drilling installation in operation at Sanga Sanga, early 20th century. The system in use (the KNPM system, also known as the East Indies flushing auger or 'Tjotok' system) continuously flushed cuttings from the hole, and was developed from earlier systems by Royal Dutch to provide a relatively light unit, more easily transportable through the jungle. The photograph (taken from next to the engine) shows the large crank wheel to the left, with the crank-shaft or 'jerk line' rising diagonally to the cable, which in turn goes up to the pulley at the head of the derrick. The cable then comes back down to where the men are standing beside the swivel head and drill poles, pounding into the ground. The record depth for a well using this system was 4,430 feet (1,365 metres), achieved with Well 21 at Puedawa, Achin, in 1916. The system remained in use in the East Indies until the introduction of rotary drilling there in the mid-1920s.

Now Kessler could execute his planned expansion. Refining capacity was doubled, and the Aru Bay bulk loading station built, to feed tank installations in the company's two main destination ports, Hong Kong and Shanghai.[53] In 1896 Deterding was lured away from the Nederlandsche Handel-Maatschappij (NHM) and appointed to expand the sales network. To Kessler, Royal Dutch had to be an integrated company with its own transport and marketing network. For that reason he rebuffed the Samuels, who approached him in 1894 and again during 1896-97 with varying proposals ranging from direct oil sales to some form of cooperation.[54] By 1898, the company had new tank installations in Shantou, Bangkok, Singapore, Karachi, Madras, and Calcutta, with Bombay, Fuzhou, Xiamen, Tianjin, Hankou, and Zhenjiang under construction. A tanker fleet of 28,000 tons, enough to ship a million cases a month, was bought and chartered to serve the establishments. Nor was the staff forgotten. The temporary dwellings, designed to last two years in tropical conditions, were gradually replaced by proper houses, and in June 1895 a bonus scheme was introduced for the Europeans.[55] Despite these improvements, conditions remained severe, and Pangkalan Brandan continued to suffer occasional outbreaks of serious labour unrest.[56] It was only in 1909 and after some pressure from Loudon that Sunday became a regular day off and that married men were allowed to have their wives join them.[57]

Consolidating the gains

Meanwhile Royal Dutch had taken steps to safeguard its strategic interests. Zijlker's original concession of 1,000 hectares was gradually expanded to an area covering just over 175,000 hectares, including all the land adjoining the Aru Bay.[58] The cost of even nursing concessions to the status of proven fields deterred the mostly speculative concession holders in Langkat from exploitation, and the change in the Royal Dutch's fortunes meant that large tracts more or less fell into its lap.[59] By contrast, the Bukit Mas concession, immediately adjoining Telaga Said and deemed to be very promising on account of the ubiquitous surface seepages, required some intricate manoeuvring to forestall a sale to the British merchant house Jardine Matheson.[60]

Rising prosperity also attracted Standard Oil's interest in a takeover of Royal Dutch. In September 1897, the board turned down an offer as financially not very attractive and unfair to the government after the loyal support received.[61] Decisiveness soon gave way to an uneasy feeling, however. What if Standard simply launched a hostile bid? A precipitous fall in the Royal Dutch share price immediately following the refusal fuelled the board's fear into a panic. With hindsight this appears nothing so much as the market correcting an inflated price. The board saw things differently, suspecting an organized raid of the kind Standard Oil was known to practise on its US rivals.[62] A few months later, Standard's friendly takeover bid for the Moeara Enim oil company, which held a good concession on southern Sumatra near Palembang, raised chauvinist

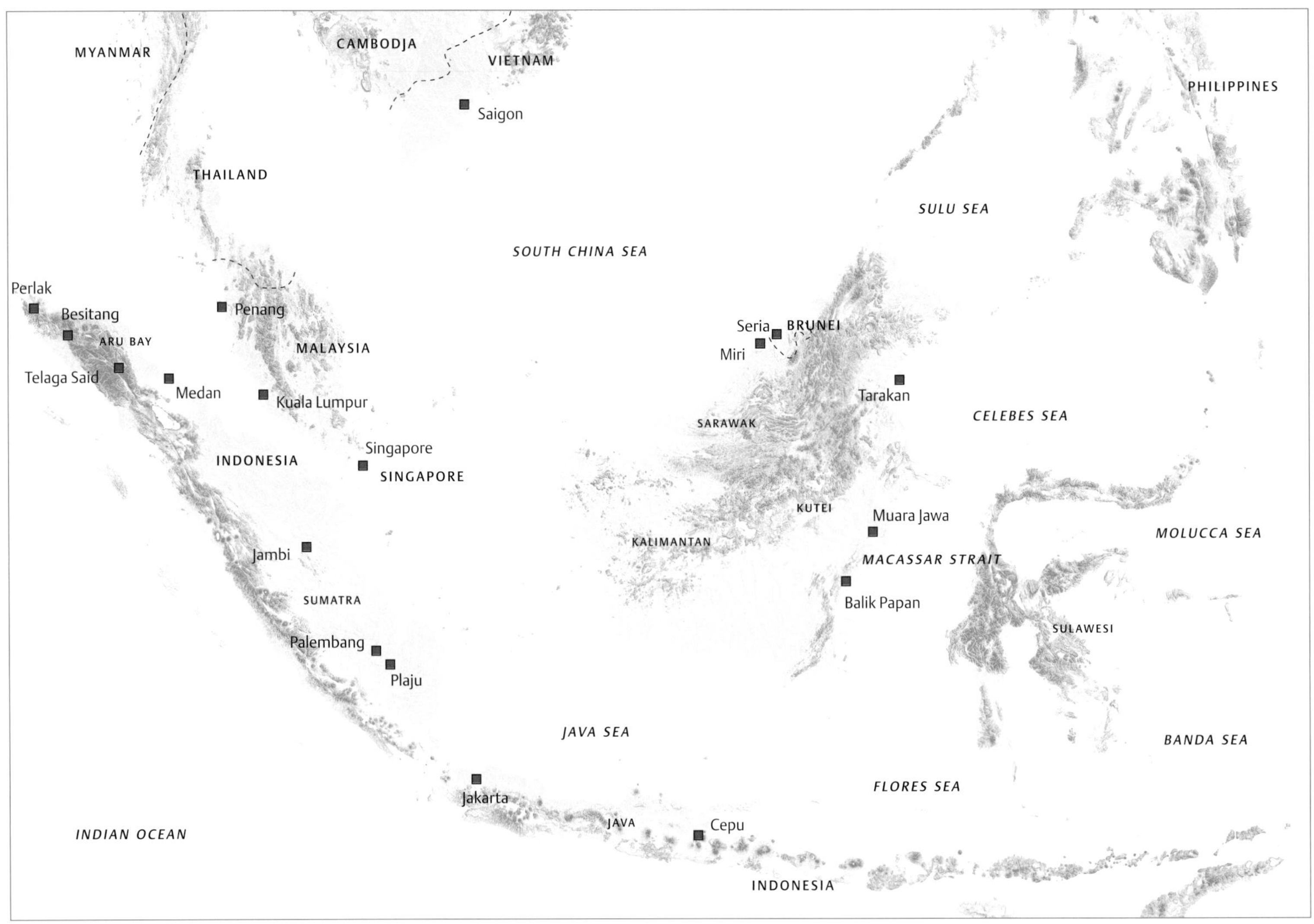

[33]

sentiment to a frenzy. Suddenly the whole industry seemed at risk from foreign predators. The Moeara Enim deal was scuppered when the Minister for the Colonies J. T. Cremer, after a visit from Kessler, dropped a heavy hint to the Moeara Enim board that permission to transfer the concession to a foreign concern might well be withheld. This threat could not have been carried out without incurring serious diplomatic repercussions, but no more than a hint was needed, or indeed even wanted by those board members opposed to the bid.[63] Meanwhile the Royal Dutch board had drafted a proposal to change the articles of association. The right of appointing board members would be transferred to the carefully selected holders of specially created 4 per cent preference shares. The proposal aroused fierce opposition from the press, shareholders, and legal specialists as being injurious to shareholders' rights, its creation of two classes of shareholders alien to the Dutch system of corporate governance. Even so, after two tumultuous general meetings devoted to the subject the shareholders voted in favour, setting a precedent in Dutch company law which was to shape investor relations into the twenty-first century.[64]

Thus, by 1898 Royal Dutch had overcome its teething troubles, and a bright future beckoned. Selling five million cases of kerosene a year, it was a force to be reckoned with on the Asian market, more so than the Dordtsche with its exclusive concentration on Java.[65]

Key locations in the former Netherlands East Indies, the hub of Royal Dutch's operations in exploration, production, marketing and sales during its early years.

On the Telaga Said concession, the company possessed eighteen richly producing wells. A huge gusher on Bukit Mas struck in February 1897 had suggested that there were unlimited reserves on the adjoining concessions as well. At the same time, the growing sales network turned Royal Dutch from a local producer into a truly international concern. The decision to locate the sales and transport department under Deterding not in Penang but in The Hague underlined the ambition for wider horizons, very similar in fact to those entertained by Samuel & Co.

From trinkets to tankers Universally admired for his profound knowledge of shipping and of oil, Fred Lane (1851-1926) was a businessman second to none, versatile, inventive, a born entrepreneur with a gift for developing business opportunities, and a man with a mind of his own. The commercial world of his day saw frock coats and silk top hats as the hallmarks of respectability, but Fred Lane wore suits and a bowler hat, signalling himself to be both independent-minded and aware of changing fashions. His discretion, charm, and persuasiveness enabled him to cross divides, and thus to fix transactions for economic rivals while remaining on good terms all around. His nickname 'Shady Lane' unjustly hints at dubious schemes and hidden agendas, whereas contemporaries unfailingly praised his honesty and uprightness.[66]

His firm Lane & Macandrew, set up in the early 1880s, was originally a shipping broker's business. In that capacity Lane served the Paris Rothschilds, organizing their kerosene transport from Batum on the Black Sea to ports along the Mediterranean. These operations developed into a deep involvement with Russian oil. In 1884 Lane & Macandrew were the first to charter a tanker to carry bulk oil to Britain. This astute move heralded a new phase in the European kerosene trade, until then dominated by American kerosene imported in cases of two five-gallon (US) tins. Eliminating the cost of casing and tins, bulk transport lowered the unit cost of transport, thus giving Russian exporters a clear advantage over US producers in the European market.[67] Moreover, bulk trading led to a concentration amongst the local importers, the numerous independent general merchants and agents making way for large and dedicated oil marketing firms such as the Kerosene Company Ltd., set up by Lane and the Rothschilds in 1888, and Standard's Anglo-American Oil Company Ltd. and Deutsch-Amerikanische Petroleum Gesellschaft, both founded two years later.[68] This consolidation gave producers a better grip on sales and prices, a vital step in the competitive race for market share.

Lane also began to explore the Asian markets for Russian kerosene. From 1885, he organized shipments of case oil to various ports through a variety of London merchant houses. The first consignments went as part-cargo; in 1888 the first full cargo of case

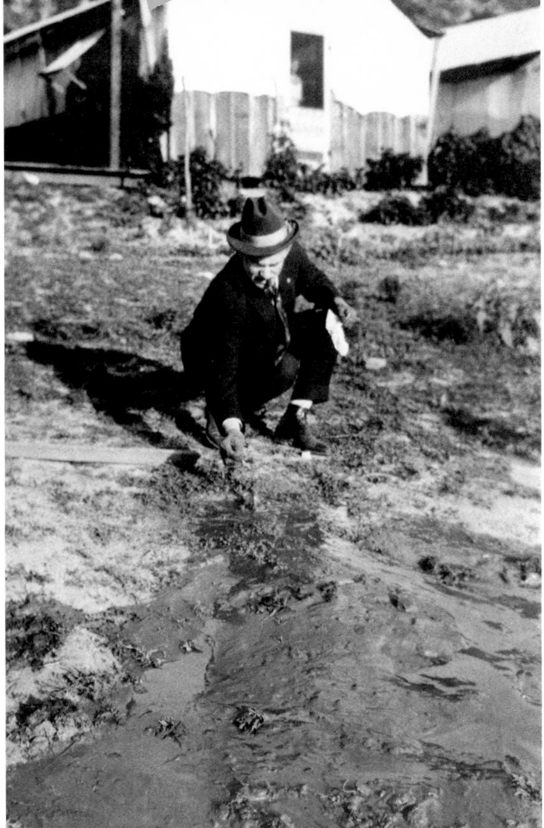

[34]

Recently found in an album from 1916, this photograph shows gas, water and mud seeping from the ground 'about 400 feet from Taylor Well number 1'. Evidently in playful mood, Fred Lane is reaching out to light the gas.

oil reached Singapore.[69] With Asian demand rising rapidly, selling lots of kerosene proved to be no problem at all, but establishing a regular marketing network remained an elusive goal in the face of Standard's well-entrenched agencies. The obvious solution, switching to bulk shipments, was not as simple as it had been in Europe. To start such a venture required familiarity with trading conditions and ties with local firms across Asia. Success would depend on scale, establishing operations in several ports at once, to pre-empt attempts to thwart the scheme by starting a price war. This in turn meant substantial investments in tanker ships and on-shore installations, which neither Lane nor Bnito, the company into which the Rothschilds had meanwhile reorganized their Russian oil interests, were prepared to make. Lane first approached the well-known firms of Wallace Brothers and Jardine Matheson & Co., who handled Bnito's case oil trade to Asia, but failed to arouse their interest for the idea of bulk transport, both firms having grown weary of Standard's price wars. He then turned to M. Samuel & Co., where he found a warm welcome.[70]

Most records of M. Samuel & Co. have been lost over time, so we have to make do with an outline of the firm and its operations, rather than a precise reconstruction. Having been established by the brothers Marcus and Sam Samuel in 1878, this firm had begun to make a name for itself as general traders to and from Asia, and notably Japan, in a wide range of goods, from machinery and tools to rice, coal, fabrics, bric à brac and, in combination with Wallaces and Jardine Matheson, case oil. Exotic shells were a special item, inherited from their father's business, and the firm's premises at 31 Houndsditch, just on the border between London's City and East End, included a workshop where these shells were glued as ornaments on a variety of seaside souvenirs, such as round lacquer boxes.[71] Though not ranking on a par in name or financial means with older established merchant firms such as Jardine Matheson or Butterfield & Swire, M. Samuel & Co. possessed just the sort of assets which Lane and Bnito needed. Experts in trade and shipping with Asia, the Samuels had a network of trusted trading partners and knew most ports from personal experience, for in the course of

[35]

[36]

their careers both brothers had travelled extensively. Moreover, the Samuels had earned a sound reputation in the oriental trade and their Yokohama affiliate Sam Samuel & Co. already imported Bnito's Anchor brand case oil into Japan.[72]

Marcus (1853-1927) was the firm's front, a rotund but somewhat withdrawn man with keen sense of decorum, a born merchant quick in spotting business opportunities, perhaps to the point of impetuosity. He craved to be recognized as the man of importance he was, aspiring to public office and relishing the ostentation that commercial and social success brought him. Yet behind the stern exterior he remained a warm-hearted man, very affectionate to his family and friends. Brother Sam (1854-1934) could not have offered a greater contrast. A lifelong bachelor, he appears to have remained a traveller at heart following the decade he spent living in Japan building up Sam Samuel & Co. into a first rank foreign trading house, with subsidiaries in Yokohama and Kobe. Consequently, he did not care much for the carriages and elegant houses of his brother, though he was happy enough to share them as a guest. Unburdened by a craving for acceptance, he enjoyed his own quiet social success in the full style of a backbench Member of Parliament, creating occasions where people could meet and reach solutions, rather than make public stands. Voluble and generous to a fault, as spontaneous and outspoken as Marcus was taciturn, Sam provided the partnership's balance, cautioning, procrastinating when his brother wanted to power ahead, and as a consequence the brothers had heated debates over every major business decision, until reaching agreement and then huddling together close, arms around each others' shoulders.[73] With both brothers in their late thirties and at the height of their powers, M. Samuel & Co. was a firm to watch.

The Samuel brothers, Marcus (left) perhaps about 30 years old, and Sam (right) in 1922.

Hatching a coup Preparations for the venture got under way in 1890 with Lane accompanying first Marcus and then Samuel Samuel on a trip to Baku and Batum. From Russia, Sam Samuel travelled on to Asia, presumably to win the support of the firm's business associates for the plans. Meanwhile the Samuels must also have commissioned the engineer Fortescue Flannery to design a tanker ship with two special features. The tanker had to have an efficient system for flushing the tanks after discharging the kerosene, so they would be clean enough to hold general cargo on the return trip. With the oil trade still very much a one-way business, tankers stood to lose their cost advantage over normal freighters unless freight costs could be spread over the full voyage. More importantly, the ship had to meet stringent safety

[37]

The adventurous Abrahams brothers, Mark (standing) and Joe, photographed about 1890, a few years before being sent out to Kalimantan to prospect for oil.

M. SAMUEL & C°. MANAGERS OF
THE "SHELL" TRANSPORT & TRADING COMPANY, LIMITED.
TELEGRAPHIC CODES USED
A.B.C. 4th Edn. Scott's Code 6th Edn. 1885 & 9th Edn. 1896. A.I. Code Watkins & Engineering Code.

TELEGRAPHIC ADDRESS.
LEUMAS, LONDON.
TELEPHONE No. 4362.
– – 4499.

16, Leadenhall Street,
London, E.C.
April 28th '99.

M. S. Abrahams Esq.,

28 April 1899: the start of one of the many letters sent from the Samuel brothers to their nephew Mark Abrahams, showing the company letterhead.

precautions, so the Suez canal authorities would allow her to pass. Until then most tankers were still converted all-cargo carriers, and after a string of fire accidents the authorities had banned them from entering the canal zone. Standard Oil had just seen its application for tanker passage rejected. Since only ships classified by Lloyd's of London as a first class risk were allowed to pass, Flannery's plans would have to be accepted by Lloyd's and then submitted to the canal authorities for approval. This procedure was well under way by May 1891, and in August the authorities consented.[74]

Meanwhile the Samuels had ordered six tanker ships from William Gray & Co.'s West Hartlepool yard. Two trusted nephews, Mark and Joe Abrahams, were sent out to Asia with instructions to obtain sites and oversee the building of the necessary tanks and installations for receiving and distributing bulk kerosene in fourteen ports from the Bay of Bengal up to Japan, starting with Singapore and Calcutta. In December, the Samuels reached agreement with Bnito. As Bnito's sole sales agent for bulk oil east of Suez, Samuel & Co. undertook to buy a certain minimum quantity of oil every year for a period of nine years. This supply contract formed the basis for a trading arrangement between Samuel & Co., Lane & Macandrew, plus seven of Samuel's Asian trading partners, who agreed to share the profits and losses of the jointly conducted kerosene trade, bought and transported by Samuel, and sold by the Asian houses.[75] As an informal arrangement, it was typical of the agency system worked by nineteenth-century trading houses, and which enabled people like the Samuels to build operations on a far bigger scale than their means really permitted.

Merchant firms such as Samuel & Co. kept costs down by operating with a minimum of staff and office space. The Houndsditch offices housed no more than the Samuels, the bookkeeper-cashier, a senior clerk, a few clerks, and an office boy. Ancillary tasks were farmed out: warehousing, pick up or delivery, buying and selling bills, chartering ships, settling insurance claims, were all arranged through brokers such as Lane & Macandrew for shipping, or the flamboyant Frank Daniels for bills of exchange, swanning into the Houndsditch office every day in a silk hat, frock coat with a fresh gardenia on the lapel, and white spats.[76] The high degree of specialization and concentration in the City of London, with a huge range of dedicated services available literally within a square mile, fostered such arrangements. Offices were in walking distance of each other, so people could meet frequently for a quick brief or to catch up on developments. Consequently, neither the trading arrangement nor the Tank Syndicate into which it was later consolidated, brought great organizational changes to Samuel & Co. The firm acted as overall manager, taking care of shipping,

[39]

A small ship that started a great business revolution: built by William Gray in West Hartlepool, the 5,010 deadweight ton SS *Murex*, the Samuels' first ship, is seen (above) in a modern painting and (below) in cross-section, revealing her oil tanks and the cofferdams separating them from the forehold and engine room. Next to the cross-section are the funnel design and house flag of S. Samuel & Co, used from 1892 to 1907.

sales, strategy, and finance, keeping accounts, and charging interest, commission, and costs for services rendered to the syndicate. Lane & Macandrew ran Samuel & Co.'s ships; Flannery, Baggallay & Johnson, Flannery's engineering firm, acted as chief engineers, supervising all technical shipping matters, down to signing repair bills.[77]

The financial construction of the joint kerosene trade shows a similar ingenuity in making a little go a long way through devolution, for the Samuels built the venture on credit rather than on capital. Gray & Co. offered very long terms of payment on the tankers. The overseas partners had to invest several thousand pounds sterling of their own money in tanks and installations, whereas the Royal Dutch agents managed installations owned by the company. Whessoes, the firm supplying this hardware, also agreed to extended terms of payment, enabling Samuel & Co. to charge its associates a commission for arranging the credit. The conditions agreed with Bnito stipulated payment for kerosene delivered in three to four month bills, for which the proceeds of sales were remitted by telegram. An average round trip from Batum to Asia took between three and four months, so Samuel & Co. had no need at all to fund the Syndicate's trade. The heavy

[40]

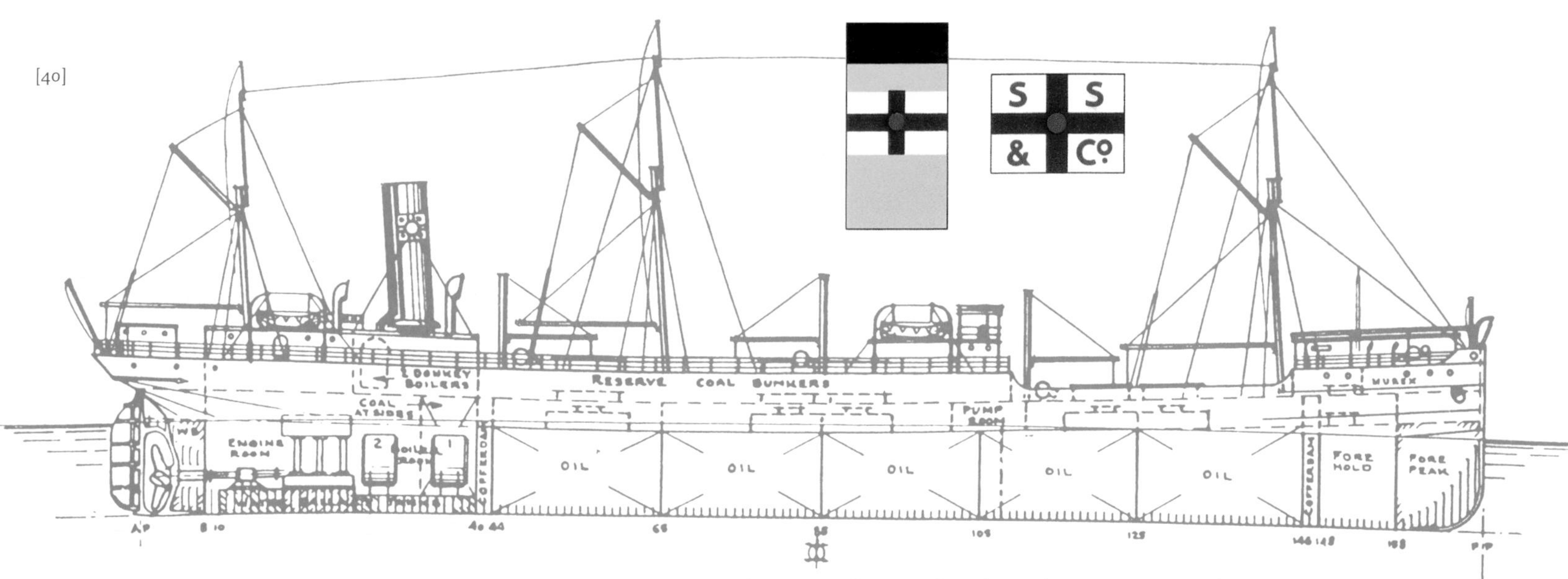

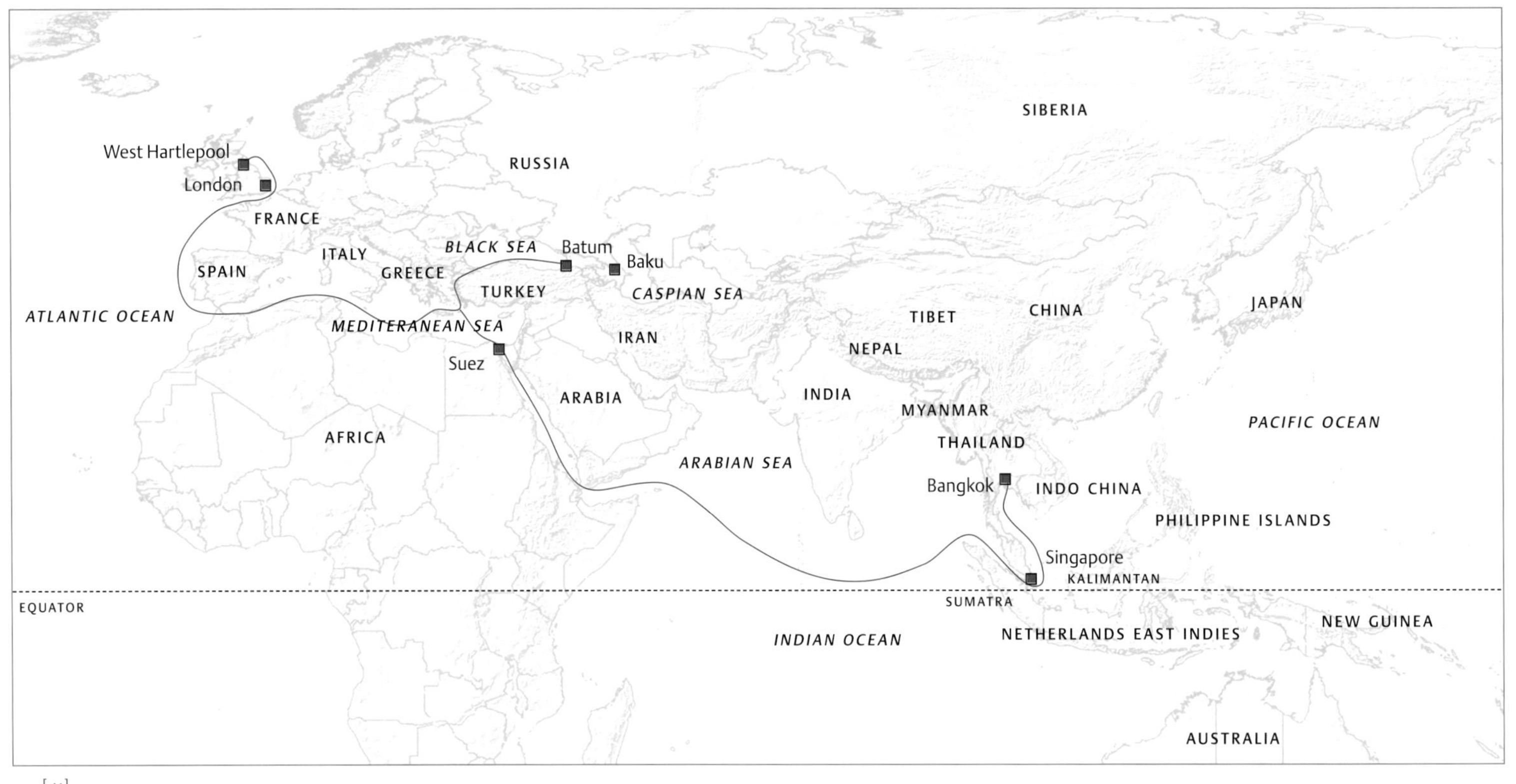

[41]

reliance on credit did have the drawback of making the Tank Syndicate very dependent on a strong cash flow, but circumstances rather favoured Samuel & Co. in this respect. Firstly the Asian demand for kerosene increased by leaps and bounds, on average by about three million cases a year between 1889 and 1893.[78] Secondly, the venture started during a period when the Russian oil industry experienced one of its recurring crises. From 1890 to 1894, kerosene prices at Batum dropped by more than two-thirds, increasing the cost advantage of Russian over US kerosene in Asia.[79]

Rumours about Samuel & Co.'s preparations immediately stirred up resistance from business interests connected to the case oil trade, ranging from Standard Oil to British shipping agents and Welsh tin plate manufacturers. Prolonged efforts to get a cancellation of the canal authorities' permission failed, however, because the British Government refused to be drawn in the controversy.[80] Local businessmen in Asia were occasionally more successful in thwarting the Syndicate's plans. Kessler and Stoop persuaded the colonial government to issue a ban on the erection of tank installations in its main ports.[81] Cloaked in the official fear of a threat posed by the Samuels' impending marketing power, this subtle but effective piece of colonial preference was yet another example of the government's willingness to waive the open door policy when it came to protecting its favourite industry. Retailers quickly cashed in on the government ban by forming cartels and raising retailing prices.[82]

In the end, exclusion from the Dutch East Indies meant an annoying setback for the Samuels' venture, but not a serious blow, given the buoyant demand on the Asian market. The passage of Samuel & Co.'s new tanker SS *Murex* through the Suez canal in August 1892 and her subsequent arrival in Singapore marked the completion of a great commercial coup. Russian kerosene rapidly established itself as a tough rival to Standard's Devoes brand, outselling it in India within a year. By 1895, Russian exports of kerosene to Asia matched those of the US.[83] Even the tactical mistake of not providing cans at the retail end, committed in an effort to reduce the sales price to the absolute minimum, proved to be no more than a temporary handicap to sales, and when

From West Hartlepool to Bangkok via Batum: the historical first voyage of *SS Murex*, 1892.

corrected it surprisingly turned into a marketing asset. Marcus Samuel simply chartered a ship with Welsh tinplate and ordered Mark Abrahams, still travelling around in Asia to construct the remaining bulk stations, to start building canning works everywhere, typically leaving him to sort out the design and the details of obtaining sites, machines, tools, and staff. Abrahams' inspired choice to have all cans painted bright red gave the Russian kerosene a distinctive product identity, which considerably helped to win consumers, if only because the fresh red stood out against the blue Devoes cans, as often as not battered and rusty after sailing the 13,000 miles or so from New York.[84]

The nature of the business

The kerosene coup helped Samuel & Co. to achieve notable prominence in the City, but that was certainly not due to oil alone. The firm's commodity trade boomed as well, notably during the Sino-Japanese War of 1894-5. Samuel & Co.'s key role as a supplier to the Japanese government led to M. Samuel & Co. being entrusted, in 1897, with floating the first Japanese loan in London, marking Samuel's ascent to the elite ranks of London merchant banks.[85] The Samuel brothers acquired great wealth, controlling assets of at least £ 1.8 million pounds, not far off Jardine Matheson's figure, but built up in a much shorter time.[86] With Marcus having served as Alderman of the City, and moving towards the prestigious position of Lord Mayor, social recognition had arrived as well.

However, the sparse available data suggest that we must qualify the long-term consequences of the *Murex* trip. For the early years of Samuel & Co.'s oil business, we really have only the fleet tonnage and the number of passages of the firm's ships through the Suez canal to get any idea of size.[87] Multiplying the number of trips which each ship made with its capacity for carrying oil, gives us an estimate of the firm's kerosene sales (Figure 1.2).[88] The graph shows the rapid expansion of the business to good effect;

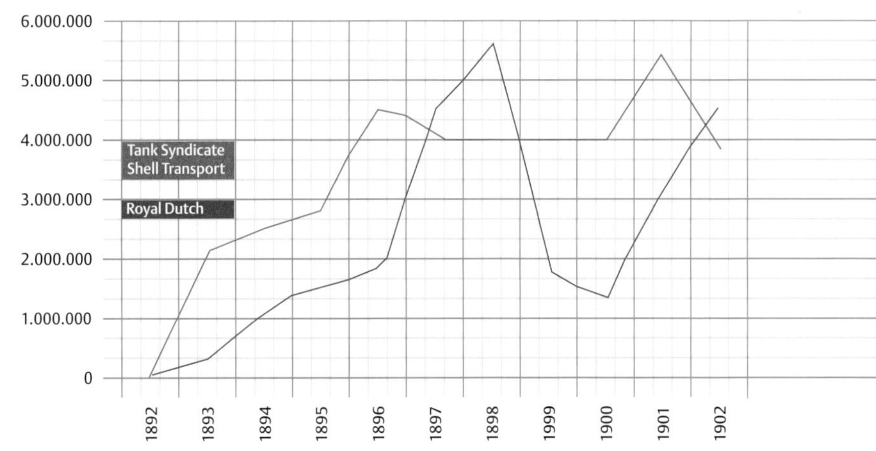

Figure 1.2
Estimated number of cases of kerosene sold by the Tank Syndicate/Shell Transport and cases produced by Royal Dutch, 1892-1902.

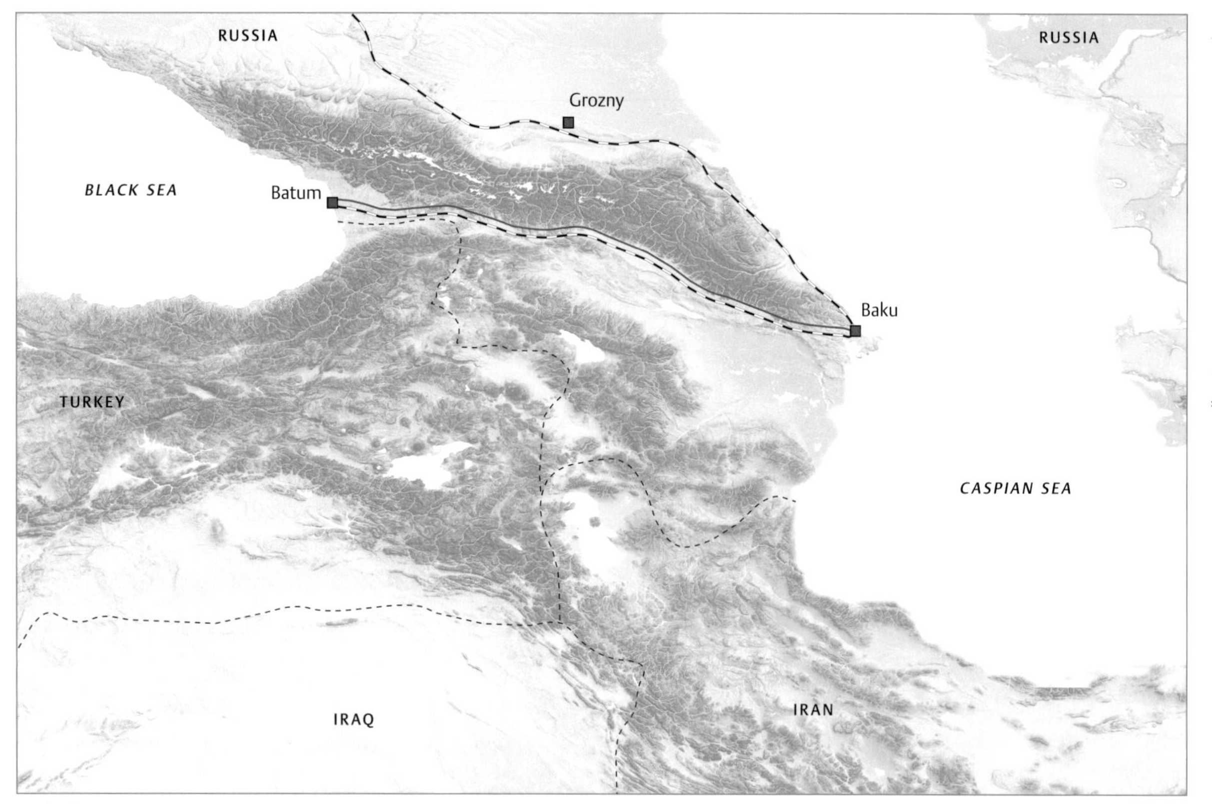

The Caspian oil-producing region of Baku and the Black Sea exporting port of Batum (modern Batumi) were the heartland of the 19th-century Russian oil industry.

— Pipeline
— Railway

[42]

estimated sales rose from just over 100,000 cases in 1892 to a peak of 4.4 million four years later, or about 40 per cent of Russian exports in a normal year.[89] That is to say, the competition from other Russian exporters remained alive and keen.[90] Case oil exports were by no means wiped off the market, as the constant pressure on prices from which the fledgling Royal Dutch suffered testifies.[91]

Moreover, from 1896, Samuel & Co.'s oil sales stagnated around 4 million cases for another four years as a consequence of the erratic nature of the Russian oil industry. Batum kerosene supplies and prices were notoriously unpredictable, not least because the Russian government constantly manipulated the freight rates on the Transcaucasian railway to suit its fiscal needs and wider economic purposes. Consequently, producers switched from kerosene to fuel oil and back as circumstances dictated. This resulted in severe shortages alternating with gluts.

One such drop in supplies began in 1897, cutting across Samuel & Co.'s expansion. As a result the firm's shipping profits dropped sharply. Profits on a single trip could vary between £ 1,855 and £ 5,781 within a single year.[92] Fluctuating supplies, the belatedly recognized need to supply tins, and the uncertainty of return freights all reduced the advantage of bulk over case oil to better stowage, less leakage, and the cost of the case. Moreover, it remains very much a moot point whether the benefits of Flannery's flushing system outweighed its cost.[93] In addition, rival tanker operators appeared on the scene; one of them, Alfred Suart, possessing no fewer than sixteen tankers of various types, and his own production in Baku to boot. Meanwhile Samuel & Co.'s kerosene sales had also been overtaken by those of Royal Dutch. During the years 1899-1900 the two companies probably remained neck-and-neck, for, as we will see below, Royal Dutch's production suddenly dropped, but Kessler kept sales going with his own imports from Baku.

We may thus conclude that the coup enabled the Samuels to establish themselves firmly in the Asian kerosene trade through a rapid build-up of scale, rather than a price advantage. However, this was not achieved overnight, and Samuel & Co.'s competitive position was less solid than the initial success of the coup might seem to suggest. By 1897 the firm's oil trade had been overtaken by

Balachamy oil field near Baku, about 1890. The Samuel brothers and their 'Tank Syndicate' – precursor of Shell Transport – first shipped Baku oil to the Far East in 1892.

[43]

Royal Dutch. This explains the brothers' exasperation with the Russian oil industry in general, and with Bnito in particular, whose terms and conditions proved disadvantageous whatever the situation in Batum. Samuel & Co. could not reap the full benefit of falling prices at times of oversupply, such as during the 1893-4 crisis when Russian kerosene sold at prices undercutting Bnito by more than a quarter. In times of shortage, the firm could not get sufficient kerosene to keep all its tankers going, because Bnito carefully rationed bulk supplies so as not to harm its own case oil sales, marketed across Asia under the Anchor brand.

The hassle with Russian supplies made the Samuel brothers thoroughly aware of their exposure in the middle of the supply chain, tied in at either end.[94] What if a rival tanker operator linked up with an independent Russian producer and entered the Asian market? Or if Bnito linked up with Standard? In either case, the business would find itself in a precarious position, sitting on expensive assets, and dependent on a single supplier to feed a rather loosely constructed transport and marketing network. Consequently, Samuel & Co. strove to consolidate their greatest intangible asset, the marketing network, by transforming the trade association into the Tank Syndicate in 1893, thus securing the formal commitment needed to turn the operations into a more lasting business.[95]

As the Tank Syndicate's sales grew, so did competition. Royal Dutch began selling in the Straits, and then in all major ports, confronting the Syndicate with the need to tighten up marketing coordination. Consequently, on 18 October 1897 the Syndicate was incorporated as the joint-stock company Shell Transport & Trading, with an initial capital of £ 1.8 million. This step was clearly taken not so much to attract fresh capital, as to achieve better integration, and because the members wanted to reduce their exposure to the risks of their growing and very volatile oil trade. About £ 370,000 of shares were sold to the public. The Syndicate members plus Lane & Macandrew together received £ 230,000 in £ 100 shares for their assets in steamers and tank installations, the Samuels together kept £ 1.2 million plus commanding voting rights to control the company.[96] The Samuels also held on to the oil trade of their Japanese firm, which did not join Shell Transport but remained their

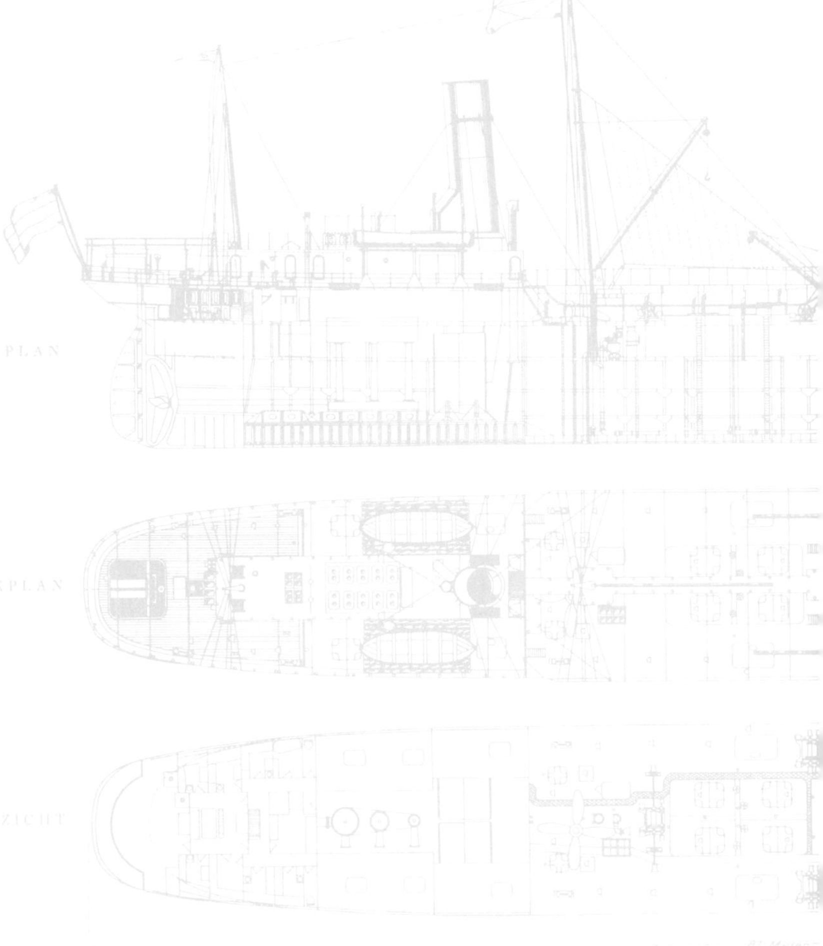

private property.[97] Marcus Samuel became chairman, the Syndicate members and Lane took seats on the board, and M. Samuel & Co. managed the company as before, with largely the same staff except for the appointment of a company secretary. The firm moved its offices from Houndsditch to a more respectable address in the heart of the City, at 16 Leadenhall Street. Circumstances helped to put Shell Transport and its chairman quickly and firmly in the public eye. In February 1898 a Royal Navy warship ran aground in the Suez canal, and was towed free by a Shell tanker. By waiving the navy's offer of financial compensation for the salvage, Marcus Samuel earned public respect and a knighthood.[98]

Shell Transport and Royal Dutch clearly recognized each other as a competitor of similar strength and intentions. Moreover, Sir Marcus and Kessler had a great liking and respect for each other, so in the early summer of 1898 Lane had little difficulty in brokering a marketing agreement between them, covering the whole of Asia and giving Crown Oil cases a small margin over two Shell Transport tins.[99] At the same time, the Samuels sought to reduce their total dependency on Bnito by opening alternative sources of supply. The Dutch East Indies was the obvious area to look to, but none of the producing companies there could oblige. Determined to get into production, the Samuels decided in September 1895 to back a wildcat operation in Kalimantan (Borneo) on a concession named Kutei. Thus began a new and dangerous adventure for the company. More or less simultaneously, Royal Dutch also had to negotiate a very tricky phase in its development.

Two crises, 1: Royal Dutch

Overconfidence lay at the heart of Royal Dutch's crisis. Telaga Said production passed the 500,000 cases a month mark late in 1897, inspiring Kessler to instruct Deterding to charter tankers for transporting a million cases a month.[100] The widespread oil seepages, and then the gusher, had convinced everyone that Bukit Mas harboured rich reserves. The company was in no rush to develop them under the circumstances, mainly because the potential volume from Telaga Said already exceeded the available shipping capacity, so a number of wells had to be shut in. Some exploration had to be done to fulfill the concession obligations, so Loudon sent crews with manual rather than steam-driven drilling equipment into the field, being both cheaper and slower.[101]

The first signs of trouble had already appeared, however. Ten wells drilled on Bukit Mas had produced no more than traces of oil

[44]

[45]

A little flattery: one of Royal Dutch's earliest ships, the *Sultan van Langkat*, was named in honour of the local Sumatran ruler. The photograph of her at anchor in Aru Bay is backed by diagrams showing her structure in cross-section and her deck plans.

and gas.[102] When the Standard Oil experts Lufkin and Fertig paid a courtesy visit to Langkat in April 1897, the Bukit Mas gusher, reopened to impress them, was dead. The two men also commented that the Telaga Said wells, huddled together on a narrow strip of land measuring no more than 100 hectares, were drawing from a single reservoir and thus likely to run out all at once. Neither Loudon nor Kessler could believe them.[103] Somewhat later during the same year, Adriaan Stoop had brushed aside Kessler's optimistic forecast of Royal Dutch producing seven million cases in 1898 with the remark that sooner or later all wells stop producing.[104] Kessler noted nothing untoward on his inspection visit during the winter of 1897-98. Indeed, the wells produced better than ever: 530,000 cases in February, almost 700,000 cases in March, inching closer to the magic million mark, so salt water gushing from a new well, a sure sign of exhaustion, was disregarded. Rumours about water in the storage tanks put a damper on the celebrations surrounding the keenly awaited arrival of the company's new tanker, the SS *Sultan van Langkat*, on 17 April.[105] Now the stops could be pulled out. All wells on Telaga Said were opened; production peaked at 840,000 cases in May. Then the flow dropped sharply, 460,000 cases in June, 300,000 in August, 200,000 by December. Loudon, who happened to be on leave in Europe, returned to Langkat as fast as possible. As the news spread, the Royal Dutch share price went into a spiral, to 400 per cent of par in July, and to 250 by October, terminating any chances to place the 1.5 million guilders of preference shares. Even Kessler hesitated before committing no less than 700,000 guilders of his capital in a desperate bid to rescue the issue. The rest was sold to the *commissarissen* and trusted business friends.[106]

The initial response to the crisis was a natural one, but bound to fail under the circumstances. Kessler downplayed the seriousness of the crisis, telling the Annual General Meeting at the end of June that bringing the company's huge reserves into production would solve the problem. The extensive drilling programme ordered failed to yield any results at all. As it turned out, Zijlker had had a triple stroke of luck. His main well at Telaga Tunggal, which came in during the second campaign in 1885, had been drilled at the crest of an anticline, in the centre of a fairly narrow elliptical area delineating the main oil-bearing strata, and had passed through a secondary oil reservoir right into the main pool. Drilling outside the narrow strip, or deeper, simply made no sense, but recognizing the location and properties of the anticline,

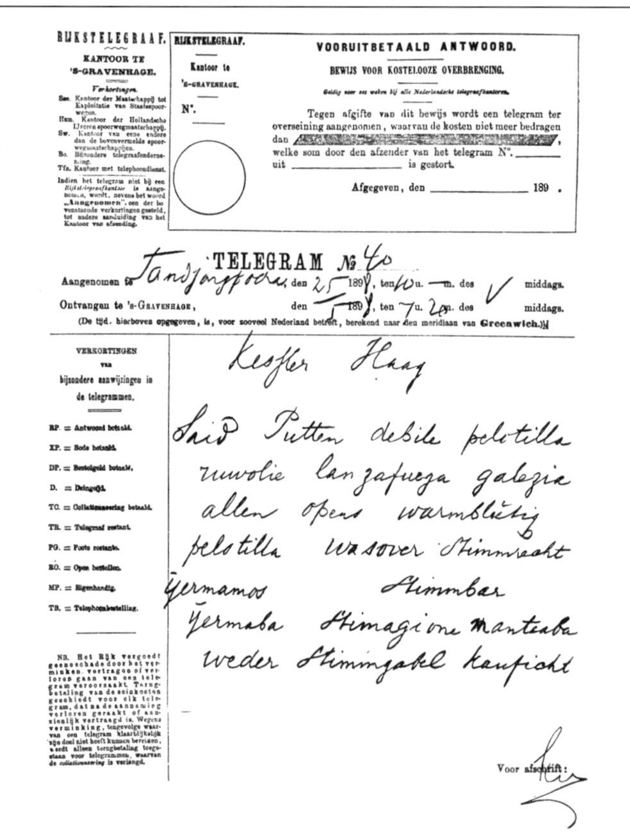

RIJKSTELEGRAAF.
KANTOOR TE
'S-GRAVENHAGE.

RIJKSTELEGRAAF.
Kantoor te
'S-GRAVENHAGE.
N°.

VOORUITBETAALD ANTWOORD.
BEWIJS VOOR KOSTELOOZE OVERBRENGING.

Tegen afgifte van dit bewijs wordt een telegram ter overseining aangenomen, waarvan de kosten niet meer bedragen dan ______, welke som door den afzender van het telegram N°. ______ uit ______ is gestort.

Afgegeven, den ______ 189 .

TELEGRAM № 40

Aangenomen te Tandjongpoera, den 25/5 1898, ten 10 u. — m. des ✓ middags.
Ontvangen te 's-Gravenhage, den 25/5 1898, ten 7 u. 20 m. des ✓ middags.
(De tijd, hierboven opgegeven, is, voor zooveel Nederland betreft, berekend naar den meridiaan van Greenwich.)

VERKORTINGEN van bijzondere aanwijzingen in de telegrammen.

Kessler Haag

Said Putten debile pelotilla
ruvolie lanzafuega galezia
allen open warmblutig
pelotilla was over Stimmrecht
Yermanos Stimmbar
Yermaba Stimagione manteaba
weder Stimmgabel kaufticht

Voor afschrift:

Tandjongpoera
25 Mei

Said
putten
Debile = in the course of a few days
pelotilla = production
ruvolie
lanzafuega = much less than expected
galeria = give
allen open
warmblutig = plenty of water
pelotilla = production
was over
Stimmrecht = 18000 barrels
yermanos = day before yesterday } 23 Mei
Stimmbar = 12000 barrels }
yermaba = yesterday 24 Mei
Stimagione = 10500 barrels } 25 Mei
manteaba = this morning }
weder
Stimmgabel = 14000 barrels
Kaufticht = I will keep you well informed

25 May 1898: a coded telegram (above) and the key to its translation (above right) bring to The Hague the catastrophic news that production in Royal Dutch's only well is falling rapidly.

The basic geology of petroleum: crude oil, migrating drop by drop through porous rock towards the surface, was found to accumulate under caps of impermeable rock. Such caps were often found in upward-folded 'anticlines', which could offer a simple clue for early prospectors.

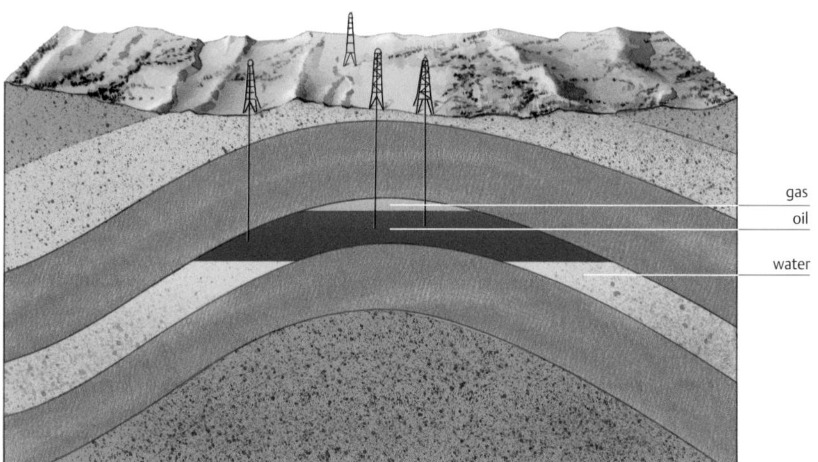

and thus the suitable drilling sites, required geological knowledge. Fennema had correctly identified the course of the anticline in his report, but Kessler and Loudon had rather relied on the know-how of their American crew, who continued to direct operations, now with rather less luck than before.[107] Nor did the increasingly desperate campaigns succeed in striking oil on any of the other concessions, since the famous seepages actually demonstrated that the oil had already escaped from the underlying strata, steeply folded, therefore fractured and porous.[108]

The board recognized the mistakes fairly early, and took effective steps to remedy them. To counter the constant rumours about the company's predicaments, the board abandoned its secretiveness and began to issue regular newspaper statements with production data.[109] During the autumn it became patently clear that further drilling along the established lines and in the available concessions was not going to produce results.

[48]

Test pits were dug close to one another in lines, with each pit being up to three feet square and between 10 and 20 feet deep. Geologists would go to the bottom of each of these claustrophobic holes to make their observations and assess the possibility of oil being present.

Consequently, the concessions and wells were all written off at a stroke, leaving a paltry profit. To save cash, contracts for packaging materials were cancelled or sold on, and ships were chartered to others. An offer from Samuel & Co. to buy the whole or parts of the enterprise was turned down. Kessler remained convinced of the company's future as an independent and integrated business, and was determined to keep the parts together for as long as possible.

One way of doing this was to copy the Shell Transport concept and feed the sales network with Russian kerosene. In February 1899, Royal Dutch sent a circular to its Asian sales agents announcing these plans.[110] Kessler obtained supplies through forming a syndicate with Ogilvy, Gillanders & Co., a London merchant firm affiliated to Royal Dutch's Calcutta and Madras agents. From 1 July 1899, the syndicate began trading under the name Eastern Oil Association, selling kerosene under the name Cross Oil, transported with tankers chartered from Alfred Suart & Co. and the Royal Dutch fleet. Ogilvy, Gillanders & Co. became general and shipping managers. At Kessler's express insistence, Royal Dutch took charge of sales. The contract bound the members for a period of two years, and envisaged sales of between 60,000 and 110,000 tons (2-3.7 million cases). Russian kerosene proved to be a very welcome expedient for Royal Dutch. Revenues exceeded income from Crown Oil by the last five months of 1899, and in 1900 the ratio was almost 2.5:1.[111] Despite the earlier good intentions on openness, the board preferred to remain vague about the arrangement. The annual reports for 1899-1901 mention the company's Russian kerosene trade, but make no reference to the Association at all, as if it was more an embarrassment than a godsend.[112] Shell Transport, which might have used the company's predicament as a unique opportunity to drive a close rival from the

[49]

[50]

[51]

[53]

The geologists' way of life in the Far East was primitive but exotic and sometimes exciting. In the main picture the bearded Dr H. Hirschi is seen seated at the centre of a large group of colleagues and workers, somewhere near Pangkalan Brandan in 1903. The upper row of pictures shows (from left to right) a geologist's house at Soengei Kerkei, 1906; oil explorers in the Dutch East Indies being transported in litters, about 1900; exploration workers being transported by boat, 1905; and another house – somewhat more sophisticated than the first – for a geologist.

market, chose not to, and even agreed to let the price agreement between the companies stand for Cross Oil. In a typical grand gesture, Sir Marcus opined that the Asian market was big enough for both. With consumption running at an estimated 37 million cases a year, he was of course right, but this remark does underline his preference for gentlemanly capitalism over all-out competition.[113] The gracious example was lost on Royal Dutch. Cross Oil opened the opportunity for price competition on its home ground in the Dutch East Indies, for which the company had not wanted to use its premium Crown brand.[114]

To solve the production problem, the board commissioned a survey from two geologists with extensive field experience in oil. This decision was remarkable both for its novel approach and for its later consequences. Geologists were rarely consulted by the oil industry, the technical side of which was dominated by mining engineers such as Stoop, chemical engineers like Dr. Paul Dvorkovitz and Sir Boverton Redwood, and drilling experts trained in the field. All regarded geologists, perhaps out of professional jealousy, as a waste of money, their knowledge a paper science not worth bothering about. Thus when late in 1897 the Deli Maatschappij board wanted a geologist's opinion about the Langkat concessions to be prospected together with the Dordtsche, it sought it behind Stoop's back.[115] The Royal Dutch board probably knew about this survey through its close connections with the Deli Maatschappij. Kessler went himself to Strasbourg to recruit candidates from the university there.[116]

Though born from despair, the decision to try a new and unfashionable approach in prospecting for oil reflects a marked change in Royal Dutch's operations, from drawing on imported know-how to accumulating first-hand knowledge based on science and technology within the company. By picking teams from his drilling crew to help conduct the surveys, Loudon ensured that the new insights quickly found a receptive audience. The geological report, issued in April 1899, contributed little to reviving the company's immediate fortunes, since it confirmed that neither Telaga Said nor Bukit Mas offered much prospect of finding oil. Yet the geological survey did provide a landmark in establishing a clear,

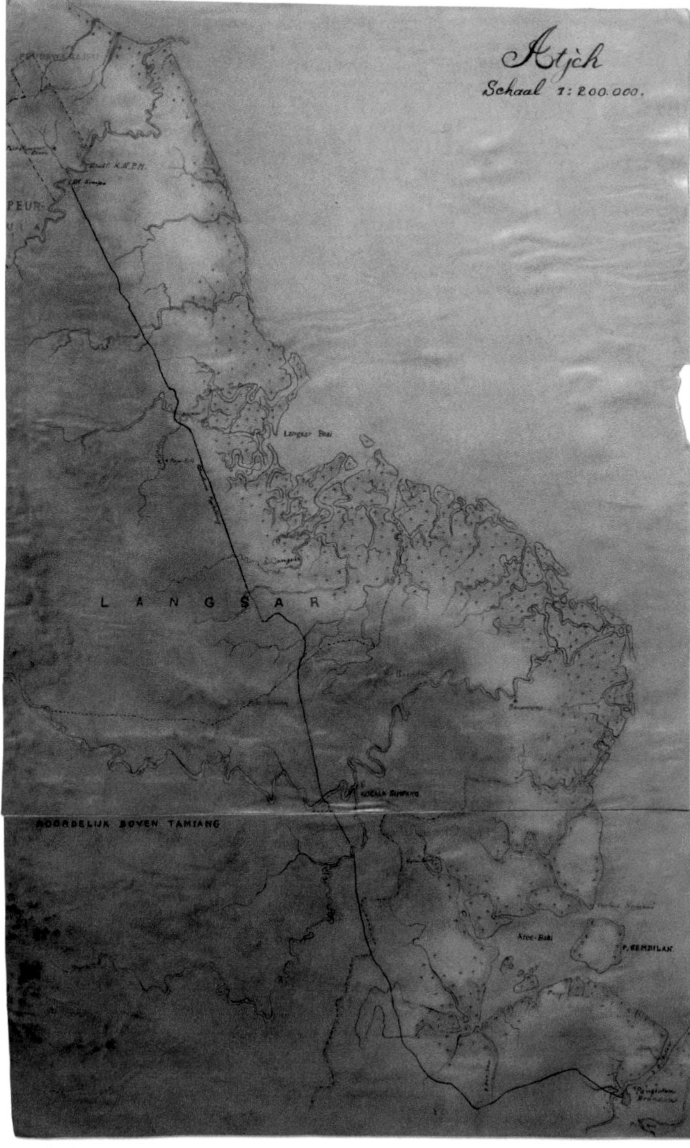

[54]

systematic manual for future campaigns, ending the happy-go-lucky days, though not of course the role of good fortune.[117]

In deciding to apply the new approach to the Perlak field, Loudon made one of the most fortunate decisions in the company's history. The area itself had only recently been brought under formal Dutch control as part of the savage war between the colonial army and the Sultan of Aceh, so the colonial government welcomed an increase of the Dutch presence there through the establishment of companies like Royal Dutch. Since under the terms of surrender, mining concessions and the maintenance of law and order continued to be the domain of the Rajah of Perlak, getting access to the oil fields required delicate manoeuvring between this prince, who fully supported exploration, and local chiefs with an allegiance to the Sultan in his determined fight against the Dutch onslaught, all the while keeping in close touch with the Dutch Governor of Aceh, Van Heutsz, to secure his administrative and, if necessary, armed support.[118]

A man of great tact and natural charm, Loudon was just the man for such occasions. The son of a former Governor-General of the Dutch East Indies, he had the advantages of a very familiar name and an intimate understanding of colonial circumstances and sensibilities, but he achieved what he did through sheer hard work, and through his gift of creating sympathy for what he wanted simply by the respect he showed to whomever he was dealing with. A geological survey of the Perlak concession undertaken in August 1899 revealed a field of unusual promise. Loudon decided to forge ahead with exploratory drilling without waiting for the formal permission required, rightly assuming that Van Heutsz would sanction him. After a final round of negotiations between Loudon, the Rajah of Perlak, and a local chief of hostile intent, drilling began in late December 1899, and within a week a powerful gusher confirmed both the great value of the systematic approach, and the optimistic expectations based on it. On the Amsterdam stock exchange, the Royal Dutch share price, lingering in the 175-200 per

Contemporary map of the Aceh region of north-eastern Sumatra, showing the location of Perlak.

In 1902 one of the producing wells at Perlak suddenly spouted out an enormous quantity of liquid mud, 'as may be seen on the derrick here'.

[55]

cent band for months, shot up to above 300 per cent of par at the news.[119]

The Perlak field proved to be a very rich field indeed, with several oil strata on top of each other, and very favourably situated, too, for the crude could be transported by pipeline to Pangkalan Brandan, about 130 kilometres away. It would take another four years before production passed the 1898 peak of 5.4 million cases, but the discovery did restore Royal Dutch's future as an independent and integrated oil company. The downstream organization had sustained the business through the crisis, but without production Royal Dutch was a lame duck. The event also heralded two further key operational changes. First, from now on exploration would be preceded and directed by the company's own geologists. One of the first recruits, the Swiss geologist Josef Erb (1874-1934), ascended to the very top of the company, demonstrating just how much Royal Dutch stood open to managers with a background in science, of whatever nationality. Teams were sent out all over the Indonesian archipelago to scout potential exploration sites. In 1901, the company employed no fewer than eleven geologists, some of them on a temporary basis, and local managers reported their results in detail to The Hague.[120] Second, Perlak crude turned out to contain 27.6 per cent of very light gasoline, or naphtha as it was then called, with just the right

[56]

Josef Theodor Erb, Royal Dutch's first chief geologist, travelled the world on the company's behalf and from 1921 to 1929 served as a managing director.

properties to be sold on the growing European market. Until then Royal Dutch was a single-product business, making and selling lamp oil. Langkat gasoline was regarded as a hazardous nuisance and burnt in pits.[121] The small volume of fuel oil and tar produced made transport and marketing uneconomical in contrast to, for instance, Dordtsche, which could sell its wax and lubricating oil more or less on the doorstep. Moreover, the company would have had a hard time competing in European markets with kerosene against US or Russian producers, both well-entrenched and with the same cost advantage Royal Dutch had in Asia. Perlak gasoline changed all that, giving Royal Dutch a large volume of a very competitive product to gain entry to new markets.

Strategic vision mattered as much as product availability in effecting this fundamental reorientation. As early as 1896, Loudon had begun selling gasoline to India and commented on the great potential for this product in Europe, but Kessler wanted to put all efforts into producing kerosene.[122] The decision to locate transport and sales under Deterding in The Hague already marked a clear shift in the company's focus. During the company's supply crisis, Deterding made a thorough study of the European markets, spotting the huge potential for selling gasoline, and planning the next moves. After the Perlak success, Loudon was recalled to The Hague as well to help manage the growing volume of work, brought about by the transformation of what used to be an administrative head office into the dynamic centre of rapidly expanding operations. And then suddenly Kessler died on 14 December 1900 on his way back from an inspection trip to Sumatra, where he had fallen ill. He was really Royal Dutch's founding father, the man who created the business and sustained it against the odds, perhaps even against economic common sense. His absolute firmness of purpose had built solid foundations, but one wonders whether he would have been the man to mastermind the metamorphosis which followed the production crisis. This now fell to Deterding, appointed to succeed him by an extraordinary general meeting in January 1901.[123]

[57]

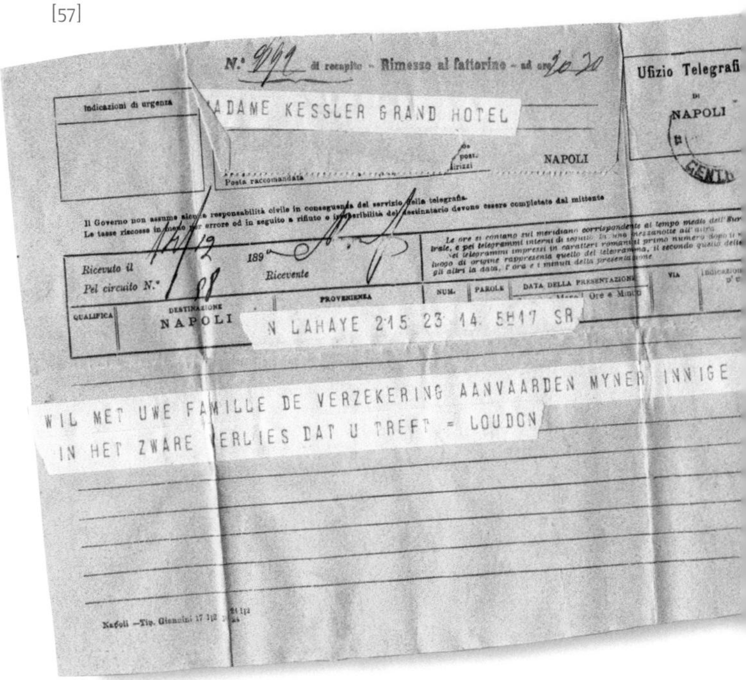
Ufizio Telegrafi
NAPOLI
ADAME KESSLER GRAND HOTEL
NAPOLI
N LAHAYE 215 23 14 5H17 SR
WIL MET UWE FAMILLE DE VERZEKERING AANVAARDEN MYNER INNIGE
IN HET ZWARE VERLIES DAT U TREFT = LOUDON

The main picture, an attractive family grouping of the Kesslers all together in the summer of 1900, shows (from left to right) daughter Go (Margo), August Kessler, Jan, Guus, Boelie, Margo Kessler, Dolph and An. But Kessler was suffering great stress, and in his touching letter (right) he says he knows he is ruining his health through working so hard, but cannot stop himself because 'one slight setback would be enough to let the whole thing collapse'. Above is Hugo Loudon's telegram of condolence to his widow.

10

26 Juli 3.

Geachte Heer Wakkie,

Ik ben werkelijk verlegen er voor dat gij mij zoo trouw schrijft en ik Uwe brieven zoo slecht beantwoorden kan. Maar U hebt werkelijk geen begrip van mijn zorgen. Lauwheid, onkunde, onverschilligheid, verval, wanorde, ergernis allerwege en dat alles moet met een handjevol geld in orde gebracht worden, terwijl bovendien de zaak moet worden uitgebreid wil men rondkomen.

Het is zwoegen en tobben en cijferen en overleggen van het oogenblik, dat ik mijn oogen opendoe tot op het oogenblik, dat ik, doodmoe, en in een staat van wanhoop 's avonds op moet houden omdat ik eenvoudig niet meer kan. Ik ruïneer mijn gestel en mijn zenuwen, maar ik mag nu niet terug, ik moet vooruit. Den 24sten kwam ik terug van een 1½daagschen tocht n/Belawan Bindjey Lan Boentoe Tandjong Poera om hout te koopen, waarmede die ellendige zaag ons vastzet en daar verrast mij de Hr. Dijkstra met de boodschap, dat hij den 28sten vertrekt, want dat hij zich zoo gevoelt, dat hij een instorting vreest, die levensgevaarlijk zou zijn; zijn zenuwgestel is geschokt; hij is doodaf, de minste inspanning bezorgt hem heftige hoofdpijn, hij kan niet denken, niets doen en voelt zich op den vooravond van een ernstige ongesteldheid. Overmorgen vroeg vertrekt de Hr. Dijkstra naar Java en het werk van iemand wiens diensten met f.1200-'s maands betaald worden komt nu nog op mijn schouders. Hoe ik het stel en uithouden zal mag de Hemel weten! Ik telegrafeerde naar Java om assistentie want het personeel (technisch) dat ik hier heb is eenvoudig akelig. Gij ziet in welken toestand ik verkeer; ik zal mij trachten er doorheen te slaan. Die financieele zorgen drukken mij ook zoo! Eén kink in de kabel en alles zakt in elkaar en al mijn inspanning is tevergeefs en ik krijg nog bovendien de schuld van een toestand, dien ik alleen met inspanning van ongewone krachten heb kunnen krijgen zooals die nu is. Als men onder dergelijke omstandigheden nog met Atjehers, Residenten, ... etc. etc. ergernis heeft, dan kunt U zich voorstellen, hoe mijn gemoedstoestand is en daarom ben ik U des te meer dankbaar voor de flinke manier waarop U voor mij in de bres springt en met mij samenwerkt om hier van de zaak het goede te maken wat er wel inzit, maar alleen met buitengewone krachtsinspanning onder de gegeven omstandigheden er uit te persen is.

Ontvang een warmen handdruk van

Uw toegen.

get. J.B. Aug. Kessler.

Dr Paul Dvorkovitz, geologist, in 1912.

[60]

Two crises, 2: Shell Transport Royal Dutch's crisis was an acute one, from which the company eventually emerged intact, strengthened, and with a new purpose. By contrast, Shell Transport suffered a process of slow attrition, which was insufficiently diagnosed and dealt with until the board had very few options left. Shell Transport's crisis began when, in its search for production in the Dutch East Indies, it backed a wildcat operation on the east coast of Kalimantan, where the retired mining engineer J.H. Menten (1832-1920) had secured three huge concessions from the Sultan of Kutei: Louise, Nonny, and Mathilde.[124]

Kutei was an even less hospitable region than Langkat had been when Royal Dutch began its operations there: a similar mix of jungle and swamps, but much thicker and wetter. The landing site at Muara Jawa was just over six kilometres away from the spot chosen for the test well, dubbed Sanga Sanga, but it was impossible to get there in a direct line, deep jungle necessitating a detour of four hours by river boat and then four hours on foot. The planned site for the refinery was Balik Papan, on a beautiful bay suitable for loading large tankers, but nearly one hundred kilometres south of Muara Jawa, again impossible to reach overland. Communications with London were handled by A. Syme & Co., a Tank Syndicate member and Samuel's agents in Singapore, which was 1,200 miles away, and the packet boat which came by only once a fortnight. Everything – getting information, supplies, labour, ordering equipment or obtaining missing parts, sending reports, keeping check of progress– took huge amounts of time, so the whole enterprise depended on the man in charge.[125]

Mark Abrahams (1862-1943) was a good choice for the job. A skilled engineer, he had proved his mettle in building the Samuels' tank installations, first in Singapore and then across India. Abrahams also knew how to make men work for him by setting an energetic example himself and treating everyone with due consideration, so his staff remembered him with warmth and gratitude.[126] Perhaps a heroic figure like Kessler might have succeeded quicker than Abrahams did, but the problems dogging the venture originated in Samuel & Co.'s underestimating its scale and complexity. The Samuels treated Sanga Sanga simply as another one of their businesses. The firm's export department kept the administration; the chemical engineer Dvorkovitz was hired as technical consultant to oversee plans, order equipment, and check bills, with a Canadian technician sitting in on meetings as a special adviser.[127] No one except Menten knew anything about Kalimantan, and Menten knew very little about oil, his career having been in coal. A mining engineer commissioned by Samuel & Co. had reported favourably on the concessions. However, the firm had omitted to employ an expert like Fennema to survey the site and plan the necessary works in detail on the spot. Nor did the Samuel brothers care to have the oil quality properly assessed. The contract with Menten stipulated that the oil found had to be of Baku quality in order for Samuel & Co. to accept transfer of the concession.[128] Samples had duly arrived in London, but the results were either misinterpreted or, in the rush to acquire production, simply disregarded. The oil found at Sanga Sanga was totally unlike any other, but the project went ahead on the assumption that it was similar to Baku crude.

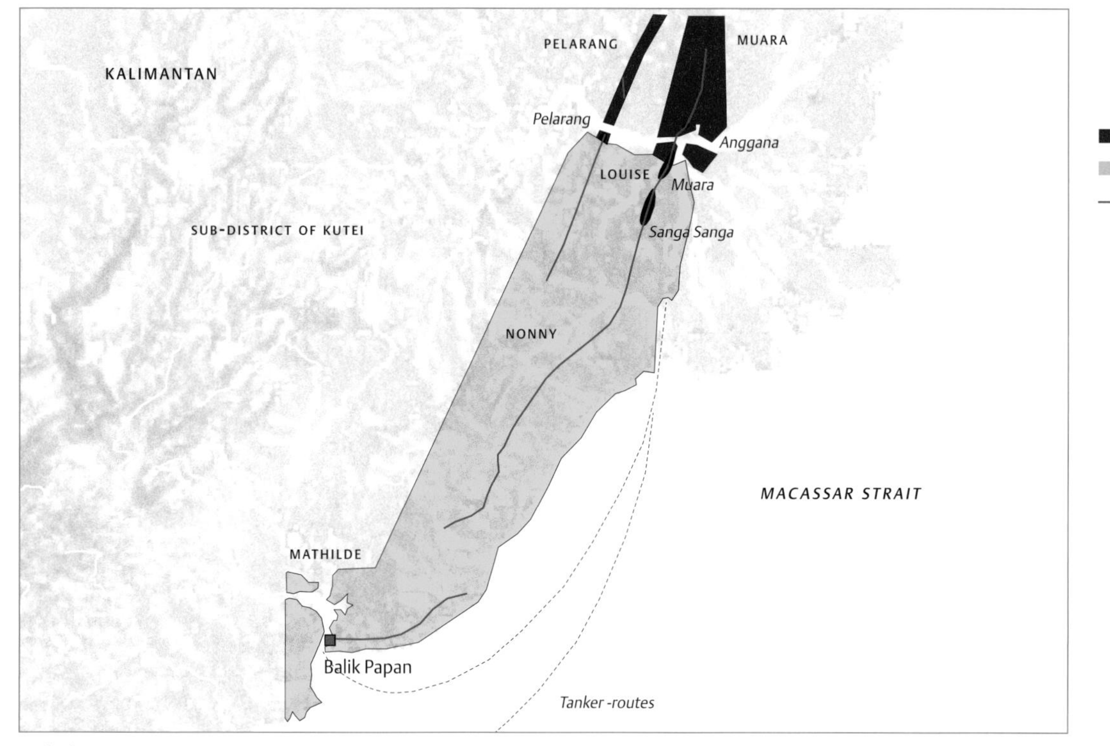

[61]

A map of East Borneo showing transport routes and anticlines.

- Royal Dutch
- NIIHM
- Anticlines

Consequently, plans for the refinery were drawn up, all equipment ordered, a drilling crew hired, everything loaded onto a steamer bound for Singapore, all by people who knew nothing about local circumstances, or about the qualities of the crude to be produced and refined. The Samuels had overall responsibility, without really knowing what happened or what needed to be done. They gauged progress by what they knew from their Baku trip, resulting in ludicrously unrealistic expectations of a few months before the venture would produce kerosene. When that did not materialize, they blamed Abrahams for not trying harder, constantly peppering him with petty queries, instructions, and rebukes.[129] A recipe for disaster.

After a crash course in the essentials of oil production in Baku, Abrahams threw himself with vigour and enthusiasm into the work. Things started auspiciously. A test well drilled in February 1897 appeared to confirm the value of Menten's concession, which was now bought by Samuel & Co. The firm then sold it, at a very handsome profit of £ 83,000 to a Dutch company, the Nederlandsch-Indische Industrie- en Handel Maatschappij (NIIHM), set up by the Samuels to comply with the requirements of the colonial mining law. The NIIHM formally owned the concessions, but Samuel & Co., and from January 1898 Shell Transport, held the majority of the shares and managed the Kutei venture.[130] However, when the two steamers crammed with equipment arrived on the Kalimantan coast in October 1897, a nightmarish drama began to unfold with dreadful similarity to what happened in Langkat during the years 1890-92: the slow progress of works for a lack of labour, the

[62]

Messrs. Syme & Co. (6). Concessions.

TRANSFER OF CONCESSIONS:

We have been expecting to hear that full sanction has been given to this, and we should much like to be able to inform the shareholders at the Meeting that all the property has been properly transferred to the Company. This is so with regard to all other than the Concessions property.

S.S."BROADMAYNE:

This steamer has now arrived and is discharging her cargo. We find that there is a considerable percentage of water although we are not yet in a position to say how much this percentage is. The presence of water changes the colour of the oil until settlement, into a brown soapy colour, and the water seems to settle out of the oil very slowly indeed.

The Captain reported that when the well No.12 was struck the pressure at every other well in the district fell considerably and production fell. We note that you refer to this peculiarity in your favour under reply. Our information is that this is usually where the primary deposit has not been struck and, seeing the area which is effected in this way, the probability is that there is a correspondingly large deposit of oil below. It appears to us uncertain that we have touched this primary deposit in No.12 well, which may probably be at a still greater depth; it will therefore be a necessary eventually to deepen almost all our present wells, a course which we recommended in a recent cable, and which we see Mr. Townsend

[63]

107/1

small production is only a matter for the moment. In this respect, we cabled you yesterday that a large production is of vital interest to us and asked what is required in order to increase this production. We have already referred to the matter of the larger bore-holes and shall be in a position to place the material on order on the best and quickest markets on receipt of Mr. Abraham's reply, as to our enquiry as to the heaviest piece of material he can conveniently handle. You will quite appreciate that the present production is entirely inadequate and, notwithstanding our constant urging, the progress of the borings is, to our mind, unaccountably slow. Mr. Samuel will no doubt give his attention to this matter but it certainly appears to us from the experience of what others have done and are doing that with 12 rigs and 12 boremasters much faster progress should be made. Your telegram of the 24th inst. shows the slowness of the borings to be about the same as that to which we referred in a special letter recently, an average of 10 to 15 ft. per well per day, - reckoning merely those wells which are referred to in the cable. When we come to look at the date of commencement of work and the depth at which it reaches on a particular day, the progress is nothing like so fast as this.

Weask you to draw Mr. Samuel's attention to this point and if the boring logs are properly kept on the Concessions and are available to him he should be able to see at a glance how lamentably slow this work is. An improvement in this respect is absolutely necessary if we are to obtain

Extracts from a letter dated 28 April 1899 from Marcus Samuel to his agents in Singapore, Messrs Syme & Co, show his concern over falling production in Kalimantan.

Workers pause for a photograph during the construction of Shell's Nonai depot in Japan in 1905.

problems of hiring new hands, the chaos of rusting or rotting equipment, the sheer hard work under appalling circumstances, the battle with heavy machinery in the driving rain, day after day, the rising floods, the missing, ill-fitting, and damaged parts, the need to do everything at once so as to avoid stagnation, all compounded by the logistic nightmare of coordinating complex works on three sites without proper means of communication between them.

Even so, Abrahams managed to get things going. By September 1898 he had forty-eight Europeans working on the concessions, plus 550 Chinese, Javanese, and Indian labourers.[131] New wells came in at Sanga Sanga, and a gusher was struck near Balik Papan. The refinery came on stream late in 1899, so Abrahams completed it in a term similar to Stutterheim and Kessler, but arguably under less propitious circumstances, and at six times the cost. Royal Dutch's total outlay on Langkat amounted to one million guilders by December 1892, whereas the Kalimantan enterprise had cost Shell Transport £ 550,000, or 6.7 million guilders, by December 1899.[132]

Unfortunately the refinery proved totally unsuitable. In fairness to Dvorkovitz, the oil was unlike any other known until then. The Sanga Sanga concession produced two types of crude: a very heavy oil, the lightest fraction of which was still too heavy for making kerosene, and a lighter oil which contained so much paraffin that it solidified at 35°C. This latter type contained, in addition to a small gasoline fraction, about 60 per cent of distillate suitable for making kerosene, but due to the very high content of aromatic hydrocarbons this lamp oil was smelly and discoloured with a marked tendency to smoke when burned.[133] The kerosene could only be sold when thoroughly mixed with other, neutral lamp oil in a ratio of 20:80. It could not be used on the enterprise itself, so cases of Standard's Devoes were brought in, to the Samuels' great indignation.[134] To solve the problem, Abrahams began experimenting with novel methods such as washing the kerosene with sulphuric acid to improve its quality.[135] The heavy crude could be used as fuel oil without further treatment, however, and from time to time a few hundred tons of it were loaded by a flexible pipeline straight into small tankers for sale in Singapore.[136] The shipping line KPM, which operated regular services throughout the Indonesian archipelago, also became a customer for fuel oil.[137] The light oil struck in February 1899 posed no less serious production problems. It was really an emulsion of 50 per cent crude oil mixed with 50 per cent water and dirt, with a remarkably high gasoline fraction equally saturated with aromatic hydrocarbons. The consequent long settlement times translated into protracted stagnations of production, and even then the high water content and other impurities caused damage to the stills and brickwork, resulting in constant breakdowns. Moreover, because of a high wax content, the light crude solidified at low temperatures, and thus could only be used in tropical climates.[138] London excelled in sending paper advice from experts, Dvorkovitz proposed one modification after another, but in the end it all came down to Abrahams and his refinery staff battling it out in a huge and expensive process of trial and error.

It was a classic business morass, absorbing so many resources that the board could not muster the courage to stop the bleeding and write it off. Only liquid fuel oil sales offered some respite, about 80,000 tons being sold in 1902.[139] Kalimantan oil also proved suitable for making solar oil or gas oil, used as a coal substitute in the production of household gas. News from the sites flowed more freely than kerosene; the Standard Oil agent in Singapore kept New York up to date with step-by-step reports.

Sir Marcus put on a brave public face, boasting that the Kalimantan concessions harboured enough liquid fuel to power all the world's ships. Privately, he admitted to feeling sick at the sight of a letter from Singapore with news from Abrahams.[140] In a fit of despair he offered Kessler Shell Transport for sale in January 1899 but Royal Dutch, deep in its own crisis, could not take up the offer.[141] Some alternative supplies were secured by buying kerosene from Moeara Enim, its south Sumatran neighbour Moesi Ilir, and from Dordtsche.[142]

From 1900 the situation at Kutei began to improve. Production rose steeply from 40,000 tons of crude in 1900 to 420,000 tons in 1905, split roughly half and half between light and heavy oil. Shell Transport promoted liquid fuel with zeal, but with modest success. In March 1899 Sir Marcus started a marketing campaign with a public speech extolling the advantages of oil over coal for shipping. The considerable publicity generated failed to boost sales, bunker stations being omnipresent and coal cheap, and it took another decade to win over the navy.[143] The refinery continued to cause no end of problems as well, so output fluctuated wildly. In 1902, a Royal Dutch inspection team sent to Balik Papan as part of negotiations to turn the combined operations in Kalimantan over to a joint venture with Shell, considered the installations to be hopelessly out of date, and very dangerous, the tightly packed tanks posing a huge fire hazard.[144] And even at 1905 production levels the venture ran at an estimated loss of nearly £ 10,000 a month, though the previous year had shown a small profit.[145] It was now a very substantial enterprise, employing eighty-three Europeans and 2,819 Asians with the usual variety of ethnic backgrounds: Tamils, Chinese, Sikhs, Javanese, Buginese, Bandjerese. A police force of ninety-two officers guarded the sites, and there were three hospitals with a total of 240 beds, though according to the chief medical officer the climate was not notably unhealthy.[146] Employment conditions appear to have been better than in Langkat, with Sundays off, and additional holidays for separate religious denominations. Abrahams also showed himself more relaxed than the Royal Dutch management in allowing an unstated number of women to come and live on the sites, though most men lived a bachelor existence.[147] Balik Papan was the social centre, with a club, a library, a stable with riding horses, and two tennis courts for the Europeans, and shops, a mosque, a Chinese temple, opium dens, and a gambling casino for the Asian workers. The site now had a telephone system, electric light, and a direct telegraph cable to the outside world, but no direct connection to Muara Jawa, and the crude from Sanga Sanga still arrived in barges.[148]

The Kutei venture was a crucial watershed for Shell Transport. A running sore drawing on profits and cash, it undermined the company's financial position. During 1901, the situation spun out of control, the debt to equity ratio rising to 80 per cent.[149] This was clearly unsupportable. The issue of £ 1 million of 5 per cent cumulative preference shares brought much-needed relief. Kutei was also instrumental in another sense. The cost and frustration of going upstream convinced the Shell Transport board to seek expansion downstream. This was a precarious strategy, for Shell Transport continued to make most of its money in shipping. In 1899, kerosene sales generated £ 153,000 gross, against shipping £ 220,000; two years later, the figures were £ 96,000 against £ 155,000.[150] Success would thus depend on whether the revenue from increased shipping would balance the cost of competing in new markets. With careful calculation and planning the strategy might still have worked, but the downstream expansion during the years 1899-1906 shows a pattern of opportunistic actions born from flights of fancy and impulsive zest, rather than a detached analysis of situations and opportunities. Two examples will suffice. In 1900 Shell Transport launched its first attempt to break into the European market for gasoline. The first setback occurred when the Suez canal authorities refused passage to tankers with gasoline to pass, forcing a re-routeing around the Cape at considerable extra expense.[151] On arrival in London, the gasoline was stored in the large tank which the company had secured in Thameshaven. Meanwhile Standard Oil had tied up the retail trade with long-term contracts. Sluggish sales combined with high transport costs forced Shell Transport to give up, consigning its acquired tank installation to idleness. The same year Joe Abrahams was sent to Australia to set up a country-wide network to distribute Kutei kerosene and liquid fuel. By the summer of 1902, he had spent the authorized £ 100,000 on installations in Sydney, Melbourne, and Adelaide, but since no supplies from Kalimantan were forthcoming, it was money down the drain. London then sent supplies from Batum, but Russian kerosene proved unsuitable for the lamps used in Australia, so sales remained low.[152]

Companies apart The two companies emerged from their respective crises in a markedly different shape and position. At first sight, Shell Transport seemed to have done rather better (Table 1.1).[153] It was still by far the bigger in 1902, and the company had also grown faster since 1898: assets had risen by almost 90 per cent, against nearly 50 per cent for Royal Dutch, sales figures an estimated 70 per cent against 55 per cent. However, during 1890-7, and again from 1902-6, Royal Dutch grew faster than its British rival, and consequently the size difference between them rapidly narrowed. The Dutch company was also far more profitable (Figure 1.3).[154] At Shell Transport, net return on assets peaked at 9.5 per cent in 1899, and then dropped sharply to less than 5 per cent until 1906, averaging 5.1 per cent over the 1898-1906 period. By contrast, Royal Dutch's net return on assets had been in double figures during 1895-7, topping 32 per cent in 1895. Once the worst of the production crisis had passed, net return on assets quickly soared back to 7 per cent and way beyond, for an average of 13 per cent over the period. Shell Transport's net profits on sales were slim indeed, about five old pennies (two decimal pence) per case,

Table 1.1
A comparison between Shell Transport and Royal Dutch, 1898 and 1902.

	1898		1902	
	Shell Transport	Royal Dutch[1]	Shell Transport	Royal Dutch[1]
Assets	£ 2,332,902	£ 1,645,636	£ 4,368,533	£ 2,421,924
Capital+reserves	£ 1,800,000	£ 1,338,209	£ 3,400,000	£ 1,348,357
Debt	£ 463,563	£ 279,094	£ 882,424	£ 371,438
Gross profits	£ 217,096	£ 120,448	£ 219,566	£ 433,839
Gross profits/assets	9.3 %	7.3 %	5.0 %	17.9 %
Net profits	£ 115,027	£ 26,591	£ 108,236	£ 179,931
Net profits/capital	6.4 %	2 %	3.2 %	13.3 %
Dividend	6 %	6 %	2.7 %	35.9 %
Fixed assets/capital	111 %	86 %	101.3 %	75.3 %
Cases sold	5,994,000[2]	5,479,694	8,910,000[3]	8,580,000[3]

[1] Guilders converted at the rate of 12,11 guilders to the pound
[2] Estimated as 80 per cent of potential fleet capacity
[3] Turnover November 1900 - October 1901

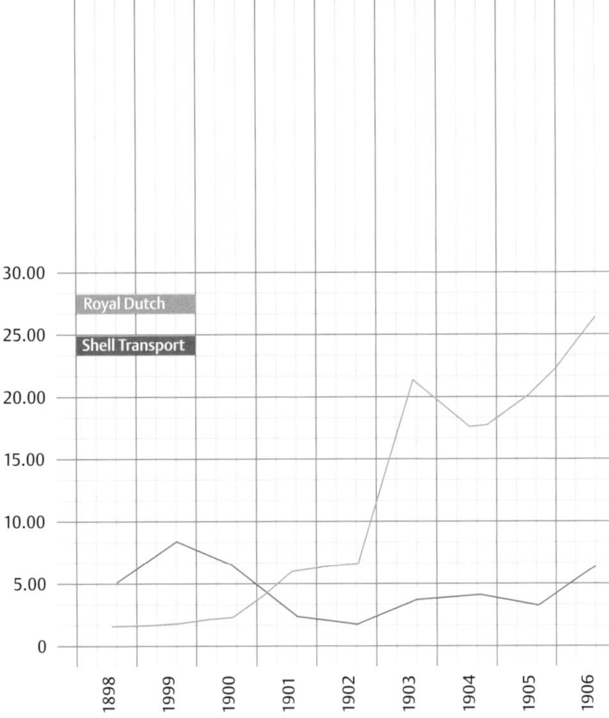

Figure 1.3
Net return on assets at Shell Transport and Royal Dutch, 1898-1906.

assuming the estimate of about 6 million cases as turnover. Royal Dutch's profits were more than double that, 0.6 guilders a case or one shilling (five decimal pence). This huge difference was not primarily a result of distance to source, rather it was inherent in the nature of the respective operations. Shell Transport was and remained primarily a shipping line. For the Samuels as well as their Asian agents, kerosene remained only part of their business.[155] During 1894-7, Samuel & Co.'s tanker business averaged profits of £ 110,000 a year. If shipping revenues for 1898 were in line, as the firm expected, then the company's profit on selling kerosene cannot have contributed more than about £ 100,000 to gross profits.[156] By contrast, Royal Dutch was an oil company, managing an integrated supply chain from drilling to retailing, and managed by dedicated oil men.

Second, margins on selling Russian kerosene in Asia were considerably lower than on Langkat products. When in February 1899 Royal Dutch considered importing Russian kerosene to sustain its sales network, the board gave agents a list of potential economies to compensate for that fact.[157] This was not just a matter of lower production costs, proximity to markets, or Royal Dutch profiting from low royalties.[158] Royal Dutch was simply far better in selling oil than Shell Transport. In 1900, with Langkat production running at a quarter of the 1898 peak, the Eastern Oil Association enabled Royal Dutch to achieve a gross return on assets of almost 15 per cent while selling Russian kerosene, double the 1898 figure on assets of similar magnitude, and more than Shell Transport ever achieved. When Asian prices plummeted in 1901, Royal Dutch's gross return rose by 24 per cent, but Shell Transport's dropped sharply. For Deterding, trading was really the heart of the business, combining high profitability with low risk.[159]

The benefits of integration showed themselves to best effect in economic efficiency. In 1902, Shell Transport's assets were 80 per cent bigger than those of Royal Dutch, but the company sold only 3 per cent more cases. Shell Transport had a far more extensive marketing network consisting of numerous ocean depots feeding elaborate up-country installations, but failed to capitalize on it.[160]

The company's chronic under-utilization of assets also highlights how dangerous Kutei was for the business. Factoring the Kalimantan venture out of the assets and the profits raises gross profits to 6.6 per cent of assets for 1902, so it was a considerable drain on already anaemic revenues.

Three further aspects bear pointing out. First, Shell Transport's reserves remained very slender, about £ 400,000 pounds on assets of over £ 4 million in 1902. This amount was built up by profits realized on issuing shares above par, and not by retained earnings. Over 1898-1902, the company's pay-out ratio, i.e. dividends to net earnings, was 101, so the board rewarded shareholders more than their due. Moreover, the reserves were sunk in the business and not held in cash or liquid investments, so Shell Transport really had no proper buffer at all. By contrast, Royal Dutch possessed a huge buffer, 9.7 million guilders invested in government bonds and on-call money, on 6.5 million guilders of capital. The company's pay-out ratio was 78 per cent, reflecting its cautious policy of expanding through retained earnings. Second, Shell Transport carried rather more debt, some of it hidden in overvaluations of assets. The balance sheets show creditors always outstripping debtors six, seven, or even twenty times, so the company habitually kept its debtors on a short lead, but paid late itself, resulting in a regular credit equalling the size of the cash flow. Conversely, Royal Dutch liked prompt payments, so creditors and debtors usually balanced. Finally, Shell Transport was both less liquid and less solvent. Cash rarely amounted to more than 2 per cent of assets, and fixed assets outstripped capital, i.e. some of the company's cash or debts were tied down, further straining the cash flow. By contrast, Royal Dutch liked liquidity. The company kept over a quarter of assets in cash, and capital and reserves showed a wide margin over fixed assets, giving the board considerable room for manoeuvre.

Family firm versus corporate organization The two companies also displayed fundamentally different managerial styles. Shell Transport did not have either a full-time manager, or dedicated staff. The Samuels, old-fashioned commercial men alien to the world of managerial corporations, still took most decisions between them. Stately, plump by now, hard workers, but neither they nor their office manager Henry Benjamin, Sir Marcus's brother-in-law, could devote their full time to the company. The firm of M. Samuel & Co., probably twice the size of Shell Transport, demanded their attention, and in addition the brothers spent considerable time on other pursuits, Sir Marcus on advancing his social career, Samuel on travelling.[161] The board met weekly and received full details about sales, stocks, progress in Kutei, correspondence, and available cash. Committees drawn from members advised the board on specific matters, but members appear to have exercised overall supervision rather than control.[162] Shell Transport's administration was scattered over M. Samuel & Co.'s various departments. The company secretary had very much a secretarial position and not a central coordinating role, so no one amongst the staff of about fifty people would have been able to get an overall view of the business. Key staff had been sent to Kutei, depleting the stock of experience, and new recruits such as the supremely able Bob Waley Cohen, who joined as an unpaid trainee in 1901, arrived too late to change the company's fateful course.[163]

[64]

[65]

Jonkheer Hugo Loudon (right) and Abraham Capadose (far right) together formed an effective team at the head of Royal Dutch.

By contrast, Royal Dutch had had two full-time managers since Deterding's appointment as sales and transport manager in 1897. Just after the turn of the century, the head office in The Hague employed a staff of thirty people, with a further 278 Europeans and 6,682 Asians employed in the Dutch East Indies.[164] After Kessler's death, Loudon was promoted to director, and in 1902 Abram Capadose, until then a non-executive director and a regular stand-in for Kessler and Deterding when they were away on business, joined the team. The production crisis reinforced Royal Dutch's corporate preference for calculated risks through detailed planning and costings. This respectable but staid principle might easily have created another Dordtsche, quietly, imaginatively, and profitably milking its comparative advantage for all its considerable worth, only to find itself caught out in the final battle over market control. It was Deterding's energetic yet patient unfolding of a world-spanning ambition which launched Royal Dutch on its spectacular course, from the groundwork prepared by Kessler.

Deterding liked to style himself as the epitome of a modern corporate manager. He claimed to attain strategic objectives with policies framed by abstract economic principles and executed on the basis of formal analysis and planning. The actual record of his performance and achievements is a mixed one, however. His economic precepts were fairly basic: supply consumers from the nearest source, reduce the waste of needless competition by dividing markets on the basis of access to the nearest supplies, and balance supply with demand to achieve remunerative prices along the supply chain. Under these all-purpose maxims Deterding frequently demonstrated a profound commercial opportunism, sometimes inspired by emotion and petty vindictiveness, which he passed off as intuition to impress his colleagues.[165] Moreover, the term 'remunerative' cannot by any stretch of the imagination be made to fit Royal Dutch's spectacular profits during the early twentieth century. True, Deterding did nurse grandiose visions for his company which, combined with a great facility to reduce complex problems to a few key numbers and ratios, greatly assisted Royal Dutch's expansion. He knew at any moment where Royal Dutch stood, and thus what the company's best options were, though he never really understood why Standard Oil would not accept the Group as an equal and come to a long-term understanding about market shares, probably because, in the absence of published figures about Standard, he failed to grasp the company's sheer size, with a net value nearly four times the Group's total assets in 1907.[166]

Deterding was more seriously handicapped by his fiery character, and by visions bordering on whimsy, indeed fantasy. He was subject to sudden flights of fancy, attempting, not always successfully, to bamboozle whomever he was talking to by throwing figures on a scrap of paper and hotly defending them as the true keys to the problem at hand. Some people were bedazzled by what they regarded as the manifestation of genius, but close associates like Lane grew weary of what they saw as a party trick, and learnt to counter Deterding's assertions with sober arguments derived from deskbound reflection.[167] Lane also learnt to work with Deterding's difficult personality, writing in 1913 that 'His mind is so active, so suspicious and so ready to take offence, that it is always advisable if possible to keep him outside the sphere of action until the last moment, and persuade him to leave himself into the hands of someone in whom he has the utmost confidence and knows his weaknesses, and is able to lead him on sound and equitable lines and prevent his bursting away in some petulant mood and out of pique adopting a regrettable course. (...) I know how unreasonable Mr Deterding frequently is in a negotiation, and it needs someone who knows him well to handle him'.[168] Deterding owed his success in building the Group to having a circle of trusted men around him who knew him well enough to accept his obvious failings, which included a marked lack of discipline and patience for the humdrum administrative side of a large enterprise. Deterding expected his staff to deal with that. The solid organization built by Kessler could cope with this cavalier attitude, but the early collaboration with Shell Transport nearly came to grief over it.

Deterding was the antithesis of the Samuels in other respects, too. Only just in his middle thirties when he succeeded Kessler, Deterding radiated charm, energy, and health, plus a passion for vigorous exercise. His commitment to business, to oil, left him no time for the gentlemanly capitalism of Sir Marcus. Deterding possessed few interests outside oil and outdoor sports, which he pursued with the same determination. As general manager, he wanted to lift the company to the status of world player. That would require a quantum leap forward, such as even Royal Dutch's rapid growth would not provide, and the political support of a great power. The Eastern Oil Association was a welcome stopgap but, as a chance coalition of independent companies without assets to complement the Royal Dutch's business, it was unsuitable for a lasting partnership. During the early months of 1901, the Batum market again suffered one of its gluts, so kerosene could be had freely and cheaply. Having set his sights on Shell Transport as the obvious candidate with which to achieve his ambition, Deterding terminated the Association in the summer of 1901, though taking care to transfer the chartered Suart tankers to Royal Dutch.[169] He now had all the figures concerning Shell Transport's situation and performance: Batum kerosene prices, freight rates and shipping costs to Asia, sales prices and costs across Asia, so he must have been thoroughly aware of the huge unused economic potential in the company's assets. He also knew the problems and potential of Kalimantan crude. After company geologists had pinpointed the best drilling site, Royal Dutch had succeeded in bringing the Kutei Lama concession, just across the river from Muara Jawa and abandoned by both Stoop and Abrahams after several fruitless exploration attempts, into production.[170] With Royal Dutch's profits soaring, Deterding could afford to wait. The Samuels, however, found themselves increasingly pressed for time, as their attempts to drag Shell Transport free from the Kutei mire and on to a new course misfired one after another.

Merging by proxy

Shell Transport's downstream expansion began during the summer of 1899, when Sir Marcus suddenly embarked on a frantic expansion, buying huge quantities of kerosene for future delivery in Batum, ordering four new tankers of 9,000 tons capacity each, chartering several more, building a new huge tank terminal in Tianjin, planning installations in Australia, New Zealand, and along the east coast of Africa down to the Cape. This surge of investment was probably powered both by expectations about imminent deliveries from Kutei, and by a desire to build muscle for the impending negotiations with Bnito about a new contract, which was finally signed in October 1900. The terms left Shell Transport free to sell kerosene and other oil products in Europe, provided they did not come from Russia, and also allowed the company to buy from other Russian producers for sale in Asia, without having to take a minimum quantity from Bnito. In return, Shell Transport gave up its exclusive agency east of Suez, enabling Bnito to begin marketing its Anchor brand.[171]

Whatever the motives, Sir Marcus's expansive strategy soon went sour. During the last months of 1900, freight rates and kerosene prices collapsed, leaving Shell Transport sitting high and dry. In China alone, agents had 60,000 tons or the equivalent of two million cases in their depots when the outbreak of civil war brought sales to a near standstill. Within a few months prices had sunk below cost, ripping up the agreement with Royal Dutch in the process. Indian sales, already under pressure from increasing competition by Burmah Oil, suffered badly when the Eastern Oil Association began importing cheap kerosene there, bought at half the price of Sir Marcus's forward purchases the year before. The Shell Transport board chose to ignore the consequences. Its dubious practice of valuing stock at cost, and not at market price, prevented the heavy loss from weighing down the 1900 accounts, and the AGM voted in favour of the proposed 10 per cent dividend paid out of what were really fictitious profits.[172]

The sharp drop in Far Eastern prices inspired Lane and Deterding to start exploring the options for a comprehensive marketing alliance. Success would depend on bringing together the main Russian exporters and the Dutch East Indies producers, under the circumstances a seemingly impossible undertaking. Deterding's cancellation of the Eastern Oil Association had antagonized Royal Dutch's former associates in Batum.[173] Nor was the company very popular with the East Indian producers. Royal Dutch rather fancied itself as the anointed leader of the Dutch East Indies' oil industry, half expecting other colonial producers to fall in behind. This regal posturing had put an end to the government-sponsored talks about

Shell storage and buildings in Tangku, near Tientsin in China, originally presented a neat and well-organized scene (below), but during the Boxer Rebellion of 1900-01 many of the Group's installations were destroyed in battles between armed Chinese troops (right) and the Boxers (far right).

[67]

[68]

cooperation following Standard's aborted takeover of Moeara Enim in 1898, and had led Moeara Enim, Dordtsche, and Moesi Ilir to seek marketing agreements with Shell Transport.[174] Relations between the companies had further deteriorated with Deterding's aggressive marketing of Russian oil.[175]

Finally, the Samuels were certainly concerned with the situation on the Far Eastern market, and notably with the resumption of sharp competition in China. However, they had just launched Shell Transport on a totally new venture in the US. In January 1901, a spectacular gusher was struck at Spindletop, near the city of Beaumont on the south-eastern coast of Texas. The Samuels immediately saw a golden opportunity for getting their own supplies and sent Benjamin over for negotiations with the biggest producer, Guffey Petroleum. In June, Sir Marcus announced with typical modesty 'one of the most important contracts ever concluded in the history of the mercantile world', binding Shell Transport to buy half of Guffey's production, or a minimum of 100,000 tons of oil a year, at a fixed rate of $ 1.75 a ton, the two companies sharing the profits on sales.[176] Three of the four new tankers, lying idle because of the slump, could now be put to immediate use, and the company ordered four additional ones of 10,000 tons each to carry the anticipated huge trade. With deliveries imminent, a sales organization had to be built. In late summer, Shell Transport established a marketing bridgehead on the European continent by taking over the Hamburg oil trading firm of Gehlig, Wachenheim & Co. The following year, the firm was reorganized as the Petroleum Produkte Aktien Gesellschaft, usually referred to as PPAG or Petrodukt, with Deutsche Bank and Steaua Romana oil company each taking a one-third share.[177]

Spindletop changed Shell Transport's entire outlook: at a stroke, the company became a global concern, and US supplies even showed potential for outpacing shipments to Asia, then running at 270,000 tons a year.[178] And yet, for all its boldness reminiscent of the 1892 *Murex* coup, the Guffey deal was a classic example of the Samuels' precipitous business style. No oil experts or geologists were sent to Texas, the Shell Transport board agreed to sign the contract on the strength of Benjamin's favourable report alone. Spindletop produced a crude similar to Kutei, very heavy and well suited for liquid fuel and gasoil, but yielding only small quantities of inferior kerosene. The market for liquid fuel, though growing, remained very small, Shell Transport having difficulty in selling the 60,000 tons or so coming from Kutei.[179] Moreover, the harbour at Port Arthur was not deep enough for the new tankers, ordered before anyone had taken the trouble to inspect local conditions. The ships could only take half their cargo in port; the rest had to be loaded from lighters in the Gulf of Mexico, which took several days.[180] Clearly the Guffey deal had been a matter of closing first, considering details later. The same was true for the Gehlig Wachenheim takeover. This firm marketed Russian kerosene from its own refinery in Baku. Now the new Bnito agreement left Shell Transport free to sell oil in Europe, provided that it did *not* come from Russia. Buying Gehlig Wachenheim thus constituted a formal breach of contract, and Sir Marcus's awkward efforts to placate an enraged Bnito show that this had been a clear oversight.[181] For the moment, however, Shell Transport appeared triumphant, and in the euphoria the forging of an alliance on the Asian market lost some of its urgency.

Since an agreement between the two main sellers in Asia would likely bring the others around, Lane and Deterding began with opening discussions between Shell Transport and Royal Dutch, having cleared the ground by effecting a truce in China.[182] Sir Marcus, just then deep in talks with Standard Oil about a possible takeover, took a very passive interest, leaving the initiative to Lane and Deterding, who explored one scheme after another.[183] By early November 1901, an understanding had been reached for transferring sales to a joint marketing organization, with the roughly equal sales figure of about 270,000 tons serving as the basis for a 50:50 ownership split.[184] This company, later dubbed the 'British–Dutch', would also have the authority to restrict production as circumstances dictated. It would not have its own

Spindletop operations: From the famous wells on the salt dome near Beaumont, Texas (left), oil was heated (centre) to make it easier to pump to refineries. The first refinery on the US Gulf Coast (right) was built in 1901 in a cow pasture near Port Arthur.

[69]

[70]

A map of the Gulf Coast of the USA and Mexico.

▲ - Oil fields

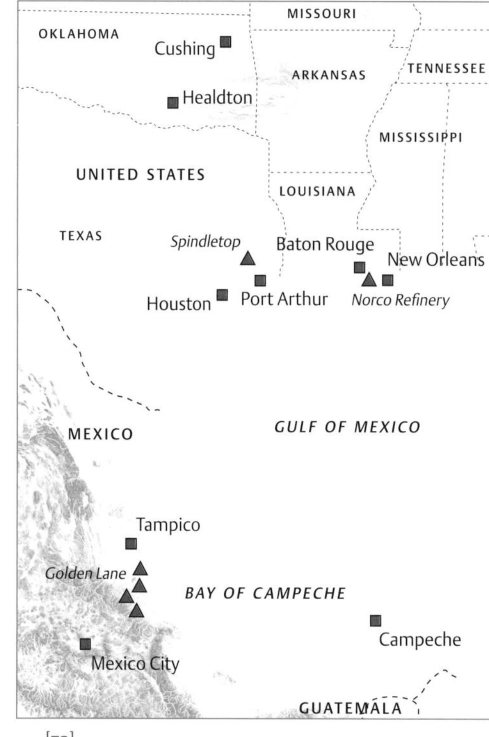

[72]

[71]

installations or ships, but lease them from Shell Transport and Royal Dutch, Deterding having to agree to lure the Samuels with high charter rates of seven shillings a ton, which threatened to render the whole venture uneconomical.[185] With characteristic circumspection, Deterding had meanwhile moved to widen his options. Through Lane, he obtained an undertaking from Bnito that Royal Dutch would be supplied on better terms than Shell Transport, thus securing supplies if the talks foundered and the Batum glut turned to a drought.[186] And he had found tentative support amongst the Dutch East Indies producers. The Moeara Enim manager Jan Willem IJzerman (1851-1932), unhappy with the Shell Transport contract, had already signalled his readiness to switch to Royal Dutch. Moesi Ilir would likely do the same, and IJzerman urged the Stoop brothers to reconsider their position as well.[187]

With the negotiations between Shell Transport and Standard casting a long shadow, the Royal Dutch board met on 3 December 1901 to consider two proposals drafted by Lane and Deterding, one for setting up a syndicate of Dutch East Indies producers, the other for the British-Dutch. The board was in a quandary, clearly

Strong shadows on the storage tanks show a bright sunny day some time in the early years of the American oil industry – but any beauty in the scene could only be seen by the most hardened industrialist.

01|70

unenthusiastic about the proposed terms for the British-Dutch, notably the high charter rate, yet apprehensive of being outmanoeuvred by events. Preferring a meagre deal giving the prospect of linking up with the seemingly imminent Standard-Shell Transport combination, to rejecting the terms and standing alone, the board accepted the drafts.[188] On 5 December Benjamin arrived in New York, knowing that Shell Transport had a provisional agreement with Royal Dutch. Until then the talks between Standard and Shell Transport had explored the options for a joint venture, with Standard managing the business and Shell Transport holding a minority stake in return for its assets. Presumably the option of Standard taking a share in Shell Transport had also been explored, for on his return from New York in October, Sir Marcus had sought legal advice on how to retain control over the company after a takeover.[189] No concrete sums appear to have been discussed between Sir Marcus and the Standard board in New York. When Benjamin wired around 23 December that Standard had offered £ 40 million for an outright purchase, his news came out of the blue.[190]

A lot of money, but certainly not a gigantic offer considering Shell Transport's position. Forty million dollars translated into some £ 8.2 million, about twice the company's 1901 assets, or over twenty times the 1900 earnings.[191] Standard could easily have afforded more. Its 1901 dividends amounted to nearly $ 47 million on assets worth $ 211 million.[192] Fed by close intelligence about the Kutei fiasco, the board clearly had a sceptical view about Shell Transport's balance sheet and earnings potential. Though he and Sam Samuel stood to earn nearly £ 4.8 million together (roughly £ 300-400 million today) on their shares in the company, Sir Marcus must have felt that the company was worth more. His brother and Benjamin disagreed, urging him to accept, but at that moment the market supported Sir Marcus. Shell Transport shares sold at around

275-300 per cent of par in kerb trading, and *The Statist* even valued them at 455 per cent.[193] Shell Transport's market value thus ranged between £ 6 million and £ 9 million.

Yet the deal probably foundered on Sir Marcus's wish to keep control. Already a very wealthy man, he valued social acceptance more than money. An alliance would have enhanced his status; selling out could have damaged it, perhaps even beyond repair. This was not a risk that he, as the Lord Mayor of London designate, wanted to take. Moreover, though the balance sheet looked decidedly weak, the future appeared splendid indeed, giving no pressing reason to sell. On 27 December, Sir Marcus signed the draft agreement with Royal Dutch, and turned down the Standard offer. Typically, the board was not consulted about either decision. With the two Samuels still owning 58 per cent of the shares the outcome would not have changed.[194]

December 1901 was Shell Transport's crest of the wave. Once it had passed, Deterding immediately set to work recovering lost ground.[195] The scope for reinterpretation remained huge indeed, no more than a statement of intentions having been agreed. However, the ensuing tortuous negotiations were not about details, but over control of the joint venture. Within months the Samuels found themselves cornered. The rejected deal with Standard made a bad impression, turning the sale of £ 1 million worth of preference shares, issued to relieve the dreadful balance sheet, into a struggle.[196] The 1901 profits were very poor indeed, down 30 per cent on the year before. Freight rates and kerosene prices continued to sink, further undermining the company's position. Shell Transport could ill afford to procrastinate, and when Lane suggested splitting the management of the British-Dutch, with Sir Marcus becoming chairman and Deterding general manager, Sir Marcus agreed.[197] A wise decision. Sir Marcus could never have mustered the necessary dedicated attention, least of all during his twelve months in office as Lord Mayor which lay ahead from November. However, by waiving responsibility, Sir Marcus really signed away the day-to-day management of Shell Transport's core operations. With the crucial point of management solved, the two companies signed a new collaboration agreement on 17 May, just ahead of talks in St Petersburg about a Russian kerosene export cartel. The agreement was to take effect from 1 July for the duration of twenty-one years. The Syndicate of Eastern Producers was set up on the same day and for the same term, without Dordtsche. The Stoops preferred to remain outside of the Syndicate which they could as yet well afford, since they already had accepted a price agreement for Java.[198]

For Shell Transport the matter was now brought to a conclusion, but for Lane and Deterding the British-Dutch agreement was only the beginning. From the outset, they had seen the deal between Royal Dutch and Shell Transport as the necessary first step in building a much wider alliance, including the major Russian producers. Lane's first scheme of September 1902 had projected a tripartite company formed by Royal Dutch, Shell Transport, and Bnito.[199] When the St Petersburg negotiations failed, they travelled to Paris and persuaded Bnito to join the British-Dutch instead. Sir Marcus would have nothing of it. To him, there was no need to bring in Bnito. With both Shell Transport and Royal Dutch having contracts, supply was assured. With Bnito, the alliance threatened to become a producers' marketing association, relegating Shell Transport's narrower shipping interests, Sir Marcus's overriding concern, very much to the sidelines. More importantly, the accession of Bnito would rob Shell Transport of its very substantial Russian kerosene trade, and reduce its benefits from the alliance to renting out its tankers and installations. Also, the NIIHM, Moeara Enim, and Moesi Ilir were to join the association as producers on their own account, so the company's Asian operations would be reduced to liquid fuel, solar oil, and gasoline sales, none of them particularly voluminous or flourishing. Consequently, Sir Marcus rebuffed Deterding's suggestion to bring liquid fuel and gasoline also under the joint marketing agreement, even though liquid fuel was hardly profitable and gasoline a licence to print money.[200] By now Shell Transport badly needed the

[74]

Oil tanks made of wooden staves held together by iron hoops were still in use as late as the Cushing Field boom period of 1913-15.

alliance, however. The company's weak financial position and 1901 performance had become public knowledge, and nasty rumours about the Spindletop crude, the first cargo of which was due to arrive in July, were doing the rounds.[201] On 20 June, Sir Marcus agreed to let Bnito become the third partner in the alliance, now dubbed the Asiatic Petroleum Company. Each partner took one third of the initial £ 600,000 capital. Efforts to bring in the other Russian majors at the start failed, though some were to join later under supply contracts.[202]

A fighting partnership The basic structure of the Asiatic contract was simple enough: an agreement for the joint transport and marketing of kerosene, fuel oil, and gasoline in the so-called Red Line Area, the area delineated by a line running from Alexandria south to the Cape, east to the 180th meridian to include New Zealand, then north to the tip of Kamchatka, north-west along the northern tip of Sakhalin, from there returning south-west to Suez.[203] Asiatic was to act as the exclusive sales agent for the participating producers, deriving its income from sales commissions plus 25 per cent of the net profits. However, uniquely for an agent, Asiatic had the authority to lay down the quantities and product specifications it required, thus creating the ability to manage markets and attain Deterding's hallowed goal of profitable stability. With Asiatic supplying just under 37 per cent of the estimated 42 million cases a year on the Asian market, plus two extensive sales networks, such a stabilization now appeared within reach. Consequently, prices in Asia were put up immediately, and Lane had a talk with Macdonald to ensure that Standard would follow suit, and the gentlemen at 26 Broadway did.[204] Singapore kerosene prices went up by as much as 40 to 50 per cent in 1902-3, including those of Standard's prime Devoes brand. This collusion led to a formal verbal agreement about market shares in 1906, in which Asiatic remained very much the junior partner. Standard dominated the market with 60 to 70 per cent of sales, and even the best Langkat kerosene could only compete with Standard's inferior brands. Moreover, the success of the cartel depended on the American oil market, not on Asiatic's strength.[205] Even so the joint marketing through Asiatic was a desirable goal for all participants. In September Deterding moved to London to manage the joint marketing from an office at 21 Billiter Street. Work discipline appears to have been rather stricter there than at Samuel & Co.[206] To reinforce the management in The Hague, the Royal Dutch board appointed Loudon and Capadose as managers, and created the title of Director-General for Deterding. The principles to which Asiatic subscribed looked simple enough, and the axiom adopted, that no participant was to enjoy an

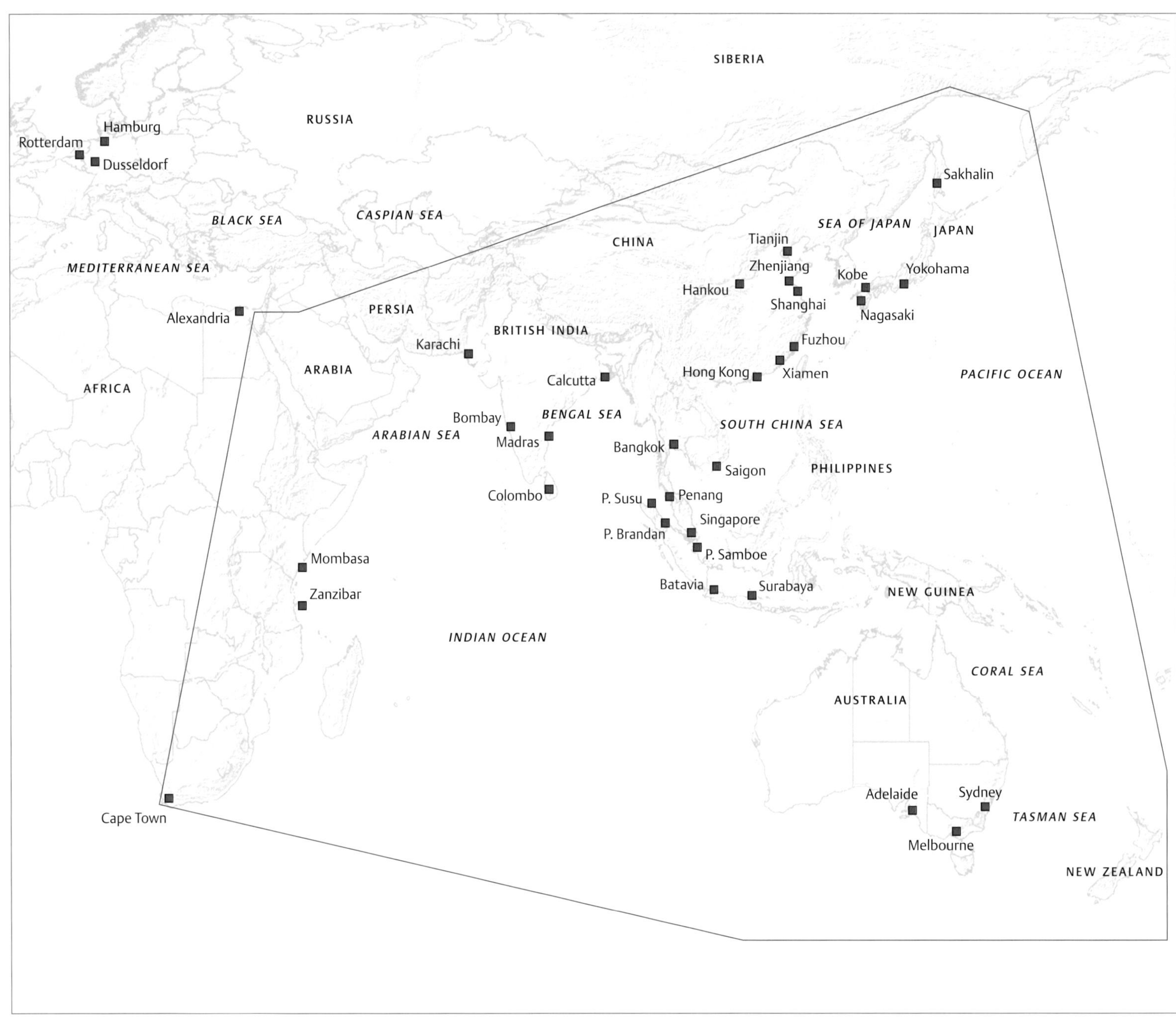

[75]

Map of the Far Eastern market around the end of the 19th century, including the Red Line area,assigned as operating territory to Asiatic.

advantage over another, appeared to safeguard everyone's interests. Yet, from the start, the combination was fraught with conflicts. It took another year of bitter disputes before Asiatic was formally constituted. Meanwhile Deterding had to work literally on his own personal account, for Asiatic had an executive committee, a letterhead, and a special name for addressing telegrams, but no formal legal existence, so Deterding could not open a bank account in the company's name.[207]

The form of the association itself, a partial alliance between equals whose other interests did not necessarily tally, almost guaranteed friction. Surrendering control over assets and operations to Asiatic required confidence in Deterding's intentions, which he himself undermined by failing to establish proper administrative procedures for the initial joint marketing operations. Consequently, he could not present anything in the way of formal accounts. The accounts for operations during July - December 1902 were finalized only in August 1904, causing much embarrassment for Shell Transport and Royal Dutch, which badly needed the data for their own annual reports. Matters improved only slowly.[208] The real causes of friction lay deeper, however. As with any cartel, the partners in Asiatic had one common goal — stabilizing prices on the Asian market at a higher level — but a whole range of potentially conflicting interests, viz. the form of the association itself, tanker charters, sales policy, price policy, the rationalization of the marketing network, and supply schedules.[209]

The form of the association. Lane, who represented the Rothschild interests on the Asiatic board, wanted it to be an independent company, with its own aims and strategies. Consequently, he opposed Deterding's proposal to share out all profits and argued for building up reserves. Sir Marcus for once agreed with Deterding, however, that Asiatic ought to remain an alliance without its own identity or funds, and they carried the day.[210] The debate about Asiatic's status was to surface again in the negotiations about the merger between Royal Dutch and Shell Transport in 1905-6.

The tankers. The rate of seven shillings a ton paid to Shell Transport was way above market rates. To narrow the gap, Deterding insisted on shifting some of this cost back to Shell Transport. The Samuels were incensed at what they regarded as an unfair attempt to renege on the agreement, but with Bnito supporting Deterding, they finally had to give way.

Sales policy. Shell Transport was built around bulk sales, but Deterding wanted to continue with case oil, to suit Bnito and to keep the door open to the Russian independents. It also gave him an opportunity to feed sales using cheaper ships. The Samuels of course took this as a deliberate attempt to bypass their tankers and installations.

Pricing. For Deterding 'profitable stability' meant maintaining prices to protect producers' margins, sacrificing volume if necessary. However, the revenues of Shell Transport and its agents depended on volume, so the Samuels' priority was boosting turnover, not price.

The rationalization of sales. To eliminate the agency problem and achieve economies of scale in marketing, Deterding wanted to build a sales network dedicated to oil. To this end he had already appointed the first of Royal Dutch's own sales agents, in Penang.[211] Network rationalization started in the summer of 1903, when Asiatic appointed the agents of Shell Transport and Royal Dutch in China and in India as joint agents for the whole of their respective

[76]

Life in south Sumatra: Perhaps unconsciously demonstrating their imperial attitude, two Europeans pose above the head of a local during the construction (above) of a new cracking installation at Plaju in 1910. A general panorama from 1904 (above right) shows the well-developed complex at nearby Baguskuning, with the hospital at Plaju in 1900 (near right) and living accommodation at Baguskuning in 1904 (far right).

[77]

[78]

[79]

countries.[212] Both steps threatened to cut off Shell Transport from its base, i.e. the agency network of general trading houses for whom kerosene was a sideline but who, together with the Samuels, were still the majority shareholders. Attempts by Sir Marcus to keep Shell Transport's former agents aligned with him rather led to a serious row in the Asiatic board.[213]

Supply schedules. Rationalization of logistics yielded a substantial part of Asiatic's gains and profits, partly achieved by merging the producers' Asian fleets into a single company, the Nederlands-Indische Tankstoomboot Maatschappij.[214] However, producers resented the subordination of production to marketing, which meant having to accept Asiatic's strict shipping schedules and instructions about what to produce and when to deliver.[215]

Moreover, Asiatic pooled all oil products, necessitating increasingly stringent product specifications.[216] This gave rise to tremendous rows over quality standards and penalties for substandard products. Notably the NIIHM suffered. In 1905 the company was heavily fined over the continuing poor quality of kerosene from the Balik Papan refinery, resulting in a cut of its kerosene profits by £ 100,000 on revenues of £ 325,000. Royal Dutch and Moesi Ilir also received penalties, but much smaller ones.[217] As a precaution, the Syndicate of Eastern Producers established its own laboratory near Singapore for regular quality checks on shipments. On the express instructions of Sir Marcus, the NIIHM voted against and refused to pay its share.[218] No surprise then that Standard Oil's European representative Macdonald reacted diffidently to Deterding's triumphant announcement of the British-Dutch alliance: with so many contentious issues built in, he must have expected this pantomime horse to break apart at its first hurdle.[219]

In these rows Lane as a rule sided with Deterding against Shell Transport, perhaps because he had developed a strong antipathy to Sir Marcus, but also because the Rothschilds were rather suspicious of the Samuels' intentions.[220] The exception to this rule was the debate about the status of Asiatic, in which for once Deterding and Sir Marcus agreed. Though usually finding himself in a minority, Sir Marcus showed his true mettle now: impetuous on the attack, he was tenacious and stubborn in defence, skilfully limiting the damage which he had himself inflicted on his company. Shell Transport could no longer afford to lose the alliance, however. First Sir Marcus's keen efforts to convert the Royal Navy from coal to liquid fuel received a serious setback when an official trial in June 1902 went disastrously wrong. Consequently, shipowners remained reluctant to convert to oil, too. By 1914 only 3 per cent of the world's merchant shipping tonnage burned oil.[221] And then the Spindletop venture foundered. During June, news had percolated to Europe about Guffey's poor kerosene, perhaps also about the first tell-tale signs of water.[222] In August production from the wells, hailed as the world's richest ever only the year before, dropped sharply. Suddenly Shell Transport faced another huge waste of assets: three large tankers, four even bigger ones to come, a European distribution network in the making, but no oil. To compound the problems, just when his company urgently needed all the attention it could get to keep floating, Sir Marcus became Lord Mayor of London, a largely ceremonial function of great honour and respect, but one which left its occupant with little time for other pursuits during his year in office. Sir Marcus immersed himself in the job. It was a month before he could spare even an hour for Shell Transport. Exasperated at what he regarded as the culminating example of years of gross negligence, Fred Lane resigned his seat on the board in December.[223]

Under the pressure of events, Shell Transport relented. In May the contractual details had at last been straightened out sufficiently for Asiatic to be formally constituted and start operation on 1 July. Under Deterding's management, sales rose sharply. Shell Transport and Royal Dutch together had sold 17.5 million cases in 1902; Asiatic sold 18-20 million during its first two years. In 1905, which was a comparatively poor year, Asiatic still handled 14.5 million cases, and at a much higher profit than the constituent companies had achieved before, unit returns averaging one shilling and sixpence (seven decimal pence; cf. Table 1.1 above).[224] Asiatic paid 49 per cent dividend for 1904, 22 per cent dividend for 1905, and 40 per cent for 1906.[225] At Royal Dutch, net return on assets shot up from 7 per cent in 1902 to over 20 per cent during 1903-7, partly as a result of the company having started to sell gasoline in Europe.[226] The board's approach in entering this market radically differed from Shell Transport's efforts through PPAG. For a start, Royal Dutch began selling gasoline, rather than kerosene, since this offered higher margins on a far less competitive and rapidly growing market. By entrusting sales to Asiatic, the gasoline could be shipped in tankers returning to Batum from Asia, neatly solving the return freight problem and cutting costs all round. Instead of attacking with fanfares and press statements, as Sir Marcus had done, Deterding opted for an indirect approach. He supported an initiative of Heinrich Späth, the manager of a Ludwigshafen gasoline refinery, to set up a German company, Benzinwerke Rhenania, in Düsseldorf, just over the Dutch border on the Rhine, to refine gasoline brought by lighters from storage tanks in Rotterdam.[227] Royal Dutch did not even figure in the Rhenania deed of association. Capadose and Loudon were named as the owners, Späth acted as manager, Wilhelm Rudeloff became chairman, the Rotterdam shipping company Phs. van Ommeren & Co. organized the shipping operations. Building began in September 1902 and the refinery came on stream six months later. The new company shook the German gasoline cartel, which had declined to support Späth's plans earlier, into concluding a marketing agreement. The cartel abandoned Standard Oil and adopted Royal Dutch as its exclusive supplier in return for a geographical division of the market. By 1904 Rhenania controlled 90 per cent of German sales, forcing Standard to come to an agreement.[228] A similar tactic was followed in France. Under supervision of the French firm of A. André Fils, gasoline refineries were built in St Louis du Rhone and Rouen forcing the cartel, until then entirely reliant on Standard, to do business on terms set by Deterding.[229] To maintain Royal Dutch's edge and obtain gasoline sources close to Europe, Deterding set out to secure concessions in Romania, for practical reasons initially for his own account.[230]

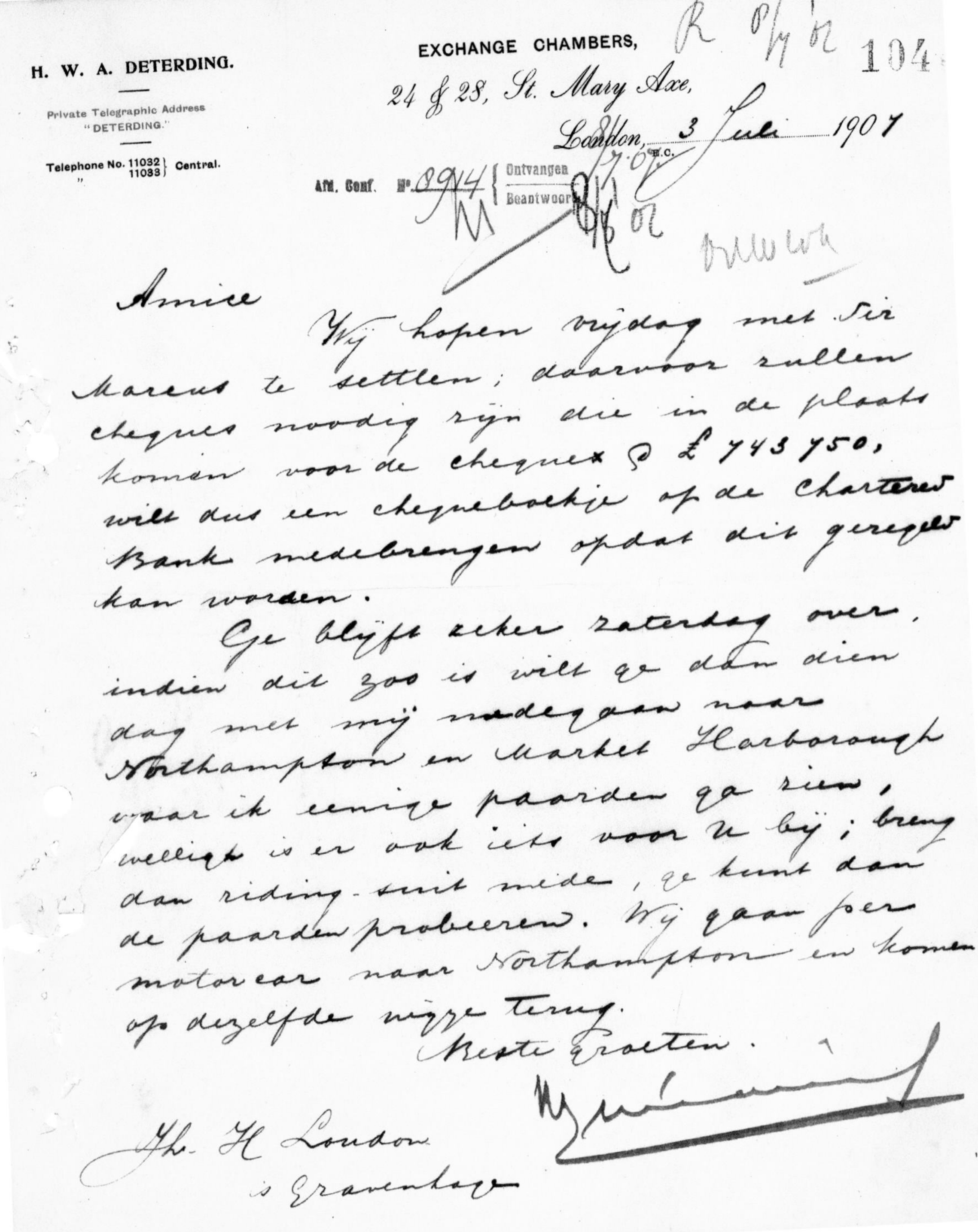

H. W. A. DETERDING.

Private Telegraphic Address "DETERDING."

Telephone No. 11032 / 11033 Central.

EXCHANGE CHAMBERS,
24 & 28, St. Mary Axe,
London, E.C. 3 Juli 1907

Afd. Conf. No. 8914 Ontvangen / Beantwoord

104

Amice

Wij hopen vrijdag met Sir Marcus te settlen; daarvoor zullen cheques noodig zijn die in de plaats komen voor de cheques @ £ 743 750, wilt dus een chequeboekje op de Chartered Bank medebrengen opdat dit geregeld kan worden.

Ge blijft zeker zaterdag over. indien dit zoo is wilt ge dan dien dag met mij medegaan naar Northampton en Market Harborough waar ik eenige paarden ga zien, wellicht is er ook iets voor u bij; breng dan riding-suit mede, ge kunt dan de paarden probeeren. Wij gaan per motorcar naar Northampton en komen op dezelfde wijze terug.

Beste groeten.

Jhr. H Loudon
's Gravenhage

In the final stages of negotiations, Waley Cohen's matter-of-fact telegram of 15 February 1907 to Sir Marcus Samuel (above right) contrasts with Deterding's effusive congratulations on 23 April (below right) concerning the 'amalgamation' of their companies. On 3 July (left), writing to his colleague Hugo Loudon, Deterding predicted that they would settle with Sir Marcus on Friday 5 July, and asked Loudon to take his chequebook with him to pay for the balance.

[81]

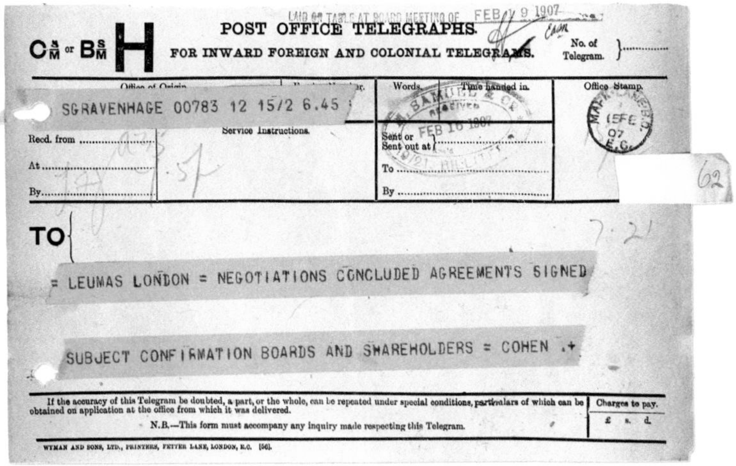
POST OFFICE TELEGRAPHS.
FOR INWARD FOREIGN AND COLONIAL TELEGRAMS.
LAID ON TABLE AT BOARD MEETING OF FEB 19 1907
SGRAVENHAGE 00783 12 15/2 6.45
TO
= LEUMAS LONDON = NEGOTIATIONS CONCLUDED AGREEMENTS SIGNED
SUBJECT CONFIRMATION BOARDS AND SHAREHOLDERS = COHEN .+
N.B.—This form must accompany any inquiry made respecting this Telegram.

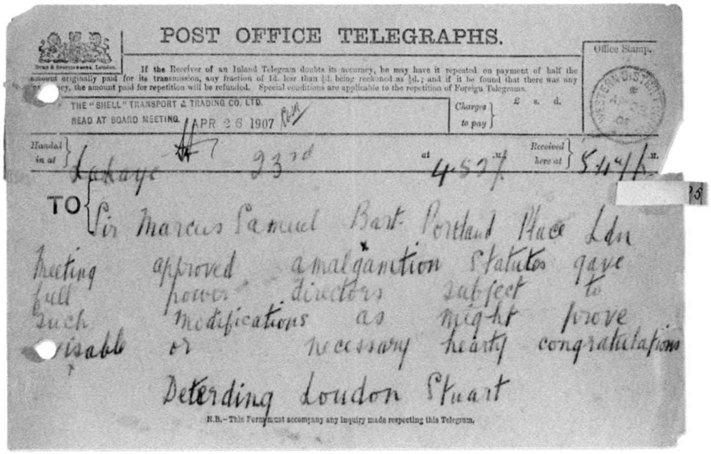
POST OFFICE TELEGRAPHS.
THE "SHELL" TRANSPORT & TRADING CO. LTD.
READ AT BOARD MEETING APR 26 1907
Lahaye #7 23rd 4.52 5.41
TO Sir Marcus Samuel Bart Portland Place Ldn
Meeting approved amalgamation statutes gave full power directors subject to such modifications as might prove advisable or necessary hearty congratulations
Deterding Loudon Stuart
N.B.—This Form must accompany any inquiry made respecting this Telegram.

[82]

Merging for real By contrast, Shell Transport had to embark on a prolonged damage limitation exercise. The company's performance improved, but it came nowhere near that of Royal Dutch. Gross revenues were considerably higher than in 1901, and the third best since 1898. Return on assets remained paltry, however, at around 4 per cent. The financial press now began to grow restive, questioning the company's performance, position, and accounting practices. One paper even called for Sir Marcus's resignation.[231] The board had learnt some lessons, keeping a close watch on costs, and discussing the building of a wax plant at Balik Papan on the basis of an extensive cost-benefit analysis.[232] Attempts to stem the losses did not produce sufficient results. The tankers built for Guffey's oil were converted to cattle carriers, but barely paid for themselves in that way.[233] Talks about Royal Dutch taking over the management of Balik Papan and turning around the refinery foundered. To Sir Marcus, surrendering control over Shell Transport's Kalimantan assets meant giving up his liquid fuel ambitions.[234] For his part, Deterding was in no hurry. Getting hold of the refinery offered considerable economic advantages, but he preferred a waiting game, keeping pressure on Shell Transport.[235] Close controls on the PPAG did not succeed in checking the drain there. The company suffered from continuing price wars, and lost more than a pound sterling on every ton of kerosene sold, making a loss total of £ 112,000, in 1904.[236]

PPAG became Shell Transport's undoing. Deutsche Bank, determined to establish a home market for its Romanian oil interests, wanted to expand the company further and beat Standard's price competition with scale. By contract, the PPAG shareholders were bound to supply equal amounts of capital. When the bank proposed raising the capital to RM 12 million, Shell Transport could not match and had to sell its stake and the tankers

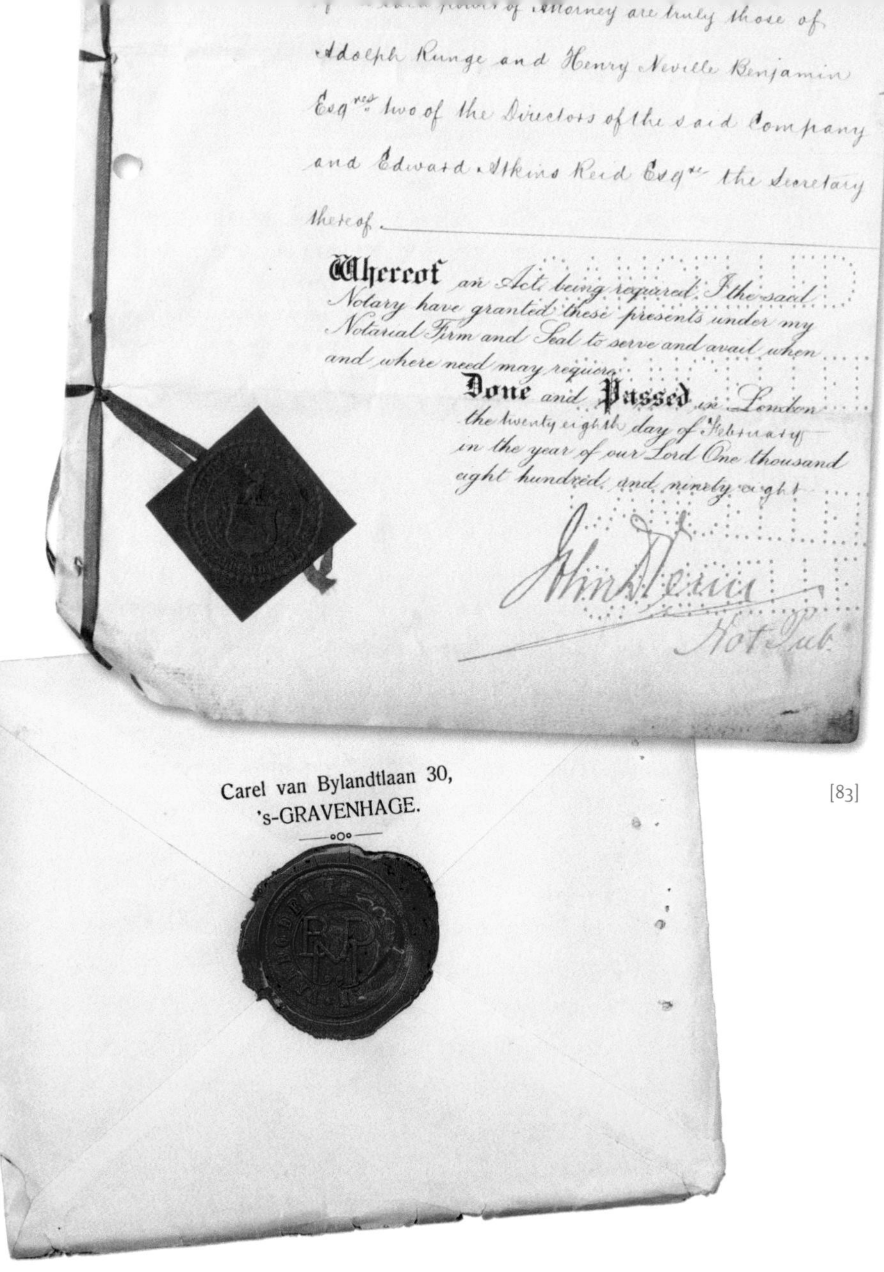

... of Attorney are truly those of Adolph Runge and Henry Neville Benjamin Esqres two of the Directors of the said Company and Edward Atkins Reed Esqre the Secretary thereof.

Whereof an Act being required I the said Notary have granted these presents under my Notarial Firm and Seal to serve and avail when and where need may require.

Done and **Passed** in London the twenty eighth day of February in the year of our Lord One thousand eight hundred and ninety eight

Not. Pub.

[83]

[84]

The legal seals of Bataafsche Petroleum Maatschappij (above) and Shell Transport and Trading (top).

employed in the PPAG operations. At a stroke, the company lost both its European sales network, and a substantial part of its transport business.[237] Now Shell Transport's habit of window dressing finally caught up with it. The company had debts of £ 580,000 which it could not meet, hidden in overvaluations of tankers and money sunk in the NIIHM.[238] In early December 1905, Sir Marcus had a first talk with Deterding about a merger between Shell Transport and Royal Dutch.[239] The British company was still twice the size of the Dutch concern by assets, but virtually bankrupt. Deterding knew it. He had been waiting for this moment since 1902. Royal Dutch could not afford an outright takeover; nor did Deterding want it. To build a global oil concern, he needed Shell Transport as a British company, rooted in the City of London, closely linked to people with access to His Majesty's Government, able to pull strings which a company based in The Hague could never hope to reach. As Deterding put it to the Royal Dutch board, it would be impossible 'to have such a large company as Royal Dutch and Shell Transport combined (...) be operating under the Dutch flag. We are in our business far too dependent on the British government because we have to sell a large part of our products in English colonies'.[240]

The merger would have to take a shape similar to Asiatic, but now forming a fully integrated oil company, from well to consumer, and with Royal Dutch holding 60 per cent of the shares in the projected operating companies. For a while Sir Marcus held out to obtain a 50-50 split. In March 1906, his back to the wall, he consented to the principles of the amalgamation after Deterding had offered to let Royal Dutch take a 25 per cent interest in Shell Transport, while letting the Samuels retain their special voting rights. Taking this quarter stake gave Royal Dutch an additional 10 per cent of the Group for a total of seventy.[241]

Chapter 1

On 12 September, the two companies signed a provisional agreement about the merger between them, but at that stage a crucial problem remained to be solved: the joint venture with Bnito in Asiatic. During an early stage of the merger talks with Sir Marcus, in January 1906, Deterding had considered taking over Bnito at the same time. The Paris Rothschilds did want to sell, but preferably to Deutsche Bank, because with a sale to Royal Dutch they would still be linked to the company in the eyes of the Russian government through their share in Asiatic.[242] In the end, Deterding probably could not mobilize the funds needed.[243] He then joined with Sir Marcus to argue that, after the merger, Asiatic would lose its purpose and should be transformed into a simple sales department.[244] There was some truth in this argument. With or without a merger between Royal Dutch and Shell Transport, Asiatic was losing its original status as an equal partnership. Bnito's consignments to Asiatic declined, both in relative and in absolute terms. Royal Dutch had overtaken Bnito's production in 1905 and would soon produce twice as much crude.[245] During part of 1906, Bnito even suspended all supplies. Yet Bnito continued to get its one-third of Asiatic's stupendous profits. Therefore, unless fundamentally modified, Asiatic would live off the production interests of the projected combine. Though motivated with an earnest desire for corporate simplification, the attempt to liquidate Asiatic really aimed to cut out the Rothschilds, who had now become a burden. Or, as Deterding put it to Loudon in March 1906, 'I believe that we stand on the threshold of tremendous years, especially if we also succeed in liquidating Asiatic'.[246]

Lane and the Rothschilds vigorously opposed the attempt. Asiatic's profits were Bnito's lifeline. Its Russian fields were nearing exhaustion and the company urgently needed new investment. Bnito's profits, never very spectacular anyway, had dwindled to almost nothing. Since 1900 Bnito had paid dividends only twice, 10 per cent in 1900 and 5 per cent in 1903.[247] The Asiatic contract put them in an ostensibly strong position. It had been the original intention, Lane argued, that Asiatic would become an integrated oil company of the combined interests, not just a transporter and distributor, but also a producer, taking up new concessions as and when they became available. That long-term intention had been the reason for setting the company's lifespan at twenty-one years. Shareholders could be bought out at par in 1923 and he could see no reason for changing the arrangements now.[248] Privately, he admitted to the weakness of the position. No contract could change the fact that Asiatic would now be run by the majority shareholder who regarded the minority interest as parasitic and who would thus be bent on reducing its profits by fair means or foul.[249] The Rothschilds agreed and contemplated selling out if the price was right, but considered it unlikely that Deterding would be unwilling or indeed able to pay the £ 1.2 million which they wanted.[250] With neither party able or willing to move, the only solution lay in a limited restructuring of Asiatic, redefining costs, charges, prices, and revenues so as to ensure full accountability and transparency to all parties of the company's business. This painstaking work took until March 1908 to complete and resulted in an impossibly complex accounting structure, opaque even to clever insiders such as Lane.[251]

The failure to liquidate Asiatic must have vexed Deterding rather, for three reasons. First, he continued to regard the profit share of the Rothschilds as unjustifiable. Second, the marketing administration remained unnecessarily tortuous and expensive, thus preventing the Royal Dutch-Shell Transport combine from reaping the full benefits of integration in what he regarded the core function of the business. Finally, he would not have the full control over the business which he wanted. At the time Deterding does not appear to have shared these disappointments with anyone, but they fed a rancour against the Rothschilds which would erupt with surprising force during the early 1930s (see Chapter 7).

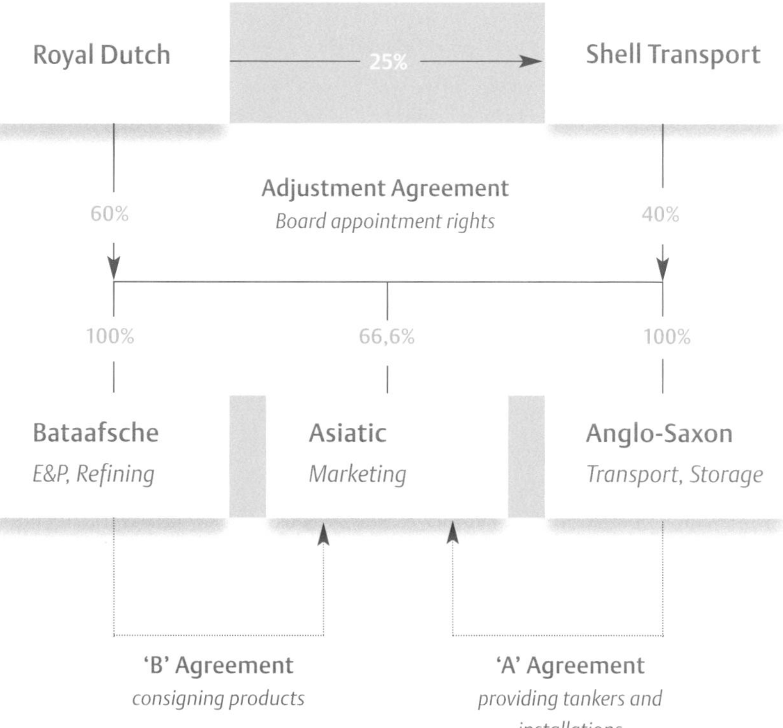

Figure 1.4
The structure of the Group in 1907.

Constructing the Group The continuation of Asiatic added further complications to the merger talks between Shell Transport and Royal Dutch. First, this necessitated splitting the business functions into separate operating companies, instead of merging them into the single company with separate departments in London and The Hague as originally foreseen.[252] Two operating companies were set up, Anglo-Saxon Petroleum Company in London and Bataafsche Petroleum Maatschappij (BPM) in The Hague, both owned in the agreed 60:40 ratio by Royal Dutch and Shell Transport, now transformed into holding companies. Anglo-Saxon was to own all ships and marketing installations and provide shipping and storage services to Asiatic under an agency contract termed the 'A' Agreement. Bataafsche would own and manage all E&P and refining operations and consign the products to Asiatic under the so-called 'B' Agreement. As before, Asiatic received two per cent commission on sales plus 25 per cent of the net profits, the rest being split by the three shareholders. Bataafsche received the rest, regardless of whether the company had actually produced the oil sold or not: if Asiatic traded outside oil, Bataafsche still profited. An Adjustment Agreement laid down further details about board appointment rights to Bataafsche, at that stage considered to be the central company of the business. As assigned, the appointment rights emphasized a clear concern for keeping Bataafsche fully under Dutch control, for it was not an outright 60:40 split. Royal Dutch appointed six directors to the board of Bataafsche and Shell Transport three, with a tenth director appointed by Royal Dutch but with the consent of Shell Transport. Moreover, until 1940 the managing directors of Bataafsche were always Dutch. No special agreement was deemed necessary for Anglo-Saxon, presumably because the negotiators regarded it as a subordinated shipping company. Having bought 25 per cent of Shell Transport shares, Royal Dutch also received the right to appoint two directors to the Shell board one of whom, Deterding, acted as manager. Conversely, Shell Transport did not have the right to appoint directors to the Royal Dutch board. Fred Lane served there from 1910 to 1926, presumably as representative of the Paris Rothschilds who were the largest individual shareholders of Royal Dutch, but otherwise no foreigners sat on it until the 1960s. Consequently, Royal Dutch had a dominating managerial presence of more than the 60:40 ratio in what became known as the Group.

Second, the splitting of business functions over different companies, the presence of an outside shareholder in one of them, and the consequent need for punctilious accounting and restrictions on the flow of sensitive business information, required using formal and strict procedures for intercompany dealings. A typical Group culture emerged with as its main axes the relative independence of subsidiary companies, and a careful allocation of all costs to the business functions or companies responsible for creating them. This form of corporate organization created a huge administration to which the overlap between two head offices contributed further, but it also provided a sophisticated and methodical cost control to a degree which would have been more difficult to achieve between separate departments of a single company.

By December 1906 the Royal Dutch and Shell Transport boards had agreed to let the merger take effect from 1 January, 1907 (Figure 1.4), though numerous issues remained to be resolved. On 15 February, the negotiating teams reached formal agreement about the last of a long set of amalgamation schemes in The Hague, but the main documents were only signed on 5 July, and the 'A' agreement took another eight months of negotiations, until March 1908. The whole process was never concluded with a joint formal ceremony to mark the momentous occasion. The respective directors, boards, and shareholders all approved the merger separately, as if they continued to operate independently, and not as parts of a larger entity. Thus the Group came into existence without fanfares or a proper date for its inception, and the constituent companies carried on their business as usual.[253]

[85]

The rough construction of a typical derrick somewhere in the jungle is easily seen here, with its four legs being trees cut from the forest.

Conclusion

The formation of the Group and notably Royal Dutch's leadership in it was so incongruous as to inspire disbelief. To the public eye, Shell Transport remained a British company, and it was automatically assumed to be the senior partner.[254] The Group carefully nursed this misapprehension, by retaining Sir Marcus as figurehead, and by adopting the Shell brand name and the pecten logo for Group-wide products and operations. Historians have offered various explanations for the unlikely outcome. Gerretson extolled the success of Royal Dutch, emphasizing on the one hand Deterding's genius, and on the other the company's rapid adoption of gasoline as a key product to conquer new markets. By contrast, Henriques described Shell Transport's trajectory in the very personal terms of Marcus Samuel, first nursing the company to greatness, then losing his touch and finally losing out to the somewhat insidious tricks of Deterding.

These explanations are only partially true, and thus necessarily incomplete. Deterding was definitely a dynamic manager, but Royal Dutch owed its success just as much to Kessler's earlier work in building a resilient organization with a remarkable facility for finding the right people to manage its operations, and not only at board level. Gasoline was crucial in facilitating Royal Dutch's diversification into new markets and products, and unit revenues were far higher than those of kerosene. Yet in 1906 kerosene still generated almost 2.5 times as much revenue.[255] Finally, it did not take much perfidy to wrongfoot the Samuels. Shell Transport was badly managed from the start. Successive mistakes only served to hasten the disastrous consequences of imprudent financial policies which would have caught up with the company sooner or later. The decisive difference between Shell Transport and Royal Dutch resided in their respective origins, Shell Transport as a shipping line, Royal Dutch as an oil producer. Only integrated oil companies could build the competitive advantage needed for long-term success. Both the huge disparity between Shell Transport's and Royal Dutch's profits, and the ultimate fate of the other Indonesian producers, testify to that. Creating an integrated supply chain from the middle, a hard act under any circumstances, would have required far more financial and managerial resources than Shell Transport could muster. Conversely, Royal Dutch simply had no other option given its location, and once production got underway the company could capitalize on its rapidly growing resources to create integrated operations. Kessler fully realized the vital importance of integration and clung to it even when it would have made sense to relent, first in 1892-94, and again during the production crisis. Thus the market forces which created the merger also explain the ostensible incongruity in the 60:40 split.

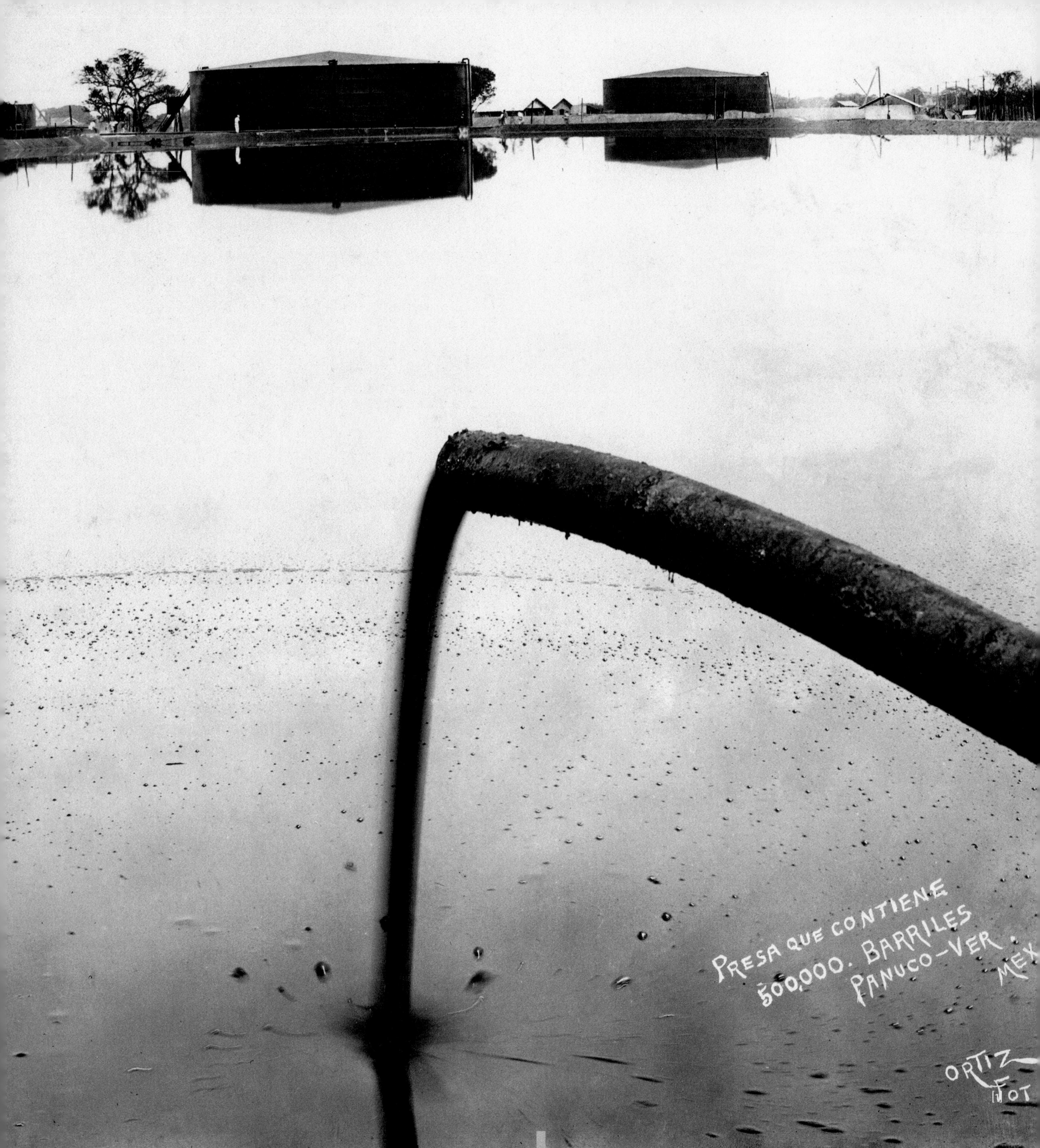
PRESA QUE CONTIENE
500,000. BARRILES
PANUCO-VER.
MEX
ORTIZ
FOT

Chapter 2

An enterprise waiting to be made, 1907-1914

Between the merger of 1907 and the First World War, the Group developed at a precipitous pace. The business owed its success partly to the way in which it was managed, with a close-knit team of managers steering the growing empire into the right directions. The merger also changed the relationship with the main rival, Standard Oil Co., and indirectly sparked an oil price war during 1910-11. This was followed by a dramatic expansion of the Group into new production areas, and by large investments in the development of new know-how and technologies. The growth spurt of 1910-14 also led to increased tensions within the organization, however, and inspired structural changes resulting in the near-complete merger of the management of the two parent companies.

End of the line: in Mexico in 1915, a pipe delivers La Corona's crude into an open-air tank holding half a million barrels of oil.

Five firms formed the core of the Group: the two parents Royal Dutch and Shell Transport as holding companies, the newly created Bataafsche Petroleum Maatschappij and Anglo-Saxon Petroleum Company as operating companies, and the Asiatic Petroleum Company as the shared marketing company.

Managing a transnational enterprise The bold and dramatic amalgamation was, at the time, probably one of the biggest mergers in history, certainly outside the US. It had to be organized without much external expert advice, because no investment banks or consultancy firms could help to guide the process. Moreover, it was a transnational merger, involving the companies of two countries with different business cultures, legal systems and languages, plus a third, formidable partner from yet another country, the Paris Rothschilds. There simply existed no precedent for such an amalgamation, and it is small wonder that it took two years to establish the organization and to sort out all the details. Yet, by March 1908 all matters had been finally settled, and the Group had been created. The big question was, of course, would it work?

Deterding was the undisputed leader of the Group. He was not only Director-General of Royal Dutch and managing director of Asiatic, his two former strongholds, but also managing director of Anglo-Saxon, of Bataafsche, and director of Shell Transport. His commanding position was fully acknowledged by Sir Marcus Samuel who, reflecting on the merger at the Shell Transport Annual General Meeting of shareholders (AGM) on 30 June 1908, stated that 'In Mr. Deterding, the Managing Director (...) we have a gentleman who is nothing less than a genius (...) I consider it most fortunate in the interests of all that we have secured his services, together with those of Mr. Hugo Loudon, Dr. AJ. Cohen Stuart and Mr. R. Waley Cohen, who are jointly the Managers of the business, and devote their best abilities ungrudgingly to it.'[1]

This quote is of interest for two reasons: first of all, *de jure* Deterding was *not* the managing director of Shell Transport, yet he was presented as such by Samuel, and he would be present in this capacity at all AGMs of Shell Transport in the years to come. Officially, Marcus Samuel, Sam Samuel, W. F. Mitchell, and Walter Samuel were the managing directors, and Deterding was no more than a non-executive director. As it was, Sir Marcus's statement served to buttress the fact that Deterding was the managing director of the Group.

The second, and perhaps more important, point is Samuel's reference to a team of directors around Deterding, formed by the Deterding's two fellow directors of Royal Dutch, Hugo Loudon and A.J. Cohen Stuart, plus Robert Waley Cohen. These four men had to fill in the Group's structure, held together by partially overlapping directorships between the five core companies and the steadily expanding number of operating companies. Consequently, they were either managing directors or directors of two or more of the core companies, as well as directors of operating companies. The composition of the Bataafsche board shows the overlap principle to good effect, with the three managing directors plus three non-executive directors or *commissarissen* of Royal Dutch, three members to be appointed by Shell, and finally a tenth member, a British subject to be appointed by Royal Dutch on condition of approval by Shell, giving Shell the practical right to appoint four of the ten members. As the Group expanded, the overlapping directorship principle was extended to the operating companies set up or taken over. Such companies were controlled through the appointment of two or more Group directors to their boards, with administrative departments in The Hague or London effecting the management.

The managerial team comprised a good variety of different backgrounds, experience, and talent. Loudon became the managing director and chairman of Bataafsche, the central operating company of the Group; he also ran Royal Dutch's day-to-day business. It is significant that he was really the only director to remain in The Hague, the other three managing directors having their base in London. The lawyer Arnold Jacob Cohen Stuart (1855-1921), who had succeeded Abraham Capadose in 1906 as managing director of Royal Dutch, also became managing director of Anglo-Saxon, while remaining closely involved with Royal Dutch and Shell Transport as well.[2] Robert Waley Cohen (1877-1952), a graduate in chemistry from Cambridge, was by far the youngest of the team.

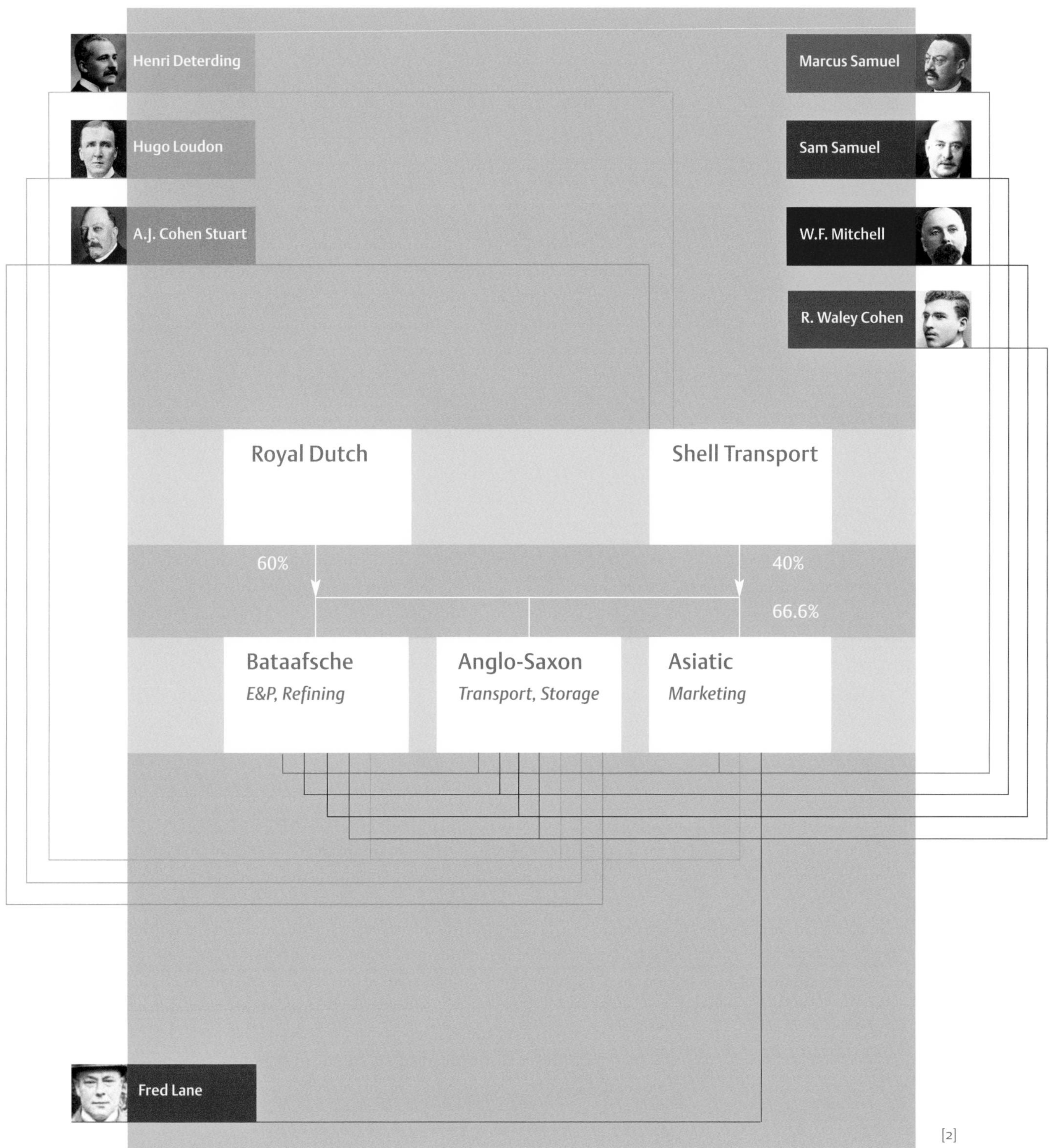

Henri Deterding
Hugo Loudon
A.J. Cohen Stuart
Marcus Samuel
Sam Samuel
W.F. Mitchell
R. Waley Cohen
Royal Dutch
Shell Transport
60%
40%
66.6%
Bataafsche
E&P, Refining
Anglo-Saxon
Transport, Storage
Asiatic
Marketing
Fred Lane

[2]

He had joined Shell Transport as a trainee in 1901, rising very quickly indeed to become a director of the company in 1906. His conduct of the merger negotiations with Loudon and Cohen Stuart established his reputation on both sides of the North Sea.[3] Waley Cohen was the only British member of the team, a great admirer of Deterding, at least at this stage of his career. His selection came as something of a surprise. As Henriques described it in his biography of Waley Cohen: 'On the Board of the Anglo-Saxon, the Royal Dutch were to have four directors, of whom the first three had been mentioned by The Hague as Hugo Loudon, Henri Deterding and Cohen Stuart. The fourth nominee of the Royal Dutch was still to be awaited (...) At the Shell Board Meeting on 2 July 1907 (...) a letter was read from the Company's solicitors (...) reporting that the Royal Dutch Co. nominated Mr. Waley Cohen as fourth Director of the Anglo-Saxon Petroleum Co. Ltd. (...) As a gesture by the Royal Dutch, the appointment was chivalrous, dramatic and quite unexpected. As a commercial measure, it was far-sighted and expedient.'[4]

With this elegant gesture, Waley Cohen joined the managerial team. His appointment also served to underline that the Group would not be an exclusively Dutch affair, in which British partners were to play a subordinate role, but that Deterding wanted to recruit all talent, old and new, into the management.

This four-man team formed the heart of the Group during its formative years. Deterding was the creative genius, but he tended to lose interest in a dossier once the main decisions had been taken, or when subsequent progress was slow. As often as not Waley Cohen, his deputy in Asiatic, or Loudon, had to do the better part of negotiations, or to get things moving again. Moreover, Deterding's genius was commercial, not technical, and he made

[4]

Henri Deterding

Success does not always breed admiration, and just two years after the Group's formation, there were public complaints about the 'colossal fees' of the Royal Dutch directors. Newsclip from the *Financial News*, 24 June 1909.

[3]

ROYAL DUTCH PETROLEUM: DIRECTORS' COLOSSAL FEES.

ANNUAL MEETING, HOWEVER, WILL VOTE UPON A PROPOSED REDUCTION TO MAXIMUM OF 30,000FL. EACH.

[CORRESPONDENCE OF THE FINANCIAL NEWS.]

AMSTERDAM, June 20.—The Royal Dutch Petroleum Company has just issued its annual report for the fiscal year 1908, and in view of the magnitude of this undertaking and its close relations to the Shell Transport and Trading Company it may be well worth while to consider in detail the figures submitted to the stockholders. The Royal Dutch Company holds 60 per cent. of the capital of the Bataafsche Petroleum Maatschappy and of the Anglo-Saxon Petroleum Company, the remaining 40 per cent. being held by the Shell Company. The capital of the Bataafsche is 80,000,000fl., on which a dividend of 17½ per cent. was paid for 1908, after writing down all expenses of exploitation and exploration and all capital outlays for improvements and for new properties, and after a liberal allocation to the insurance fund. The Anglo-Saxon paid a dividend of 10 per cent. on its £4,000,000 shares, also after liberal outlays for depreciation. Consequently it has not been thought necessary that the Royal Dutch and the Shell should make any reserves for depreciation themselves. Besides these dividend payments, the Royal Dutch Company also collected 20 per cent. dividends on its holding of Shell ordinary shares, which amounts to 25 per cent. of the latter's outstanding common stock. The total profit of the Royal Dutch was 13,473,000fl., out of which 4 per cent. dividend is paid on its 1,500,000fl. preferred and 28 per cent. on its 40,000,000fl. common stock. The preferred dividend is limited to 4 per cent., but the preferred shareholders control the company, as they have to approve

... will now be reduced to a maximum of 30,000fl. for every member; but the five present members will be compensated by getting each 120,000fl. of Royal Dutch shares, which are worth at present more than 500,000fl. Although dividends will, of course, have to be paid on this additional stock at the full rate, the change will nevertheless mean that the stockholders will in future receive about 1¼ per cent. more in dividends than at present, figuring on the same net income, while the difference will, of course, get greater or smaller, according to an increase or a decrease of the profits in future.

The International Roumanian Petroleum Company will be reorganised soon. It is intended that the refinery "Aurora," of which all shares were in the hands of the International Company, will be worked in future by the Steaua Romana, while the International Company will sell its oil under contract to the same concern. The nominal value of the company's shares will be reduced from 1,000 florins to 300 florins, in order to cover the losses sustained under the former management, which for a number of years declared dividends at the expense of the depreciation funds, so that the property deteriorated.

On the Stock Exchange dealings in oil shares were very brisk. Indian sugar shares also were strong, gains of from 5 to 20 per cent. being recorded. Tobacco stocks remain dull, on the expectation that the remaining part of last year's crop will sell at exceedingly low prices. So far the prices obtained have been considerably above expectations; but there remains much inferior tobacco to be marketed, and the Austrian régie, which is the chief buyer of this product, has changed its modus operandi, by entrusting all its orders to a special agent; so that the competition for this class of tobacco has disappeared to a large extent. Good tobacco is still obtaining high prices, the United Lankat Tobacco Company still leading in the prices made for its production. The tendency for mining stocks has grown weaker, especially for Great Cobar, on the disappointing result of last year's operations. The decline might have been still sharper if the quotation had not been held up by bear repurchases. As to American securi

[5] Hugo Loudon

[6] A. J. Cohen Stuart

[7] Robert Waley Cohen

no bones about delegating technical issues to the two scientists, Loudon and Waley Cohen. Similarly, nearly all negotiations with the British government were conducted by Waley Cohen and/or Sir Marcus Samuel. The team members met at the regular meetings of the different Group boards, discussed strategy and tactics there, often taking decisions jointly; votes appear to have been taken rarely.[5] The intention had been to have the team consist of the managing directors of the five companies, but the rules of appointment were not strictly adhered to. When, in March 1909, Waley Cohen had to resign as Bataafsche director because Dutch law required that a majority of directors be Dutch, 'he continued to attend all meetings of the Board in The Hague, was entitled to receive all information that was available to the Directors, could intervene in the actual meetings and in fact could do everything as if he were still a Director, except vote'.[6] During the initial years this relaxed attitude to formality was probably beneficial, because it created a great flexibility, but as time went on and the Group expanded ever further, the lack of a clear managerial structure coupled with procedures based on a formal assignment of responsibilities and tasks to individual directors would create increasing friction.

The guiding team spirit is perhaps best demonstrated by the distribution of the directors' fees, which added up to a huge income. A memo dated 10 July 1907 laid down that 'all salaries, commissions and fees received by the four Managing Directors were to be pooled and divided in the portions of 33⅓ per cent to Deterding, 26⅔ per cent to Loudon, and 20 per cent each to Cohen Stuart and Bob'.[7] This very generous gesture from Deterding cemented the personal relationships between the four men. The sums involved were substantial. The managing directors of Bataafsche and Anglo-Saxon together received, in addition to remuneration from other directorships of core Group companies, 4 per cent of the total dividends paid out by the two companies, with another 1 per cent going to the non-executive directors.[8] On top of this the managing directors of Royal Dutch received 3 per cent of the dividends paid out by the company.[9] The lack of financial details about Anglo-Saxon and Bataafsche prevents us from giving figures about the remuneration of Group directors during these years, but the directors' conditions of pay attracted a great deal of attention in the years to come. Financial analysts judged that the managing directors were 'perhaps, the best-paid officials of any corporation all over the world'.[10]

As chairman of Shell, Asiatic, and Anglo-Saxon, and a director of Bataafsche, Sir Marcus Samuel was in fact also part of the Group's inner managerial circle, but initially he did not function as such, since he found it difficult to accept the merger, which he considered to be a defeat. Samuel withdrew to The Mote, his Kent country house, 'living in the country and going up [to London] once a week'.[11] In the eyes of the British public, however, he was and remained a Napoleon of oil, and his seclusion did not last long. Loudon in particular proved instrumental in wooing Sir Marcus back to the business of the Group.[12] Indeed, Loudon helped to cement the managerial team together, leading one writer to comment that 'the real credit for this remarkable marriage of strong and diverse personalities within the Group, and of British and Dutch characteristics, must go to Hugo Loudon'.[13] Sir Marcus, for example, always consulted Loudon on his chairman's speech for the AGM, every year a key public relations exercise for the Group.

Though the monthly Bataafsche meetings in The Hague became a regular fixture for all directors, London definitely was the managerial centre of gravity. Since his move to Britain in 1902 to manage Asiatic, Deterding had gradually adopted the lifestyle of a British squire, buying a country house and organizing shooting parties, at times together with Sir Marcus's son Walter Samuel.[14] He relished living in Britain and recommended it to his Dutch colleagues. One of them, Cohen Stuart, followed in his footsteps, since his main duties as director were also in London. The Hague directors, and notably the Royal Dutch board, sometimes complained about being relegated to the sidelines, London having all the information and often taking the lead in major decisions, a complaint that would lead to some bitter debates after 1912.

The four-man team, with Deterding as the unchallenged leader, was also a good compromise between British and Dutch styles of management, between the collective responsibility and consensus-oriented governance of a board, as traditionally favoured by the Dutch, and the British tradition of a strong CEO. Shell Transport had been managed by the Samuel brothers without much reference to their fellow directors, but from 1907 the board assumed the central managerial role. By contrast, the Royal Dutch board had originally been closely involved until Deterding's command of the business had gradually eroded their position. Consequently, the scope of Deterding's responsibilities worried the Royal Dutch board, fearful of operations becoming too reliant on a single man. For that reason Royal Dutch had appointed, in 1902,

[8]

Marcus Samuel

Capadose and Loudon to run The Hague's operations when Deterding moved to London. The team solution failed to put the directors' minds at rest, and in 1913, with the Group's honeymoon years definitely over, the board again voiced its concern.[15]

For support, the managerial team relied on a group of middle managers, technical experts, lawyers, and other external advisers. In 1907 such services were still mostly contracted out, but on a regular basis leading to increasingly strong bonds with some experts. August Philips (1864-1954), for example, a corporate lawyer in Amsterdam, functioned as main architect of the merger agreements on the Dutch side; he subsequently took part in nearly all important negotiations conducted by the Group.[16] Geologists were also still employed for specific terms or projects. The most prominent of them, Erb, a stalwart of Royal Dutch since that company's 1899-1900 production crisis and subsequently hired for one assignment after another, did not get a formal job with Royal Dutch until 1911, when he was asked to set up a geology department of which he became head two years later.[17] Both Philips and Erb moved from freelancer to permanent employment by the Group, to become directors of Bataafsche in 1921.

Others remained freelancers, amongst them Fred Lane, who had played such an important role in the years leading up to the

[9] Sam Samuel

[10] W. F. Mitchell

[11] Walter Samuel

merger. He owed his allegiance first and foremost to the Paris Rothschilds, whom he represented on the board of Asiatic and, from 1910, also on the board of Royal Dutch. However, Lane also continued to mediate and to negotiate on behalf of the Group. A similar role was played by Calouste Gulbenkian (1869-1955), the uncrowned king of the oil trade in much of Europe and the Middle East. He became the eyes and ears of the Group in the Middle East, in Russia, and in France, negotiated with American and Mexican partners, tried to organize the issue of Royal Dutch shares in Paris, and helped to manipulate the London stock exchange when share prices were considered to be too low, always of course on a strictly confidential basis, with usually only Deterding in the know.[18] Gulbenkian's status as a trusted ally of Deterding is emphasized by his many directorships of Group operating companies. In addition to the fees which he drew from these appointments, he usually received handsome commissions for the deals which he had helped to broker but, though so close to the inner circle, he never became a member of it. Whether Gulbenkian did not aspire to become a Group director, or was deemed unsuitable for admission, is not really known; it looks as if Deterding deliberately kept him on the sidelines, probably because he feared that otherwise his leadership would be challenged.[19]

[12] Calouste Gulbenkian

Controlling a rapidly expanding business The system of overlapping directorships meant that all lines came together at the top in the hands of the four-man team, burdening them with many and varied responsibilities. A key managerial problem was the Group's sheer spread of operations and the inherent challenges of pulling the business together which forced the managing directors to travel often and far to inspect the marketing in, for example, India and China, or the drilling and refining in the Dutch East Indies, or new concessions in Romania, Trinidad, California, or Egypt. Having witnessed Samuel & Co.'s distance mismanagement of the mounting problems in Kalimantan, Deterding knew the crucial importance of having a steady flow of independent information on developments of the Group's operations around the world, supplied both by the managers on site, but also by passing visitors, who were entitled to propose modifications to the managers and, if necessary, impose them on the spot. The practice of regular inspection trips had been started at Royal Dutch by Kessler, who regularly visited the Sumatra operations and the marketing agencies across Asia, and was continued by Loudon and others. After the amalgamation the directors and chosen experts made many similar trips. Deterding himself made a famous one in 1908, a true 'Grand Tour' taking in all major existing and potential interests of the Group in Asia, including the Dutch East Indies, Japan, China, and the United States. He was accompanied by Dolph Kessler, JBA's eldest son, who had just joined the Group as a management trainee and acted as Deterding's assistant.

[14]

[13]

[15]

[16]

The Group staff also had a pronounced variety of nationalities. British and Dutch managers dominated at the top, and a Swiss geologist and an Armenian oil merchant were amongst the regular consultants. The staff in the London and The Hague offices was still quite small, no more than a few dozen clerks, secretaries and accountants.[20] German clerks were considered to be very trustworthy and efficient, and thus popular with managers, both in The Hague and in London. When, in 1907, the management of the Dutch tanker fleet was transferred to London, Deterding specifically asked for the accountant Viehoff to move to London. Former ship captains moving into fleet management added to the office staff variety. In 1907 the news of desk job opportunities in the London office created an enthusiastic response from Group captains, 'because it is a great attraction for every Captain to get a job on shore, as nearly all of them grumble without exception at being on sea, especially on a Tank steamer where they have very little time off'.[21] Unfortunately for the applicants there was only one opening, which was given to Captain Van Rijn.

Ter herinnering aan onze groote wereldreis
January 8 – June 1 1908
Dolph Kessler

[17]

In 1908 Deterding made a world tour of the new Group's facilities. Accompanying him as personal assistant, Dolph Kessler took many photographs, starting with the touring party on board the *Empress of India* (far left and top left). The main picture shows Deterding's first wife Catherine Neubronner being carried in a litter by Chinese workers. Deterding and his wife have the no doubt fascinating experience of inspecting stills (top right) and visit well number 4 in the Graciosa field (near left).

[18]

Sarawak, the north-western part of Borneo (present day Kalimantan), was arguably one of the most inhospitable places on earth. It was governed by a succession of 'white' Rajahs, the Brookes, who had close links with the British Empire. Part of their free trade policy was to encourage British investors to explore the mineral wealth of the region. Since oil was found in surrounding areas – in East Borneo and Brunei – getting a concession to explore oil in Sarawak was one of the obvious goals of the Group. In 1909 negotiations began, which were concluded in 1910. At the AGN of 1910 Marcus Samuel could state 'that we paid nothing whatever for that concession. The Rajah thought ... that it was better to give these concessions to a company like this, who will develop
[19] them satisfactorily and spend any amount of money that may be required for the purpose'.

Exploration was an immediate success: in 1911 oil was found at Miri, 'an almost unknown, primitive jungle village'. In 1950 B. Bromfield, who was one of the members of the first team that began operations there, wrote down his recollections of these first years:

'We lived in leaf bungalows..., hastily erected in clearings, with the jungle so close that wild boar were shot from the verandahs... There were no telephones or electric light, and we used oil lamps... Opening up the jungle released hordes of mosquitoes, and there were numbers of black cobras and pythons, but happily it was normally the most innocuous snake with secreted itself in the rafters over your head and dropped into the chair on your verandah. The Dayaks feasted on the snakes and regarded them as a most coveted luxury.

There was no white woman in Miri for more than two years, and we were a bachelor community of eight Europeans in all. Most of the evenings were devoted to playing bridge, but at times we would be lured to the kampong to join in the Malay dancing, with its haunting music, and perhaps to beat a tom-tom for a hour or two. The Malay girls, beauties some of them, were all discreetly kept behind curtains....

Weeks would pass without our seeing a ship – there was the feeling that we were utterly cut off from the world – and it was indeed a red-letter day when a glimpse of smoke on the far-off horizon heralded the arrival of mail from home. All too often, however, no contact could be made with the shore

NEW BORNEO OILFIELD.

FIND IN BRITISH TERRITORY.

From a Correspondent.

SINGAPORE, Mon

After passing through two good oil p
Sarawak, oil has been struck in large qu
tities at a depth of 860ft by the bore-masters
operating on account of the Anglo-Saxon Petro-
(Ltd.) of London. It is re-

0]

over the raging bar, and the ship sailed on to its destination, to land the mail on its return weeks later, if we were lucky.

In my earliest days in Miri there were no amenities of any kind. We spent much of our leisure shooting crocodiles in the river and hunting deer or tusked wild boar, and these made welcome additions to our table and a change from the monotonous buffalo meat.....'

These memories are obviously coloured by the passage of time, but do give an impression of life in such a distant place.

[21]

From over a century Sarawak was a private kingdom governed by 'white Rajahs', members of the Brooke family. Sir James ruled 1842-68; his nephew Sir Charles Anthony Johnson succeeded him, ruling until 1917; and Sir Charles's son Sir Charles Vyner Brooke continued until 1946. As monarchs they could issue their own currency, such as the coin (left) from the reign of Sir Charles. It was during his reign – as reported in Britain by the *Daily Telegraph* (left) on 17 January 1911 – that the Anglo-Saxon Petroleum Company struck oil in commercial quantities in the kingdom. Preparing to develop the discovery, Chinese workers (above left) haul a boiler towards its destination near the previously obscure village of Miri.

Employment and labour conditions in the Dutch East Indies

In terms of the number of people employed, Sumatra, Borneo, and Java comprised the centre of gravity of the Group. By 1913, the first year for which a reliable figure is available, Bataafsche employed 23,167 Asian employees and 825 Europeans in the Dutch East Indies.[22] Royal Dutch initially employed American and Canadian drillers, but strove to replace them with Dutch and German ones. The company had sent a group of young trainees to Sumatra to acquire those skills as early as 1898, and these men became the core of the technical staff of the different fields. During the production crisis, Loudon had also taken care to have them accompany the geologists so they would come to understand that part of operations as well.

Labour conditions in the Dutch East Indies were on the whole quite harsh, and it took a long time before Royal Dutch accepted the need for basic improvements such as Sundays off and family reunion.[23] On the other hand, wages appear to have been fairly good, particularly for the European staff, that is to say, better than at Shell Transport in Kalimantan. A comparison of pay conditions in the run up to the merger concluded that 'As a general rule, the servants of the Royal Dutch Co., and especially their administrators, doctors, drillers and others on the field, receive a higher level of remuneration than similar employees of the Shell Co.', the difference resulting mainly from a piece work system, i.e. bonuses received for achieving specific results.[24] Drillers, for example, received 600 guilders per month at Royal Dutch, against 5 gold dollars a day when employed by Shell, slightly more than half what was paid by Royal Dutch. Morever, at Royal Dutch drillers could earn 'drilling premiums amounting to considerable sums for expeditious and successful drilling'.[25] The two companies paid their

[22]

In 1913 works manager W. P. H. du Pon received a retirement gift (above) from the Sultan of Langkat. The stocky Sultan personally inscribed his photograph with the words 'As a memento to my friend Du Pon and his wife', exemplifying the Group's good relations with local rulers.

Without prejudice: a letter (below) from Deterding to Hugo Loudon, 20 April 1909, concerning salaries of employees in the Far East. These ranged from 50 guilders a month for very junior local workers to a maximum of 600 for laboratory technicians from Europe ('which might', said Deterding, 'be considered as not very highly paid').

DE BATAAFSCHE PETROLEUM MAATSCHAPPIJ, 's-GRAVENHAGE.

68

AB.

CONF.

30 April 1909.

No. 266.–

WITHOUT PREJUDICE.

Amice,

Ik heb nog te beantwoorden Uw schryven van 27 April No.701.

Cranies beginnen op een aanvangssalaris, dat veel minder bedraagt dan f.150.-, gewoonlyk bedraagt dit f.40.-, f.50.- à f.60.- Er zyn echter onder deze, Chineezen en Javanen of Maleiers, zeer handige menschen, die in werkelykheid meer werk presteeren dan de Indische jongelui, van wie het minste salaris ongeveer f.125.- is. Noch de Chineesche noch de ~~Indische~~ Inlandsche Kranis deelen in de uitbetaalde gratificaties, wel de Indische jongelieden, die als stillmen etc. fungeeren.

De klerken zyn verder alle jongelieden uit Europa uitgezonden.

Wat betreft Uwe verdere opmerkingen omtrent de salarissen, kan ik U mededeelen dat voor het Laboratorium gebruikt worden jonge technologen, die meestal uitgezonden worden op een salaris van f.350.- en f.400.-, voor wie dus het maximum-salaris van f.600.- niet bizonder hoog kan worden

[23]

managers a similar salary, about 1,200 guilders or £ 100 per month, but Royal Dutch also paid a bonus of 1.5 cents per unit on the first million units of kerosene manufactured, going down to 1/20 cent on each unit beyond the 8 million threshold. In addition, the administrator received 6,000 guilders for each field brought into production.[26] If successful, managers of drilling operations and refineries had well-paid jobs, but then demanding ones, too, requiring considerable managerial and diplomatic skills, the latter very necessary in maintaining relations with local rulers and representatives of the colonial administration. Some local rulers, including the Sultan of Langkat, had the right to royalties from production and would, from time to time, demand advances on royalties due, forcing managers to walk a tightrope between commercial common sense and the need to maintain amicable relations. Deterding disliked the cash bonus system, largely because he felt that bonuses quickly lost their incentive power.[27] Individual bonuses also created administrative complications and undesirable pay differentials between similar ranks in different parts of the organization, restricting flexibility in moving staff around. In 1912 the Group, after considerable reflection and against the resistance notably of employees in the Dutch East Indies, set up the Provident Fund to harmonize conditions, gradually replacing the specific bonuses with a profit share paid into member accounts.[28] The fund did not eradicate all pay differentials, however. Performance bonuses remained for the upper reaches of the managerial hierarchy, and initially only white employees with a permanent job qualified for membership, though the coverage was gradually widened.

Below the Dutch administrators and clerks, and the expatriate drilling personnel, there was a middle layer of 'Indische' employees, i.e. those born in the Dutch East Indies of mixed parentage. They tended to have regular jobs in semi-skilled work, for example as stillmen, and be eligible for bonuses, but they were not promoted into the higher echelons of the organization.[29] Asians formed by far the biggest group of labourers: Chinese, Javanese, and Bandjerese, some Tamils, and Sikhs who, 'with their martial turbans and beards' worked as guards.[30] The supply of unskilled workers remained problematic under the indentured labour system, dependent as it was on coercion to enforce the contracts. Moreover, the consequent wastage kept skills levels and productivity unnecessarily lower than required for work in the oil fields. For that reason Bataafsche changed tack in 1905 and began to recruit free labour from Java. This policy proved highly successful and already in 1910 about 90 per cent of the labourers were no longer indentured.[31] Meanwhile other aspects of labour conditions had also begun to improve, Loudon achieving, after some debate, a free Sunday for everyone and an eight-hour day for the European staff in 1909.[32] Medical services were also expanded and by building better staff quarters and other amenities, Bataafsche began to create the conditions for a normal family life for both the Javanese and the European staff.

Exploiting synergies The success of the merger would not, of course, be proven by the functioning of the organization, the fact that somehow the Group partners were able to work together, but by the creation of synergies, the new business adding up to more than the sum of the different parts. The two companies had looked ideal partners to all close observers: Royal Dutch with its great operational skills, in which Shell Transport had failed so miserably in Kalimantan; Shell with its large tanker fleet, an extensive, if underused, marketing network, and the easy access to the corridors of world power as a British, London-based company. It now came down to realizing those evident synergies.

One of the major jobs requiring immediate attention was the need to stop the bleeding in Balik Papan. Royal Dutch was acutely aware of the technical shortcomings of Shell Transport's operations there. Following the formation of Asiatic in 1902, Deterding and Sir Marcus had talks about setting up a joint venture to run the combined operations in Kalimantan. Royal Dutch would take over management of the refinery, in return for which the company

[24]

[25]

The Group began building houses of uniform design in 1896, initially with roofs of corrugated iron, but when the wives of employees came out to join their husbands they complained about the heat of the houses. Improved ventilation was installed and a younger generation enjoyed far better facilities, as shown by this dwelling at Pangkalan Brandan in 1913 (below left) and houses for Javanese workers in the village of Aloer Ganding, 1912 (below right). Despite the improvements, Du Pon was disdainful of the suburban appearance: 'Does that look like a Javanese village with its lovely intimacy in the shadows of high and flexible bowing bamboo and thickly leafed fruit trees?'

would get preferential rates for refining the crude from their neighbouring fields, which would then be shipped to Pangkalan Brandan for processing, at the same refinery. This project ultimately failed because Sir Marcus did not want to hand over control, but there are reasons to believe that Deterding, on finding out how bad the situation at Balik Papan and the financial position of Shell Transport really were, stalled the talks, preferring to let Shell weaken further through this drain rather than obtain transport cost savings for Royal Dutch. This was a shrewd assessment of the situation, because the huge debt of the Nederlandsch- Indische Industrie en Handel Maatschappij (NIIHM), the Balik Papan operating company, to Shell Transport helped to bring Sir Marcus's company to the brink of bankruptcy.[33] Early in 1907, with the merger negotiations nearly over, Deterding still warned local managers not to give advice to the technical staff at Balik Papan about restructuring the refinery, because this advice was 'a very valuable asset that we bring into the amalgamation, and not before that'.[34]

After the merger Dutch engineers immediately took over the sites and set to work modernizing the installations, with the most urgent repairs and modifications alone costing an estimated eight million guilders.[35] Inspecting Balik Papan on his world trip in 1908, Deterding waxed enthusiastic about the potential of operations there. He was convinced that the Kalimantan fields were the richest of the Dutch East Indies and considered the natural harbour at Balik Papan ideally situated. Crude from other fields could easily be shipped in for processing and then exported to China, Japan, and Indo-China.[36] Operational results did not immediately come up to expectations, however. In January 1909 Borneo production began to decline due to water seepage into the wells, attributed by some of the English staff to the introduction of drilling techniques from Sumatra, which they deemed unsuitable for the soil conditions in Borneo. In less than a month, however, engineers had thoroughly analyzed and solved the problem, as they had done in a similar situation in Perlak, enabling Deterding to write to Loudon that he

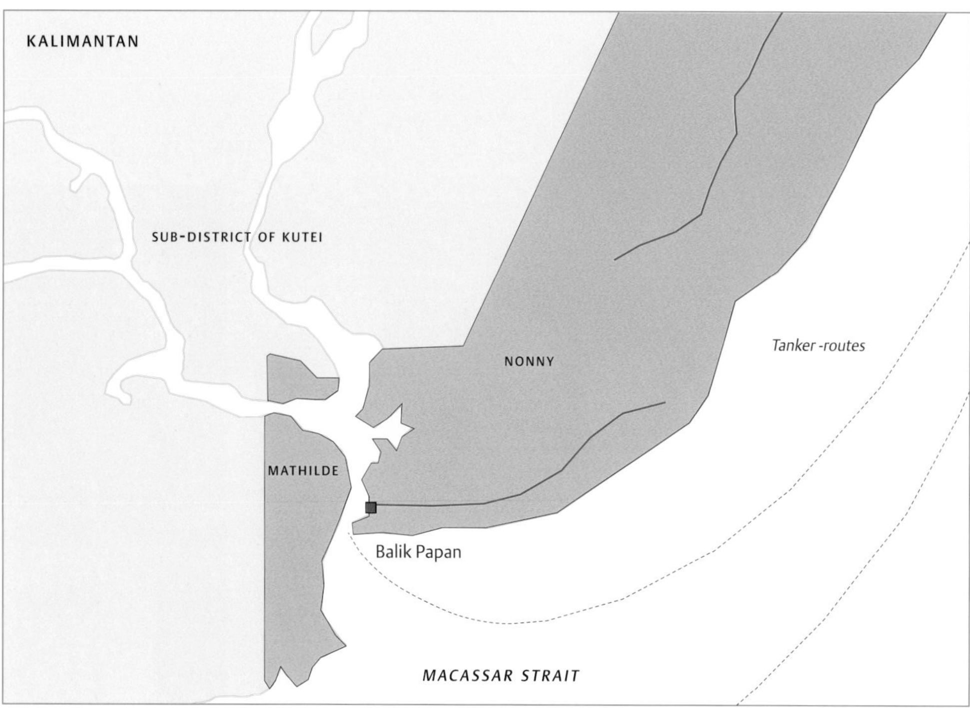

The natural harbour at Balik Papan.

[26]

would make an English translation of the report for Sir Marcus and invite him to address the criticism from the English staff on the technical prowess of their Dutch colleagues.[37] With the production problem solved, Borneo quickly became the Group's most important source of crude oil in the Dutch East Indies, supplying about one-third of the region's total crude output in 1912 and 1913.[38]

For a long time, however, the specific characteristics of the crude and the resulting poor quality of the products made from it continued to pose a formidable puzzle. The high paraffin content was put to good use by building a paraffin factory, but it took longer to find a solution for the aromatic hydrocarbons. Bataafsche's laboratory, which was then still attached to the gasoline installation at Rotterdam Charlois, found a commercially viable solution in the form of the process developed by the Romanian chemical engineer Lazar Edeleanu. This process extracted the aromatic hydrocarbons by washing the kerosene with sulphur dioxide. After having experimented with other

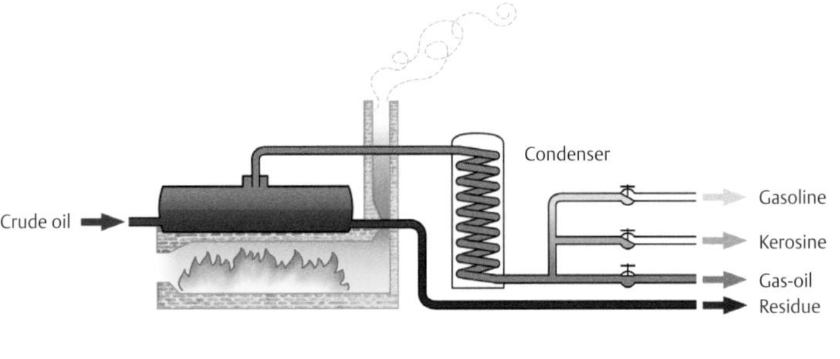

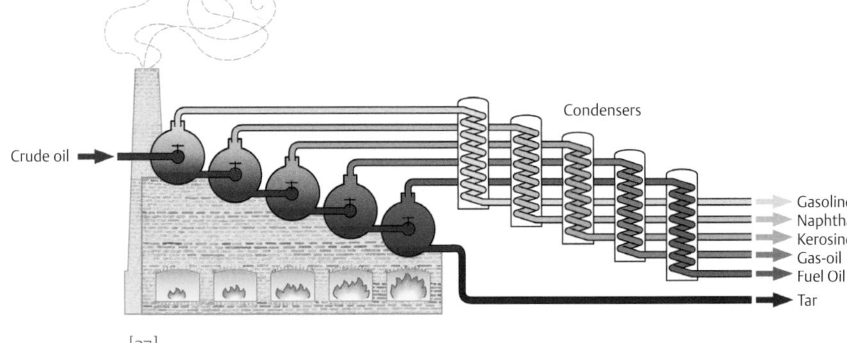

[27]

The distinction between Batch and Bench stills: A Batch still (above) heated crude oil to the boiling point of gasoline, distilled that off, then increased the temperature for the next product. The more efficient Bench still (below) passed crude oil over progressively hotter fires in a continuous process, each still in the series separating one fraction.

The exterior of the refinery at Balik Papan (below left), and its interior battery of boilers (below right), both seen in 1914.

[28]

[29]

solutions (see below) the Group bought a licence in 1911 and immediately began building an installation in Balik Papan, but it took another five years before it came on stream, primarily because it was the first large-scale plant of this kind.[39] The Edeleanu process would subsequently become one of the Group's great technical and commercial successes, and Balik Papan its model of state of the art operations. Recognizing the need to provide the expanding operations with better technical support, Bataafsche set up a technical department in The Hague, charged initially with setting standards and norms for all installations and subsequently also entrusted with designing plants. This department became a major asset in the Group's worldwide operations.

Deterding also used the need for rearrangements after the merger to reorganize the administration in the Dutch East Indies. Balik Papan was integrated into the administrative and accounting system of Bataafsche's 'Indische administratie' and the local head office was moved from Pangkalan Brandan to Batavia. This was done for both economic and political reasons. With the rising importance of operations outside of Sumatra, the Group needed a separate and a more centrally located head office to keep the business together and to cultivate relations with the colonial government. The Group now had a near monopoly on oil production in the archipelago and with the takeover of Dordtsche Petroleum Company, the last major competitor, its position appeared unassailable. This situation began to invite insidious comments. With new and promising concessions coming up for tender, including the highly regarded Jambi fields, Deterding wanted Bataafsche's managers to be close to government officials, just to make sure that the company would get its way.[40]

Reorganizing marketing and shipping The merger also enabled Deterding to take the reorganization of marketing and shipping, begun immediately after the formation of Asiatic in 1902, a full step further. Shell and Royal Dutch still competed in some markets with their own brands sold through their own channels.[41] Shell's more extensive marketing network, the core of Asiatic, was still largely in the hands of general merchant firms selling a range of products as well as oil, who were more interested in volume than in price. Visiting Penang during the 1908 trip, Deterding noted that 'Our agents are fairly active but confine themselves to dealing with one dealer only so as to enable them to give only some of their time to our business. Consequently they know practically nothing of the trade in the interior.'[42]

London had limited power over such agents and could thus not set marketing strategies or prices. Attempts to increase market share by cutting prices sometimes only resulted in the agents simply pocketing higher margins; marketing agreements with rivals such as Standard Oil Company could not be enforced in the face of these independent agents.

[30]

From India to Japan, and (overleaf)...

[31]

...from Thailand throughout south-east Asia, agents' advertisements for the Shell and Crown brands appeared in many languages.

During his trip to India in 1904, Waley Cohen had already noticed 'how sadly the marketing organisation of the Shell (...) had gradually slackened'.[43] At the same time he had suggested the solution: Asiatic should circumvent the merchant houses and edge 'as close as possible to the consumer', which in practice meant 'get closer to the retailer'.[44] In large parts of Asia European merchant firms sold their goods in bulk to Chinese or Indian wholesalers, who distributed to retailers. Cutting out these merchant houses would be revolutionary enough, but Waley Cohen and Deterding wanted to go further: they also wished to bypass the Chinese and Indian middlemen by setting up a comprehensive marketing organization of their own. The formation of the Group created an opportune moment for starting this process, since the merger required a revision of contracts. In December 1907 Asiatic notified its main overseas agents that agreements would not be renewed, and began setting up its own marketing organization.

The reorganization of marketing in China provides a good example of the new policy. With about 400 million inhabitants and hardly any electricity, China was by far the biggest market for kerosene in Asia, and access to it had been a major goal of both Shell and Royal Dutch from the start. Traditionally, Standard Oil dominated the market, but the high price of its case oil imports left enough room for new entrants. The Shell agent in China was the German firm Arnhold, Karberg & Co., a member of the Tank Syndicate and subsequently a founder and shareholder of Shell Transport.[45] From 1892 onwards the firm had built bulk oil installations and tinning plants in the main Chinese ports, which it had transferred to Shell Transport in return for shares while continuing to manage them on behalf of the new company. In 1897 Royal Dutch had begun to import kerosene in bulk for distribution by its agent Meyer & Co., another German firm.[46] In 1903, the two marketing networks merged when Asiatic appointed Arnhold Karberg and Meyer & Co. as its joint agents for China.[47] Three years later Asiatic set up its own office in Hong Kong and started to build an up-country marketing organization bypassing both the German firms and the large Chinese wholesalers who had dominated distribution from the port cities to the hinterland. These firms often clubbed together to get the lowest possible price from foreign importers and keep their own margins high.[48]

Starting an up-country sales organization was a novel approach, for most European businesses did not venture out of the ports where they had the protection of treaties between the Chinese Empire and the major European states. Until then only one foreign company, the cigarette manufacturer British American Tobacco (BAT) in 1905, had set up a distribution network in inland China. BAT had patterned its network on the main navigable rivers, but the Group went beyond that and also used the railways to feed wholesale and retail distribution facilities in remote corners of China.[49] Within a year the country-wide organization of depots and agencies was ready; Asiatic cancelled the agency contracts with its German agents from 1 January 1908.[50] Europeans, mostly Englishmen, staffed the supervising ranks of the organization, from the lowest rank of agency inspectors to branch office manager.[51] The Chinese up-country market was strategically important, because the Group could sell its low-quality Borneo kerosene there. Sales of the quality brands from Sumatra were concentrated in the coastal region, where they competed head-to-head with Standard Oil's Devoes brand. Standard soon followed Asiatic's example, setting up an inland distribution network from 1910 and switching to bulk imports the following year.

Shell entered the Australian market using agents Burns Philp and Co, whose name appears (below) in about 1900 at the base of the storage tanks in the Group's first depot at Clyde, New South Wales. However, the Group soon established its own distributor, British Imperial Oil Company, whose name appeared on these storage tanks in Sydney (below right).

The substitution of independent agencies with salaried agents was also introduced on Java, in British Malaya, Australia, and Europe.[52] In 1908 the Penang agent, Martijn, was told by Deterding 'that we are going to withdraw his agency. I understood that as soon as we take our business in our own hands Martyn will discharge his employee for petroleum Mr Arthur Oechele and, as we want for ourselves a man fully acquainted with local conditions, I have engaged Mr Oechele for 6 months provisionally on a salary of £ 500 a year'.[53]

In Australia marketing was taken over by the British Imperial Oil Company, set up in 1905 as a subsidiary of Asiatic, and the role of the independent agents was 'greatly reduced'.[54] For the British market Asiatic was tied to its contract with British Petroleum, the joint venture with Deutsche Bank, due to expire in 1916.[55] In the Netherlands, however, the contract with the gasoline distributor, E. Suermondt & Zoonen, could be terminated in 1907, and the Group established its own sales organization, NV Acetylena. Gasoline sales, and from 1910 kerosene sales as well, rose substantially under the new management.[56]

[32]

[33]

PETROLEUM.

Groote Petroleum Maatschappij zoekt met het artikel bekende

VENTERS

voor den verkoop van petroleum met handketelwagen. Brieven met volledige inlichtingen onder No. 985 aan het bureau van dit blad.

[34]

[35]

[36]

[37]

[38]

[39]

Clockwise from top left: Responding to newspaper advertisements, street vendors of Shell products used a wonderful variety of means of transport. Buffalos were favourites in the Philippines and camels were commonplace in Australia; donkeys featured widely in the Middle East, and man-powered freight tricycles in Malaysia. Dray horses in Australia could haul barrels of oil just as well as the more traditional barrels of beer.

Further cost savings could also be realized by integrating the management of the two tanker fleets. In 1906, during the initial merger negotiations, Cohen Stuart and Benjamin had exchanged notes about the consequences of fleet rationalization for the staff in The Hague and in London. They fully expected an 'immediate or gradual dismissal of a great portion of the personnel', but that fear soon proved unfounded since the Group's continuous growth actually created more jobs.[57] More importantly perhaps, the merger led to a better utilization of the fleet. Labour conditions on board the ships were also harmonized, and a detailed system of staff management, including a schedule for regular assessment and promotion, introduced.[58] The fleets could not be entirely merged into a single one, however. Royal Dutch wanted to keep its largest tankers under the Dutch flag, thus sacrificing economic efficiency for reasons of prestige.[59] Moreover, under colonial regulations the inter-island shipping within the Indonesian archipelago had to be done by a company domiciled in the Netherlands, so part of the fleet had to remain Dutch. Soon, the management of the Dutch fleet returned to The Hague, although in close coordination with London.

Reaping the benefits Modern analyses of the success or failure of mergers concentrate on what happens to share prices and the market capitalization of the companies concerned. If a merger produces the expected synergy and strategic advantages, this should translate into sharply rising share prices during the months and years following the merger.[60] Making such an analysis for the formation of the Group is difficult simply because the whole process lasted so long. Negotiations began during the autumn of 1905 and the Royal Dutch board first discussed the subject in December, but the merger was only concluded in July 1907. Rumours were rife on the London and Amsterdam stock exchanges, nudging share prices upward (Figure 2.1). Share prices of Shell Transport, which fluctuated between 100 and 120 per cent during 1903-5, began to climb slowly from November 1905 onwards, when news about the negotiations will have started to leak out. The rise clearly accelerated during the second half of 1906,

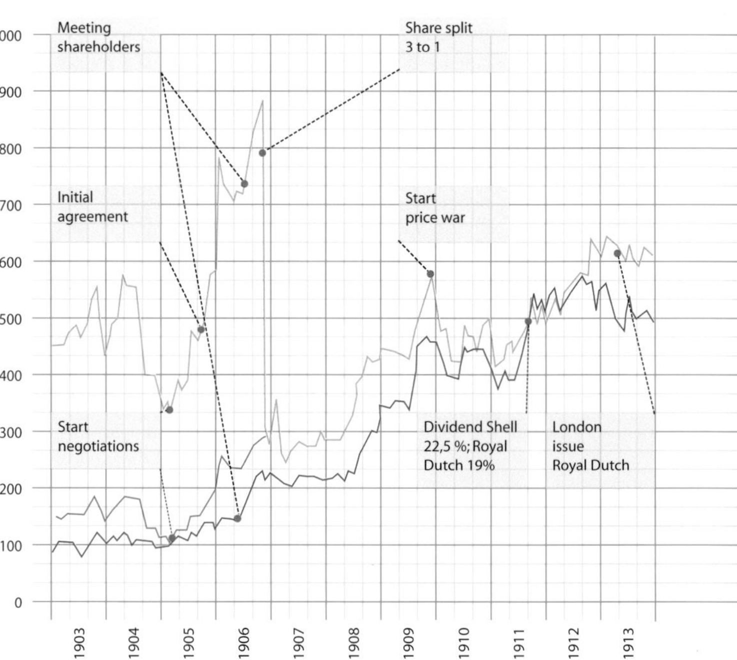

Figure 2.1
Share prices of Royal Dutch and Shell Transport (as a percentage of nominal value), September 1903 - Juli 1914).

Advertisement for shareholders in Royal Dutch and Moeara Enim, 1904. 'The undersigned warn all shareholders in Royal Dutch not to take fright from the shares' price drop which occurred during the last few days following MENDACIOUS news spread by a powerful bear syndicate. Therefore we give shareholders, in their own interest, the serious advice not to sell their shares. Signed by Many Shareholders.'

when the merger had been done and only the formalities needed finalizing. Another steep rise occurred early in 1907 following the acceptance of the definite plan by Shell Transport's General Meeting of Shareholders on 1 January. Overall, between October 1905 and June 1907 the Shell Transport share price doubled. We must also take into account that the intrinsic value of the shares was rather less than 100 per cent, since Shell Transport's net debts amounted to £ 580,000 or almost 30 per cent of the nominal value of the shares. Consequently, the Shell Transport shareholders profited substantially from the merger. The Dutch investing public had even more reason to be grateful, however. Royal Dutch share prices, which hovered around 350 per cent in the summer of 1905, attaining a low of 320 per cent in October, responded immediately and dramatically to all the steps in the process of amalgamation. During the first stage, October 1905 to May 1906, share prices increased from about 350 to almost 500 per cent. During the second stage they skyrocketed to almost 900 per cent during the spring of 1907, touching an all-time high of 906 in May, so they almost tripled in value.[61]

Rising share prices created an interesting differential between the capitalized value of the two companies. In the autumn of 1905, Shell Transport and Royal Dutch were considered equally valuable by the investing public, at around £ 3.1 million They then diverged sharply as Royal Dutch share prices rose and Shell Transport stagnated. In May 1906 Royal Dutch had gained a third in value, against Shell Transport less than 20 per cent, for a total of £ 4.2 million against £ 3.6 million By May 1907, with the merger almost done, Royal Dutch had become about 50 per cent bigger, with a capitalized value of £ 8.2 million against £ 5.6 million for Shell Transport, or close to the 60:40 per cent distribution of the new Group. Thus investors appear to have liked the merger and to have expected more benefits for Royal Dutch than for Shell Transport. However, given Shell Transport's poor financial condition, which only became public knowledge when the negotiations were in full swing, one might argue that the rewards for Shell Transport shareholders were bigger than those for Royal Dutch shareholders.

After such a surge one cannot expect shares to continue their upward march. Moreover, the 1907 financial crisis intervened to break the optimism of investors. And yet two constituent companies of the Group maintained their highly increased capital value during the relatively difficult years of 1907 and 1908, quite an achievement in itself. From 1909 Royal Dutch and Shell Transport share prices resumed their buoyancy, recording total gains in the order of 60 per cent. It would be difficult to find a merger which has been equally successful in rewarding shareholders.

Koninklijke Petroleum-Maatschappij.

Ondergeteekenden waarschuwen alle Houders van Aandeelen in bovengenoemde Maatschappij zich niet te laten verontrusten door de daling der koersen in de laatste dagen, welke gebleken is het gevolg te zijn van leugenachtige berichten, verspreid door eene uitgebreide contramine-partij. Wij geven dan ook den Aandeelhouders in hun eigen belang den ernstigen raad, zich thans niet van hun Aandeelenbezit te ontdoen.

(17422) **VELE AANDEELHOUDERS.**

Aandeelen Moeara Enim.

Op de verlaagde koersen wordt aankoop dezer Aandeelen, met het oog op den intrinsiek goeden toestand der Maatschappij, **dringend aanbevolen.**

(17466) **INSIDER.**

Ramp.

Te Bandoeng is het bericht ontvangen, dat den 18en April te Pladjoe (Palembang) vier tanks, gevuld met croosine en petroleum, in de lucht zijn gevlogen.

De brand duurde voort tot den volgenden morgen.

Vrouwen en kinderen vluchtten op de schepen; inlanders vluchtten de rimboe in.

Een boormeester, een assistent-scheikundige, twee inlanders en vier Chineezen zijn gedood.

In de locale markt blijven de Petroleumaandeelen nog steeds de aandacht trekken.

Nu de speculatie eenigszins in deze afdeeling scheen te hebben uitgewoed, trad een koersherstel langzamerhand in. Daarop volgde echter weder een teruggang, die het gevolg was van het ontbreken van zaken in dezen hoek.

Blijkbaar kwam de reactie na de drukke dagen en het publiek, door adviezen die hemels-

In hoeverre deze geruchten waar zijn, is nog niet officiëel bekend. Wel hadden zelfs de buitenlandsche couranten de geruchten overgenomen, doch het fijne van de zaak wist men nog niet.

Of het eene absolute amalgamatie zal zijn, dan wel eene overeenkomst als de Sumatra Palembang is nog niet uitgelekt. Wel scheen een groot deel van het beurspubliek veel voor het eerste

[40]

Royal Dutch

Royal Dutch (corrected for share split)

Shell Transport

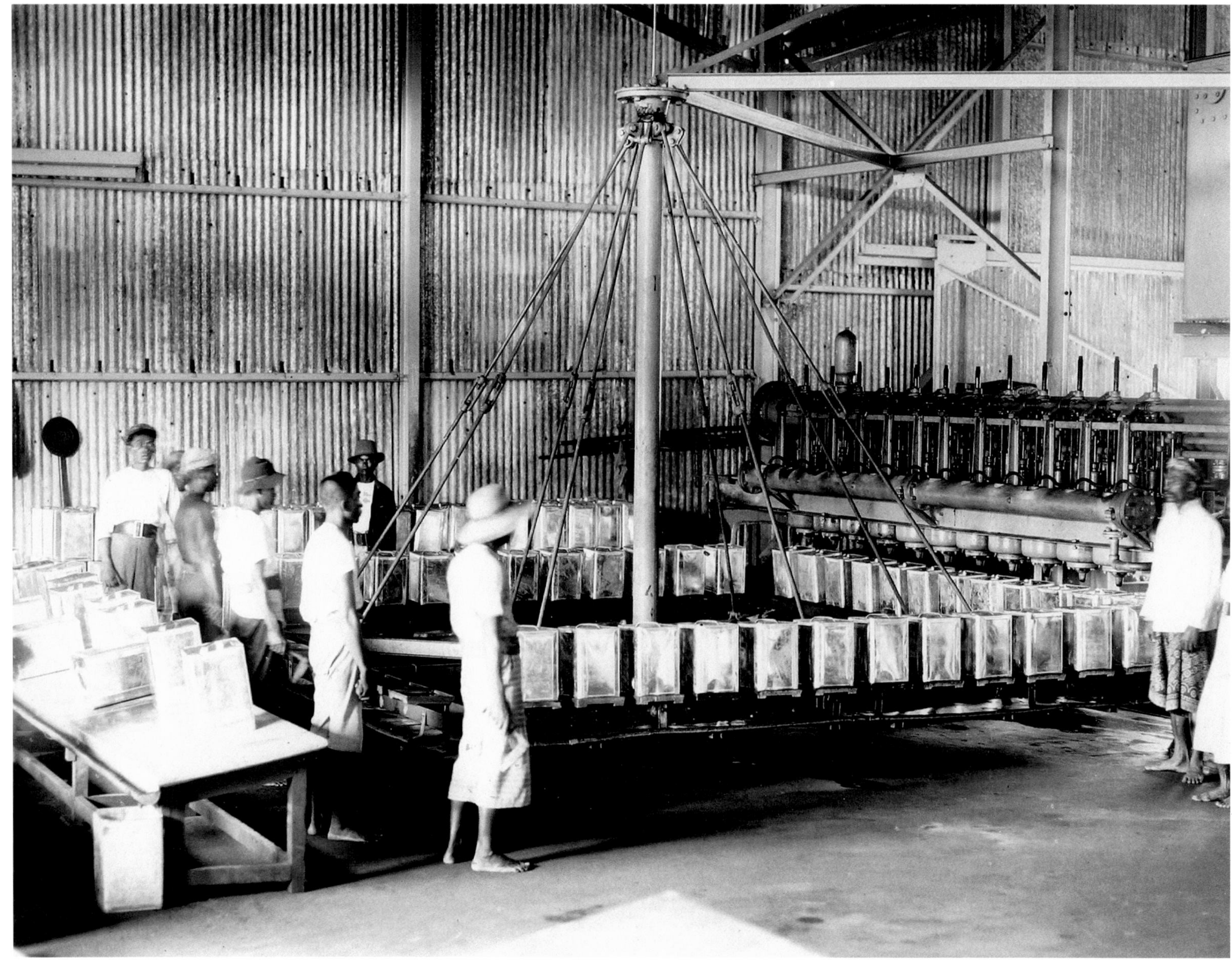

[42]

For retail sale and non-bulk transport, kerosene was often packed in tins containing five US gallons each. The tins were then packed in pairs in wooden cases – hence the term 'case-oil'. On the previous pages, in a photograph taken at Plaju some time between 1913 and 1915, a solitary European oversees the semi-automated filling of dozens of tins. The same operation is being performed at Pangkalan Susu (above left) in 1916, and the manufacture of tins is seen (above right) in Pangkalan Brandan in 1916.

[43]

Playing a global chess game: relations with Standard

In March 1907, during the last phase of the negotiations with Shell, Deterding travelled to New York for talks about cooperation and market sharing at the Standard Oil head office on 26 Broadway. The invitation had been arranged by Walter Teagle, Standard Oil's coming man and, as manager of its British marketing organization Anglo-American Oil Company, the officer responsible for European operations. During negotiations between Anglo-American and Asiatic about market sharing in Europe and in Asia Teagle and Deterding had come to know and like each other. Deterding had found Teagle very receptive to his ideas about collaboration between the two oil companies being more constructive and profitable than Standard's aggressive policy of destruction through price wars. Now Teagle wanted Deterding to convince the Standard board of his vision. It was Deterding's first visit to New York and the first meeting between the leaders of Royal Dutch/Shell and Standard Oil. As such it was a moment of recognition for what Deterding had achieved, building an oil company of sufficient power that he could talk as an equal to the Standard Oil board. Yet the occasion must have been a disappointment to him, for he made very little headway. Having listened to Deterding Vice-President John D. Archbold, Standard's acting CEO, politely rejected his ideas by telling him that policies concerning Europe were delegated to 'the 14th floor' and that he had to try to convince that committee.[62]

Even so relations between the two companies improved markedly, building on the rapport between Teagle and Deterding. Moreover, Standard wanted to collaborate because the company had a shortage of gasoline and needed Sumatra supplies from Asiatic, particularly for the German market.[63] Consequently, the market share agreement for the European gasoline trade was continued for another two years. From 1907, however, the situation began to change. New refining techniques in combination with a rapid expansion of US crude oil production created a growing gasoline surplus there and subsequently also in Europe, leading to renewed tensions which sparked a price war during 1910-11.

Deterding had apparently failed to read the signs during his talk with Archbold in 1907, for the following year, on his world tour, he paid another visit to 26 Broadway to discuss collaboration. This time the reception was much cooler. Teagle was absent or, according to Deterding, kept away from him. He did not get access to board members and was relegated to meetings with W. E. Bemis, in charge of Standard's China trade.[64] After complaining about the way he was being treated, Deterding was finally received by two directors, A. C. Bedford and H. H. Rogers. They told him that Standard was prepared to make a 'firm offer for the whole combine' of $ 100 m. giving a premium of about 70 per cent on the Group's current value.[65] This offer shows how the American company viewed Deterding and the Group at that time: as a sound business worth a substantial takeover premium, but not a potential partner for marketing arrangements on equal terms such as Deterding wanted. Deterding dismissed the offer out of hand and took only Loudon into his confidence about it. He thought it was 'not wise for Standard to buy us. I dream and (if we have success in Romania, California or Russia this will soon be realized) I believe that our goal is to be the *only* opponent, and therefore silent partner of the Standard. There is no place for a monopoly on this earth, but there is for two great companies working wherever this is possible'.[66]

AUSTRIA-HUNGARY
RUSSIA
Brasov
Galati
Buzau
Campina
Ploesti
Bucharest
ROMANIA
Constanza
DANUBE
BLACK SEA
Giurgiu
BULGARIA

[44]

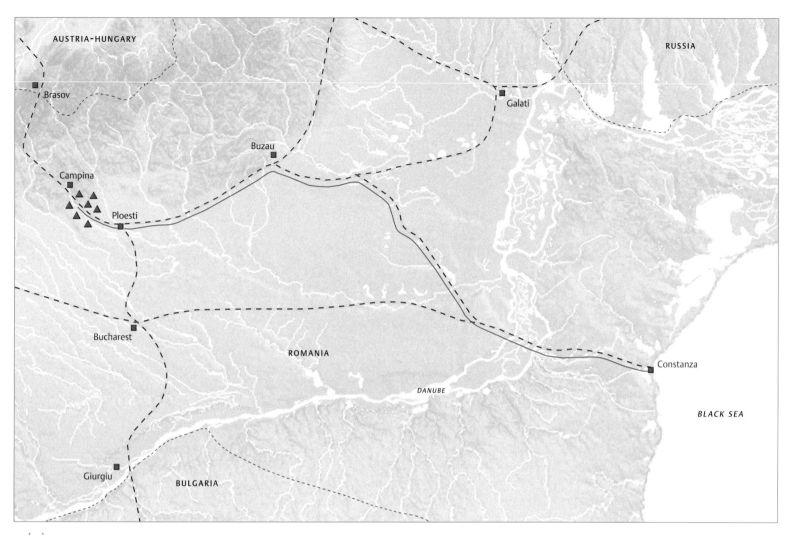

Map of Romania

— – Pipeline
▬ – Railway
▲ – Oil fields

While Dolph Kessler was working for Shell in Romania, his mother paid a three-week visit in the spring of 1906. Photographed in a mountain meadow (right), Dolph is standing on the right with C. W. Pleyte, a veteran of Group operations in the East Indies, in the centre, and Dolph's mother Margo and sister Go seated. The third standing figure has not been identified.

[45]

29
28
27

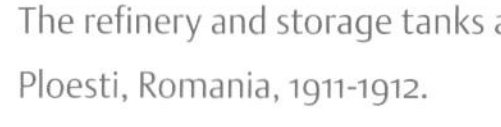
The refinery and storage tanks at Ploesti, Romania, 1911-1912.

As Deterding wrote to Loudon, widening the supply base was a logical next step in the struggle with Standard. The Group initially concentrated much of its efforts on Romania, for a number of reasons. First, the Russian oil industry was going through a deep crisis during the early years of the twentieth century, partly brought on by the Russo-Japanese war of 1904-5 and the subsequent political unrest which rocked the empire. Consequently the supplies which had fed Shell Transport's expansion on the eastern market were drying up. Second, the Group needed a gasoline supply closer to European markets, because freight costs made the imports from Sumatra uncompetitive there in comparison with Romania. Most important, however, Standard had already started to invest in Romania with the aim of solving its recurrent gasoline shortages. Royal Dutch had started to gather intelligence about Romania in 1900, when Kessler had visited the country. He had been followed by other experts surveying opportunities.[67] The first approaches by Asiatic to the Romanian oil industry, in 1903, were a direct response to Standard Oil's activities there, but they had been limited to supply contracts with Romanian oil companies.[68] In 1906 Deterding decided that it was time to move ahead; not wishing to wait for board approval, he secured some concessions on his own account.[69] To follow up this initiative Royal Dutch sent two of its foremost experts, C. M. Pleyte and Erb, to buy more properties and start up operations. These two men personified one of the Group's strategic advantages, a pool of technical and administrative experts who could be entrusted with the difficult task of starting from scratch in unfamiliar territory. In November 1906, Royal Dutch set up the Consolidated Dutch Oil Company (Geconsolideerde Hollandsche Petroleum Company) to hold the properties acquired. Two years later the Group bought a local refinery named Astra and reorganized this into a production company.[70] In 1910

The refinery and storage tanks at Ploesti, Romania, 1911-1912.

Astra merged with Regatul Roman, which possessed important producing fields but almost no refining capacity, to form the Astra Romana, which became the biggest producer in Romania, crude output rising from 85,000 tons in 1909 to 520,000 tons in 1911. Pleyte's efforts turned the Romanian operations into models of efficiency. When the always critical Deterding visited Romania for the first time in 1911, he wished 'that all employees in the Dutch East Indies (...) would spend one day in Romania, in order to see how the business is run here'. He considered the key to success was that 'the Director and the chief engineer visit the sites regularly and are personally in contact with all members of staff'.[71] Thus the rapid expansion into Romania succeeded in parrying Standard's move in that direction.

Teagle's growing influence helped to keep relations between the Group and Standard good throughout 1908 and 1909, leading to further market sharing agreements and extensions of existing ones. In January 1909, for instance, the two companies concluded a price agreement for Thailand and Indo-China, followed by a new agreement over the Indian market four months later, after which Deterding 'at once wired to India to raise prices to the old maximum'.[72] In August 1909, foreshadowing the famous Achnacarry cartel of 1928, Deterding even proposed a comprehensive agreement covering gasoline in Europe and kerosene in Asia, 'to limit his trade in the future to the trade of the 1 July 1908 to 30 June 1909'. Writing to the Paris Rothschilds, Fred Lane showed himself puzzled about the intentions behind this deal, because 'one can never tell what is inside Deterding's head'.[73] In Europe the agreement with Standard contributed to the overall profitability of the Group, according to Lane: 'In Belgium and Holland, instead of making a loss since the alliance we have made a profit (...) The moment we lose that friendly relationship the profit is wiped out and turned

[48]

Mobile drilling rig at Ploesti, Romania, 1912-13, capable of fast drilling to 600 metres' depth.

into a loss.' In Asia the alliance stabilized market share: 'Roughly speaking, as between Standard and ourselves, we have 36% and the Standard 64% over the whole of the East', Lane estimated.[74] Deterding may have intended this agreement simply as an experiment, for, corresponding with Cohen Stuart, he anticipated 'a blazing row with the Standard' to cause 'a bad year' for the Group.[75] A world recession during 1908-9 created oil surpluses both in Asia and in the United States. In June 1909 negotiations between the Galician oil producers and Standard broke up without agreement after the Austrian government had intervened. Deterding predicted that this would lead to a price war in central Europe.

At the same time he was negotiating with Deutsche Bank, which controlled the second largest producer of oil in Romania, Steaua, about the formation of a syndicate to combine the European oil interests of the two companies, including the Kasbek Syndicate and its fields in Grozny owned by Deutsche.[76] The bank had come to realize that oil was not its core business, and needed to curtail its commitments to finance another major project, the famous Baghdad railway. The proposed combination with Deutsche did not materialize, however, but the Group did buy the Kashbek Syndicate and thereby acquired large properties in a new oil region.[77] Standard considered the imminent combination between Deutsche Bank and the Group as a grave threat to its position on the German gasoline market and the company tried to intervene.[78] Deterding refused to reconsider, however, so relations broke down into open conflict: the great oil war of 1910-11 had begun.

It was war on nearly all markets, with the UK as the most important exception, because Teagle and Deterding succeeded in maintaining some form of truce there.[79] Oil prices dropped sharply, depressing profits; as a result, share prices fell by about

[49]

'It's an ill war that blows nobody any oil' - Cartoon of the oil price war from *The Illustrated Finance*, 1910.

25 per cent in the five months after the beginning of the price war in August 1910.[80] The drop in profits forced a dividend cut in 1911. Royal Dutch lowered its dividend more than Shell, causing a degree of friction amongst managers.[81] Though the Group lost revenues, other companies bore the brunt of the price war. Standard aimed to hit the Group in its most profitable market, the Dutch East Indies, but it was Dordtsche Petroleum company which suffered most. After the Group's takeover of the Shanghai Langkat company in 1910, Dordtsche was the last independent producer on Java. The company sold nearly all of its products there and the sharp drop in prices brought it to its knees. Relations between Dordtsche and Royal Dutch had been tinged with an edgy rivalry since the days of Kessler. Nor did the managers, in particular Adriaan Stoop, get along with Deterding, so they would not turn to Royal Dutch for help. In November 1910 rumours spread that Dordtsche wanted to sell out to Standard Oil, and the prospect of having the American trust as his close neighbour in the Dutch East Indies started to worry Deterding. In an attempt to win over Dordtsche, he proposed a joint lobby for increasing the tariff on oil products imported into the Dutch East Indies, which would

definitely have damaged Standard.[82] Such a policy was not feasible, however, so Deterding and Loudon had to wait and see how the negotiations between Dordtsche and Standard Oil would develop. They failed because Teagle considered Stoop's price too high.[83] Dordtsche therefore had to turn to the Group, which, in contrast to the Americans, could pay the Dordtsche shareholders with Royal Dutch and Shell Transport shares. So instead of weakening the Group's position in the Dutch East Indies, Standard's offensive there only reinforced it.

Prudent financial management gave the Group sufficient resilience to conduct a holding operation during the price war. Turning the tables on Standard would be more difficult. The key to Standard's power was the American market. The company could fight its overseas price wars on the strength of profits from high domestic prices. Moreover, after the collapse of Russian oil output, the US again became the single most important producer of oil in the world, its share in world output rising from 42 per cent in 1900 to 65 per cent in 1910. A large part of this impressive growth came from California. The supply elasticity of the other oil producing regions was much lower; in spite of the massive investments in fields and equipment by the Group, the Dutch East Indies hardly produced more than 3.5 per cent of world output. So a logical move in the chess game was getting access to the US. This had been on the Deterding's mind since the merger, and on his 1908 world trip in he visited the Graciosa oilfields in California, which had been offered for sale to the Group. He was greatly impressed 'by the possibilities of this country, the fields being of enormous extent (...) but (...) worked in a somewhat rough and ready fashion'.[84] But he also understood the extent of the risks. Drilling was very expensive, because oil was only found at depths of 3-4,000 feet. Many fields had very short life cycles, output often declining after only one or two years. In May 1908, after receiving a detailed assessment report Deterding had had to decide that it was still too early to go to the US. In 1909 the geologist P. Kruisheer was again

I was greatly impressed by the possibilities of this country, the fields being of enormous extend. At the same time I must admit that it stryck me that the fields ~~have been~~ are operated worked in a somewhat rough and ready fashion, but if this is really so it proves how extensive the field must be because if after a superficial survey no dry holes have been struck, what would have been the production of the field if a careful exploitation had been made. All the oilwells are very deep (3 to 4000 ft) as they are being drilled by the Pensylvanian system, with rppe and bit, it takes about 1 year to finish a well and each well costs at least $25000.
Most of the wells are pumpers and are pumped by the workingbarrel system so that the capacity is limited; and I can hardly say how much this production could be increased by working the wells in another way. At present, if after pumping a well begins to spout; the working barrel cannot be take away and the produwtion is therefore limited.
All the oil is of a heavy grade, asphaltum base, and contains roughly 20-25% kerosina, 6-11% benzine.

Shell moves into the USA: A letter (left) to Hugo Loudon from Deterding, written on 14 May 1908 during his world tour, in which he extols the possibilities of the oil fields around an Francisco. Four years later, in 1912, the Group opened its first depot in Seattle (right).

CANADA
Vancouver
Richmond Beach
Seattle
WASHINGTON
MONTANA
Portland
IDAHO
Eugene
OREGON
UNITED STATES
CALIFORNIA
NEVADA
Sacramento
Martinez
Stockton
San Francisco
Oakland
Fresno
Valley Pipeline
Coalinga
Belridge
Mt. Poso
Coalinga
Bakersfield
McKittrick
Midway Sunset
Los Angeles
Ventura
San Diego
Long Beach
PACIFIC OCEAN
MEXICO
"SHELL"
MOTOR SPIRIT & OIL
STOP FOR GASOLINE
SHELL MOTOR SPIR

[51]

[52]

dispatched to America, this time to inspect fields in the Midwest, including Oklahoma. Again the conclusion was negative.[85]

The price war changed the priorities. The growing oil production in California made Standard Oil stop its purchases of Sumatra gasoline for that market, confronting Asiatic with the need to find another outlet for the production of Shanghai Langkat which it had sold to Standard. The obvious solution was to start selling the gasoline in California. Within a few months this decision was taken and implemented by sending a tanker over and selling the gasoline more or less on the quayside. The operation was a resounding success and by July 1910 Loudon and Deterding wrote jubilant letters about it to each other.[86] In January 1911 two employees, J. C. van Panthaleon van Eck and F. P. S. Harris, were sent to Seattle to start a marketing organization for the Pacific North-west and the San Francisco Bay area. The new company sold its gasoline under the Shell brand, but Van Eck and Harris chose to name it the American Gasoline Company, since there already existed a Shell Petroleum Company in California, run by two brothers of that name.[87] Simultaneously Deterding set out to purchase producing properties both in California and in the Midwest. Two of the Group's geologists, Erb and Kruisheer, crisscrossed the continent to scout for concessions and companies. In December 1910 the Group received an offer to buy concessions in the Ventura field, destined to become famous during the 1920s.

Meanwhile the balance of oil power was turned upside down. On 15 May 1911, the US Supreme Court ordered the dissolution of the Standard Oil Trust. This was a decision of profound importance for the oil industry as a whole; at a stroke Standard's dominance of markets in the US and around the world was broken up. The old Standard Oil now became Standard Oil New Jersey, usually referred to as Jersey Standard or simply Jersey, with the parts of the company split off usually named after the state in which they conducted their business. Over time the impact of the dissolution would be mitigated because, as Deterding and others expected, the successor companies continued to collaborate fairly closely. Immediately following the verdict, however, the Standard companies were too busy organizing themselves and were thus happy to accept a truce in the price war. Moreover, market conditions had changed once again. Rising demand in America turned the surpluses of 1910-11 into shortages; early in 1911, Standard increased its Java kerosene prices, other markets soon followed.

The Group's expansion into the United States did provoke a like-for-like counter move. In April 1912 Standard Jersey set up a Dutch subsidiary, Nederlandsche Koloniale Petroleum Maatschappij or NKPM, to apply for concessions in the Dutch East Indies and more specifically the prized Jambi concessions. The Group had much feared such a step and Deterding was enraged, in particular by the NKPM's propaganda offensive against Royal Dutch which, for example, argued that Bataafsche was not really a Dutch company. The Royal Dutch board immediately reacted with a lobby 'to urge [the ministers concerned] that neither in Jambi nor in other reserved fields concessions will be given to any other company but Royal Dutch and definitely not to companies under the direction of Standard Oil Company'.[88] It would take another decade and large amounts of money before Jersey would get some production in the Dutch East Indies, but in the short run this move had a very high nuisance value. The appearance of an alternative bidder put the authorities in the Netherlands and in Batavia in a quandary, forcing them to postpone a decision again and again. Jambi would remain on the agenda of both companies for some considerable time to come.

A dramatic expansion Overall the Group may be hailed as the winner in the chess game with Standard Oil in 1910-11, with the consolidation of its position in the Dutch East Indies, and the acquisition of extensive properties in Romania, Russia, and the United States. This was only the beginning. The Group continued buying producing fields and concessions at an amazing tempo all over the world: more in California and in the Midwest, in Russia, Egypt, Mexico, Trinidad, Venezuela, the Ottoman Empire, and Sarawak (British Borneo), to mention only the most important acquisitions. As a consequence, oil output rose dramatically. During the first years after the merger, Group production remained more or less constant, but between 1909 and 1913 oil output almost tripled, from 1.3 million tons in 1909 to almost 4 million tons in 1913. Consequently, the Group's share in world output grew from 4.3 per cent in 1910 to 9.1 per cent in 1914, entirely due to new producing territories as the share of the Dutch East Indies stagnated at about 3.5 per cent (Figure 2.2).

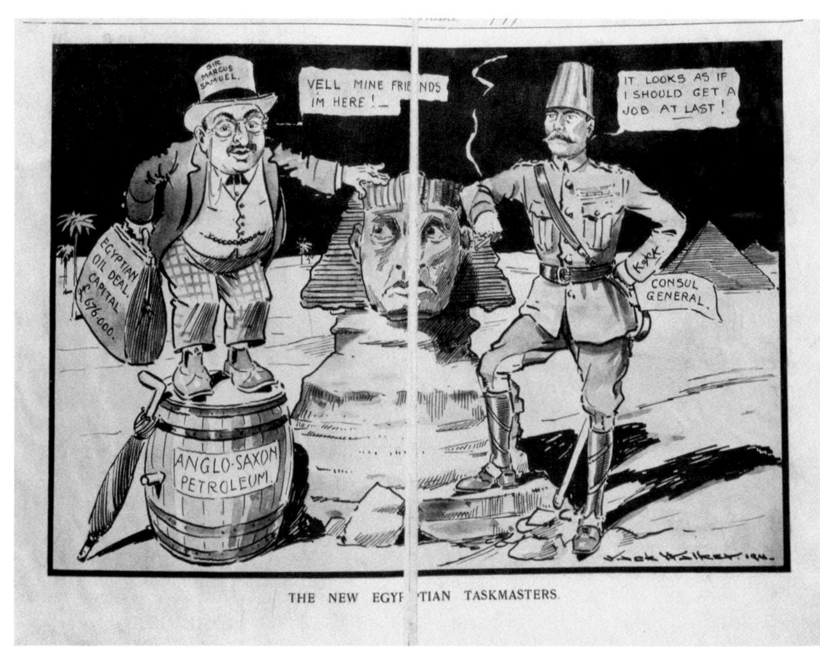

[54]

Cartoons often verged on being anti-Semitic. Above right, in an unknown publication from 1911, Samuel and Anglo-Saxon are seen as the new taskmasters of an apprehensive Egypt, while the Consul General hopes for employment at last.

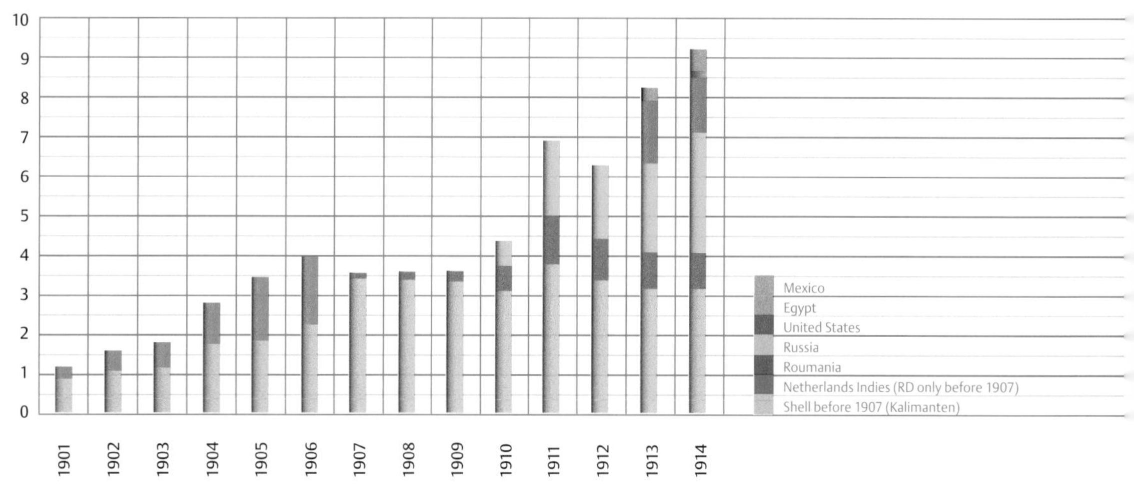

Figure 2.2
Crude oil production by the Group as a percentage of world output, 1901-1914.

Chapter 2

[55]

[56]

[57]

It is impossible to detail here all the fascinating stories about these new ventures or acquisitions, let alone the equally interesting stories about failed projects. Three big schemes succeeded each other rapidly. The first one has already been mentioned, the takeover of the Kasbek Syndicate from Deutsche Bank in 1910. Two years later the Group also obtained, through its collaboration with Deutsche Bank, a 25 per cent stake in the newly established Turkish Petroleum Company (TPC), which was to buy Deutsche Bank's concessions in Mesopotamia, now Iraq. The Group would also act as manager of the new company.[89] At this early hour, Deterding had clearly already understood the strategic importance of the Middle East. The second scheme was the purchase, in 1912, of the oil interests of the Rothschilds in Russia, their controlling stake in Bnito, directly followed by the purchase of a majority interest in another Russian oil company, Mazout. With these acquisitions and the Kasbek Syndicate, the Group at once became the biggest producer in the Russian Empire. Attempts to reach a closer understanding with the Nobel group, the other major producer, failed however.[90]

The third and potentially most spectacular scheme was the attempt, in 1913, to take over Gulf Oil Corporation, the biggest independent in the United States with a production of about 1.2 million tons. Bnito and Mazout were tiny by comparison, with a total production of about 550,000 tons. The actual negotiations were carried out by Gulbenkian, in daily contact with Deterding, and Kuhn Loeb & Co. on behalf of the Mellon family, which controlled Gulf Oil. Two issues dominated the talks, price and structure of the transaction. The Mellons asked $ 53.5 million and preferred a clean deal, i.e. a cash purchase of Gulf at market value. Deterding was prepared to pay $ 45 million but he wanted to form a joint venture combining Gulf with the Group's American and Mexican properties.[91] As manager the Group would receive 25 per

La Corona's Mexican operations around 1915: Left, at top, a riverside wharf with a small crane for loading and unloading building materials and other cargo; centre, the construction of a refinery building; and bottom, the pressure valves at the head of a well, for controlling the flow of crude oil. In the main picture, two gentlemen of the company have brought three daring lady visitors a short way out, to pose precariously on the very narrow bank separating two enormous open reservoirs of crude oil.

[58]

cent of the profits, while holding only 16.7 per cent of the shares.[92] This was unacceptable to the Americans. Moreover, Kuhn Loeb doubted the advantages Gulf Oil would get from such a deal, telling Gulbenkian 'confidentially, that the hallmark of the Royal Dutch Company does not enhance the value of the Gulf Co. in their eyes, for they have as much confidence in the Mellon management as in anybody else's in the world'.[93]

Share issues, and in particular the issuing of stock in exchange for the shares of companies taken over by the Group, played an important role in financing the dramatic expansion. Deterding was sailing very close to the wind, however. Starting greenfield investments on four different continents at the same time was very capital intensive, whilst they could not be expected to generate much revenue for a number of years either. To maintain investor confidence the Group also had to keep dividends high. The rapid evolution of demand enabled Deterding to sustain this juggling act. Rising demand for first gasoline and then liquid fuel revolutionized the oil market, until then dominated by kerosene.

Riding the markets The Group had ample supplies of both, but whereas gasoline found a ready market, fuel oil sales remained sluggish for a long time. Marcus Samuel never tired of extolling the qualities of fuel oil over coal for the propulsion of ships, and the Group itself gradually converted all its tankers to burn it; still, outside Russia, where liquid fuel was used extensively on the railways and inland shipping, demand remained limited. For all the obvious advantages in labour savings and increased speed one of the biggest potential customers, the British Admiralty, hesitated to switch to fuel oil. Both Sir Marcus and Deterding collaborated closely with Admiral Sir John Fisher, an ardent advocate of fuel oil within the Admiralty, to win support, but nonetheless failed to make much headway.[94] The lack of enthusiasm within the Admiralty was partly a matter of access. Britain had all the coal it needed, but fuel oil sources were invariably overseas and in foreign hands. Institutional inertia also played a role, however, and in that respect the British Navy did not stand alone. The US Navy was also slow to convert to fuel oil, despite having access to the biggest oil production in the world.[95]

Mestres Tank farm near Tampico
La Corona Tankfarm & National Line Bridge at a dis
Kilo. 9.

[59] [60]

From 1910 the British Admiralty began to adopt fuel oil, however. For prestige reasons Sir Marcus was keen to push Shell as a major supplier, and the Group happened to have large stocks, Deterding complaining about an 'enormous excess supply of liquid fuel' at Balik Papan in 1911.[96] And yet the Group had difficulty in getting access to this market. It was partly a matter of price; freight costs made Group supplies from Borneo uncompetitive with fuel oil from, for instance, Iran, where the Anglo-Persian Oil Company (APOC) had recently discovered large fields of heavy oil eminently suitable for liquid fuel. However, the Admiralty also mistrusted the Group for being foreign controlled, an attitude which Sir Marcus's inept advocacy of his cause helped much to foster. Moreover, APOC manager Charles Greenway, keen to create room to move for himself since he had slender capital resources and entirely depended on Asiatic for distribution, emphasized the 'Shell menace' at every opportunity.[97] Efforts by Waley Cohen and Marcus Samuel to collaborate with Greenway failed. Greenway skilfully exploited the 'Shell menace' to win the Admiralty over to secure its own fuel oil supplies by taking a £ 2.2 million stake in

Daily Express 14/3/12
40S A TON
KING COAL
OIL FUEL
HARRISON
MASTER OIL FUEL: "Yes, your Majesty, I'm small at present, but I'm growing rapidly."

[61]

More glimpses of Mexico around 1915.

In this cartoon, published on 14 March 1912 in the British newspaper *Daily Express*, the massive King Coal shows his high price – forty shillings a ton. Realizing that competition from oil might be no laughing matter, he glares down at the presumptuous youngster beside him. 'Yes, Your Majesty,' says Master Oil Fuel, 'I'm small at present, but I'm growing rapidly.'

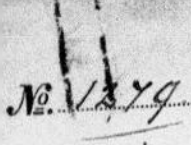

№ 1379

CERTIFICAAT VAN DEUGDELIJKHEID.

IN NAAM VAN HARE MAJESTEIT DE KONINGIN DER NEDERLANDEN.

De Hoofdinspecteur voor de Scheepvaart verklaart, dat

het Nederlandsche tank MOTORSCHIP Vulcanus
onderscheidingssein Q.B.M.T.
gebouwd van staal te Amsterdam in het jaar 1910
groot 1178,82 bruto register tonnen (van 2,83 kubieke meter)
met een Diesel- motor van 400 E. paardekracht
benevens de motor en de uitrusting van dit schip door de ambtenaren van de scheepvaartinspectie zijn onderzocht, in zeewaardigen staat verkeeren en voldoen aan de bepalingen van de Schepenwet, Wet van den 1 Juli 1909 (Staatsblad N°. 219), en van het krachtens deze Wet uitgevaardigde Koninklijk Besluit.

Op grond hiervan wordt door hem dit Certificaat uitgereikt, dat geldig blijft tot den 1e Januari 1912.

'S-GRAVENHAGE, den 19e December 1910.

De Hoofdinspecteur voornoemd,

[signature]

[62]

[63]

APOC.[98] The agreement was signed in May 1914, much to the dismay of Sir Marcus. In defending the agreement in the House of Commons Winston Churchill, then First Lord of the Admiralty, pulled out all the stops, heavily emphasizing the excessive prices supposedly charged by the Group and, with scarcely disguised anti-Semitic undertones, contrasting the Englishness of APOC favourably with the Group's uncertain nationality.[99] The Navy would now convert to fuel oil, but with APOC as its preferred supplier. Sir Marcus and Sam Samuel were deeply disappointed. Deterding's reaction is unknown, but he would have had reasons to be disappointed as well: despite having moved close to the corridors of world power, the Group was still not regarded as really British.

In the end, this defeat hardly mattered, for the Group did profit from the growing demand for fuel oil. On the eve of the First World War, liquid fuel made up an estimated 40 per cent of product sales by volume, turning a low-value by-product into a main revenue earner. Meanwhile the Group had moved on to the next technological challenge, diesel engines, commissioning a trial vessel from the Werkspoor shipbuilding and engineering works in Amsterdam. The ship, christened *Vulcanus*, was launched in September 1912 and marked a major event for the Group.[100] Though suffering from teething troubles, diesel engined ships were destined for a great future, from which the Group stood to benefit handsomely.

MS *Vulcanus*, the world's first ocean-going motor ship, was built for Royal Dutch/Shell in 1910 and soon demonstrated the economy of her diesel system. She could sail for 88 days without refuelling, and compared to steam-driven ships of similar size, she used only 20 per cent of the fuel by weight, and employed only five engine-room staff instead of sixteen. Left (above), her original certificate of seaworthiness, and (below) the ship herself – a revolution in the industry. Moreover, with her high forecastle and poop deck, and her bridge amidships, she was an early model of the 'three-island' superstructure that became the norm for tankers.

Lifted by the surging demand for all products, Group profits soared. Between 1907 and 1911 total net profits had fluctuated around £ 1.7 million, with a slight dip of £ 1.4 million in 1911.[101] In 1912 profits more than doubled to £ 3.4 million, reaching £ 4 million in the next two years. Bataafsche generated by far the most profits, contributing 70-75 per cent to the total dividend income received by Shell Transport and Royal Dutch. Most of these profits came from Indonesia, because the dividends from other sources, i.e. Romania and Russia, remained small.[102] Moreover, these figures seriously underestimate the true profits, for Group managers liked to build hidden reserves in the form of generous depreciations and large contributions to the central insurance fund. Judging by the calculations which Cohen Stuart made for 1907 and 1908, comparatively lean years, very, very large sums disappeared into the hidden reserves, for factoring the sums back in boosted the estimated profits of Bataafsche by 42 per cent for 1907 and by 67 per cent for 1908.[103] The Group's global expansion was thus clearly propelled by Bataafsche's substantial profits in the Dutch East Indies.

Building the skills base The Group's growth spurt was not just a matter of geographical expansion, but also a process of graduated skills, deepening by carefully considered investment in know-how and technology. The creation of a separate geological department by Erb has already been mentioned; the need for such a department had become urgent because, with the chief experts often away on surveys abroad, there was nobody left in The Hague to read the reports coming in and brief the directors about what had been found in, say, Russia or a particular concession in the Midwest. During the summer of 1913 Erb, in between two visits to the Middle East, organized a small department to solve this bottleneck.[104] The new department devoted most of its initial efforts to the exploration of new fields, but gradually the application of geological know-how to maximize the production of existing fields became more and more important.

[64]

Extracting wax from crude oil led to a useful additional business, the manufacture of candles. 'Made in Sumatra' became a selling point, with Royal Dutch's 'Crown' symbol featuring often in labels and advertisements.

The Royal Dutch director J. E. F. de Kok became personally involved in the Group's manufacture of candles and nightlights. In his memorandum (far right) he insists that it is essential to get an idea of the quality of candles sold in other countries, such as England, where the Price and Palmer brands were preferred; and the charming illustrations in the centre remind us that before electricity, candles were used for purposes now long forgotten – no doubt with a terrible strain on the eyes.

[65]

[66]

FIETSKAARSEN.

Wij leveren speciale Fietskaarsjes, Nr. D 1, 20 m.M. dik en 75 m.M. lang, terwijl het cylindrisch gedeelte 60 m.M. lang is. De brandtijd is ca. 2½ uur.

Een kistje houdt 25 pakjes à 12 stuks in.

Ten einde het spoedige doordrukken der kaarsjes in de fietslantaarn te voorkomen, bevindt zich op elke kaars een blikken ring met een diameter van ongeveer 9 m.M., waardoor de opening van den rand van den houder der lantaarn, die de kaars moet tegenhouden, verkleind wordt. Op elk pakje is voorts eene gebruiksaanwijzing geplakt.

13

PIANOKAARSEN.

Op een huiselijk muziekavondje brengt kaarslicht op de piano ongetwijfeld eene intiemere stemming teweeg, dan gewoon kunstlicht. Wij fabriceeren speciale pianokaarsen, ook in ROOD, van de volgende afmetingen:

Nr. C 1 — 18 m.M. × 202 m.M.
Nr. D 6 — 20 m.M. × 120 m.M.
Nr. D 7 — 20 m.M. × 215 m.M.

Benoodigt men een afwijkenden diameter, of wenscht men eene andere lengte, dan zal het niet moeilijk vallen eene keuze te maken uit onze overige talrijke soorten.

23

[67]

PROEVEN OVER DE VERVAARDIGING VAN NACHTLICHTJES.

Alvorens tot het maken van nachtlichtjes over te gaan hebben wy eerst de in den handel zynde merken geanalyseerd. In Holland zyn de voornaamste kwaliteiten die van Verkade, Gouda en de Kabouterlichtjes. Om tevens een denkbeeld te verkrygen van hetgeen in het buitenland op dit gebied geleverd wordt zyn door ons tevens onderzocht de Price en Palmer lichtjes, welke in Engeland preferent zyn.

I. Gouda Nachtlichtjes.

Prys per doos van 12 stuks (branddur 6 uur) 25 cents.
Er zyn lichtjes van 6,8 en 10 branduren en wel 2 kwaliteiten n.l.

1.) die welke gebrand worden in papieren hulsjes.
2.) die welke gebrand worden in blauwe glaasjes, waarvan de verkoopsprys 7 cent per stuk bedraagt.

Het materiaal dezer lichtjes bestaat uit paraffine van 47 à 48° stolpunt, zonder eenige toevoeging. De pit bestaat uit 3 × 4 draden en is met was omgeven van 66°C., waarvan het zuurgetal practisch nul en het verzeepingsgetal 11 is; het schynt een mengsel te zyn van byenwas (echte of substituut) en carnaubawas.

[68]

Chapter 2

In 1907-08 Dr Lazar Edeleanu of the Romanian National Chemical Laboratory developed the process that later bore his name and proved of great value to Royal Dutch/Shell. The essence of the process, shown in the diagram below, was to improve the quality of crude oil by washing it with sulphur dioxide. Tests by De Kok and Pyzel validated Edeleanu's experiments, proving that inferior Borneo kerosene could be made equal to the best, and in 1911-12 the Group bought the patents for several areas. In 1912 the Ploesti plant was ready, and a small pilot plant at Monheim in 1914, but the war delayed development and it was not until 1918 that the first commercial plant (below right) was ready at Balik Papan.

[69]

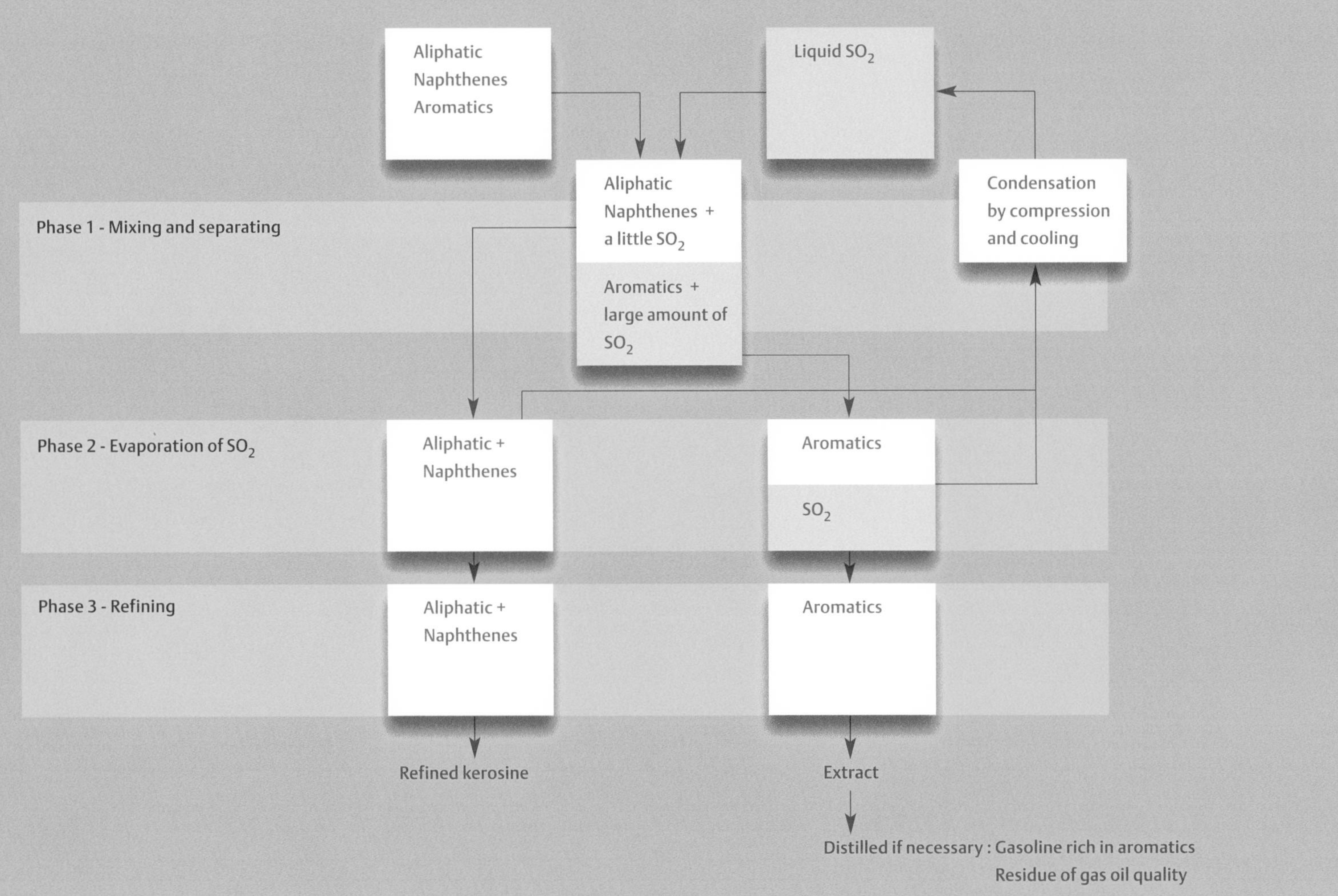

Earlier that same year, the decision had been taken to build a new laboratory in Amsterdam, Royal Dutch's third, after Sluyterman van Loo's makeshift facilities at the company's head office, and the first proper laboratory set up as adjunct to the Charlois gasoline bulk installation in 1902. The importance attached to in-house research was a pertinent aspect of Royal Dutch policy almost from the start: the company had employed Sluyterman van Loo since 1895, long before the board considered hiring geologists. By contrast, Shell Transport relied on outsourcing any research the company wanted to get done. One such project, commissioned from H.O. Jones and H.A. Wootton of Cambridge University's chemistry department, provided the first clue to solving the Borneo kerosene problem by demonstrating that the cause lay in the specific composition of the crude, i.e. its high content of aromatic hydrocarbons such as benzene and toluene.[105]

Jones and Wootton's discovery set the agenda for much of the research being done prior to 1914 at the Charlois laboratory. One task of great commercial urgency was finding a way to improve the very poor quality of Borneo kerosene and make it saleable. The laboratory pioneered a process for extracting the aromatics, but when this proved too slow for commercial purposes engineers turned their attention to two patented processes,

[70]

[71]

Construction of the Balik Papan Edeleanu plant begins.

sulphonation, based on washing the kerosene with sulphuric acid, and the Edeleanu process already mentioned.[106] In 1911 the Rotterdam laboratory invented a better sulphonating process by applying heat and pressure and Bataafsche decided to build a sulphonating factory in Balik Papan. Not everyone was convinced, however. Dan Pijzel (1877-1972), internationally better known as Pyzel, contested the outcome of the experiments. After starting his career building installations for Dordtsche, Pyzel had joined Royal Dutch in 1907, becoming its chief engineer in 1911.[107] Helped by J.E.F. de Kok (1882-1940), who had become the laboratory's head in 1910, Pyzel demonstrated the commercial superiority of the Edeleanu process, convincing the Bataafsche board to build another factory using this.[108] In February 1914 the sulphonating factory, which had cost one million guilders, was finally closed.[109] One lesson drawn from this episode was the need for experimenting with pilot factories on a semi-industrial scale, which could not be done at the Rotterdam laboratory. Consequently a new laboratory was built on a site bordering Amsterdam harbour and formerly used by Dordtsche Petroleum, and this opened in February 1914.[110] The search for extraction processes ran side-by-side with a project to extract aromatics from Borneo gasoline, conducted by the suitably Anglo-Dutch tandem of R.E. Robinson and W.C. Knoops. They had perfected a commercially viable process in 1908, paving the way for the production of mononitrotoluol at Rhenania's Reisholz works.

Testing the span of control One reason why the Gulf Oil deal failed was a lack of money for an outright takeover. As it was, the bankers on the Royal Dutch board as well as the Shell Transport directors considered the amount at stake too great.[111] Since as yet no market for Shell or Royal Dutch shares existed in the United States, an acquisition there would have had to be financed by raising cash, whereas the Group paid for most of its other takeovers with stock, a cheap way of financing given the high share prices in London and Amsterdam (Figure 2.1). In this way large blocks of shares found their way to private investors; as a consequence of the Bnito takeover, for instance, the Paris Rothschilds became the largest individual shareholders in Royal Dutch.

The Group's dependence on the London and Amsterdam stock exchanges irked Deterding. He felt that Royal Dutch in particular suffered from restraints imposed by the relatively small size of the Amsterdam market. Moreover, spreading share ownership over more stock exchanges lowered the risk of speculative raids on Royal Dutch, a risk of which Deterding, himself a keen speculator, was acutely aware. Paris, the dominant stock market on the continent, was the obvious place for an introduction of Royal Dutch shares, the more so since the powerful position of the Rothschilds would likely smooth the way. 'You must have the Paris market' Gulbenkian urged Deterding in April 1913, with the negotiations on the Gulf deal in full swing.[112]

[72]

... The more I think about this business the more I am convinced that YOU MUST HAVE THE PARIS MARKET FOR ROYALS and the more I become confirmed in my view that you must not allow Shells to come here alone and thus increase their speculative value. I would like to do the Shell and Royal business in a loft digni-fied manner not by roundabout crooked dealing. I do not think it would be advisable to mix other people in the business on your side – you know to whom I am referring – you should have the matter absolutely and entirely in your hands...

The extract from Gulbenkian's letter of April 1913 to Deterding (above) and his telegram to Deterding six months later (below) show his emphatic insistence on the value of the initiative – a view which Deterding wholeheartedly endorsed.

VERKORTINGEN
voor bijzondere aanwijzingen in de telegrammen

NB. Het Rijk vergoedt geene schade door het verminken, vertragen of verloren gaan van telegrammen veroorzaakt.

Ontvangen te 's Gravenhage, den ... 191.., ten ... u ... m. ... middags door ...

YOUR OPINION ON POINT I RAISED WILL MAKE USE YOUR OPINION WITH DISCRETION STOP FEEL VERY HAPPY BECAUSE CONSIDER INTRODUCTION ROYALS PARIS GREAT EVENT IN ITS HISTORY STOP SHALL I RESERVE ROOMS FOR YOU AND LANE AND DOCTOR PHILIPS WHOSE PRESENCE NECESSARY = GULBENKIAN +

[73]

This ambition brought Deterding into conflict with Sir Marcus Samuel. As Figure 2.1 shows, Shell shares had done a little better than those of Royal Dutch after 1911, reflecting the fact that Shell had paid higher dividends than Royal Dutch during the price war. Deterding had meanwhile put a stop to Sir Marcus's freedom of action by forcing him to give up Shell Transport's independence over dividend policy and new share issues.[113] His wings clipped, Samuel made another attempt to fly. He still owned a substantial amount of Shell shares and wanted to boost their value by having them introduced on the Paris stock exchange. It was an unequal contest. The Rothschilds, 'the almost almighty power (...) in all matters connected with finance in France', sided with Deterding and favoured the introduction of Royal Dutch shares, while Gulbenkian took care to isolate Sir Marcus when he visited Paris to find bankers who would support the introduction of Shell Transport shares.[114] Six years after the formation of the Group the two leading directors of the parent companies could still be seen pursuing their own, rather than common, goals.

Issuing shares in Paris proved to be difficult, however. The government, which had to give permission, demanded that two Frenchmen be appointed to the Royal Dutch board, a demand that after some negotiating was reduced to a single one appointed by the state. Taxation on share issues and on dividend payments also turned out to be heavy and the French members of the syndicate could not reach agreement about who would get a seat on the board. When all was settled the issue failed, according to

[74]

[75]

[76]

A selection of Shell advertisements from 1910-1916.

"SHELL"
Motor Spirit
SHELL
SHELL MOTOR SPIRIT
THE SPIRIT OF THE AIR

02|143

Deterding because Banque de Paris et des Pays Bas, a member of the syndicate, did not live up to its promises.[115] Having foreseen this possibility, Deterding had drawn up an alternative plan and secured the support of N. M. Rothschild & Sons for an immediate issue in London, where the Royal Dutch shares were an instant success. The Samuels, who saw their market invaded, were furious and upset: 'They term your action unfair, discourteous, unfaithful', Gulbenkian wrote to Deterding in November 1913.[116]

For all his obvious skills and successes, Deterding increasingly encountered resistance against his bold plans, notably on the board of Royal Dutch. Having grown used to getting his way, one way or the other, he found it difficult to cope with setbacks and to accept criticism, and when stressed by work he became irritable and easily lost his temper. This began to become something of a problem during 1912. Standard Oil's move to establish NKPM vexed Deterding considerably, which translated into occasional outbursts against colleagues and staff. During the last months of 1912, with several big schemes on his mind, he lost his temper in a discussion with IJzerman, a respected oil expert and director of Royal Dutch. After returning to London he apologized to him, but in a way that was quite typical: 'In the first place I am more sorry than I can express that, even if after all everybody would agree with my argument, I have really been too brusque, and as a result of my temper I do not have many friends, and can therefore badly miss one of the few. I do hope that you will forgive and forget my brusqueness.'[117]

During 1913 the critics of the Group's precipitous expansion became more vocal. The Royal Dutch directors, including Lane, protested that it was financially impossible to sustain the hectic pace; at the same time clear signs appeared that the organization could not keep up with the rate of growth. The managers of the Kasbek properties, for example, could not supply the required detailed information about the drilling operations there, nor stamp out the inefficiency and theft which drove up costs.[118] At a board meeting in May 1913 Royal Dutch directors, anxious to keep track of the Group's commitments following recent acquisitions in Russia and the US, had criticized Deterding about his habit of rushing decisions through without fully informing them. In response to the complaints, the board had then resolved to send the board minutes of the main companies to all Group directors, but the proposed full audit of the Group's position and obligations by an industry expert was finally dropped.[119] To some extent the Anglo-Saxon directors shared the concerns of their Dutch colleagues about the Group's precocious and uncoordinated growth, asking Deterding in October 1913 for a full list of subsidiaries.[120] Instead of tackling these evident failings, Deterding turned his attention to Bataafsche's organization in the Dutch East Indies, which he took to task for being too bureaucratic and too slow in securing the Jambi concessions from the colonial government in Batavia. He wanted to abolish the regional head-office in Batavia, and run the operations directly from The Hague.[121] This decision only served to increase the strains within the managerial team. More or less at the same time two members, Cohen Stuart and Loudon, tendered their resignations as managing directors of Anglo-Saxon and Bataafsche, although they continued as managing directors of Royal Dutch. We do not know the motives for these resignations. Both men were in their fifties, so age should not have been a factor. It appears more likely that stress took its toll. Cohen Stuart had medical reasons for wanting to step down.[122] As for Loudon, his relationship with Deterding never really recovered from one of the latter's outbursts in the winter of 1912.[123]

Thus the team of managing directors appeared to have overstretched itself. In November 1913 the Anglo-Saxon board devoted its attention to the management crisis. Probably sparked by a surprising proposal from Sir Marcus, there was a 'considerable discussion', a strong term in the always sober minutes, 'as to the advisability of the shareholders taking their proportion of the shares held by the Company in subsidiary Companies instead of providing Capital for the Company to pay off its indebtedness incurred for payment of such shares'.[124] This radical proposal threatened to break up the carefully crafted structure of the Group, distribute the worldwide interests to shareholders, and remove the control over the operating companies from the team of managing directors to the holdings. It was nothing short of a coup attempt, perhaps inspired by Sir Marcus's earlier defeats over the Paris share introduction and Shell Transport's dividend policy. Deterding was apparently caught unawares, very much unlike him, and he could not sweep the idea under the table. The meeting decided to call for a special combined session of Royal Dutch and Shell Transport directors, to be held on 11 December, when everybody would be in London for the official opening of the new head office at St. Helen's Court. Unfortunately no minutes survive of this unique meeting, as far as we are aware one of the rare occasions on which the two boards met in joint session.[125] We can glimpse something of the outcome, however, in the minutes of other boards. The break-up proposal disappeared without a trace and more conventional ways to finance the operating companies were agreed, i.e. a new share issue by Anglo-Saxon and a bond issue for Bataafsche.[126] The meeting must also have settled the outstanding management issues. Pleyte and H. Colijn were to succeed Loudon and Cohen Stuart at Bataafsche, and Waley Cohen became managing director of Anglo-Saxon, confirming his position in the ascendant.

With the December 1913 meeting the period of internal struggle over managerial control ended, leaving Deterding in command of the situation. The struggle originated partly in unfinished merger business, partly in the growing pains of rapid expansion, partly in Deterding's increasing tendency to act on his own and inform his colleagues later, and partly in his fiery temper. The Group stood on the threshold of a more mature phase in its existence in which the former two factors no longer played a role. By contrast, the latter two would become more and more dominant.

A chest of Shell products provided as part of the Group sponsorship of the British expedition to the Antarctic in 1910, led by Captain Robert Falcon Scott CVO RN (1868-1912).

[78]

[79]

Conclusion

By any standards the new Group was an outstanding success during the first seven years of its existence. The capitalized value of its stock increased from slightly more than £ 6 million in 1906 to £ 55 m. on the eve of the First World War. Crude oil production increased from about 1 million tons in 1906 to 4.7 million tons in 1914. Whereas in 1907 the Group had been dependent on a few, very productive fields in the Dutch East Indies, in 1914 it produced on four continents.[127] Moreover, the company had completely reorganized its transport and marketing, built a strong knowledge base in geology and engineering, and was moving to the forefront of new technology in drilling and refining. Finally, the Group owned the largest tanker fleet in the world, which was being modernized through the introduction of new technologies such as the diesel engine.

The spectacular success of the Group had many fathers. First of all, oil was a booming business: demand for gasoline and for fuel oil expanded at an incredibly rapid pace, and the Group was better positioned than Standard Oil to profit from changing market conditions. Second, oil was a relatively young business. The opportunities for exploring and exploiting new fields were abundant and, having invested in the necessary know-how, the Group only needed to pick what it considered the best. Standard Oil had a disadvantage here, having traditionally concentrated on transport and refining, leaving exploration and production to others. Most of all, however, the Group profited from the fact that it was an integrated company controlling the entire value chain of oil. The many oil companies which the Group took over had, with one or two exceptions, concentrated on production and neglected to invest in transport and marketing. Dordtsche was a fully integrated, well-run and technically proficient company but, content to restrict itself to working the Java market for its considerable worth, it could never hope to match Royal Dutch for marketing power. The Mellons also wanted to sell Gulf Oil because the company lacked the clout to access markets dominated by Standard and Royal Dutch Shell. Moreover, during their formative years in the Dutch East Indies both Royal Dutch and Shell Transport had assembled very solid human resources, with Deterding himself as a fine example, to power the global expansion after 1907. New acquisitions could be reorganized rapidly, new drilling sites developed, new refineries built. Again, the contrast with Standard Oil is illuminating. The company clung to its centre of operations in the American North-east and was ill prepared for a world in which oil production was decentralized, in which strategic possession of producing fields came to matter more than control over transport and marketing. Finally, the Group sought out best practice technology wherever it was developed and modernized its plants and drilling sites whenever a better technology became available. This worldwide exchange of cutting edge technology, partly developed by the Group itself, would become a key success factor in the years to come.[128]

These seven years of plenty also welded the Group together as a single corporation. For a long time managers continued referring to the Group as an 'amalgamation' or a 'combination of interests', suggesting a somewhat loose arrangement between companies which maintained their separate identities. In his autobiography, Deterding used these terms synonymously.[129] The convoluted management structure of overlapping directorships also reflected a sort of alliance rather than an integrated company. Yet there can be no doubt that it was a full merger. The two holdings kept their separate legal existence and also kept some leeway in financial policy until the 1930s, but the operating companies worked as a seamless unit, and the managerial team considered and treated them as such. The skirmishes between Samuel and Deterding over dividend policy, over share introductions, and over the ownership of operating companies were really no more than rearguard actions. The stupendous commercial success cemented the various parts firmly together; but Group managers failed to take this transformation to its logical conclusion and change the management structure to suit a more mature enterprise. This was to have lasting consequences.

[1]

Chapter 3

Surviving the crucible, 1914-1919

Having just taken steps to ensure a closer cooperation between its constituent companies, the Group found its integrity sorely tested during the First World War. Nationalism split the operating companies into different categories with opposed interests, warfare cut supplies from core producing areas, and the breakdown of regular consultations between London and The Hague strained the organization nearly to breaking point. However, the Group's unorthodox corporate structure now proved remarkably flexible and resilient, enabling the business to survive the challenges more or less intact. In London, managers strove hard to portray the Group as an ardent and selfless supporter of the Allied cause, but its position was in fact equivocal. Emphasizing its status as a neutral, Bataafsche continued to maintain close relations with the subsidiaries in Germany and Romania. With the coming of peace a new threat arose. To secure supplies, the British Government attempted to achieve some measure of control, as part of a deal to acquire access to Middle Eastern concessions. Group managers in both London and The Hague were initially prepared to give in to the demands, but the threat was finally averted.

Minutes of a Meeting of the Board of Directors of the Asiatic Petroleum Co. Ltd. held at St. Helens Court, Great St. Helens, on Wednesday, 5th August 1914.

Present:-

Mr. H.W.A. Deterding
Dr. A.J.C. Stuart
Mr. R. Waley Cohen
Mr. H.N. Benjamin

In attendance

The Secretary

In the absence of the Chairman Mr. Deterding took the Chair under Clause 81 of the Articles of Association.

438 Provident Fund. A letter from the Secretary of the Provident Fund dated 31st July, was laid on the table, and a copy of the Regulations ~~of the Regulations~~ of the Fund in Dutch and a translation of the same in English were signed by two Directors and the Secretary.

439 Employees joining the Colours. A memorandum from Mr. Engel with regard to notifying the Press of the action of the Asiatic and the Anglo-Saxon, was laid on the table and it was decided that the Press be notified unofficially that the two Companies had allowed their employees to join the Colours guaranteeing to them to keep their positions open and to pay them their full salaries for the next three months.

A further memorandum from Mr. Engel re out-of-pocket expenses of employees owing to holiday arrangements being cancelled, was laid on the table, and it was decided that everything that was fair should be done in the matter, that each case should be treated on its merits and that a list of all cases should be submitted to the Directors.

Employees due for leave. It was decided that instructions

[2]

Extract from the minutes of an Asiatic Petroleum Board meeting, showing the decision to offer paid leave to military volunteers, regardless of nationality.

Almost as usual The first practical war-related question confronting the directors of the Anglo-Saxon Petroleum Company concerned the arrangements for staff joining the colours. On 29 July, six days before Britain declared war on Germany and Austria-Hungary, but with mobilization proceeding apace all over Europe, the board resolved to offer military volunteers three months' leave on full pay with the promise of a return to their old job at the end of the war. Keeping true to the company's international orientation, the offer was extended to all employees regardless of nationality. As it turned out, Anglo-Saxon and its subsidiaries employed only two Germans and one Austrian. The latter had already gone back to his native country to join up, and he received his salary until November. Of the two Germans, one was dismissed without further ado, probably because he had only been employed for a short time. The other worked as a geologist with the Anglo-Egyptian Oil Company, which was considered sufficiently remote to keep him on the staff, provided he would make arrangements to receive his salary on a Dutch rather than a German bank account. The agencies with German trading houses were formally terminated at the same time. In at least one case, that of Wilhelm Rudeloff in Hamburg, it simply meant a transfer of operations from Asiatic to Bataafsche. Rudeloff's marketing business had very recently, in June 1914, been reorganized into a limited liability company with regular facilities to discount customers' bills. This had enabled him to remit sales revenues of seven million marks, about 4.1 million guilders or 330,000 pounds, before the close-down of international payments. Asiatic's German employees overseas were only dismissed with three months' salary in June 1915.[1]

The outbreak of war also thwarted the tentative moves towards greater integration between the Group's constituent companies. In June 1914 the Anglo-Saxon board suggested calling another joint meeting of the Royal Dutch and Shell Transport boards such as the one in December the year before to consider the Group's financial position, but nothing came of it.[2] Indeed, a sharply contentious issue threatened the Group's very integrity as an economic unit. The Dutch army had seized the gasoline stock of 17,000 tons at the Rotterdam-Charlois Benzine Installatie, but on 1 August the War Ministry agreed to release some of it for sale in return for an undertaking from Bataafsche to keep 10,000 tons of gasoline ready for army needs whenever required. With war between Britain and Germany a matter of days or even hours away, The Hague immediately sent a lighter with 2,300 tons of gasoline to the Reisholz refinery near Dusseldorf, owned by its subsidiary Rhenania. This action aroused fierce anger at Asiatic. Bataafsche had acted in breach of contract, and ought to have respected Asiatic's express instructions to terminate supplies to Germany, wired on 4 August following the declaration of war. Bataafsche responded to the criticism from London by digging up an old Anglo-Dutch bone of contention and regular cause of armed conflict between the two countries, i.e. the free trade of neutrals in war. Having had a lucky escape from being drawn into the carnage, the Netherlands now clung to a religiously-maintained neutrality. Group companies had to remain true to their host countries, the Bataafsche board argued. Staying neutral meant selling equal amounts to all belligerents, not by rights, but as a *duty*. Sending consignments to Germany would enable the company to sell similar quantities to the Allies, too, as evidenced by a 3,000 ton cargo of gasoline to Portishead (UK) by the SS *J. B. Aug. Kessler*, sent after the contested Lindavia shipment.[3]

After further animated correspondence a solution was found in early September, by excepting the Netherlands, Belgium, and Germany from the 'B' Agreement under which Bataafsche consigned its entire production to Asiatic for sale. This was a pragmatic screen, not a principled response to political circumstances. In a letter to his Paris patrons, Lane was quite open about the matter. The trade was transferred for practical reasons of communication and because, as an English company, Asiatic could no longer transact with Germany. Bataafsche would pay the revenues generated over to Asiatic in the same way as the profits from Dordtsche Handelszaken on Java.[4] The rearrangement had little practical consequence. The 'B' Agreement remained in force, so the London directors could simply redirect the product flows away from Rotterdam. This was exactly what happened. By mid-October, the Charlois gasoline installation lay idle through a lack of supplies.[5] As a result gasoline exports to Germany halted. In a letter to Rhenania manager Späth, Bataafsche director Colijn blamed low stocks, but he knew the real reason to be London policy.[6] Royal Dutch's shipping manager Phs. van Ommeren kept the Rhine tankers busy for another few months with stocks bought up here and there. The ships were soon laid up for the duration of the war, as the Allied blockade gradually restricted oil supplies to the absolute minimum needed for local consumption by neutral countries and left little or nothing for trade with the Central Powers.[7] Danish oil imports, for instance, dropped from 158,000 tons in 1913 to 18,000 tons in 1918. Sweden, Switzerland, and the Netherlands suffered similar sharp falls, though in the last case the official figures may be too low. Just before the war Bataafsche concluded an arrangement with Dutch shipping lines to carry heavy oil such as liquid fuel or filter oil in the double bottoms of general cargo and passenger ships. If continued during the war, as seems likely, such imports will have escaped notice.[8]

Deterding had meanwhile laid down guidelines for Group policy. Companies must, he wrote in December 1914, 'provided that they do not cause a *breach of neutrality* to be committed by the country in which they are established, (...) commit no act which might be injurious to the interests of Holland or Great Britain and their colonies'. The British and French Governments placed great confidence in the fairness of companies connected with the Group, and Deterding thought it would be 'a betrayal of such confidence if any injurious act were committed, however profitable it would be'. He stated his express wish to sever his links with any companies that could not toe this line, and consequently resigned from the board of the Astra Romana which, based in a neutral country but partly financed with British capital, also claimed the right to continue trading with Germany and Austria. Cohen Stuart resigned his seat on that board as well. Moreover, Deterding sought to forestall his being associated with Royal Dutch in London, to the point of eradicating his name from letters and documents which might otherwise have linked him expressly to the company and its board. Judging by the attendance of meetings, his visits to The Hague became increasingly rare, once only in 1915 and in 1916, and not at all during 1917 or 1918.[9]

The British companies took steps to ensure that their subsidiaries conducted no clandestine trade. In June 1915 correspondence intercepted by the counterintelligence services cast a definite suspicion on the Swedish and Norwegian sales companies having sold oil to Germany by diverting ships to the Baltic ports, partly disguised as impositions by the Kriegsmarine. As a result, the Foreign Office prohibited further exports to Scandinavia. Investigations conducted by a London manager sent over for the purpose uncovered further fraudulent sales to Germany and false accounting by the general manager responsible, who was dismissed immediately. To clean up the mess, the entire management structure of the Scandinavian companies was reorganized and responsibility transferred from Anglo-Saxon to Asiatic. Supplies to Scandinavia were subsequently tied to strict quotas agreed with the Foreign Office, and all agents received firm instructions to do their utmost to prevent any oil being re-

[3]

A political sketch by P. de Jong, 1917: the oil barrels are dry. Lamenting 'the end of our petroleum supply', a loaf of bread worries about the possible end of grain supplies too.

exported. Finally the Group's correspondence was placed under special observation by the British censor.[10]

The Group's opponents in Britain and abroad seized on the incident with alacrity to cast doubt on Shell's trustworthiness.[11] If such doubts lingered, it was because they had some substance in them. For though the Group's British companies could no longer be accused of direct trading with the enemy, the Group as a whole still could. At no point did the respective boards consider selling or liquidating subsidiaries based in or trading with Germany and Austria-Hungary. Deterding gave up his formal functions in Astra and avoided taking overt responsibility for decisions regarding the company, but he continued to be closely involved with its management. Moreover, Shell Transport held on to its Astra shares and continued to follow its business. By July 1916 these facts were sufficiently public to generate questions in the House of Commons.[12] The core group of managers does not appear to have seen this as a problem, perhaps because they regarded the Group as an abstract entity, not as a real economic unit. This was a convenient fiction. Throughout the war, the main companies retained their very close personal and economic ties, allocating staff, assets, and funds across the Group as and when required in the same spirit as before. Moreover, Deterding, the Samuels, Waley Cohen, Foot Mitchell, Cohen Stuart, Loudon, and Lane held on to their interlinking directorships in London and The Hague. Nor did the English directors with sons in the field, amongst them Sir Marcus and Lane, feel compelled to resign their board positions.

Even so a certain degree of disintegration did take place. The London managers suspended their monthly visits to The Hague for the board meetings of Bataafsche and Koninklijke, delegating Loudon or Philips to vote for them by proxy. They could thus claim to have no knowledge of decisions taken, though this stratagem did not relieve them of formal legal responsibility. Moreover, Bataafsche and Koninklijke continued to treat them as members of the Group's managerial team. At one point Fred Lane renounced his fee, for having attended no meetings at all, only to get an appeal from the Royal Dutch board to accept it since his valuable services to the Group did not depend on regular attendance.[13]

Until the Spring of 1915, when the Scandinavian imbroglio fully demonstrated the risks of sending sensitive information by mail, Group directors also continued to receive board minutes and other data from the various companies, and Lane passed them on to Paris. The last Bataafsche minutes sent contained a full report about Rhenania's new lube oil factory and plans for supplying it with Romanian distillate instead of products from Balik Papan and Tarakan, as planned. Interestingly, the report emphasized cost as the reason for the switch, mentioning supply difficulties due to the war only as a secondary factor. The translated text tactfully omitted any reference to the factory's location and the currency of the cost estimates, German Reichsmarks.[14]

We do not know whether such information reached directors subsequently through other channels. Shell Transport continued to receive rather cryptic agendas for Bataafsche meetings. Deterding must have had access to all operational details; Lane's correspondence with the Rothschilds also betrays a deep familiarity with them. Other directors may well have opted not to know. They would have needed to possess some mental agility to forget about the Group's subsidiaries such as the Rhenania gasoline works and the Monheim lube factory plans, which had figured regularly at pre-war Bataafsche board meetings. Moreover, Anglo-Saxon only transferred its Rhenania shares to Bataafsche in March 1916, which required the board's full consent.[15] Thus the ostentatious allegiance to the Allied cause displayed by the British companies and managers, and notably by the Group's informal head Deterding, must have been driven, at least partly, by the uneasy awareness of a rather equivocal position. It was an uphill struggle convincing sceptical officials anyway. Civil servants at the Foreign Office, for instance, remained convinced that, like British Petroleum, the joint marketing venture with Nobel and EPU which was sequestrated as partly enemy property, the Group had a large German interest. Nor did the managers make a favourable impression. One Foreign Office minute referred to Waley Cohen as 'utterly unscrupulous', adding that 'the susceptibilities of his colleagues could probably be better assessed in terms of pounds, shillings, and pence than as "national pride".'[16] In a final twist of irony, Rhenania refused to pay for the gasoline consignment, keeping the money due to Bataafsche as an advance on disbursements from The Hague which had not been forthcoming due to the spending restrictions imposed following the international financial crisis.[17]

Thus the opening months of the war severely tested the Group's organization as an assembly of holdings and operating companies, held together by a set of exclusive agency contracts, underpinned by shareholdings, and governed by a core group of managers with interlinking directorships. At Bataafsche two new managers, appointed to enable Loudon and Cohen Stuart to concentrate on the management of Royal Dutch, found themselves thrown in at the deep end. One of them, Cornelis Marinus Pleyte (1865-1951), was a dedicated oil man with a model career within Royal Dutch, at first in the Dutch East Indies and then as the pioneer manager of operations in Romania. Family ties provided him with a direct link to the colonial administration, his brother having become Minister for Colonial Affairs in 1913. This link proved very practical when Bataafsche began planning the Curaçao refinery and wanted to obtain tax concessions and other facilities from the island's government.[18]

The other appointment as Bataafsche director was a complete outsider. As a young officer in the colonial army, Hendrikus Colijn (1869-1944) had taken an active part in the often vicious campaigns to extend Dutch rule over the Indonesian archipelago, winning decorations for bravery on the island of Lombok and in Northern Sumatra. His ability to get things done caught the attention of General Van Heutsz, the Aceh governor who, on becoming Governor-General of the Dutch East Indies in 1904, steered Colijn to a senior position in the administration. In 1909, Colijn returned to the Netherlands and moved into politics, becoming Minister of War the following year. He earned his reputation as an able fixer by reorganizing the Dutch army within a remarkably short time. With his closely cropped hair and resolute bearing, Colijn cultivated his image as a man of action to the end of his life. Public office suited him very well, but even this brief excursion had exhausted whatever financial means were available to him.

[4]

The high tide of Dutch colonialism included senior officials of Royal Dutch. Above left, Cornelis Pleyte is seen in the jungle near Pangkalan Brandan, 1898; and above right, at Banda Aceh in January 1903 the Sultan of Aceh,

[5]

together with his son, promises allegiance to the Dutch government. The governor general, J. B. van Heutsz, stands to the right of the portrait of Queen Wilhelmina, with captain-adjutant Hendrikus Colijn in front of it.

This Passport is not valid for the zone of the Armies.
Ce Passeport n'est pas valable pour les zones des Armées.
British Consulate General

JE MAINTIENDRAI

N: 350.

P.P. 3,75

Signalement.
Agé de 45 ans
Cheveux, Sourcils blonds
Yeux gris
Nez régulier
Barbe et moustache blondes
Taille 1 mètre 7 décimètres 8 centimètres – millimètre
Religion protestante.

Signes Particuliers

Valable pour un AN.

Signature du Porteur.
H. Colijn.

Le Ministre des Affaires Etrangères
de Sa Majesté
La Reine des Pays-Bas, Princesse d'Orange-Nassau,
&c. &c. &c.

Prie et Requiert

au nom de SA MAJESTÉ, tous les Amiraux, Généraux, Gouverneurs, Commandants, Magistrats et autres Officiers, tant Civils que Militaires, quels qu'ils puissent être, des Princes et Etats, Amis et Alliés de SA MAJESTÉ, non seulement de laisser passer Monsieur H. Colijn, ancien Ministre de la Guerre de Sa Majesté la Reine des Pays-Bas, Directeur de la Bataafsche Petroleum Maatschappij, demeurant à La Haye (Hollande méridionale)

avec ses Hardes et Bagages, allant en Angleterre et ailleurs,

sans lui donner ni souffrir, qu'il lui soit porté aucun trouble ou empêchement quelconque, mais aussi de lui donner, ou faire donner au besoin, tout aide et secours, ce qu'Il sera prêt à reconnaître.

Donné à La Haye, le 18 Septembre 1914

Pour le Ministre,
le Secrétaire-Général,

[6]

As a director of Bataafsche, Colijn had to travel widely, but his wartime passport was specifically not valid in 'the zone of the Armies'.

At that moment Royal Dutch asked him to become manager of Bataafsche. Colijn reckoned that the remuneration offered would enable him to return to Parliament as a man of independent means within five to six years, though he agreed to stay for a term of ten years.[19] But why would the Group have wanted to give such a senior position to this 44-year-old man, who had no commercial training or even proven business aptitude? The initiative to parachute him onto the Bataafsche board probably came from Deterding, who knew Colijn from their joint involvement with a colonial property investment trust. In Deterding's eyes, Bataafsche needed a fundamental reorganization of its Dutch East Indies operations. Urgent projects such as acquiring the Jambi concessions, a personal and enduring obsession for Deterding, had made no progress at all, and he blamed slow bureaucratic procedures. With this purpose in mind, Colijn's background as former minister, fully at home in the colonial corridors of power, more than made up for his lack of business experience. Hiring him was a typical example of Royal Dutch's networking. After all, why would Deterding have overturned an earlier decision and allowed Colijn to accept a seat in the Upper House of parliament, if not because he valued the new director's political connections? In March 1914, Colijn took up his post with Bataafsche after a five-month trip visiting Russia, China, and of course the Dutch East Indies, planned and paid for by the Group.[20]

The beginning of this long and eventful association was auspicious enough. No-one but the former Minister of War could have obtained the release of Bataafsche's gasoline stock, if only because he knew that the barely motorized army did not really need it. Colijn mastered the intricacies of the oil business with remarkable rapidity, aided by daily briefings from two bright young managers on the rise, De Kok, promoted from the Amsterdam lab to lead the Technical Department in 1916, and the head of the Geological Department, Erb. As the war progressed, Colijn's stature facilitated access to diplomats, for instance to arrange for the free passage of a German engineer to Balik Papan to help solving the continuing problems with the Edeleanu installation there.[21] One interpretation credits Colijn with driving Bataafsche's policy on trade with Germany single-handedly to its logical conclusion, the partial cancellation of the contract with Asiatic.[22] The initial correspondence on the subject does indeed carry his opinionated and self-important stamp. However, as a novice in the organization Colijn could not have sustained his position, still less have achieved the waiving of fundamental contract clauses, without the full consensus and cooperation of his London colleagues. The correspondence reads as a board debate by proxy, not as a row over the unilateral declaration of independence. The solution chosen suited the Group, not Bataafsche alone.

A neutral ally

Despite Deterding's formal instructions, neither Colijn, Bataafsche, nor Royal Dutch observed a very principled neutrality. Gasoline deliveries to Germany stopped, but during the summer of 1916 Bataafsche agreed to help out Anglo-Saxon, caught short by a sudden surge in British Expeditionary Force (BEF) demand for gasoline, with 1,000 tons from supplies earmarked for the Dutch army and released by the Government at the request from The Hague.[23] Large amounts of surplus cash from the Dutch companies propped up the British war debt through investment in Treasury bills. Asiatic's payments to Bataafsche were no longer remitted to The Hague, but remained in London, thus also easing the pressure on the pound-guilder exchange rate. By December 1915, Anglo-Saxon held over a million pounds in British Treasury paper on trust for Bataafsche. The amount steadily rose to a peak of over £ 13 million, some 130 million guilders at the then current exchange rate, during most of 1918, dwarfing the amounts held by the Group's British companies. Royal Dutch also bought substantial amounts of Treasury bills, but preferred to have the paper sent over to the Netherlands by diplomatic bag, to serve as collateral when the company needed cash. During 1916 alone, Deterding helped to sell £ 7 million of Treasury bills to the Netherlands.[24] Over time, the financial expediency of funneling Bataafsche's cash flow through London had important consequences. Anglo-Saxon became the Group's treasury department, a function not foreseen in the 1907 agreements, but vital in effecting a closer integration.[25]

The famous toluol factory episode also demonstrates the Bataafsche's natural inclination towards the Allied cause. The origins of the Group's involvement with toluol derived from the efforts at the Rotterdam laboratory to extract aromatics from Kalimantan gasoline. Benzene and toluene had a ready market as raw materials for the manufacturing of dyestuffs and explosives, an industry until then entirely dependent on supplies from light coal-tar fractions produced as waste by gas factories. When Robinson and Knoops had perfected their extraction process, Balik Papan started to send 'straight run' gasoline, that is, gasoline containing all fractions boiling at 60-180° C, to a new refinery at Rotterdam-Charlois, where it was distilled into four main products:

[7]

[8]

light gasoline for motor cars; a white spirit substitute; benzene-rich gasoline; and toluol-rich gasoline. This latter product then went by lighter to Reisholz, where in 1909 a factory, built by Knoops working for Rhenania, came on stream to extract mononitrotoluol by way of a nitration and distillation process. The planned participation in and expansion of an existing explosives factory was finally rejected when the manager made a poor impression. Within two years of the first research into Kalimantan crude, an entirely new and sophisticated range of operations had been set up, foreshadowing the Group's research-led diversification into chemicals during the late 1920s.[26]

However, the venture in itself proved to be insufficiently competitive with the omnipresent gas works, so the Charlois installation remained an isolated specimen equipped with special Heckmann fractionating columns for this specific two-stage destillation process. With the outbreak of war, the works suddenly acquired an immense strategic importance. The surging demand for the toluol-based TNT high explosive could not be met by the inelastic supplies from coal. Kalimantan straight run was the obvious alternative source. The available documentation does not allow an exact reconstruction of events and decisions, but the sequence was probably as follows. During September, the Group's chief chemical engineer and manager of the Charlois installations, Sluyterman van Loo, realized the crucial importance of straight run, and he probably communicated his insights to London. At more or less the same time the cartel of French refiners led by Deutsch came to a similar conclusion. They approached Asiatic to buy 5,000 tons of straight run a month, to be refined at its Balaruc installation in the south of France under the control of the French Government, the cartel handling supplies and selling the light gasoline remainder.[27]

Since toluol offset the cost disadvantage of transport over a longer distance, this proposal offered a golden opportunity to gain a firm foothold in a market where until then the Group had been held at bay by the cartel and its main gasoline supplier, Jersey Standard. The opening was all the more important because of the expected national oil monopoly, mooted before the war in France as in Germany and now thought to be only a matter of time. The

The unintentional and terrible double meaning of 'Shell' in wartime: in 1915 French workers fill shells with TNT produced by Shell (far left), and high-explosive shells are loaded for firing (left and right).

[9]

French Government, robbed of the coal mines in the northern part of the country, was desperate for alternative supplies of toluol and eagerly accepted the terms offered, even agreeing to obtain the necessary tankers from its British ally, so the Group did not have to use its now scarce shipping capacity. The deal gave the Group a unique lead over Jersey Standard, which saw its market share in France decline from 67 per cent in 1914 to 39 per cent four years later, as gasoline imports from the Dutch East Indies rose from 5,000 to 36,000 tons. Toluol provided a similar lever into the Italian gasoline market in 1916, when the Group closed a more or less identical deal with the Italian Government. In 1917 the Group supplied 25,000 tons of oil to Italy from the Dutch East Indies, against a prewar total of only 1,000 tons; however, this volume dropped to 13,000 tons in 1918, when supplies from Egypt and the US began to meet part of Italian demand.[28]

By contrast, the British authorities needed some convincing, first of a toluol shortage threat, then of Kalimantan straight run as the best alternative, and finally of the Group's offer being desirable. The proposal differed from the French and Italian one. Anglo-Saxon offered to erect the necessary installations and run them; accepting that meant giving the Group a pre-eminent position in high explosives and in light gasoline, a little hard to swallow for a government which only months before had taken a large share in Anglo Persian to avoid becoming dependent on the Group for liquid fuel supplies. Finally the tanker bottleneck was solved by the Admiralty agreeing to take 5,000 tons of Kalimantan liquid fuel a month FOB rather than CIF to free shipping for an equal volume of gasoline. By early January the two sides had reached agreement, and The Hague sent instructions to Balik Papan for raising straight run production from 5,000 to 7,140 tons a month.[29]

Time was a crucial element in the deal. The Group offered to dismantle the now idle Rotterdam-Charlois refinery, transport it to the UK, and rebuild it there. Toluol production could then start in a matter of weeks, whereas building from scratch would take months, and drain other scarce resources. Bataafsche readily agreed to sell the installation to Anglo-Saxon. Colijn does not appear to have realized the political implications of a neutral country selling vital war equipment to only one of the belligerents, at

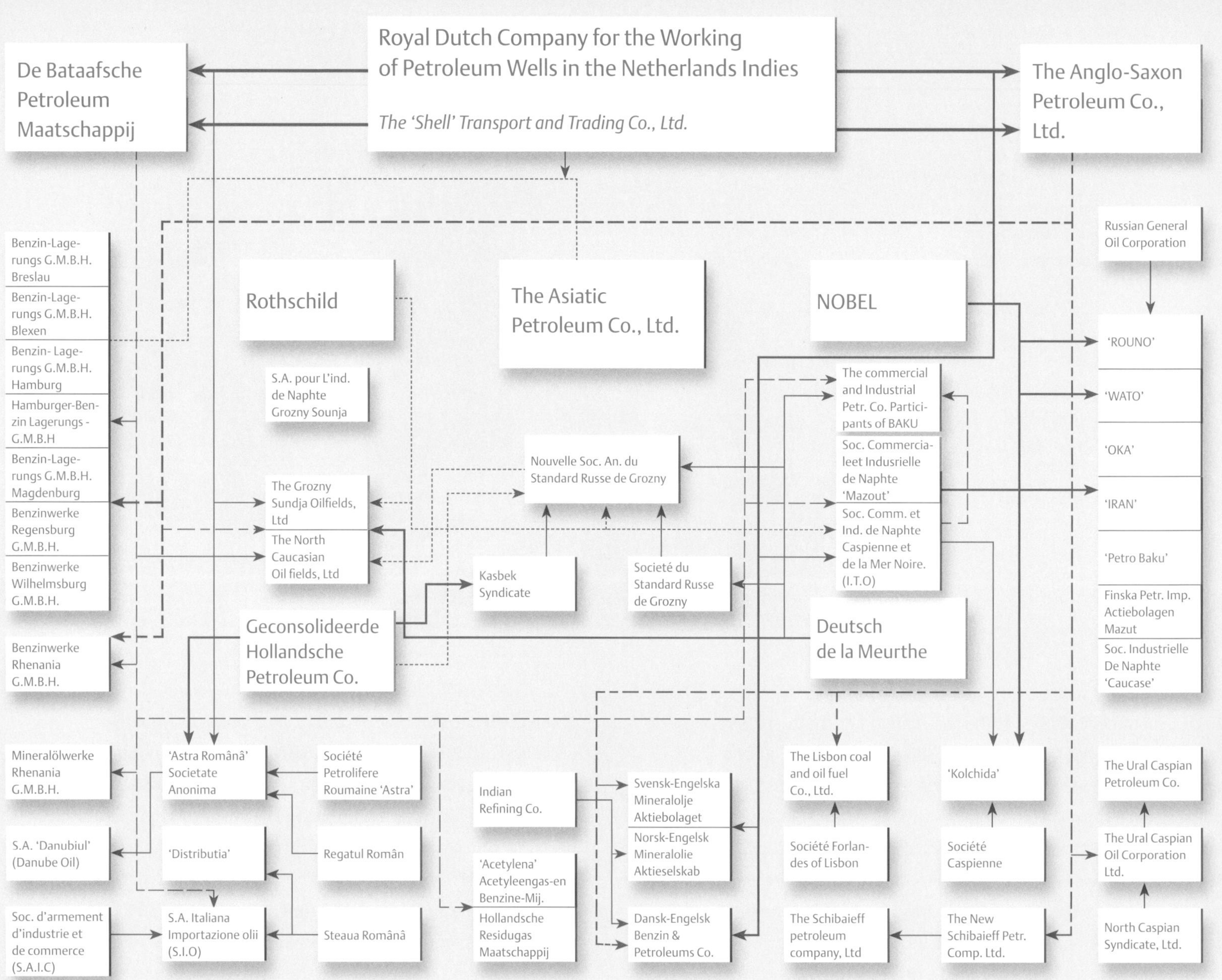

[10]

Financing companies
Producing and/or refining companies
Transport companies
Marketing companies

Connection broken off in the course of the period

Chart of the organization of Royal Dutch/Shell and the western market in 1914.

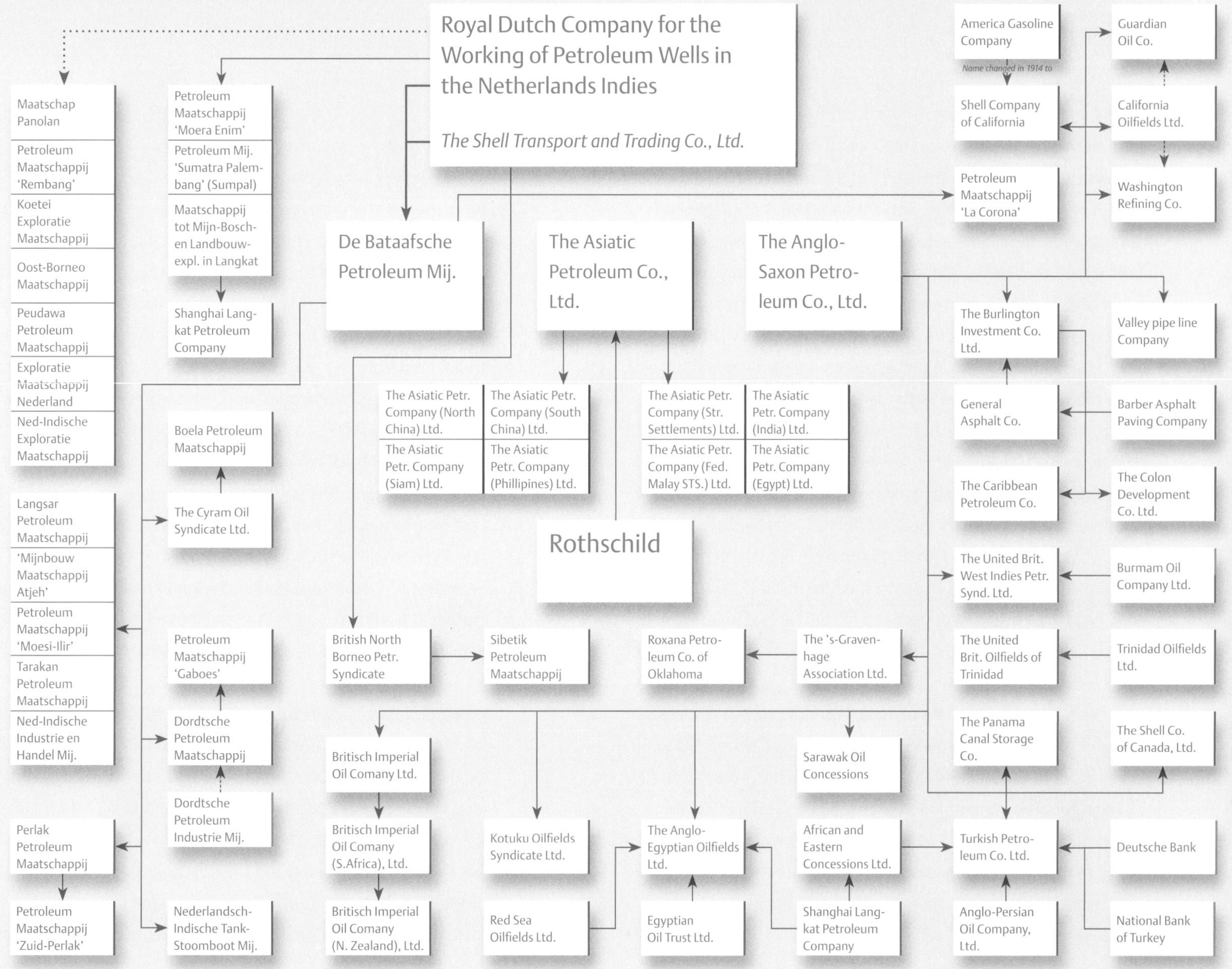

[11]

Financing companies
Producing and/or refining companies
Transport companies
Marketing companies

Associated by Exploration, production and/or marketing agreement
Connection broken off in the course of the period

Chart of the organisation and the Eastern and the American market in 1914.

least intitially, for he wrote to the Dutch Government about it on 25 January, after Waley Cohen and Deterding had urged him to do so. Nor did he wait for permission. Preparations for the refinery's removal must have been in full swing by the time Colijn approached the Foreign Minister, who gave permission on 2 February, but the freighter carrying the crates with equipment had already left Rotterdam harbour on the night of 30 January, after a clockwork operation that suggests detailed preparation and close cooperation with the Rotterdam authorities and the river police.[30] Contractors directed by Bataafsche engineers rebuilt the refinery in Portishead, near Bristol, where Anglo-Saxon had an installation for landing bulk gasoline from tankers. Toluol production resumed at the beginning of April, managed by Sluyterman van Loo, who was hired by Anglo-Saxon from Bataafsche for the duration of the war.[31] To cope with the flow of light gasoline, canning factory equipment destined for the Far East was redirected to Portishead, soon producing 8,000 two-gallon tins of gasoline a day.[32] Before the year was out, Anglo-Saxon had opened a second much bigger factory of similar design in Barrow-in-Furness, adjacent to another of its bulk gasoline port installations. Nitration installations at Chester and Oldbury extracted the mononitrotoluol, together producing 1,150 tons per week for a total of 30,000 tons during the war, about half of the British forces' needs. A special TB (Toluol Benzine, i.e. gasoline) Department administered this particular business at Asiatic, in its usual capacity as agent for Anglo-Saxon.[33]

Getting hold of this strategic resource was of considerable importance to the Allies. The Dutch Government soon had cause to regret its permissive attitude. The Hague immediately commissioned the Feyenoord engineering works to rebuild the factory, thus giving the lie to its own argument used to get export permission that the old one was idle and thus surplus to requirements. Moreover, Bataafsche drove a hard bargain when the War Ministry approached it for toluol supplies in June 1915.[34] The Germans had also learnt a lesson. Equipment for rebuilding the refinery ordered from German companies came with a clause forbidding onward sale to countries at war with Germany, so Loudon refused, with noted regret, Anglo-Saxon's request to repeat the coup.[35]

For the Group, toluol confirmed the central importance of fundamental chemical research to guide product development. This was notably different from Jersey Standard, where labs remained closely tied to refineries and entrusted with little more than product testing.[36] After Pleyte had blocked an attempt by Waley Cohen to have the work on toluol and benzol derivatives transferred from the Amsterdam lab to the UK, a joint team was set up composed of chemical engineers from Amsterdam and Cambridge University to conduct research into the various hydrocarbons present in Kalimantan crude, with a view to entering the dyestuff business. This particular direction proved a dead end for commercial reasons. Nevertheless the Group's commitment to research remained strong indeed, towards the end of the war switching towards investigating cracking processes and hydrogenation, leading to a long-standing but ultimately fruitless involvement with the Bergius process for coal hydrogenation.[37] At the same time Waley Cohen arranged funding for a research project on Kalimantan crude into a different direction. Chairing a gasoline supply commission, he met the engineer Harry Ricardo, then occupied with investigations into the tendency of engines to knock, i.e. detonate prematurely, which reduced performance and kept compression ratios low. Ricardo had already discovered that benzol reduced knocking and he subsequently found that the aromatic hydrocarbons in Kalimantan crude did as well. At the instigation of Waley Cohen, Anglo-Saxon commissioned Ricardo to research the effect of various gasoline components on engine knock. This project made a high-performance aircraft gasoline which enabled Alcock and Brown to make the first successful transatlantic flight in 1919. The association between the Group and Ricardo lasted for more than thirty years and produced further important results, notably in the field of gasoline composition and Diesel engines for road vehicles. Royalties from patents and licences helped to fund further research, since the Group channelled its part of the fees back into Ricardo's firm.[38]

Oil in total war On 21 November 1918, ten days after the Armistice, the dinner guests at Lancaster House in London, all delegates to the Inter-Allied Petroleum Conference and including Deterding, Sir Marcus, Lane, and Waley Cohen, celebrated the wave of oil which powered the Allied victory.[39] The Allied forces certainly used prodigious quantities of petroleum products, with Britain the biggest consumer by far. Total UK oil consumption nearly trebled, from 2.1 million tons in 1913 to 6.2 million in 1918. Royal Navy operations alone burned 3.7 million tons of liquid fuel during that last year.[40] France used only a fraction of the British volume, 700,000 tons in 1913 rising to around 900,000 tons five years later.[41] Russian production peaked at 9.5 million tons in 1916, but the following year ethnic tensions and revolutionary violence led to an increasingly precipitous decline.[42] However, exports from Russia and Romania came to a halt in September 1914 when the Turkish Government, hostile to Russia and therefore a natural ally for the Central Powers (Germany, Austria-Hungary, and Bulgaria), formally closed the Dardanelles to shipping, though few ships had crossed the straits since German warships had entered it in early August.[43] Total consumption by the Allies (Britain, France, Italy, and the US) amounted to about 9 million tons in 1918, 60 per cent of it liquid fuel, 20 per cent gasoline, with kerosene, lubricants, and gas oil taking up the rest.[44]

Oil was indeed crucial to the Allied strategy. Fuel oil enabled the Royal Navy to sustain the extended operations required by fighting a drawn-out war of attrition as opposed to the short war of annihilation envisaged by nearly everyone during the summer of 1914.[45] The logistics of land forces were equally geared towards a short conflict. Army supply systems were based on hauling goods from the rear by rail for onward distribution to field units by horse-drawn wagons. This worked well enough for armies on the move, living off the land whenever necessary as they had done for centuries, but trench warfare imposed totally different requirements, notably a hugely increased demand for ammunition. Thus in 1914 a division of the BEF needed twenty wagon loads of food and horse fodder a day, plus seven wagons of other supplies, including ammunition. By 1916, the non-food requirements of a division

[12]

'Germany hastens to embrace its new ally, Turkey.'

[13]

[14]

The defeat of the Central Powers in 1918 and the ensuing peace treaties reshaped the map of Europe from its pre-war form (top) to a new pattern (below), with new states, the enlargement of some of the victorious nations, and the diminution of the defeated.

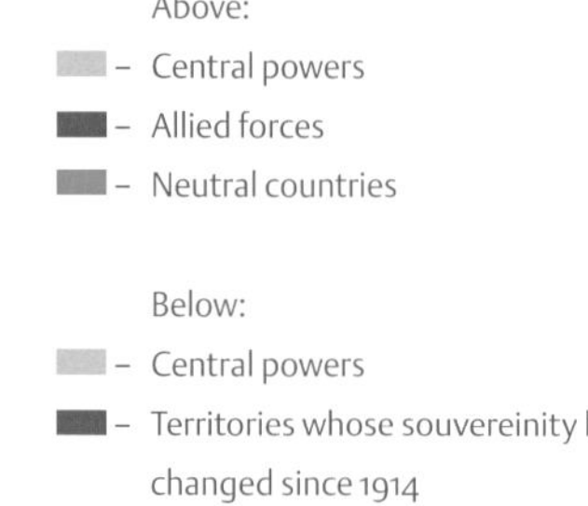

had risen to thirty wagons a day for a total of fifty.[46] Only motor trucks could provide the additional means of transport, since their productivity outweighed their cost and the need for additional fuel supplies. One five-ton lorry could replace three teams of horse wagons. The advantages of motorized transport had been clear for more than a decade, but even the Group, with its privileged access to gasoline, had only just begun to switch, ordering its first batch of Leyland trucks in April 1914 for gasoline distribution by the British Petroleum marketing organization. They were promptly requisitioned at the outbreak of war.[47]

When the expected short campaign turned into a stalemate, motor transport could be expanded fairly quickly. The supply of horses remained inelastic, since it takes five years to raise a foal. However, horse transport remained essential in the shell-cratered and often muddy wasteland front areas.[48] The Allied armies generally used motor trucks for the intermediate link between the railheads and the field depots. Some operations depended entirely on them, however. In 1916, the French could hold the fortress of Verdun only because of supplies brought up by convoys of *camions* driving day and night along the single narrow road later named the *Voie Sacrée*. Moving the same volume with horse wagons would have blocked the road.[49] Similarly, the repulse of Ludendorff's March 1918 offensive and the subsequent decisive counterattack during the summer succeeded because of rapid troop movements by truck, backed up by flexible supplies.[50] The total number of trucks in the French army rose to an estimated 92,000 in 1918.[51] At that time the British army possessed almost 828,360 horses and mules, about half of them in France, plus 57,000 trucks and tractors, and 23,000 cars and vans.[52] Motorized transport was also

'Shell at the Front!' Wartime newspaper advertising left readers in no doubt about 'what petrol is doing'.

space which newspapers have been expected to supply for announcements of an official or semi-official character connected with the war. It is felt by many that the generosity of the Press in this matter has been in no small measure abused.

Why, it is asked, should newspaper proprietors be required to supply for

'SHELL'
AT THE FRONT!

Quality tells more than calumny

The Allies

Be on the side of the Allies and use "Shell"

OBTAINABLE EVERYWHERE

WHAT PETROL IS DOING

[15]

vital where speed and a degree of comfort mattered. For that reason mechanized ambulance services, linking dressing stations at the front with the hospitals in the rear, appeared early in the war. In October 1914 the famous American Volunteer Motor Ambulance Corps alone already had ten ambulance cars stationed in France. A year later the number had risen to sixty, which together had transported 28,000 wounded, and two further US organizations ran such cars in France. The British Army possessed a total of 7,000 ambulances by November 1918.[53]

As for the new weapons of war, tanks and aircraft, both were intensive gasoline users. Military aviation developed rapidly. In August 1914, Britain possessed only 84 aeroplanes for military use but no independent staff or units flying them; four years later, the Royal Flying Corps counted 1,230 on the Western front. France had almost 2,750 front-line aircraft, and the US forces fielded another 600 for a total of 4,500.[54] Tanks were not deployed in any great numbers until the last twelve months of the war, with the first military successes achieved by British forces during November 1917. A year later, the British army had about 2,600 tanks, and the French army 3,800.[55] This new kind of armoured vehicle was essential, arguably more so than aircraft, in forcing the Allied breakthroughs during the summer of 1918. Tanks also used more fuel, 42,000 tons a month against 31,000 tons for all Allied aircraft.[56] Finally, having gasoline meant that the Allies could mechanize agriculture and thus fill part of the manpower gap created by the drafting of farm labour. To keep food supplies stable, France needed about 1,000 tons of gasoline a month by 1918 for farm tractors and combine harvesters.[57]

Access to oil thus enabled the Allies to wage a mechanized war to an unprecedented degree, from running a tight naval blockade denying supplies to the Central Powers, forging new supply chains to the front, building air forces to command the skies, to creating the weapons which finally broke the stalemate.

Consequently, the challenge for Allied oil policy was essentially a logistical one, i.e. mobilizing the resources needed to ensure sufficient supplies to meet racing demand, and to allocate the different products to the right destinations, at the right time, in the quantities and to the specifications required. This was difficult enough in itself, for no preparations had been made at all. The Army had detailed plans for obtaining the necessary horses, but not for getting motorcars or gasoline.[58] German submarines posed a lethal threat to oil supplies, for shipping was a crucial resource. Britain commanded about half of the world's entire tanker fleet, 200 ships with a total GRT of 883,000 tons in June 1914. Whereas British oil consumption tripled, its tanker capacity rose by only 50,000 tons and world tonnage by just over a quarter, most of it during 1917-1918.[59] The closing of the Dardanelles

Below, a vast collection of motorized ambulances gathers in Paris in 1915 before leaving for the front. Right, National Motor Volunteers line up for inspection of their vehicles.

[16]

[17]

KEEP it COMING
"We must not only feed our Soldiers at the front but the millions of women & children behind our lines"
Gen. John J. Pershing
WASTE NOTHING
UNITED STATES FOOD ADMINISTRATION

[18]

Above, ambulances of the Royal Army Medical Corps are seen close to the front line in France, 1916; and below, a US poster from 1918 exhorts American citizens to keep supplies coming.

[19]

imposed further constraints. Supplying Europe from Asia or the US instead of Romania or Russia tripled the distance covered and thus the tanker capacity required. France and Britain together imported some 700,000 tons of oil from Black Sea ports. Substituting this meant finding an extra 1.4 million tons of shipping.[60]

The situation was made worse by the Admiralty's requisitioning of tankers to carry fuel for the Navy, which sharply reduced the capacity available for carrying gasoline and other products. France became more or less at the mercy of its ally. The refiners' cartel possessed only a few small tankers of its own, which for reasons of convenience had been transferred to the British register of shipping just prior to the war and had now been requisitioned. Moreover, the country lacked adequate port resources to handle the large quantities needed, and the ports that possessed any installations at all lay in or very near the war zone.[61] As tanker losses to submarines mounted in 1916-17, Britain faced shortfalls in supplies first of gasoline and then also liquid fuel, resulting in the rationing of gasoline for private motorcars and some restrictions on Navy operations.[62] France introduced rationing measures in April 1917 and imposed firmer controls in December, when stocks of gasoline threatened to run out by March 1918.[63] Though often described in terms of an acute crisis, this situation was not one of actual shortages, but of an anticipated divergence between projected demand and stagnating or even falling supplies. The threat was sufficient to goad the Allies into tackling the transport bottleneck by drafting the pool arrangements and setting up the Inter-

DBl
BK
D Bl
G
BK
L.GR
BK

1]

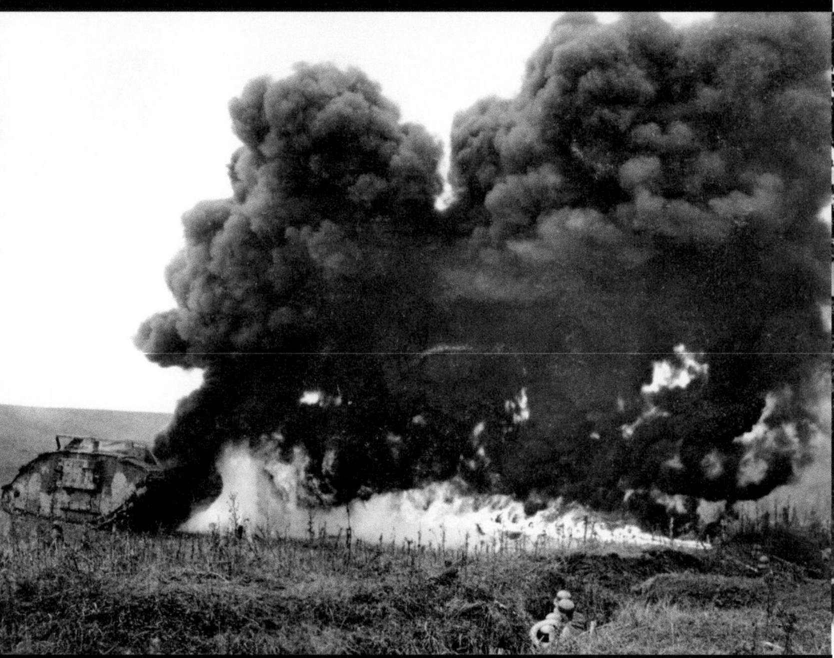

[22]

Tank warfare: left, women camouflaging a tank; above, a tank as flame-thrower, projecting a deadly blaze far ahead; right, a Frenchman and his children look at a Tank Type Mark V damaged by shell-fire, with Tank Corps engineers beginning to make repairs.

Allied Petroleum Conference whose delegates attended the victory dinner at Lancaster House.[64]

Before that celebration, in the long and often acrimonious negotiations over transport and gasoline, the British Government had the great advantage of a direct approach to the major oil companies, transporters, and distributors. Representatives from the oil trade ran the gasoline rationing scheme for the Board of Trade, and the petroleum pool board set up in March 1917 was equally composed of managers from the oil majors who could take binding decisions.[65] The Group continued to encounter some animosity, notably at the Admiralty, in the form of a resurrected 'Shell menace', now composed of its commercial power, foreign ownership, and the suspicion of dealings with the enemy.[66] Sir Marcus and Bob Waley Cohen attracted considerable personal hostility from Government officials as well, some of it in thinly disguised anti-Semitic terms. At one point Waley Cohen even received a formal ban on entering the Foreign Office, though six months later he became Gasoline Adviser to the War Office.[67] The commercial,

[23]

The wreckage of an aircraft lies in the foreground (left), with soldiers of a Belgian outpost above it. The experimental nature of early aerial warfare (right top): a bomber prepares to release his weapon, 1914; (centre) a Lewis gun mounted on a Scarff ring in one of the two Howdah gun positions, fitted to the upper wing of an F2A at Felixstowe, 1918; (bottom) Dutchmen fish the remains of a low-powered aircraft out of the sea, 1915.

[24]

[25]

[26]

technical, and managerial resources which the Group commanded were indispensable to the war effort, however. Deterding, Sir Marcus, Waley Cohen, Lane, and other managers such as the Group's Marine Superintendent Cornelis Zulver, found themselves increasingly drawn into semi-official duties, ranging from hurriedly organizing tankers to evacuate French gasoline stocks from the advancing German armies, financial engineering, official trips to the US to plead for increased oil supplies, advising on maximizing tanker utilization, to developing more efficient ways for the refuelling of aircraft, and finally building the entire infrastructure for supplying bulk gasoline to the Allied forces.[68] Initially gasoline was shipped in cans from the Group's bulk installations at Barrow and Portishead. To save shipping capacity, labour, and leakage, bulk supplies were in 1916 redirected to Rouen and Calais and handled by the Group's canning works transplanted in their entirety from Britain, production resuming five days after the last cans had been filled in Britain.[69] This system relied entirely on tins brought up to the front line by trucks; the American Expeditionary Force introduced filling stations with pumps, an innovation which was to spread rapidly during the 1920s.[70] In recognition of the services of Group managers, the British Government after the war bestowed

[27]

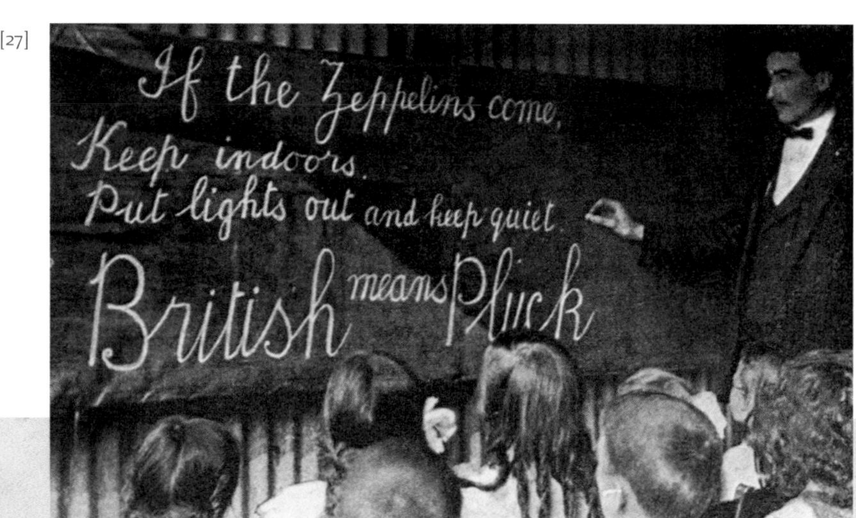

British schoolchildren are told what to do if a German Zeppelin should come (left), and (below) in 1914 the citizens of Ostend watch a British airship over the North Sea.

[28]

In 1916, trained women workers from the Shell can-making factory at Fulham in London were moved with their entire factory to France.

[29]

knighthoods on Deterding and Waley Cohen; a peerage on Sir Marcus, who became the first Lord Bearsted; an OBE on the chief engineer Aveline; a CBE on Andrew Agnew, the Group's general manager in Singapore; and a silver service with inscription on Zulver.[71]

By contrast, the French Government had no access to management resources or technical expertise at all. The oil industry consisted of the refiners' cartel which, for all its commercial acumen and downstream assets, had become an agency dependent on its suppliers. As minority shareholders in Asiatic and in the Group's Russian ventures Deutsch had direct access to the operating companies, but even they could only move resources outside France on the same commercial conditions as everyone else.[72] As a consequence, Gulbenkian came to occupy a pivotal position in Paris, for though he held no official position with the Group, he did have Deterding's ear. To facilitate his intermediation Gulbenkian received official privileges such as permission to travel on particular reserved ferries between Folkestone and Boulogne, and occasional use of the diplomatic bag for sensitive or urgent documents.[73]

From an energy point of view the war was no contest. Germany and Austria-Hungary conducted a holding operation, as remarkable for its precariousness as for its efficiency in making a little go a very long way indeed. Austria-Hungary had production in Galicia, but Germany imported about 90 per cent of its oil, most of it from the US, the Dutch East Indies, Romania, and Russia. At 1.4 million tons a year in 1913, consumption was far lower than that in Britain, but twice that in France. Most of the oil was used for lighting and industrial purposes since, for a motor industry pioneer, Germany had surprisingly few cars, about 61,000 registered vehicles in 1913, against 91,000 in France, and 209,000 in the UK.

The army used a quarter of all gasoline imports.[74] The Schlieffen Plan, Germany's sweeping attack on Belgium and France in August 1914, depended entirely on access to railways. Nor did Germany use much liquid fuel. The *Kriegsmarine* had hardly begun to convert its surface fleet and, after the strategic defeat of its High Sea Fleet at the Battle of Jutland in the early summer of 1916, gave up any ambitions to patrol the high seas anyway, ceding priority of operations to diesel-engined U-boats.[75] In March 1914, government threats of a state monopoly had led to a secret agreement with the main importers, Jersey Standard and the Group, in which the companies guaranteed 'to maintain a stock of Benzine [i.e., gasoline] in Germany for a period of years at the disposal of the Government on the distinct understanding that no Monopoly is established'.[76] As a result Germany had oil stocks of about 340,000 tons at the outbreak of war, three-quarters of it kerosene, the rest gasoline and lubricants.[77] The Allied blockade ensured that, during the war, Germany and Austria-Hungary could never muster more than the roughly 2 million tons which the two countries together burned in

Left, Belgian troops recover the town of Termonde (Dendermonde), south-west of Antwerp, after its destruction by German forces in September 1914; centre, French soldiers march through the vineyards of Champagne during the harvest of 1914; right, French marines come to Belgium's aid, 1914.

[32]

1913.[78] The Central Powers thus had to organize a scramble for resources. Private motoring was curtailed as early as February 1915 by the simple expedient of cancelling all car licences and reissuing only those considered vital. Increased gasoline supplies from Galicia and, from 1916, Romania, replaced two-thirds of Germany's overseas imports. Without the stocks and sources captured in Romania, Germany would have collapsed two years earlier. The loss of Romanian oil through the destruction of stocks and equipment by a few British soldiers was fairly quickly recovered, but this action still counts as 'without doubt one of the most important single achievements of the World War', since it hit Germany at its weakest spot.[79]

Substitutes or *Ersatz* covered an estimated third of German oil consumption by 1918.[80] The very sophisticated coal-tar industry produced toluol and benzol, used as a gasoline substitute in various blends, diesel fuel for the U-boat fleet, and about half of the navy's 100,000 tons of liquid fuel. In 1916 Bergius set up a pilot plant for manufacturing gasoline from lignite and heavy oil fractions by cracking the large hydrocarbon molecules into smaller ones in high-pressure vessels filled with hydrogen.[81] Tentative estimates put German synthetic oil production during the war at more than 2 million tons.[82] The Haber-Bosch process for producing ammonia through nitrogen fixation, perfected just prior to the war, supplied additional building blocks for high explosives.[83] Methylated spirits made from potatoes went some way to replace kerosene, though the shortage of this particular product hastened the switch from kerosene lamps to gas and electric lighting in Germany. Lube oil producers concentrated their efforts on methods for recycling oil and on finding suitable vegetable and animal fats as additives to eke out supplies.[84] Finally, German refining practices appear to have been very efficient. In 1917, French experts analysed captured aviation fuel and highlighted its purity as the reason for the far superior speed and climbing performance of German aircraft.[85] However, these qualities presumably found their origins in the benzol additive and its effect on engine knock, as Ricardo had just found out.[86]

[33]

On 21 November 1918, ten days after the Armistice came into effect, the German High Seas Fleet surrendered to the Royal Navy (left). Nearest the camera is HMS *Queen Elizabeth*. In a near-starving Germany the poster (right) appeals to citizens to collect fruit stones to be crushed for 'ersatz' oil production. A list of collection points is attached.

[34]

Despite all efforts, Germany succeeded in meeting only 85 per cent of peace-time oil consumption, with lube oil supplies attaining no more than two-thirds of requirements.[87] This put an effective check on mechanizing the army. The German army had no more than 45 battle tanks in 1918, its 40,000 trucks faced an Allied total of 200,000, and even then gasoline shortages severely curtailed the use of the trucks.[88] General Ludendorff's desperate March 1918 offensive failed to secure a breakthrough because the army did not have sufficient trucks to execute the required movements of men and supplies.[89] Fuel allotments to the air force dropped as the number of front line aircraft rose. In the spring of 1917, the Germans had 2,270 aircraft in France and could dispose of 11,000 tons of fuel a month. For his offensive, Ludendorff could field 3,600 aircraft supplied with 7,000 tons of aviation fuel a month, to attack Allied forces of 4,500 aircraft and 31,000 tons of gasoline a month. By June German gasoline supplies had dropped to less than 5,000 tons, restricting their squadrons of 12 to 15 fighters to a total of ten flights a day.[90]

The Allies' wealth of gasoline thus boosted their small numerical superiority by the crucial factor of being able to fly many more sorties. Moreover, the German shortage of suitable lubricants severely curtailed the use of the simple and effective rotary aircraft engine, leading to the early phasing out of the famous Fokker Dr 1 triplane, the Red Baron's favourite aircraft, which had it.[91] German army logistics depended on the railways and on horses to a far greater degree than the Allied armies. The railway system simply could not cope with the demands of prolonged warfare. Though possessing huge reserves of its own and in the captured mines in Belgium and France, Germany developed very severe coal shortages early during the war, since stocks could not be moved from the pit heads. As a result, supplies even dropped 15 per cent below peacetime levels, and industrial production of vital war equipment such as aircraft fell.[92] Horses also became very scarce, in 1918 forcing the German High Command to hold on to the Ukraine as a vital supplier of animals when logistic sense demanded a withdrawal.[93] Combined with the inability to mechanize farming, these shortages resulted in falling food supplies from 1916, ending in the near famine conditions which forced Germany to give up. Oil thus had a decisive impact on the course of the First World War. The Allied camp and the Central Powers occupied diametrically opposed positions, which translated into completely different policies: managing a glut on one hand, juggling to make ends meet on the other. The Group found itself caught between these extremes, exerting great efforts in generating the supplies to meet urgent demand, whilst facing severe constraints in production.

The rising spectre of nationalization

In November 1914, Deterding wrote to Loudon forecasting a glorious future for the Group. The war might lead to political change in Russia and Romania, he thought, but the overall economic outlook was very favourable, especially in Russia. The war demonstrated just how important oil already was, and it could only become even more so. Now that most horses had been requisitioned and vans cost as little as 500 dollars in the US, shopkeepers and delivery firms would all switch to motorcars, the more so since rearing horses to replace war casualties would take five years. If only politicians could be dissuaded from obstructing business by imposing heavy taxation.[94] Deterding's vision about motorcars driving the Group's prosperity came true during the 1920s, but his optimism about Romania and Russia proved to be entirely misplaced. The war profoundly changed both political and economic conditions there, leading to the extension of state controls over the oil industry in the former, and to the nationalization in the latter, the first instances of what would become a recurrent plague to the industry.

With the Dardanelles closed, the Group could no longer obtain supplies from either Romania or Russia, a serious handicap since Asiatic used to draw about 450,000 tons before the war,

[35]

The 'Red' Baron von Richthofen (1882-1918) stands with a colleague in front of his trademark red tri-plane (left), epitomizing the way of wars to come. However, despite the mechanization of warfare, during the First World War all armies fighting in Europe depended to a greater or lesser extent on horses. General Sir Douglas Haig, a cavalry officer and commander-in-chief of the British Expeditionary Force, believed strongly – but wrongly – in the ability of cavalry to break through the German lines. In the remarkable main photograph, the British cavalry seen in action in 1916 were probably among those used in the early stages of the Battle of the Somme. At far right, Canadian soldiers are seen watering their horses in 1915. Germany also used cavalry but depended much more than the Allies on horse transport, such as this supply caravan (right) resting near the Belgian town of Mouland, 1914.

[36]

[37]

[38]

mainly gasoline and fuel oil imported into Britain and France.[95] Immediately following the outbreak of war, the Romanian Government took control of oil exports in what was to be the first step in a gradual process of extending the state's grip on the industry. Until August 1916 Romania remained neutral and, like the Netherlands, claimed the right to sell to both sides. Under guidance from The Hague, Astra scrupulously sought to maintain that line, the British envoy in Bucharest, with whom Astra general manager H. Jacobson kept in close contact, accepting that this meant the company would sell to the Central Powers.[96] Logistic bottlenecks greatly reduced the scope for exports, however. During the autumn of 1914 the Allies bought up large stocks of gasoline, but they could no longer be shipped out. Nor could Austria-Hungary or Germany obtain large volumes of oil, for Serbia kept the Danube closed and the railways could not handle the tonnage formerly shipped by barge. Consequently, Romanian exports fell dramatically. Stocks built up, so companies had to run down first drilling and then production. The Government took the economic emergency as an opportunity to introduce legislation to force companies to comply with official policy, if necessary by seizing installations and products.[97] The situation then eased gradually, enabling exports to Germany to recover and rise past their 1913 level, and opened up after Serbia's capitulation in November 1915, which removed the blockade of the Danube.[98] Rising prices pushed up Astra's profits, from 13.5 million Lei (305,000 pounds) in 1913 to 23.5 million (530,000 pounds) in 1914 and 32.5 million (730,000 pounds) in 1915.[99]

When Romania finally entered the war on the Allied side in 1916, the Government sequestrated Astra's assets as enemy property. This decision was rooted less in any consideration of economic warfare than in the Government's determination to gain control over the oil industry. Astra, the country's biggest producer, was a fully Romanian company. Four-fifths of its shares were held by the Group, only one-fifth remained in German hands, and no more than two out of fifteen directors were Germans, so it took a little imagination to brand Astra an enemy company. The minister responsible, Vintila Bratianu, made no bones about the real reason for the seizure: Astra's longstanding and firm resistance to official policy.[100] Soon the course of events imposed different priorities, however. Germany, intent on capturing vital stocks of oil and cereals, attacked Romania and its army gained rapid headway. The Allies now began to exert pressure on the Romanian Government to destroy oil installations and stocks, Britain sending over a special task force for assistance. Anglo-Saxon instructed Jacobson by telegram via the Foreign Office to collaborate with the authorities and burn the stocks, though omitting to mention the demolition of field installations and equipment.[101] The overall damage inflicted on the Romanian oil industry by the destructive actions

[39]

amounted to about five months' worth of production.[102] Astra later claimed damages of 252 million lei, about £ 5.7 million.[103] A special section of the German Army Command directed the rebuilding of installations with the object of restoring production at all possible speed. By the end of 1917, Astra was back to 60 per cent of its normal volume.[104] In 1919 the Group successfully challenged Astra's sequestration under Romanian law and regained control of the company, but the continuing state involvement with the industry seriously restricted its freedom of action.[105]

In Russia the war had also reinforced calls for state control of the oil industry, notably to solve the acute shortage of liquid fuel

This military band marching through Bucharest (left) heralded Romania's declaration of war against Austria in 1916, by which time Shell people had severely damaged the Romanian oilfields (above) in an attempt to deny them to the enemy.

[40]

03|182

[41]

[42]

As president of the Third International in 1919, the Russian politician Grigori Evseevitch Zinoviev (1883-1936) addresses workers in the Baku oilfields (above). The previous year, in the upheavals of revolutions, the same fields had been set ablaze (main picture).

for industry and the railways and to combat alleged abuses by the main producers, such as production restraints to keep prices high.[106] In 1914 the Group had finally completed the reorganization of its Russian companies so as to fit the mould of other operating companies, with proper administrative controls and standard reporting procedures to the relevant departments at central offices, and the appointment of a chief engineer in Baku and a chief geologist in Grozny.[107] The war imposed serious handicaps in the form of transport disruptions, distribution bottlenecks, material and manpower shortages, and labour unrest, but even so the production of Group companies rose steadily, from about one million tons in 1913 to almost 1.6 million the following year. Profits were marginal, however; Bataafsche earned around 6 per cent dividend on the four companies which it controlled.[108] Since the profits could not be sent over, the Group used these frozen rubles to expand and consolidate its operations, centralizing the management of all companies in the marketing company, Société Mazout.[109]

Following the October Revolution of 1917 the Caucasian oil regions descended into a vicious spiral of violence driven by political, ethnic, and religious strife. Both main cities Grozny and Baku changed hands several times, Bolshevik Soviets (councils) establishing themselves there only to be ousted again by one or other of the groups fighting for control of the region: Armenians, Tartars, Kazakhs, Chechens, Turkish troops, even a small British force sent in to deny Baku to the Central Powers. On 20 June 1918 the fledgling Bolshevik regime in Moscow summarily nationalized the country's oil industry, which was at that point entirely outside its grasp. Supported by the Western Allies, a Russian general had turned against the Bolshevik Revolution and taken possession of the Caucasus, supporting the formation of new, nationalist governments and helping with the reconstruction of the oil industry. Cut off from the Russian market, the oil companies now had to find other outlets; in July 1919, the Group, Nobel, Jersey Standard, and a Russian company set up a joint export agency in Batum.[110] The counter-revolutionary movement quickly lost its momentum, however. The Red Army moved in to re-establish Moscow's power, taking Grozny in February 1920, Baku in April, Batum in July, enforcing the nationalization of the oil industry.

With the nationalization of its Russian companies the Group lost about one-third of its total crude production, a grievous loss under any circumstances, plus a very substantial investment. The shares in fifteen Russian companies were worth £ 8.1 million, about 97 million guilders. However, the assets lost were estimated at £ 33 million (396 million guilders).[111] Deterding took it as a personal affront. As years went by the experience turned him into a fervent anti-Communist, which brought serious consequences during the 1930s.

Keeping pace with demand Partly as a consequence of losing its Russian and Romanian oil, the Group could not, unlike its competitors, increase its production to match Allied demand. Crude production remained roughly stable at around 4.5 million tons during the war, peaking at 5.1 million in 1916 as the fast-rising output of the US compensated the declining production in Russia and Romania. The Group's total supplies will have been higher, but we do not know by how much. Already before the war the Group purchased some 20,000 tons of oil products a year in the US for export to the Asian market.[112] During the war a special office was set up in New York under N. G. M. Luykx to buy oil for shipment to Europe, some of the supplies coming from Roxana.[113] The volume is likely to have been considerable. Oil imports from the US into Britain and France rose from 2 million tons in 1914 to 4.7 million in 1918.[114] Assuming that the Group succeeded in maintaining an overall market share of 30 per cent in these two countries, then we would need to raise its total supplies by 800,000 tons in 1918, nearly twice the entire output of Astra Romana in 1914. This tallies with Asiatic's handling of 780,000 tons of 'outside oil' in 1918.[115]

Jersey Standard's supplies showed a much greater stretch, however. Jersey's annual refinery runs, for that company a better gauge for supplies since it bought most of its crude, rose from 38 million barrels or 5.4 million tons in 1914 to almost 60 million barrels, 8.5 million tons, in 1918. Of the Group's British rivals, Anglo-Persian more than tripled its production from 270,000 tons to almost 900,000, whereas Burmah enjoyed a moderate increase from 900,000 tons to almost 1 million.[116]

The Group's fundamental challenge during the war thus centred on finding the supplies to maintain its competitive position. Two approaches were adopted: vigorously expanding production, and developing new technology to increase refinery yields. Unfortunately a marked lack of data prevents us from tracking the Group's performance during these years in sufficient detail. In their annual accounts, the two holding companies summarized assets of Anglo-Saxon, Bataafsche, and selected other operating companies, but gave no operating data other than net profits, and said nothing at all about Asiatic. The three main operating companies did not publish annual reports and hardly anything has survived of the internal reports which the various boards and departments concerned must have received for all companies. The board minutes for such a key company as Bataafsche show large gaps from March 1915 until January 1918, rendering it impossible to follow managerial decisions at close range, all the more frustrating in the absence of formal accounts and detailed production data. Any analysis must therefore remain approximate and impressionistic.

About the operations in Germany, for instance, the Bataafsche records offer little more than a tantalizing glimpse of production and results. In addition to the Reisholz works, Rhenania possessed gasoline refineries in Hamburg and Regensburg. After the breakup of the cartel arrangements with the Vereinigte Benzinfabriken in 1913, the company had erected five bulk installations in strategic locations to supply the entire German market from its refineries. Having perfected the refining of heavy gasoline into sharp fractions, Rhenania produced both gasoline and diesel fuel. In 1912 manager Späth finally won his campaign to obtain a reduction of the special diesel duty, which boosted sales considerably.[117] Lube oil operations began in 1913, when a separate company was set up at Monheim to process the type of heavy oils from the Dutch East Indies for which Pyzel developed the double-bottom transport system.[118] The two Group subsidiaries were of crucial importance to the German war effort, arguably more so than the Jersey subsidiary Deutsch-Amerikanische Petroleum Gesellschaft (DAPG) which, a distributor rather than a manufacturer, became the official agency for the import and distribution of oil products.[119] Despite the suspension of overseas imports, Rudeloff and the two factories succeeded in expanding operations. The 1915 results were reported as very much better than the 1914 ones, and in 1916 the Bataafsche board approved 8 million marks' worth of new investment, probably destined for lube oil production.

During those years the Reisholz gasoline works regularly paid out 100 per cent dividends on its admittedly low capital of 840,000 marks. In 1917 the gasoline refinery, the lube oil works and the various separate bulk installations were merged into the

Mineralölwerke Rhenania AG with a capital of 15 million Reichsmarks, a sum designed to lower the huge dividends by spreading them over a larger capital base.[120] In 1914 a pilot installation at Monheim had shown the great benefits of applying the Edeleanu extraction process to lube oil manufacturing. This improved both the yield of the product, by as much as 20-40 per cent, and its quality. The washing with sulphur dioxide helped to remove contaminants, improved the viscosity index, that is the lube oil's resistance to flow, and flattened the viscosity curve, meaning the oil retained its resistance to flow across a wider temperature range. A full-scale plant probably came on stream in July 1915, using residue from Galicia and Romania and possibly heavy oil in barrels imported from the Netherlands. The company's pioneering work during the war was to put it at the centre of the Group's lube oil business during the 1920s.[121] Regular visits from Colijn and others ensured Bataafsche's grip on the German operations. Rhenania showed rapid growth until Germany's collapse in November 1918.[122]

Returning to the search for supplies, the Group's attempts to increase production from existing wells and concessions had only limited success. Two areas which the Group had recently entered with substantial investments, Mexico and Venezuela, failed to live up to their great promise, in the former case because of civil war, in the latter because of logistic difficulties, disputes about concessions and, when those had been resolved towards the end of the war, mounting shortages of drilling equipment. The building of the Curaçao refinery, a site chosen for its central location for processing Venezuelan crude, also suffered from long delays in deliveries of essential equipment from the US. When the installations were nearly ready, it proved impossible to obtain sufficient transport, leading to negotiations with the British Government to take crude oil delivery off Lake Maracaibo in Venezuela rather than liquid fuel supplies from Curaçao.[123] Production in Egypt, mostly heavy oil, very suitable for liquid fuel and ideally located for supplies to international shipping and military forces around the Mediterranean, achieved a peak of over 275,000 tons in 1917, but remained notoriously unpredictable for technical reasons.

Since negotiations with the Dutch Government over Jambi remained in deadlock, no new concessions could be obtained in the Dutch East Indies. Bataafsche managed to raise crude output from 1.5 million tons in 1914 to 1.7 million in 1918, to produce kerosene, gasoline, and liquid fuel in a ratio of roughly 35:25:25. After more than four years of hard work, the Edeleanu factory at Balik Papan continued to pose major technical problems, restricting production in Kalimantan, because without treatment stocks of unsaleable kerosene piled up. The installation finally came on stream in March 1916, leading to an immediate jump in kerosene output.[124] Better fractionating helped to increase the gasoline yield, but a substantial switch was undesirable as the corresponding drop in kerosene output would have made it more difficult to sustain the Group's Asian market position, which already required imports from the US. By 1918 Bataafsche consigned more gasoline than kerosene to Asiatic, and Asiatic sold more kerosene from outside suppliers, i.e. from the States, than from Bataafsche. Following detailed market research, Bataafsche also reorganized lube oil production. Until 1915 all factories in the Dutch East Indies used different specifications, different cost–price accounting methods, and even different packaging for lube oil. Following a general managers' conference, production was largely concentrated at Balik Papan using base stocks shipped in from the other factories in the area. To support sales, a Bataafsche liaison manager was appointed at Asiatic's Singapore office, and the Balik Papan manager Dubourq sent detailed charts with specifications and cost calculations to London in 1917. Two years later Asiatic already sold 365,000 pounds worth of products, 7 per cent of turnover.[125]

Only the US provided the scope for expansion which the Group so badly needed. At the outbreak of war the Californian operations had just been renamed the Shell Company of California Inc. Within the space of two years, the company had moved from selling Sumatra gasoline along the western seaboard into local production by acquiring California Oilfields for 13 million dollars. This deal gave Shell California an annual production of 4.4 million barrels, almost 630,000 tons, in the Coalinga field.[126] More or less at the same time the Group moved into the mid-continent fields.

[43]

It looks like little more than a garden shed, but in 1916 this was Shell's first experimental refinery and pilot plant at Martinez in California.

By 1914 the Roxana Petroleum Company had a modest production of 500,000 barrels, about 75,000 tons, in Oklahoma. The following year Roxana acquired extensive concessions in the Cushing and Healdton fields, and after only seven months, by December 1915 the company produced 18,000 barrels a day (600,000 tons a year), comfortably exceeding Deterding's target of 15,000 barrels within three years.[127] The points of entry differed, California starting with trade and Oklahoma with production, but the intentions were the same, i.e. to build an integrated oil business from wellhead to consumer. To that end both companies quickly moved into transport and refining. Shell California built the Valley Pipe Line to feed a

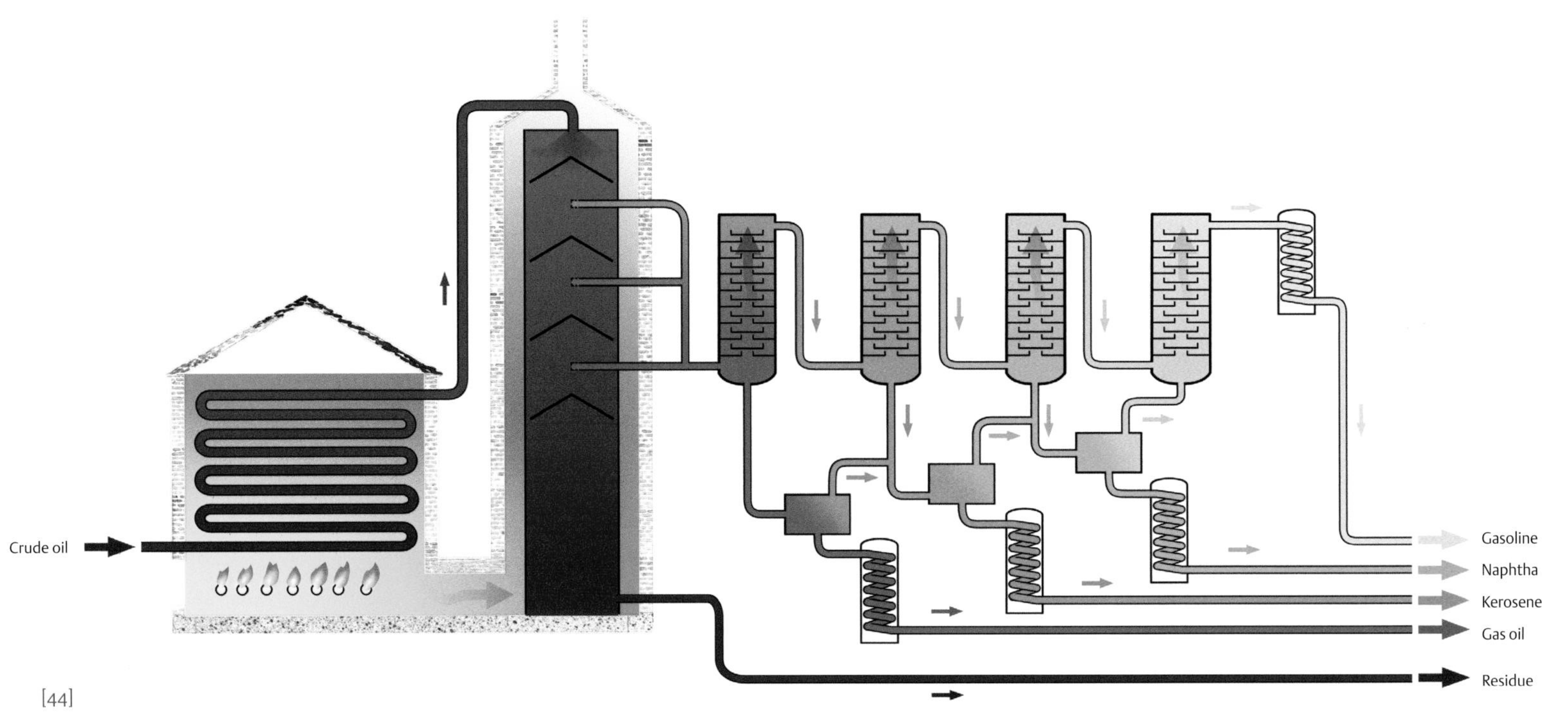

[44]

projected refinery at Martinez. Roxana set up refineries at Cushing and St Louis, and a bulk installation in New Orleans for shipping products to Europe. By 1918, the US had become the Group's second biggest producing area with 9.5 million barrels or 1.4 million tons, eclipsing Russia, where production had collapsed following the Bolshevik revolution and foreign invasions of the Caucasian oilfields, and closing in on the 1.7 million tons produced by the Dutch East Indies. The Group now produced more crude oil in the US than Jersey Standard, 27,500 barrels a day against 18,783 (1.4 million tons to 900,000), but then Standard had always relied more on buying oil rather than on producing it, so its revenues of 454.8 million dollars in 1919 dwarfed those of Roxana and Shell California, a total of 23.7 million dollars.[128]

Coming at a time when the Group badly needed production, the US operations were a heaven-sent bonanza, not just because of the new oil supplies which, in addition to oil purchases in the States, helped Asiatic to sustain the markets previously supplied from Russia and Romania. With Shell California and Roxana, the Group established itself firmly in a huge market at a very favourable moment, just before the comparatively low prices gave way to a sustained rise. By March 1916 Roxana made net profits of £ 50,000 a week, so Lane estimated that the initial outlay on concessions and equipment had been recouped in just five weeks of full operations.[129] Moreover, the Group now also gained an intimate knowledge of conditions in the US, the world's dominant market and, with almost two-thirds of the total, its biggest producer by far, enabling the fine tuning of Group operations elsewhere.[130] Thus Shell California and Roxana marked the Group's transformation from a multinational corporation into a global concern. Finally, the Group obtained access to new technology. Its first acquisition, Milon J. Trumble's refining system, was a resounding technical and commercial success.

Trumble developed his system as a topping plant, i.e. an installation to remove the light fractions or 'tops' from the crude so as to make liquid fuel, capable of dealing with the high water content common in Californian oils. Preheated crude oil was pumped through pipes winding upwards over a furnace to enter the top of a column which Trumble called the evaporator, where

[45]

[46]

The purchase of the rights to the Trumble process was a major investment for the Group. The diagram (above left) explains the principles involved. The massive brick construction (above) is the Group's original plant built in California in 1914, and the sleeker unit (above right) is its first UK installation, erected at Shell Haven in the 1920s.

vapours of water and the light fractions separated from the heavy ones, which were drawn off at the bottom. From the evaporator, the light fractions went into separators redistilling them into three fractions: kerosene, diesel fuel, and gasoline. To complete the system, Trumble added a converter for 'cracking', that is, applying a mixture of heat and pressure to rearrange the molecules of heavy oils and turn them into lighter fractions. Hot residue and the heat from vapours were used to preheat the crude running into the system, thus saving fuel.[131] Patented in 1910, the Trumble system heralded a new era in refinery design. It combined the virtues of ingenuity with simplicity and great economies. Trumble installations were far cheaper than conventional refineries with rows of numerous large battery stills, using 25-40 per cent less space, costing 50 per cent less to build, and consuming less fuel per unit of crude.[132] They were also less prone to the breakdowns through clogged pipes and the formation of coke which dogged common stills. The fractionation was better and easier to adjust to changing product specifications, whilst notably the residue had a very even quality rendering it ideally suited for liquid fuel or as stock for lubricating oils. It was this latter aspect which caught the attention of Shell California managers occupied with planning the Martinez refinery. In July 1914, Anglo-Saxon bought a licence to use the Trumble patent in the US, which was extended to the UK and its colonies three months later, after a guarantee had been obtained that the installation would produce liquid fuel to Admiralty specifications. The board immediately began considering building Trumbles in Mexico, Dublin, and Thameshaven, but only the last one was built, on an adjacent site renamed Shell Haven.[133]

Trumble's original design possessed far more potential than its inventor appears to have realized. When building began at Martinez, the Group sent over its chief engineer, Dan Pyzel, as supervisor. Responsible for the design and building of all the Group's refineries and installations, he had also pioneered the scheme for carrying oil in double bottoms.[134] On studying the Trumble plans at Martinez, Pyzel found that with some modifications the concept would be revolutionary rather than just innovative. By replacing the separator with a series of dephlegmators or fractionating columns, which Pyzel had built at Balik Papan and at other Group refineries, the system would also give a closer separation of the higher oil fractions. Rather than boiling them off, as in bench stills, these fractions were fed from the evaporator into successive dephlegmators, each cooled to the temperature at which the heaviest fraction condensed, resulting in much better control over the final products.[135] Modifying the design made no sense, however, for under the terms of the licence such alterations had to be shared with the Trumble Refining Company, the patent holders. In February 1915 William Meischke Smith, who had become Shell California's president after having run the Group's operations in China, wrote to Anglo-Saxon proposing a takeover of Trumble Refining. Deterding hesitated, but Pyzel's detailed and enthusiastic description and Meiscke Smith's arguments convinced him, and in April 1915 the company and its patents were bought for one million dollars in cash.[136] Shell California immediately built a modified Trumble at Coalinga, opening markets for both gasoline and liquid fuel which the company could not otherwise have served. In May 1915 Anglo-Saxon ordered another four Trumble units for the US.[137] By 1916 the Group had Trumbles running at eleven refineries with a total capacity of 2.7 million tons, or equal to about half of its entire production, which according to Pyzel had already generated savings equal to the amount spent on buying the patent. Discussions with the German Government over building an installation at Monheim for treating Romanian crude ended inconclusively in March 1918, presumably because of material shortages.[138]

To keep the flow running

The increasing tanker shortage of course had a serious impact on the Group's operations, much more reliant on global supply lines than, for instance, those of Jersey Standard, which drew most of its oil from the US. In May 1915 75,000 tons out of the Group's 256,000 DWT fleet, almost a third, were either requisitioned by the British Admiralty or on voluntary charters to the Government.[139] Group managers emphasized this shortage at every opportunity as a serious operational handicap. The Dutch Government discovered this during the negotiations over toluol supplies, when Bataafsche had claimed to have insufficient capacity for transporting the small volumes of straight run required and refused to arrange transport.[140] Waley Cohen complained vociferously time and again about the Group suffering an unfair disadvantage in having to cope with severe logistic constraints whilst US oil companies squandered precious sea miles on needless trips and detours.[141]

However, the tanker shortage became a pressing problem for the Group only during the last eighteen months or so of the war. Initially Anglo-Saxon even had capacity to spare. During 1913-14 Asiatic sold some 140,000 tons a year, enough to keep three or four 10,000 ton tankers busy, of products from Russia and Romania, termed 'outside oil' as coming from outside the 'Red Area', the Asian market as defined by the map attached to the 1902/1907 Agreements.[142] The closing of the Dardanelles freed this capacity for other purposes, so the Group offered tankers in charter to the British government at 'blue-book rates', which just about covered the running costs. Moreover, for a long time the Group's marine managers successfully responded to the logistical challenges in various ways, primarily of course by chartering or buying tankers as and when required. Up to January 1918, Anglo-Saxon regularly received offers from shipyards and brokers like Lane & Macandrew for tankers to buy. The board accepted some, and declined others, always on the basis of price and anticipated returns. Between January 1916 and July 1918, Anglo-Saxon lost 62,186 tons from its own fleet to enemy action, but by purchase and building succeeded in adding 112,075 tons, including three sailing ships, for a net gain of almost 50,000 tons.[143]

[47]

An excursion to view the new Panama Canal, 1914.

The Group also occasionally chartered Jersey Standard ships to carry gasoline to the UK on their homeward run from Asia, took tankers on time charter from the Nederlandsch-Indische Tankstoomboot Maatschappij during the slack months in the company's marked seasonal shipping pattern, and chartered ships from Group companies such as Corona. Optimizing shipping schedules and the routeing of supplies, and cutting turnaround times, also helped to maximize the use of available capacity. With the opening of the Panama Canal in August 1914 the Group inaugurated a round-the-world trip, from the Dutch East Indies to Europe with liquid fuel or gasoline, then to the US Gulf Coast to load kerosene for, say, China, and then back to Balik Papan or Pangkalan Brandan.

[48]

[49]

To assist such voyages fuel installations were built at both ends of the canal. One of four new tankers using diesel engines made this trip of 27,000 miles in 162 days, the fuel economy enabling the captain to sail from the Gulf of Mexico via China to Singapore without bunkering.[144] During 1916 and 1917 three new 8,000 ton tankers, originally destined for Shell California and built at a shipyard in Wilmington (US), were chartered out at very profitable rates to the French refiners Les Fils de A. Deutsch & Co. to carry oil from the US to France. Meanwhile Shell California helped itself out by chartering local transport and expanding storage. Clearly these tankers were surplus to the Group's own immediate requirements.[145]

Seen above with a non-Shell ship in 1915, the Panama Canal was opened for navigation in August 1914. As at Suez in 1892, a Shell tanker, Eburna, was [....] engines were not used.

The complaints about tanker capacity originated rather in the pressing need to sustain competitive positions, in the rising demand being for particular oil products only, and in the frustration over lacking the means for a commensurate response. The war effort required enormous supplies of liquid fuel, gasoline, and straight-run, products which formed only a fraction of the crude from which they were made. As more and more tankers were used for conveying the fractions, transporting the bulk remainder became increasingly difficult, resulting in unwanted expensive stocks and the undermining of market positions. Rivals such as Jersey Standard had fewer constraints. The company's German fleet alone had counted some 40 tankers which were reallocated to other tasks, enabling Jersey to increase its UK market share, adding to Waley Cohen's vexations. On the other hand, Jersey's Asian fleet, domiciled in Hong Kong and flying the British flag, was requisitioned by Britain. In at least one case the Group succeeded in using one of these ships to supply its Hong Kong bulk station, thus angering Jersey Standard.[146] It was the struggle over market share, too, which led Waley Cohen to propose a Petroleum Pool in 1915, when oil imports from the Dutch East Indies into Britain peaked at almost 300,000 tons, up from 112,000 tons in 1913. Freezing this very favourable status quo by a pooling scheme would have greatly benefitted the Group, for it could not sustain imports from the Dutch East Indies, which subsequently dropped to 127,000 tons in 1918, as shipping capacity dictated the redirection of gasoline to France and to other destinations. The authorities rejected Waley Cohen's suggestion, which became practicable and necessary only two years later.[147]

The practice of transporting oil in the 'double bottoms', in reality the ballast tanks, of ocean liners also added to the Group's overall shipping capacity. We do not know the full quantities of oil

[50]

Worse things happen at sea: unrestricted submarine warfare threatened all shipping on the Atlantic, regardless of nationality.

[51]

By the spring of 1918 wheeled aeroplanes were capable not only of taking off from but also of landing back on the deck of a ship, and British naval air power was becoming a useful factor in the war. Left, a Sopwith Pup aeroplane is winched up from the hold of an aircraft carrier, ready for take-off (above right) from the forward flight deck; and (right) a scouting airship (SSZ 59: Submarine Scout Zero 59, launched in April 1918) lands on the after-deck.

[52]

[53]

transported by the Group in this way before the scheme was offered to the Admiralty, nor the shipping lines concerned. It was a rather sensitive issue, if only because the oil did not figure as cargo on bills of lading and was smuggled through the Suez Canal.[148] Bataafsche had an agreement with the Stoomvaartmaatschappij Nederland since January 1914 to transport oil in this way from the Dutch East Indies to the Netherlands. Presumably during the war some oil continued to be shipped along this route in the ballast tanks of Dutch liners, circumventing the naval blockade. Owning up to it could therefore have compromised the Group very seriously indeed, but this liability was ingeniously turned into an asset by offering the concept to the Admiralty for public use, with full engineering support from Zulver and his staff at Anglo-Saxon's Marine Department. The Admiralty ordered some trials in 1915 but then hesitated, presumably because conveying oil in this way reduced a ship's overall carrying capacity by an equal weight and imposed specific demands on routeing schedules.[149] The scheme was finally adopted in June 1917. Zulver acted as technical supervisor, Anglo-Saxon received a £ 3,000 annual fee plus Zulver's salary and costs. By November 1917 400 ships had been converted and 500,000 tons of liquid fuel carried; six months later, 700 ships carried 120,000 tons a month, for a total of 1 million tons by the war's end.[150] More or less simultaneously, Anglo-Saxon proposed another Zulver plan to create capacity, i.e. by converting general freighters into oil carriers. The Admiralty adopted it in August 1916, leaving all details of management and execution to Anglo-Saxon. By September six ships had been bought for conversion, with five more under offer. Officials clearly appreciated Anglo-Saxon's expertise and entrusted a growing number of ships to its management. Running Admiralty tankers increased the Shipping Department's workload, just when the administration of the Nederlandsch-Indische Tankstoomboot-Maatschappij had been returned to The Hague to ease the pressure. Finally, close cooperation with the Admiralty and subsequently the Ministry for Shipping put the Group in an ideal position when, in 1919, the British Government began to sell off surplus tankers.[151]

Thus the tanker capacity was less a handicap on the Group's overall operations than a restraint on growth, which might otherwise have followed booming demand more closely. The shipping shortage also drove a wedge between the operating companies, however. As early as 1915 the rapidly rising market freight rates undermined the set prices adopted in the 1907 agreements for internal settlements between Anglo-Saxon and Asiatic, contributing to a growing conflict of interests between the Asiatic shareholders.

[54]

ST. HELEN'S COURT BULLETIN. 3

THE LOSS OF S.S. "ARCA."

ON this page we reproduce the photograph of Captain Welch and the Officers of the s.s. "Arca," which was torpedoed on her way home last October, and sunk with the loss of several lives, including the officers and engineers of the vessel.

Captain Welch entered the service of M. Samuel & Co. in October, 1895, and in a very short time was given command of one of their vessels. He was taken over by the Anglo-Saxon Co. in 1907, and had been in charge of various of their vessels since that date down to the time of his death. He was a valued and trusted servant of the Company, who suffer a very severe loss by his sudden end.

The Chief Officer Mr. E. H. Medland, and Second Officer Mr. C. W. Barlow, entered the Company's service early in 1917; Third Officer L. J. Rae joined in April, 1918; Chief Engineer Alfred Mavin, of whom we regret we have not at the moment a photograph, entered the Company's service in August, 1915; whilst Second Engineer Corbett Crosswell joined the Company in April, 1914; and J. M. Clarke and W. M. Pugh, Third and Fourth Engineers, in the Spring of 1918.

Notwithstanding the short length of time that the junior officers and engineers were in the Company's service they proved themselves most worthy and promising officers, and it is the Company's misfortune to lose such valuable servants in such a way.

E. H. MEDLAND (Chief Officer).

L. J. RAE (3rd Officer).

C. W. BARLOW (2nd Officer).

CAPTAIN WELCH.

CORBETT CROSSWELL (2nd Engineer).

J. M. CLARKE (3rd Engineer).

W. M. PUGH (4th Engineer).

[55] [56]

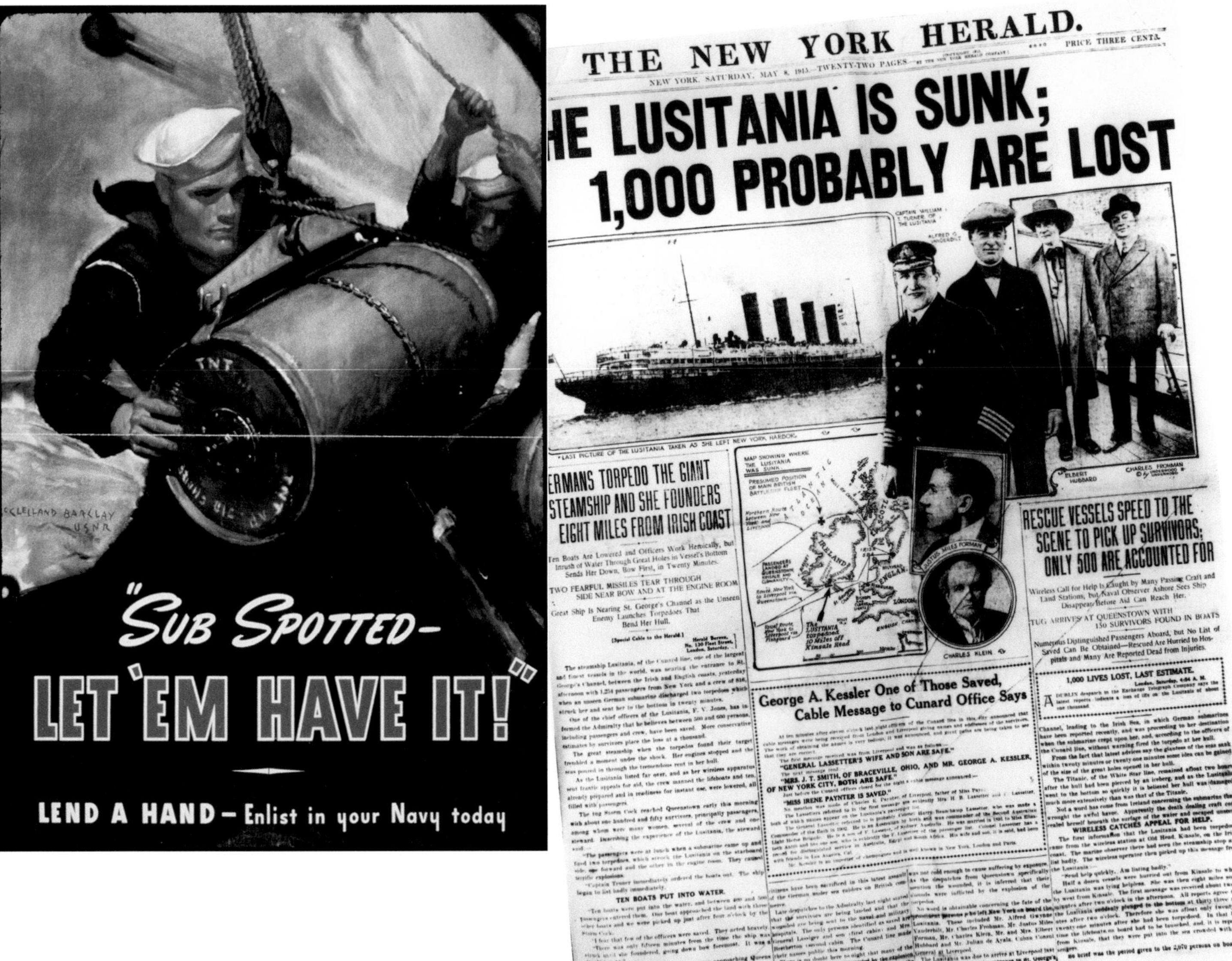

THE NEW YORK HERALD.

NEW YORK, SATURDAY, MAY 8, 1915.—TWENTY-TWO PAGES — PRICE THREE CENTS.

HE LUSITANIA IS SUNK; 1,000 PROBABLY ARE LOST

CAPTAIN WILLIAM T. TURNER, OF THE LUSITANIA — ALFRED G. VANDERBILT — ELBERT HUBBARD — CHARLES FROHMAN

"LAST PICTURE OF THE LUSITANIA TAKEN AS SHE LEFT NEW YORK HARBOR

MAP SHOWING WHERE THE LUSITANIA WAS SUNK

The LUSITANIA torpedoed 10 Miles off Kinsale Head

JUSTUS MILES FORMAN

CHARLES KLEIN

ERMANS TORPEDO THE GIANT STEAMSHIP AND SHE FOUNDERS EIGHT MILES FROM IRISH COAST

Ten Boats Are Lowered and Officers Work Heroically, but Inrush of Water Through Great Holes in Vessel's Bottom Sends Her Down, Bow First, in Twenty Minutes.

TWO FEARFUL MISSILES TEAR THROUGH SIDE NEAR BOW AND AT THE ENGINE ROOM

Great Ship Is Nearing St. George's Channel as the Unseen Enemy Launches Torpedoes That Bend Her Hull.

[Special Cable to the Herald.]

Herald Bureau, No. 130 Fleet Street, London, Saturday.

The steamship Lusitania, of the Cunard line, one of the largest and finest vessels in the world, was nearing the entrance to St. George's Channel, between the Irish and English coasts, yesterday afternoon with 1,254 passengers from New York and a crew of 816, when an unseen German submarine discharged two torpedoes which struck her and sent her to the bottom in twenty minutes.

TEN BOATS PUT INTO WATER.

George A. Kessler One of Those Saved, Cable Message to Cunard Office Says

"GENERAL LASSETTER'S WIFE AND SON ARE SAFE."

"MRS. J. T. SMITH, OF BRACEVILLE, OHIO, AND MR. GEORGE A. KESSLER, OF NEW YORK CITY, BOTH ARE SAFE."

"MISS IRENE PAYNTER IS SAVED."

RESCUE VESSELS SPEED TO THE SCENE TO PICK UP SURVIVORS; ONLY 500 ARE ACCOUNTED FOR

Wireless Call for Help Is Caught by Many Passing Craft and Land Stations, but Naval Observer Ashore Sees Ship Disappear Before Aid Can Reach Her.

TUG ARRIVES AT QUEENSTOWN WITH 150 SURVIVORS FOUND IN BOATS

Numerous Distinguished Passengers Aboard, but No List of Saved Can Be Obtained—Rescued Are Hurried to Hospitals and Many Are Reported Dead from Injuries.

1,000 LIVES LOST, LAST ESTIMATE.

London, Saturday, 4:04 A. M.

A DUBLIN despatch to the Exchange Telegraph Company says the latest reports indicate a loss of life on the Lusitania of about one thousand.

WIRELESS CATCHES APPEAL FOR HELP.

The first issue of *St Helen's Court Bulletin* appeared in November 1914. By then several hundred British Shell employees had 'joined the colours', and the magazine was created by their colleagues to keep them in touch with home life. It also paid tribute to those who had fallen, including those from SS *Arca*, the last of Anglo-Saxon's tankers to be sunk in the war (left), just nine days before the Armistice. Her entire crew of 52 officers and men died. Germany's policy of unrestricted U-boat warfare, with victims such as the Cunard liner SS *Lusitania* in 1915 (above right), encouraged the US to enter the war on Britain's side. In this American poster from 1918 (above left) a sailor prepares to launch a depth-charge filled with TNT.

Going separate ways The Asiatic contract gave the Group an option to buy out the Paris Rothschild interests at par in 1923 when, in accordance with the 1902 contract, the company would have to be disbanded on the expiry of its 21-year lifespan.[152] There was thus a natural end to the partnership, which was becoming somewhat frayed anyway as the commonality of interests which formed Asiatic's *raison d'être* began to disintegrate. The sale of Bnito and Mazout to the Group turned the Rothschilds from producers into investors. Moreover, the Group's shifting centre of gravity towards the western hemisphere and away from the 'Red Area' undermined Asiatic's position as the core marketing company. Asiatic's profits almost quadrupled from 975,180 pounds in 1913 to 3.7 million in 1918, but its sales volume stagnated at around 1.5 million tons. At the same time the Group's US production doubled from 700,000 to 1.4 million tons which, though handled by the Group, remained entirely outside Asiatic's remit.

War drove a wedge between the partners in Asiatic by creating a succession of contentious issues, beginning with the exclusion of Germany, Belgium, and the Netherlands from the 'B' Agreement with Bataafsche in August 1914. A month later gasoline sales in the UK suffered a blow with the sequestration of British Petroleum, the joint venture with Nobel and EPU which ran this business. BP's publicized German connection caused private sales to fall, but this did not prevent the company from supplying gasoline to the British Expeditionary Force in France.[153] Bataafsche immediately threatened to withdraw UK gasoline consignments from the 'B' Agreement with Asiatic.[154] In August 1915, the Bataafsche board even suggested abandonning Asiatic completely and start direct deliveries to national sales companies.[155] Since the Asiatic contract had another eight years to run this was clearly a non-starter, but the incident does show the direction of managerial thinking which the war inspired. Practical constraints imposed by the original contract added further impetus. Asiatic's capital of 1.5 million pounds was insufficient to finance its needs, so the company had come to rely on bank credit, spare funds from other Group companies, and shareholders leaving some dividends on deposit. Converting dividends into equity would have reopened the old debate about Asiatic's status, all the more difficult to conduct now that the shareholders had reversed their original positions. Deterding wanted the company to operate independently and to expand; the Rothschilds preferred to keep Asiatic as it was, a dependent and very profitable syndicate.

The war raised the stakes and simultaneously imposed a stalemate over the capital question. Spiralling costs meant that Asiatic needed more funding, but the banks demanded more safeguards for fewer facilities. At the same time the repatriation of dividends became problematic. Leaving them on deposit, the obvious option, now threatened to seal the outside shareholders into holding rising amounts of immobilized debt, solving the status question by default. In January 1916, Deterding estimated that Asiatic needed at least £ 4.5 million capital and demanded the conversion of dividends into equity. The company now put a disproportionate strain on Group funds and failed to take proper care of its own financial needs, Deterding even hinting that he considered this a breach of contract.[156]

Meanwhile two further issues had arisen, i.e. freight rates and insurance premiums. To end the continuous bickering about freight rates, Asiatic's 1908 'A' Agreement with Anglo-Saxon had laid down a fixed rate of 13s. 6d. (67.5 pence) per ton, regardless of distance, as the basis for settlements between the two companies. In March 1915, Lane & Macandrew reckoned that rising costs necessitated raising this to 16s. 6d. (82.5 pence) to give Anglo-Saxon, which had lost about £ 160,000 on its shipping operations during 1914, fair compensation.[157] Paris had hardly agreed when Lane discussed with Deterding the 'absurd' situation of freight rates from the US to Asia which, not fixed by the 1908 contract but tied to market rates, had risen to £ 15 a ton on the shortage of tanker capacity. The oil shipped had fetched only about £ 5 per ton, so Asiatic stood to lose huge sums on supporting the Group's Asian market share with US supplies.[158] This question provided the tune for a song of never-ending troubles. As a compromise solution, Deterding proposed to transfer the entire outside oil trade to Anglo-Saxon in return for a fixed commission to Asiatic.[159] Such an arrangement would have meant Asiatic giving up the oil trade between the US and Asia. After the war, Russian exports would inevitably be eclipsed by US oil, Lane argued, leading the Rothschilds to suspect that Deterding expected huge future sales and was trying to cap the level of profits in which Asiatic might expect to share. Paris deferred a decision, then accepted, but meanwhile the simple question of cost allocation had ballooned into a matter of general policy. Shifting cost meant shifting profits, exposing the Group to unnecessary tax liabilities; fiddling with freight rates raised questions about the equally artificial oil pricing system used for internal settlements.[160]

Moreover, Group managers had little sympathy for Asiatic's complaints about rising costs, since it was the most profitable of all, and by far. The company had paid more than 60 per cent dividend over 1913, 1914, and 1915. Anglo-Saxon dropped from 14 per cent in 1913 to 7.5 per cent in 1915. Over 1917 the company even suffered a loss of £ 700,000. Bataafsche paid a steady 20-25 per cent during those years, Shell Transport paid 35 per cent in 1914 and 1915, and Royal Dutch 49 per cent. The fact that a third of Asiatic's booming profits did not even benefit the Group but outside shareholders generated considerable ill-feeling, even between Lane and Deterding, who until then had been close and mutually respectful collaborators. The Group clearly wanted to cut Asiatic to size. This was neither unfair nor unreasonable. The prices at which Asiatic bought oil from Bataafsche had been fixed, like the freight rates, by the 1908 contract so as to escape the constant bickering about market values. Consequently, Asiatic enjoyed all the benefits of the war in the form of higher product prices, while Anglo-Saxon and Bataafsche carried the burdens, i.e. rising wage bills and material costs. In justification for their demands, Group managers quoted the spirit of the 1902/1907 agreements, the express intention that none of the partners was to have an advantage over another.[161] It testifies to Lane's remarkable skills that he succeeded both in convincing the Rothschilds of the fairness of this particular aspect of the joint-venture, and in single-handedly parrying the Group's intention and keeping Asiatic's profits high.

The January 1916 dispute over insurance premiums led to a clear separation of minds. Asiatic's oil account bore the risk of war damage to Anglo-Saxon's ships and Bataafsche's cargoes under an ingenious scheme devised by Lane to save cost. Until then there had been no loss, but the submarine threat greatly increased the likelihood of substantial and rising charges which Asiatic, with its slender capital base, could ill afford to bear. To keep the premiums within the Group, Royal Dutch and Shell Transport proposed Asiatic taking out formal insurance with Bataafsche at market rates, which Anglo-Saxon already did for its US-based ships *Silver Shell* and *Gold Shell*.[162] Lloyd's conditions proving onerous to Asiatic, Lane set about to get better terms and a new round of acrimonious negotiations began.[163] Something snapped during those talks. For over a decade, Deterding and Lane had worked closely together to build the Group on the basis of shared interests. Now the endless arguing, Deterding surrounded by his staff, Lane sitting on his own, led them to recognize that the goals which they pursued had become really irreconcilable, and would have to lead to a separation between the Group and the Paris Rothschilds. Each clearly signalled this to the other, and each of course carefully avoided spoiling his negotiating position by formulating the obvious conclusion.[164] Though agreements over all outstanding issues subsequently cleared the air, the die had been cast. For Paris selling out appeared increasingly desirable. Lane was in his mid-sixties now and suffering from declining health, at times having to spend weeks on end seeking seaside cures for his bronchitis. The Rothschild interests depended on his unique position in the Group and his intimate acquaintance with its affairs. Moreover, the bank recognized the fairness of the Group's claims for profit redistribution in the spirit of the agreements and accepted Asiatic paying some compensation to Anglo-Saxon. Opening the fixed prices enshrined in the contracts made Asiatic's future increasingly uncertain, however, so selling out prior to 1923 became an attractive option.[165]

Negotiations did not start for another year pending the talks with the British Government over Asiatic's joint venture BP. Since November 1916 the Group possessed its own UK sales organization, Shell Marketing, under the energetic Frank Harris, and no longer needed BP. Giving up the fight would have depressed the value of its assets, however, needlessly benefiting Anglo-Persian, the Government's intended buyer, which finally obtained the company in the spring of 1917.[166] Lane and Deterding immediately began negotiating and reached an overall agreement in January 1918. The Group agreed to buy out the Rothschilds' interest in Asiatic of £ 500,000 at par plus estimated profits until 1 July 1923, based on actual dividends since 1912. The total came to £ 3.1 million, which was paid in Royal Dutch shares. Typically, Royal Dutch and Bataafsche were informed only at the end of January, when the negotiations had been concluded, Deterding and Cohen Stuart requesting urgent assent.[167] Just at that moment the entire situation changed because of the British Government's wholesale tanker fleet requisitioning. Anglo-Saxon now claimed *force majeure* and wanted to cancel the 'A' Agreement with Asiatic. The company would have to charter its own ships back from the Government at market rates, which Lane estimated to cost nearer £ 20 a ton than 16s. 6 d.[168] He coolly played for time. When in April the Asiatic board finally met to consider the matter, Lane stated that it was really for the Group to decide now that the buy-out had been agreed in principle.[169] Thus the close collaboration on which so much of the Group's business had rested, ended in a transformation of the Rothschilds from partners into formal shareholders. The Paris bank continued to occupy a special place amongst the Group's business relations, for until the mid-1920s the manager of its oil interests, Weill, attended Anglo-Saxon board meetings about once a year. The disputes over Asiatic were by no means the only tensions within the Group, however.

Strains in the structure War made the Group in some definite respects more coherent and uniform. The Technical Departments in The Hague and London took initiatives to standardize both equipment from well head to tank installations, and designs of capital goods such as staff bungalows and other amenities. Staff policies of the operating companies converged, partly stimulated by the gradual extension of the Provident Fund to all operating companies and fleet personnel, partly by the exposure to common pressures. When Bataafsche considered the position of its employees called up by the Dutch Army, the directors followed Anglo-Saxon's earlier arrangements, and rising inflation led to similar, though uncoordinated, arrangements for wage compensation. All Group staff received a bonus to celebrate the Royal Dutch Silver Jubilee in 1915, after considerable board debate

about allocating the cost.[170] Directors were of course quick to use terms and conditions of employment in one area of operations to try and discipline staff in another. In August 1915, Deterding wrote to The Hague that the preferential treatment meted out to Bataafsche staff had to stop. This ran counter to the agreement to harmonize employment conditions within the Group, he argued; moreover, it was time for people in the Dutch East Indies to realize that profits were increasingly generated elsewhere.[171] Staff circulated between operating companies in rising numbers. There had always been a couple of troubleshooters such as Mark Abrahams and technical experts like Erb or Kruisheer sent wherever the Group needed their services. Now a regular pool of managers, among them the two Kessler sons Dolph and Guus, began to emerge, who followed a career track leading them from

In 1915 Royal Dutch commissioned a commemorative album marking its Silver Jubilee. Illustrated by the famous Dutch artist Cornelis Jetses (1873-1955), the album's intricate details show the combination of classic heroic images and exotic images that characterized the imperial mind-set, simultaneously rejoicing in power and fascinated by the culture of the colonies.

TER·GELEGENHEID·VAN·HET
ZILVEREN·JUBILEUM
DER
KONINKLIJKE·NEDERLANDSCHE
MAATSCHAPPIJ·TOT·EXPLOITATIE
VAN·PETROLEUMBRONNEN·IN
NEDERLANDSCH-INDIE
OP·16·JUNI·1915
BIEDEN·ONDERGETEEKENDEN
ALLEN·MEDEWERKERS·AAN·DE
BELANGEN·DIER·MAATSCHAPPIJ
HUNNE·GELUKWENSCHEN·AAN.
DE·MAATSCHAPPIJ
BLIJVE·STEEDS·MET·EERE·GE-
NOEMD, HARE·WERKZAAMHEID
BIJ·VOORTDURING·GEPREZEN
ALS·EEN·VOORBEELD·VAN
HETGEEN·NEDERLANDSCHE
ONDERNEMINGSGEEST·VERMAG!
HEIL·DE·KONINKLIJKE!
VIVAT·CRESCAT
FLOREAT!
C. JETSES. INV. DEL.

an initial desk appointment in London or The Hague to one regular overseas appointment after another, grooming them in international business, creating personal linkages across continents holding the Group together, and intensifying the exchange of ideas about best practices. Thus Pyzel immediately understood the huge importance of the Trumble process for the Group's very different crudes produced in Russia, Romania, Egypt, and Mexico, which he knew either from direct experience or close second-hand knowledge. At Roxana's Yarhola lease, he also designed a casing head gasoline plant similar to the one which he had pioneered in Sumatra. To support the US operations, Erb overcame the opposition of Deterding, who was anxious to guard company secrets, to set up an American geological service. W. J. M. van Waterschoot van der Gracht, a man of many talents, was sent over for the purpose. He joined Roxana and became the company's President in 1917. When Roxana embarked on a big exploration campaign in 1918-19, Van der Gracht had his men use core-drilling, a technique then new to US oilfields, which he knew from previous experience in Romania.[172]

However, at board level the Group developed some definite signs of fatigue. Whether or not the appointments effected in January 1914 eased the managerial strains evident during 1913, wartime pressures soon created new ones.[173] Moreover, the dynamics of the Group's development affected the balance of interests within the business. With the expansion in the western hemisphere and the interruption of supplies from Russia, the Group's centre of gravity shifted away from the original constituent interests captured in the 1902/1907 agreements which were essentially constructed to serve the Asian and European markets. Since the operators in the western hemisphere were run by Anglo-Saxon, Bataafsche's influence on overall policy waned. Moreover, rising sales of outside oil undermined Bataafsche's role as the Group's main producer. In 1914, Asiatic's sales of outside oil were just over half of Bataafsche sales, at 540,000 against 995,000 tons respectively. By 1918, however, the gap between them had narrowed to just 50,000 tons, 780,000 tons of outside oil against 830,000 for Bataafsche.[174] Consequently, fault lines opened up between London and The Hague, and between London and Paris.

The suspension of the monthly meetings with the London directors rapidly generated uneasy feelings in The Hague, reinforcing earlier doubts about the Group's overall management. When the customary management procedures broke up in August 1914, The Hague keenly felt its dependency on London for key operational data, which revived fears of being relegated to the sidelines. In November 1914, Deterding wrote to Loudon in response to a complaint from the latter, that he had given orders to relay all incoming information, such as the weekly reports on US operations, to The Hague.[175] Even so managers continued to feel at a disadvantage. Colijn organized a statistical department in The Hague to keep him up to date about the London operations.[176] To give Royal Dutch better control over the Group's subsidiaries, Deterding also proposed to transfer all shares in operating companies held by Bataafsche and Anglo-Saxon to the two holding companies Royal Dutch and Shell Transport in the usual 60:40 ratio. After all, he argued, Anglo-Saxon had originally been set up as a political expedient, a British company to operate the tankers and installations within the overall Dutch concern, to ensure British political and diplomatic support. For similar political reasons, the company had come into possession of shares in subsidiaries such as Anglo-Egyptian Oilfields Ltd., North Caucasian Oil Fields Ltd, or California Oilfields, which was really alien to its purpose. Moreover, despite receiving full operational details neither Royal Dutch nor Bataafsche could exert effective control since the Anglo-Saxon board minutes offered insufficient details about strategy and objectives. This defect would be remedied by the proposed transfer.[177]

Deterding had presumably intended to soothe the Royal Dutch board with an assurance of overall control, but he touched a raw nerve. With assets of almost 265 million guilders (just over £ 20 million) in 1914, Bataafsche was still by far the biggest of the

Group's three operating companies, Anglo-Saxon coming second with nearly 177 million guilders (nearly £ 15 million), and Asiatic third with 67 million (£ 5.4 million). However, Anglo-Saxon grew faster than Bataafsche, and The Hague managers clearly felt that Anglo-Saxon's leading role in developing the US operations threatened to eclipse the Dutch company. At the December board meeting, with Deterding absent and represented by his Anglo-Saxon colleague Tresfon, feelings ran unexpectedly high. Chairman Capadose firmly opposed the proposal. He felt that the share reshuffle would tilt the Group's balance further towards London. Anglo-Saxon was far too big already. Contrary to what the merger arrangements had intended, the company controlled more production than Bataafsche, and the gap would grow with the steep rise of production outside the Dutch East Indies. He had full confidence in Deterding's skills and intentions, but it was unwise to be so dependent on a single person, or to rely on the Dutch managers heading key departments in London. A transfer to Shell Transport exposed any shares to Sir Marcus's well-known penchant for window-dressing through optimistic valuations. It would also lead to the English directors forming an inner circle for policy decisions, rendering The Hague redundant.[178] Capadose's arguments appear somewhat irrational and curiously aggrieved, fired by the reshuffle but clearly inspired by an anxiety about being relegated to the sidelines. Giving 40 per cent of of Bataafsche's subsidiaries to Shell Transport mattered far more to him than getting 60 per cent of Anglo-Saxon's. Moreover, he opposed giving Royal Dutch a more direct hold over Anglo-Saxon's subsidiaries, but did not suggest taking action over the far graver issue of Anglo-Saxon becoming the Group's core operating company. If Capadose simply used the occasion to alert his fellow directors to apparent trends, his words do show a remarkable ability to bark up the wrong tree at anything which appeared to threaten the Dutch preponderance in the Group. After a long debate, the board accepted the proposal with the smallest possible majority of four to three, adding the safety-valve of a buy-back option for Bataafsche and Anglo-Saxon.[179]

Though only a paper transfer, the share reshuffle appears to have pacified the Dutch boards. During the war London inevitably became the Group's operational centre, but the surviving Bataafsche and Royal Dutch minutes do not record further principled debates about a loss of power or position until the very end. Meanwhile the Bataafsche board strove hard to turn the The Hague office into an effective organization. Staff numbers more than doubled during the war to 460 in 1918, causing severe stress to Bataafsche's managers, who in the old hierarchical structure took all decisions.[180] Colijn and Pleyte, but also middle managers such as Erb and De Kok, struggled to control basic administrative processes such as handling mail, managing staff, obtaining data or particular files, getting expert advice, interviewing a steady flow of job applicants in between. The staff department was rudimentary; a memo from April 1917 commented that the company kept stock books for nuts, bolts, all kinds of equipment, horses, and draft oxen, but no regular records on staff numbers, qualifications, or performance reviews.[181]

In the run up to the move to the proud new Royal Dutch head office at Carel van Bylandtlaan, The Hague, in 1917, Colijn and Pleyte overhauled Bataafsche's entire organizational structure. The company's management suffered badly from an imbalance between functional and regional departments, a recurrent problem throughout the Group's history. Bataafsche had separate departments for each operating company, but only two fully fledged functional departments, Geological and Technical. All other business functions were loosely grouped in staff bureaus reporting to the two managers, who therefore had to devote most of their time to administrative chores, leaving precious little for policy matters. Moreover, the departments running the operating companies tended to become little fiefdoms fiercely defended by the chief clerks responsible for them, hampering the efforts to improve Bataafsche's functional performance.

New office buildings

The growth of the two organizations made it necessary to move to bigger office buildings. In the Netherlands this took quite a while. In 1912 the decision was taken to build a new office, but the municipality of The Hague did not allow this on the site that was selected – at the Old Scheveningen Road, opposite the Peace Palace – because a residential quarter had been planned there. This induced the Raad van Commissarissen to consider moving to Amsterdam, which was considered to be 'the place where a big trading company as Royal Dutch should be located'. The proposal led to a request by 144 employees who feared the financial consequences of the move. Finally, in 1913, it was decided to buy another site, at the Carel van Bylandtlaan. Two architects, the brothers M. A. and J. van Nieukerken, were invited to design the building. They were pupils of their well-known father, J. van Nieukerken, who had been the architect of a number of neo-renaissance buildings particularly liked by the colonial elite residing in the Netherlands. His most famous building was probably the mansion Duin & Kruidberg, designed for the former director of the Deli-company and minister of colonial affairs, J.T. Cremer. The neo-renaissance building that was erected between 1915 and 1917 is a clear product of his 'school'. Implicitly this re-invention of the 17th-century Renaissance architecture stresses – as Colijn and his secretary Gerretson liked to point out in those years – the continuity between the Dutch Golden Age and the 'second' Golden Age of the early 20th century, of which Royal Dutch was considered to be the foremost example. This idea of continuity in Dutch imperial ambitions is also a theme in the well-known 'History of Royal Dutch' by Gerretson.

The design by the brothers Van Nieukerken was criticized quite a lot. For a new generation of architects

[58]

such as H. P. Berlage and his followers, who were revolutionizing Dutch architecture in these years, this was a rather 'old-fashioned' building, lacking the clarity and soberness of design that had become fashionable. After finishing the office of the Bataafsche, the brothers Nieukerken concentrated on another project for the Dutch colonial elite, the building of the Royal Tropical Institute in Amsterdam, which was partly funded by Royal Dutch and the Bataafsche.

The new building in London, St. Helen's Court, was perhaps less memorable for its architecture than for the occasion of its opening in December 1913 (see main text). An interesting detail concerns the way in which the name of the new building was selected: at the meeting of the Board of Anglo-Saxon of 13 November 1913 the name St. Helen's Buildings was proposed, but no decision was taken because Deterding was not present. The minutes of the next meeting, on 20 November, tell us that Deterding had not approved this name, 'but suggested that the new building be called St. Helen's Court ... and that this had been approved by the chairman' (i.e. Marcus Samuel). Deterding, indeed, cared for every detail. But he did not win all discussions. A proposal by Waley Cohen and him for the 'supplying of lunches in the lower basement on certain terms to the staff' met stiff resistance from Marcus and Sam Samuel, and was therefore withdrawn. Another interesting detail of the actual opening (dinner) is the list of invited guests: 'St. Helen's Court 'At Home' invitations should be sent to the Secretary of the Admiralty, the Director of Navy Contracts, the Director of New Transports and to Sir Francis Hopwood at the Admiralty, to the Secretary of the Board of Trade, to the Chairman and Directors and to Mr. McClean and Mr. Woods of the Anglo-American Oil Co., and Mr. Smith of the Tank Storage and Carriage Co.' The most important guests appear to be those from the Admiralty – Marcus Samuel was still hoping to improve relations with them. The Jersey Standard competitors were also invited (Anglo-American Oil co. was the British subsidiary of Standard), but British competitors – the directors of Burmah Oil co. and the Anglo-Persian Oil co. – were notably absent. Nor were politicians invited, although the Samuels had links to the Conservative Party – Sam Samuel was also an MP.

[59]

On 18 June 1916, the foundation-stone of 30 Carel van Bylandtlaan was laid (left) by Gerard Coenraad Bernard Dunlop, a Royal Dutch director and representative of its original backing syndicate. Still easily recognizable today, the original offices (right) have been enlarged for modern needs. In the entrance hall of St Helen's Court, Shell Transport's own new post-war offices, a part of the company's Roll of Honour can be seen.

To overcome these handicaps, Colijn and Pleyte set up separate managerial departments for administrative matters such as staff policy, budgeting, statistics, and internal communications. They then reorganized Bataafsche's operations into four functional departments, Production, Manufacturing, Storage & Transport, and Trade, supported by three executive bureaus, Technical, Purchases, and Accounts. The regional departments remained in a subordinated position, reporting not to the managers, but to a department with the inconspicuous name *Algemeen Secretariaat* or General Administration. This particular branch of the organization functioned very much like a ministerial cabinet office entrusted with confidential matters and assignments. In addition the department had a staff of assistant managers or *directie-assistenten*, who were each responsible for particular geographic areas and entrusted with maintaining the balance between regional and functional interests. The formation of the Algemeen Secretariaat was probably inspired by a reorganization at Anglo-Saxon, about which more below.[182] The function of assistant manager was also intended as a training ground, creating an alternative track towards the top for law graduates and other non-technical people. It failed to solve Bataafsche's heavy reliance on engineers and chemists becoming managers, however, which continued to worry managers well into the 1930s. The first appointments as assistant manager included F. C. Gerretson. He was soon overburdened with work, partly because from January 1918 the department had to compile Bataafsche's new and comprehensive monthly surveys of Group operations as well.

Preparations for the administrative reorganization began in October 1915 with The Hague staff visiting St Helen's Court to report on modern office arrangements and amenities at the London companies, notably document flows, recordkeeping, stores management, and the typing pool. The whole process was rounded off by the introduction of printed standard forms and filing cards for routine jobs.[183] Surrounded as it was by open fields, the new head office offered sporting facilities such as a football pitch and two tennis courts for the staff clubs.[184] From November 1917, a monthly staff magazine *Maandblad van het Personeel der verbonden Oliemaatschappijen* also helped to foster an *esprit de corps*. Colijn also wanted to reform the Group's top management. In line with his posturing as a man of action spurning idle words, he looked down on the Royal Dutch and Bataafsche boards, composed of executive and advisory directors, as talking shops. He evidently preferred the British model of joint accountability and pleaded with Deterding for a new managerial style, transferring some of the General Manager's tasks on to a new and wider team of directors meeting at least once a week rather than once a month, the common frequency in The Hague. The members of such a board would have clearly defined competences, ending the undesirable practice of allowing managers to choose their own responsibilities.[185] Deterding's reply is not known, but Colijn's proposal did not even reach the stage of board-level discussion.

How Colijn and Pleyte, grappling to steer a sprawling business through the uncharted waters of a world war, must have yearned for the kind of policy debates which they witnessed on their visits to London! The British boards, though managing companies hardly bigger or more complex than Bataafsche, were both bigger and more active than the Dutch ones. Moreover, the directors were all oil men, as opposed to the fractious mix of bankers and lawyers at Bataafsche and Royal Dutch. Meeting once and sometimes twice a week, the Anglo-Saxon board functioned as the prime forum for all major policy issues concerning the Group. Its members, and especially Sir Marcus, Waley Cohen, and Cohen

[60]

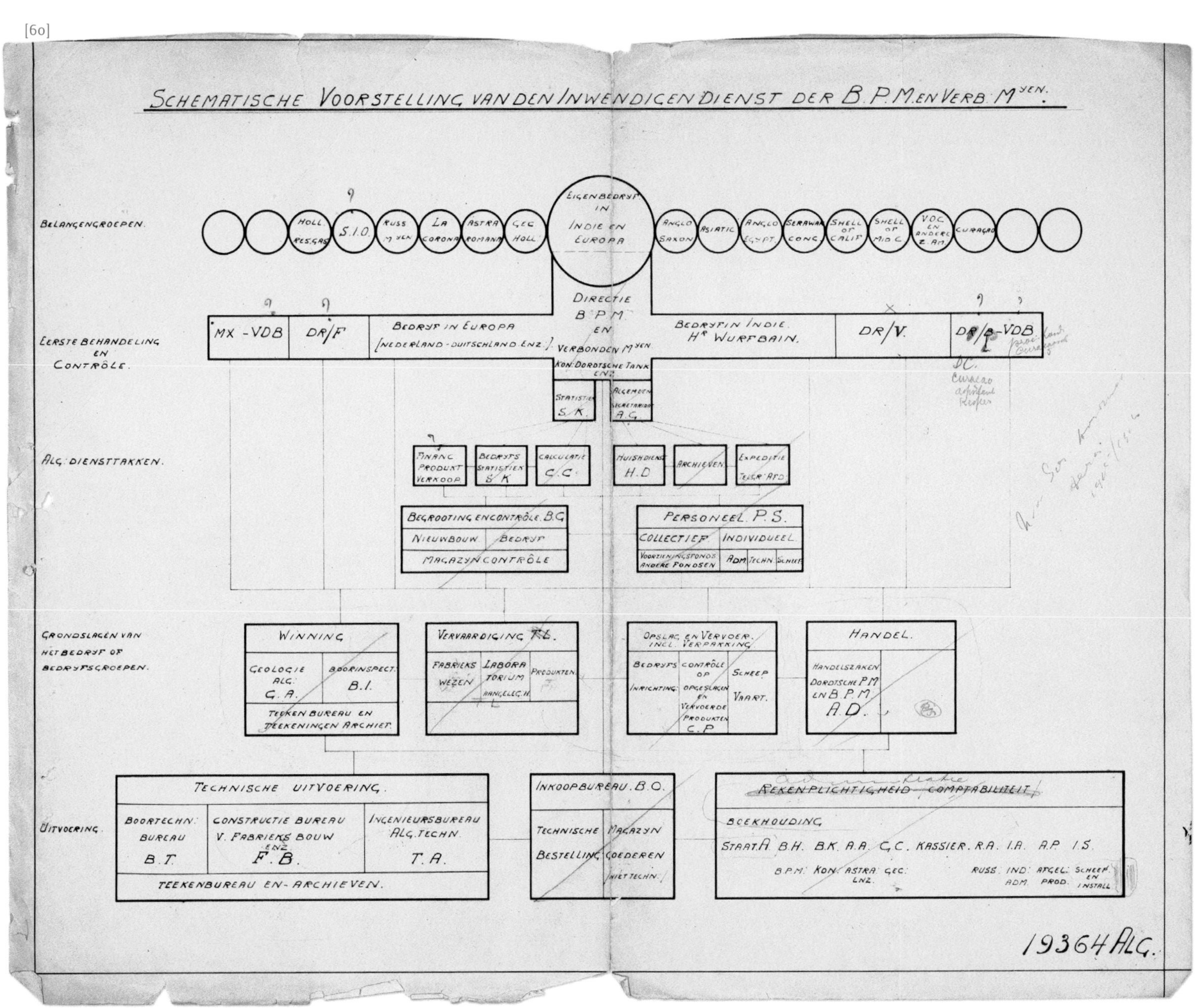

Colijn's reorganization of Bataafsche Petroleum Maatschappij, 1916.

Stuart, fully participated in debates and decisions, allowing Deterding to appear as *primus inter pares*, without whom, however, no important decision was taken. Reflecting the Group practice of respecting the separate corporate identity of subsidiaries, the board couched its directives in polite terms of suggestions and tentative proposals, but of course they were hardly less pertinent. As we will see in Chapter 5, this form of instructions from central offices later became mandatory for reasons of fiscal expediency, since it avoided the impression that the company concerned was subject to taxable controls.[186] In 1915 the Asian, Russian, and American departments, which had formerly done little more than handling correspondence, were reorganized and given similar reponsibilities to those of the Marine and Technical departments, i.e. monitoring their respective areas, drafting regular surveys on operations, reporting to the board on specific policy issues, and preparing major investment decisions.[187] The change more or less coincided with the rise of a new class of middle managers, such as the young Fred Godber at the American department, who began replacing senior clerks as departmental heads.[188] Anglo-Saxon kept some purely administrative departments such as Steamers and Accounts Steamers, but wherever possible specific tasks were farmed out to Asiatic, which managed both the various toluol gasoline installations and the new Trumble refinery at Shell Haven on the Thames estuary.[189] Consequently Asiatic, though the smallest of the Group's main operating companies in terms of assets and turnover, had the biggest staff. For the handling of documents in St Helen's Court alone the company employed 40 messengers of a total staff of 437 in March 1920, up from 276 in 1914. During that same period Anglo-Saxon's staff rose from 153 to 379. By contrast, Shell Transport had a staff of only eight people.[190]

War forged the British companies closer together in a St Helen's Court team spirit, fostering initiatives such as office staff squads working Sundays at the Fulham canning factory to increase production, sports clubs which after the war found a home at the Lensbury Club complex at Teddington, and the staff magazine *St Helen's Court Bulletin* which, launched and edited by Mark Abrahams, specifically aimed to keep up morale and forge ties

[61]

ST. HELEN'S COURT BULLETIN. 21

Miss E. TYLER.

It gives us great pleasure to publish the photograph of the Secretary to the Asiatic Petroleum Company's "War Relief" Fund, Miss E. Tyler, and we are sure it will be hailed with great enthusiasm by those members of the staff now with H.M. Forces who have been the lucky recipients of the parcels so kindly packed

and forwarded by this lady. This photograph will serve as an introduction to those fighting members of the Asiatic Petroleum Co.'s staff who have not hitherto had the pleasure or an opportunity of meeting Miss Tyler. In the early days of the war, Miss Tyler undertook the duties of Secretary to the "War Relief" Fund, the moneys subscribed being at first divided between the various National Funds, but since the beginning of this year such donations have (on the Secretary's suggestion) been set aside for the purpose of providing comforts for those of the Company's men who are serving at the front. The duty is a very onerous one, and occupies nearly all Miss Tyler's spare time, but we have to con-

business instincts and energy that she has been entrusted with the post of Secretary to the "War Relief" Fund. As is well known, she is very popular with all the members of the staffs of St. Helen's Court.

ASIATIC WAR RELIEF FUND.

Another Splendid Gift.

Yet another Christmas sees the continuance of the war, and we renew the unfulfilled hope of last year that the coming New Year may bring us an honourable and lasting peace. Until this consummation we have to shoulder the burden of our responsibilities, not the least of which is the upholding of this Fund, which is now doing so much to cheer and hearten a few of the men who are serving the Empire. So far it has been upheld with increasing rather than diminishing ardour, and we have great pleasure in thanking the staff of the Rising Sun Petroleum Co., Ltd., in Japan, for a splendid gift of £77— for parcels to the A.P. Co.'s staff.

Christmas parcels to France have all been despatched just prior to the date fixed by the Post Office as the limit to ensure their reception by Christmas Day, and we can only hope they will all arrive in good time and serve as the expression to the recipients of our desire to impart to them as much of the spirit of Christmas as it is possible for men away from home and friends and, under such trying conditions, to know.

Further contributions are gratefully acknowledged as follows:

	£	s.	d.
Staff of Suez Refinery (representing 50 per cent. of the sum already acknowledged in total by the Anglo-Saxon Comforts Fund)	4	10	3
Staff at Sandycroft	2	11	2
Mr. N. G. M. Luykx, 50 per cent. of ..	1	18	9

In addition to the Christmas parcels, the following have been despatched to the staff on Active Service:

Sergeant O. H. Bond, Private S. H. Casperd, Private H. D. Hayland, Private G. W. Bailey, Private F. Berry, Sergeant E. Brooker, Captain A. D. Clark, Private C. W. Eastman, Lance-Corporal W. G. Holder, Stoker O. Hicks, Lance-Corporal S. C. Jones, C.P.O. G. F. Jones, Corporal H. Jacobs, Second-Lieut. A. W. K. Money, Captain J. F. M. Robinson, Private A. G. Stokes, Second-Lieut. P. J. Shaw, Private S. A. Gulland, Private A. Trusson, Corporal G. H. Smith.

LETTERS TO MISS TYLER.

'Many thanks for your extremely nice parcel. I am awfully sorry I have not written to you before but still, as the saying goes, better late than never. I'm sure one could not wish for a better one. Kavanagh and I are both in the same battery, also are both telephonists, so we see a decent bit of each other. I was sorry to read about the death of Captain Maginn. I was near his battalion very often, but never had the luck to run across him.'

C. Thornelow

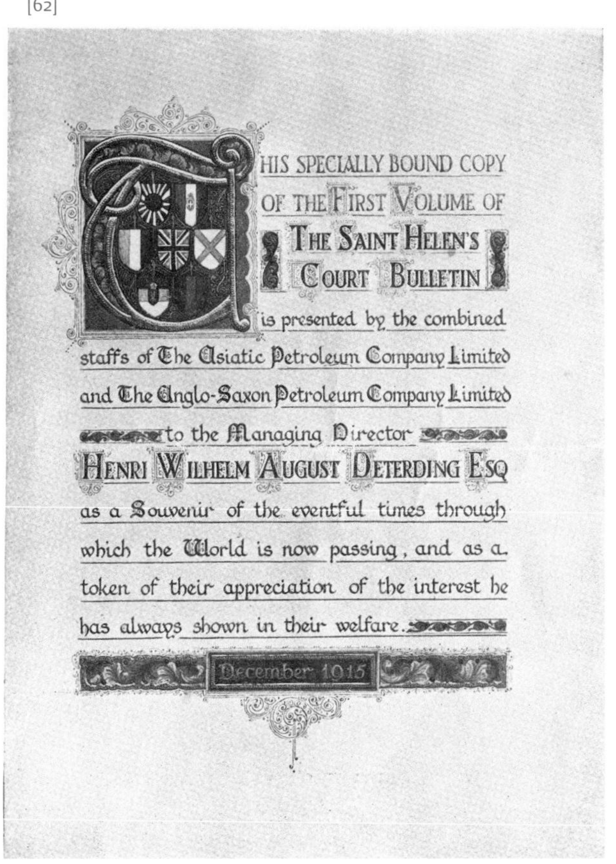

THIS SPECIALLY BOUND COPY OF THE FIRST VOLUME OF THE SAINT HELEN'S COURT BULLETIN is presented by the combined staffs of The Asiatic Petroleum Company Limited and The Anglo-Saxon Petroleum Company Limited to the Managing Director HENRI WILHELM AUGUST DETERDING ESQ as a Souvenir of the eventful times through which the World is now passing, and as a token of their appreciation of the interest he has always shown in their welfare.

December 1915

In 1914 people expected the war to be over by Christmas. In December 1915, as the second Christmas approached, staff presented Deterding with a specially-bound copy of the first volume of St Helen's Court Bulletin (above) 'as a token of their appreciation of the interest he has always shown in their welfare.' The Bulletin became a very human record of the entire conflict, and every issue included letters of thanks to Miss E. Tyler (left), secretary of the Asiatic's War Relief Fund, for the packages she arranged to be sent to staff serving in the armed forces. Early in 1918, with mingled pride and sorrow, the magazine published a Roll of Honour of all Group staff who had been killed.

Pro Patria

1914 Roll of Honour. 1917

ASSOCIATED COMPANIES

Anglo-Saxon Petroleum Co., Ltd.

NAME.	RANK.	REGIMENT.
Bird, R. T.	2nd Officer	Royal Navy
Blackwell, A. R.	Lance-Corporal	Duke of Cornwall's Light Infantry
Carter, G. S.	3rd Officer	Anglo-Saxon Co.'s Fleet
Clarke, L.	2nd Officer	Anglo-Saxon Co.'s Fleet
Corby, F.	Sergeant	Royal Field Artillery
Cox, T. R.	Lieutenant	Durham Light Infantry
Cordrey, W.	Lance-Corporal	24th London Regiment
Fraser, P.	C.P.O.	Royal Naval Air Service
Kell, A.	C.E.R.A.	Royal Naval Volunteers
Mann, H. W.	Lieutenant	Loyal North Lancs.
Munro, J. W.	Private	Essex Regiment
Oram, J. A.	1st Officer	Anglo-Saxon Co.'s Fleet
Pearce, E.	Gunner	Royal Garrison Artillery
Raffray, H. L.	Chief Engineer	Anglo-Saxon Co.'s Fleet
Scott, Nigel	2nd Lieutenant	Royal Flying Corps
Stevens, R. N.	4th Engineer	Anglo-Saxon Co.'s Fleet
Stott, E.	Captain	Anglo-Saxon Co.'s Fleet
Sullivan, W.	2nd Engineer	Anglo-Saxon Co.'s Fleet
Young, J. H.	Lieutenant	Royal Naval Reserve

Asiatic Petroleum Co., Ltd.

NAME.	RANK.	REGIMENT.
Baldry, W. G.	2nd Lieutenant	11th Essex Regiment
Barnes, F.	Private	7th Royal West Kent
Barnes, J.	2nd Lieutenant	7th Notts and Derby Regiment
Bird, E. H.	Captain	10th Queen's R.W. Surreys
Bishop, B. B.	2nd Lieutenant	Duke of Cornwall's Light Infantry
Blyth, J. R.	2nd Lieutenant	Rifle Brigade
Cowie, G. J. H.	2nd Lieutenant	10th Royal Fusiliers
Dixon, C. S.	2nd Lieutenant	2/6th London Regiment
Everett, W. J.	Rifleman	Rifle Brigade
Fulton, C. J.	Captain	2nd Batt. Royal Guernsey Light Inf.
Halliday, Victor	Lance-Corporal	King's Royal Rifles
Hughes, E. J. W.	Captain	Machine Gun Corps
Jackson, B. R.	Captain	Coldstream Guards
Jepson, A. G.	Captain	London Regiment
Jones, W. F.	Private	1st Somerset Light Infantry
Lachlan, C. G.	Lieutenant	West Yorkshire Regiment
Leete, W. J. H.	Captain	Lancashire Fusiliers
Leslie, W. R. Norman	Lieutenant	Gloucester Regiment
Maginn, P. A. C., M.C.	Captain	London Irish Rifles
Overall, C. B.	Corporal	4th Middlesex

Asiatic Petroleum Co., Ltd.—*continued.*

NAME.	RANK.	REGIMENT.
Paxton, E. G.	Lance-Corporal	London Rifle Brigade
Purser, H. A.	Gunner	Royal Field Artillery
Richardson, F. J.	Captain	South Staffordshire Regiment
Rust, P. R.	Private	13th Batt. Kensington Rifles
Wadsworth, W. C.	Corporal	17th King's Royal Rifles
Wells, N. L.	Lieutenant	Loyal North Lancs.
Whitewright, A. R.	2nd Lieutenant	Sherwood Foresters
Willers, E. W.	Private	Kensington Rifles

Anglo-Egyptian Oilfields, Ltd.

NAME.	RANK.	REGIMENT.
Chambers, P.	Private	The Rangers
Jamieson, R.	Private	

British Imperial Oil Co., Ltd.

NAME.	RANK.	REGIMENT.
Buchanan, E. F.	Lance-Corporal	Australian Expeditionary Force
Chandler, P.	Sergeant-Major	Australian Expeditionary Force
Charge, G. C.	Private	Australian Expeditionary Force
Ellis, J. O.	Private	Australian Expeditionary Force
Emery, E.	Private	Australian Expeditionary Force
James, J. R.	Bugler	Australian Expeditionary Force
Kimmond, R. M.	Lance-Corporal	Australian Expeditionary Force
Meggy, D. A.	Signaller	Australian Expeditionary Force
Momphlait, A. V.	Private	Australian Expeditionary Force
Morris, G. L.	Private	Australian Expeditionary Force
O'Rourke, T. G.	Private	Australian Expeditionary Force
Pyman, R. L.	Lieutenant	Australian Expeditionary Force
Seaman, R.	Corporal	Australian Expeditionary Force
Sives, J. P., M.C.	Lieutenant	Australian Expeditionary Force
Sweetland, F. J.	Sergeant	Australian Expeditionary Force
Walker, R. J.	Corporal	Australian Expeditionary Force
Watson, L.	Private	Australian Expeditionary Force
Williams, W. D.	Private	Australian Expeditionary Force
Wilson, K.	Sergeant	Australian Expeditionary Force

United British West Indies Petroleum Syndicate.

NAME.	RANK.	REGIMENT.
Benwell, E. H.	Private	London Rifle Brigade

NAAML. VENN. „ACETYLENA"

ACETYLEENGAS-EN BENZINE M!J

VERKOOPKANTOOR DER
KONINKLIJKE NEDERLANDSCHE MAATSCHAPPIJ TOT EXPLOITATIE VAN PETROLEUMBRONNEN IN NEDERLANDSCH-INDIË.

TELEGRAM-ADRES:
"ACETYLENA-ROTTERDAM."

TELEFOON: 4134-4155.

NED. INDISCHE BENZINE.

~~ACÉTYLÈNE DISSOUS~~

~~GECOMPRIMEERD IN STALEN CYLINDERS.~~

BRIEVEN VOOR DE VENNOOTSCHAP BESTEMD GELIEVE MEN AAN HAAR TE ADRESSEEREN ZONDER VERMELDING VAN PERSOONSNAMEN.

Gelieve in Uw antwoord te vermelden

Afd. ....

WB.

ROTTERDAM, 26 November 1918.

VAN OLDENBARNEVELTSTRAAT №49.

POSTBOX 56.

De Directie der Bataafsche Petroleum Maatschappij,
Carel van Bijlandtlaan 30,
's Gravenhage.

Mijne Heeren,

Het Personeel der N.V. "Acetylena", Acetyleengas- en Benzine Maatschappij, te Rotterdam, heeft de eer U door dezen zijn hartelijken dank te betuigen voor de prachtige gratificatie, welke Uwe Maatschappij het heeft toegedacht en welke in deze moeilijke tijden zoo byzonder welkom is.

Het stelt Uwe welwillendheid op hoogen prijs en geeft U de verzekering dat het door onvermoeid de belangen der N.V. "Acetylena" (dat zijn dus ook Uwe belangen) te blijven bevorderen, zal toonen die welwillendheid waardig te zijn.

Hoogachtend,

Het Personeel der
N.V. "ACETYLENA".

chef de bureau

Afdeeling Boekhouding :

P. van Es

P. Dekker

[64]

On the signing of the Armistice in 1918, 'in gratitude to all for so much special work during the war', the Bataafsche directors gave every Group employee a bonus of one week's basic salary for every three months of service, to a maximum of 13 weeks' extra pay. Paid in cash, this prompted many letters of thanks, including one (left) signed by the whole office of NV Acetylena for such a 'splendid' gift – 'particularly welcome in these difficult times.'

between the home front and employees uprooted by the war through military duty in France or elsewhere.[191] Its tone was therefore markedly different from the *Maandblad*. At board level, the shared premises at St Helen's Court and the practice of scheduling meetings on the same or on consecutive days facilitated quick consultations and decisions. Shell Transport directors Foot Mitchell, Sam Samuel, and Walter Samuel regularly attended Anglo-Saxon boards, tightening the overall coordination between the companies and creating a common bond from which even Fred Lane, an insider knowing all directors well and having frequent meetings with them, felt excluded.[192] With their London colleagues almost gone native, the directors from The Hague stood at a greater distance than he. Initially Colijn and Pleyte visited London once or twice a year, but neither did so from the spring of 1916 until October 1918. Loudon must have gone over as well, but there are no London minutes showing his presence; Capadose went at least once, and during the summer of 1915 Deterding urged The Hague to send over middle managers as well. This included overseas managers such as Dubourq, the Balik Papan manager, who visited London in 1916 discuss lube oil product specifications with Asiatic.[193] The directors from The Hague found themselves facing a barrage of proposals from St Helen's Court, firmly established on the high ground of full and informed discussions. Struggling to catch up and guess the intentions, they were condemned to remain out of touch with the climate of opinion because telephone lines were cut, mail and telegraph deliveries fell into disarray, and courier services by ferry became increasingly dangerous and erratic. By the time reports from Anglo-Saxon's Russian or American departments had percolated to The Hague, London would already have formed an opinion and initiated a response.

In short, The Hague had become the flywheel to London's engine, generating the cash to keep the machine turning, but often seen as a drag on acceleration. The 1916 New York share issue demonstrates just how far Royal Dutch directors, though fully informed and consulted at every step, remained out of touch with events of central concern to their company. In February 1916 Colijn reported to the board that London planned to sell Royal Dutch shares in New York, following talks conducted by Gulbenkian with the Paris agents of Kuhn Loeb & Co.[194] Deterding had now outlined the scheme, in which American certificates would be issued by the Guarantee Trust Co. for Dutch shares deposited there, and asked for approval, subject to the outcome of further detailed negotiations. The shares would not be listed on the New York stock exchange, but sold privately. After some discussion the board accepted the principle, but raised a number of objections as to timing and conditions. In April, Colijn returned from London, clearly invigorated by a peptalk from Deterding. His detailed explanation of the Group's finances at once made the board give up its resistance to Deterding's idea of paying the 1915 dividends in shares. Moreover, Colijn emphasized the Group's dynamic development and the urgent need for more capital in the US, in Mexico, in Venezuela, for ships, for pipelines, for a refinery in St Louis.[195] The board again approved the proposal for the New York issue. Raising money in the US for local expansion made sense, but no one appears to have asked for a specification, nor indeed how far the proposed 5 million guilders, the equivalent of about $ 2 million, would last.

Talks with Kuhn Loeb continued throughout the summer and autumn. When the two sides neared agreement in early November, Deterding suddenly revealed his intentions by proposing to use the money for buying UK Treasury notes issued in New York. The capital requirements in the Group's US operations were not quite so pressing after all, he argued; for the moment it would be better to put the money into first-class marketable securities, which could be sold or lombarded as and when the companies needed cash. The Royal Dutch board was painfully surprised. Only a few weeks before Deterding had underlined the volume of anticipated investments, and now there was money to spare. Nor did the purchase of yet more UK government paper find much favour. The Group already possessed too much of it: and what if the Netherlands should suddenly find itself at war with Britain? With arguments over a wide range of topics raising tempers on both sides of the North Sea, this possibility was unlikely but not entirely fictitious. Cancelling the deal appeared no longer an option, if only because the terms looked tempting indeed. Royal Dutch was to sell 7.4 million guilders of shares for a total of $ 13.6 million, giving the company an estimated 15 per cent premium.[196] The directors sent a terse telegram to London strongly objecting to buying more British paper, wanting instead to have the money put on call on collateral of prime securities, or used to redeem Bataafsche bonds in the Netherlands.[197] Deterding now had a free hand. He arranged a complex transaction in which Royal Dutch placed $ 14 million on-call with J. P. Morgan in New York for use by the British Government, against a portfolio of various foreign securities held in trust by the Bank of England. In January 1917, the board accepted the construction. Coming on top of the large investments in Treasury bills, this financial engineering on behalf of the British Government helped Deterding to earn his knighthood.[198]

Like the share reshuffle, the New York issue was paradigmatic for relations between London and The Hague during the war. A structural asymmetry of information put the Royal Dutch and Bataafsche directors more or less at the mercy of the Anglo-Saxon team, turning the formal control vested in Royal Dutch into something of a charade, and the consultative procedures designed to safeguard it into subtle games to make the Royal Dutch and Bataafsche directors agree to things they were really against. This is the background of Waley Cohen's chilling observation to Philips in 1923, that to him control was a matter of sentiment, and 'if by transferring control to the Hottentots we could increase our security and our dividends I don't believe any of us would hesitate for long'.[199] Clearly Waley Cohen was impervious to notions of corporate governance. With his formidable intellect and drive, he knew he would get his way whatever the formal arrangements. The Hague did succeed in winning some arguments, such as the thwarting of Waley Cohen's attempt to have the aromatics research transferred to the UK, but in matters of general Group policy, London called the tune.

Tempted by Middle Eastern oil Consequently, The Hague could hardly muster effective resistance when London proposed trading Royal Dutch's control over the Group for a stake in Middle Eastern oil. The Group's constructive contributions to the Allied war effort gradually wore down the initial hostility nursed by some departments and officials, but those very achievements reinforced the arguments, first heard to defend the Government's stake in Anglo-Persian during the spring of 1914, to bring the oil industry under political control for strategic reasons. In 1916 a Board of Trade memorandum emphasized secure oil supplies as a core Imperial policy aim.[200] The memo accepted that this would require control over Shell Transport, which meant lifting the company out of the Group and merging it with another British oil company such as Burmah to form one single and powerful all-British concern. In talks at the Board of Trade, Sir Marcus and Waley Cohen had apparently suggested as early as June 1915 that such a scheme would be acceptable to The Hague, though there are no records documenting any formal consultations about the matter.[201] To Waley Cohen the question of control was an academic one after all. Sir Marcus simply sought to bolster Shell Transport's status as a British company; since 1902 he had developed a habit of responding to taunts about Shell Transport's doubtful parentage by offering the Government a seat on the board, only to meet with steady refusal because of the Royal Dutch link.[202]

The Board of Trade proposal foundered on conflicting departmental policy perceptions and commercial rivalry, but the quest for ways to secure control over Shell Transport continued. In May 1918 the Government appointed the Petroleum Imperial Policy Committee chaired by Lord Harcourt to investigate how the British Empire could best secure its long-term oil supplies.[203] Whitehall patience with Anglo-Persian's tiresome scheming against the Group was wearing thin, and the company's continuing poor performance ruled it out as a serious partner for any strategic policy, so the committee's initial explorations naturally focused on the Group. Talks with Deterding exploring the options for a combine with other oil companies led nowhere. Deterding insisted that any such business had to be both commercially sound and free from state interference and shareholdings. The Harcourt committee did not want to trade a potentially very lucrative stake in the Anglo-Persian for vague voting rights.

Just when the Harcourt committee became stuck in the same mire as the Board of Trade before, international events moved to render it the sought-for leverage. In October 1918, Turkey signed an armistice with the Allied Powers, and the following month British troops occupied the Mosul region of Mesopotamia, securing control of what was considered one of the most promising oil fields around. Though the fate of the former provinces of the Ottoman Empire was to be settled at the Paris peace conference which followed the Armistice in November 1918, Britain now had the strong claim of possession, fortified by French support expressed by Prime Minister Clemenceau to Lloyd George in December.[204] As a consequence of the agreement concluded in March 1914 between the Government, the Group, and Anglo-Persian, oil exploitation would fall to the Turkish Petroleum Company (TPC). Anglo-Persian had half of the TPC shares, the Group 25 per cent, and the rest was held by the public trustee for enemy property as belonging to Deutsche Bank. Instead of Shell Transport merging with Anglo-Persian, the Harcourt committee now proposed reconstituting TPC, giving 40 per cent to Anglo-Persian, 40 to Anglo-Saxon, and 20 to the British Government, with the management entrusted to Shell Transport for the first seven years. Both Anglo-Saxon and Shell Transport would become fully British companies, and control over the Group's Romanian, Venezuelan, and Mexican companies would be transferred to them.[205]

The proposal was both practical, commercially sound, and supremely tempting. Just as in the struggle over BP, denying opportunities to the competition appears to have been as impor-

Extrait de..........
Adresse : JOURNAL DU CH
BOURGES
Date : ....
15 MAI 191
Signature : ....
Expos

Don de 1.000.000 de fran

Alexandre Ribot a reçu de
enkian, le délégué à l'étranger
tion pétrolifère Royal Dutch,
gé par M. Deterding, le direc
de cette Association, de
résident du Conseil la somme
on de francs que ce grand
pays a décidé de donner à la
our aider au soulagement, sous
rmes, des misères résultant de
allemande dans les régions év
vahies.

In 1917 French newspapers were full of admiration for Deterding, who had given a million francs to the French government to help relieve the suffering caused by the German occupation. The same papers reminded readers that this was 'not the first good action of M. Deterding', and that he had 'already marked his affection for our country' by undertaking to pay, until the end of the war, the expenses related to 200 beds reserved for French wounded in a Dutch hospital.

tant as securing them for the Group. With expansion in the US, Mexico, and Venezuela proceeding painfully slowly because of material shortages, the Group did not urgently need new concessions, but opting out meant giving them away to Anglo-Persian or Burmah, something which Deterding, keen to thwart the rival companies' expansion, was not prepared to consider.[206] Moreover, the Mosul fields had the benefit of proximity to both European and Asian markets, offering a direct competitive threat to the Sumatra fields.[207] With typical circumspection, Deterding moved to hedge his bets. Gulbenkian received instructions to work on the French Government and get it to claim the Deutsche Bank share, with the proposal to vest it in a French company managed by the Group. More or less at the same time, Lord Cowdray signalled his willingness to sell Mexican Eagle to the Group. Thus when in January 1919 Deterding and Lord Harcourt initialled a provisional understanding about transferring control over Shell Transport and Anglo-Saxon in return for access to Iraq oil, Deterding had several options open.

In February Colijn, having been briefed by Deterding in London, sounded out the Royal Dutch board about the proposal in terms which suggested that at that stage nothing had been signed or initialled.[208] Though a Bataafsche director, Colijn did not then have a seat on that board, which underlines just how much he had become a pivotal figure in the Group, more so than Stuart, presumably because Deterding's peremptory style, anathema to someone like Loudon, suited his own military persona.[209] After much discussion the Royal Dutch board declared itself prepared to surrender control over Shell Transport and Anglo-Saxon, and to transfer the Russian, Venezuelan, Mexican, and Romanian operating companies to Anglo-Saxon, but not Asiatic.[210] This key condition would ensure, as it had done in 1902-1907, the Group's continuing grip on overall operations, and since it ran counter to the purpose of the argeement that safeguard had to go. Discussing a full proposal in March, with Deterding present, the Royal Dutch board attached seven further conditions of secondary importance, one claiming a tax advantage, another the extinction of Sir Marcus's special voting rights in Shell Transport. One final clause stipulated that agreement was to come into force as and when the British Government would have given the Group full access to all oil fields in Asia Minor.[211]

These conditions did not materially alter the apparent fact that the Royal Dutch board had accepted a full surrender. Had it really? The Harcourt committee certainly thought so, but the agreement initialled by Harcourt and Deterding in May provided little more than the kind of paper controls which Waley Cohen ridiculed in 1923: the companies concerned to become or remain British-registered, the majority of their boards made up of British-born British subjects, and the appointment of a special nominee with a veto over proposed new board members and over alterations to the agreement. Both Anglo-Saxon and Shell Transport already complied with the first two conditions, and other operating companies concerned could easily be made to do so, in as far as they did not already. The third one did not pose much of a handicap either, since the agreement's preamble specifically excluded interference 'with the commercial policy or financial or business

Un don d'un million pour soulager les misères

M. W. Deterding, directeur d'une importante société industrielle hollandaise, vient de donner au gouvernement français une somme d'un million pour aider au soulagement, sous toutes ses formes, des misères résultant de l'occupation ennemie dans les régions évacuées ou envahies.

Cette somme devra être distribuée par les soins du groupe parlementaire des départements envahis à la tête duquel se trouve M. Cuvinot, sénateur.

M. Deterding a tenu déjà précédemment à marquer son affection pour notre pays en prenant la charge jusqu'à la fin de la guerre de l'hôpital néerlandais du Pré-Catelan où deux cents lits sont à la disposition des blessés français.

---

16 mai 1917

...male, le stock des vins de ... à 2 sh. 6 d. la bouteille sera ... trois mois.

... cette restriction ? — ajoute-t-il. ... tonnage ? Mais les bateaux qui ... charbon à Bordeaux, au lieu de ... vide, peuvent en rapporter du ... se surcharger.

... qui porte 6.000 tonnes de char... emporter 7.000 barriques repré... millions de bouteilles !

... nt du tonnage n'existe donc ...

... our empêcher l'exportation des ... anniques ? Mais les finances ... s et les finances françaises ... ires ; et, comme l'ajoute M. ... , il y a intérêt, pour les deux, ... uissions envoyer les poteaux de ... la Grande Bretagne a besoin, ... qui sont un aliment de pre... , au lieu d'être obligés d'en... or.

... haitons que la diffusion de ces ... irrésistibles fasse comprendre ... et alliés que nos « clarets » ... hez eux à quelque sympathie, ... la même sympathie dont ils ... sés sur notre sol.

... ET BIJOUTIERS MARRONS

... MONDE INTERLOPE DE NICE

... guerre, on ne s'embête pas à Nice ... de des grands hôtels et des bijou... tiers.

...rimée par des chagrins de famille et par ... tat de santé, une femme fort honorable, ... certain âge, était allée passer quelques ... es sur la Côte d'Azur et s'était instal... ns un des meilleurs hôtels de Nice. Quel... ours à peine après son arrivée, le direc... rant de l'hôtel trouvant insuffisant, sans ... les revenus de sa profession, harcelait ... nte des plus vives sollicitations pour la ... aux mains d'une troupe de fournisseurs, ... iers, modistes, fourreurs, bijoutiers. C'est ... de l'un de ces derniers que l'opération ... s activement menée, avec succès, d'ail... Un collier de perles représenté par le di... de l'hôtel comme ayant une valeur d... 20.000 francs était « laissé » à 13.000. ... ais on n'avait vu d'aussi magnifique oc... ! »

... urée et circonvenue, la dame fit l'achat du collier qu'elle paya au bijoutier en un chèque de 13.000 francs sur la Société Générale.

Quelques jours après, l'acheteuse reprenait le chemin de Paris et apprenait avec stupéfaction, d'une personne compétente en matière de perles, qu'elle avait été audacieusement volée. Bien mieux, il ne pouvait pas y avoir l'excuse d'un acte commer... al fait par le bijoutier

LA GÉNÉROSITÉ AMÉRICAINE

Les Dons Stillmann et Deterding

Les Américains ne se lassent pas de prouver à la France leur sympathie sous les formes les plus tangibles.

Nous signalions, il y a quelques jours, la générosité de M. James Stillman qui, non content de mettre à la disposition des blessés français son superbe hôtel du Parc Monceau, avait remis au Président de la République un chèque de 500.000 francs.

Ce magnifique geste vient d'être répété par M. H.-W. Deterding, Directeur Général de l'Association Pétrolifère Royal Dutch, qui a chargé son Délégué, M. C.-S. Gulbenkian de remettre à M. Ribot, Président du Conseil des Ministres, la somme d'un million de francs pour venir en aide aux victimes des régions évacuées ou envahies.

Ce n'est pas la première bonne action de M. Deterding. Sans qu'il ait été fait beaucoup de bruit autour d'une autre de ses générosités, nous nous permettons de rappeler que l'Hôpital Néerlandais du Pré-Catelan, où 200 blessés reçoivent dans le plus merveilleux cadre les soins les plus dévoués, a été complètement pris à la charge du grand philanthrope jusqu'à la fin de la guerre.

Le plus bel éloge que l'on puisse faire de cet homme de bien est qu'il est fils de ses œuvres et que son immense fortune est uniquement le fruit de sa vaste intelli... ... n travail acharné. C'est un exemple à suivre !

Sans mettre en doute aucunement la valeur intellectuelle de la demi-douzaine de nos « Rois du Pétrole » qui ont si largement profité de la guerre depuis trois ans, nous devons regretter qu'ils soient de mentalité assez obtuse pour ne pas comprendre qu'il faut savoir faire la part du feu et que quelques centaines de mille francs sur tant de millions volés pourraient leur servir plus tard de circonstances atténuantes lorsque sonnera l'heure des règlements de compte avec tous ceux qui ont profité de la dureté des temps et de la domestication des volontés pour imposer au public l'achat à vingt sous de ce qui en valait quatre. Avis à nos grands pétroliers !

François Massol.

LE COMMERCE FRANCO-CHINOIS

... été attirés par l'appât des ho... mais lorsqu'ils surent quel ét... veau journal, ils se sont reti... les autres. »

L'accusation était formelle. ... pas seulement la politique ru... milieux financiers français.. ... mutisme absolu sur cette ... grave ?

Il y a lieu de supposer qu'e... pé aux trois ministres qu... pas tolérer que, sous le ... des manœuvres de trahiso... même dans les hautes sph... bituellement les représenta... dénoncées comme allema... protestations francophiles.

En tout cas, et sous ... cette bizarre situation, nou... droit que ces collaborateurs e... blissements mis en cause p... accusatrices faites en pleine ... à fonctionner à Paris et qu... être de prendre à leur égard... réclament les soucis d'une ... lance des menées allemandes ... France par l'intermédiaire ... russes ou français.

Un homme nous paraît ... fournir de gré ou de force au ... du Gouvernement français le ... que nous sommes en droit d... M. Mathieu, à la *Banque Inter...* Laffitte, l'ineffable successeur ... riste Raffalowich qui nous le... de beaux chiffres pour nous ... réel fait par la famille Nicol... milliards que le peuple franç... au peuple russe pour la réfe... la construction de son artiller... de ses armées.

A défaut de l'intervention ... çaise dans les agissements d... ques russes dénoncées comme ... pleine Douma, il appartient ... surveiller ces nids de vipère ... acte louche, de les faire supp...

LE SUCCES DE L...

des nouvelles obligation...

C'est avec un vif empressem... teurs de Bons Municipaux on... pel de la Ville de Paris en ... leurs titres contre des Obliga... 5 ans 5 1/2 0/0 de l'Emprunt ... Jusqu'à présent, plus de 330 ... souscrits par la fidèle clientèl... palité : c'est un nouveau témoi... fiance dans les sages mesu... s'est toujours efforcée de pren... commun.

Nous ne nous avancions d... prévoyant un grand et légitim... émission, qui a su obtenir par ... conditions attrayantes en plu...

... La ... tra... des ... our ... ou ... ndi... ouf... ns; ... ues

... l'*Information* un article que l'on aimerait voir reproduire par tous les journaux qui se piquent de s'intéresser aux problèmes vitaux de l'alimentation dont l'importance... et l'acuité n'ont pas cessé de s'accroître depuis le début de la guerre.

management of the companies concerned' from the controls imposed. The committee had found it impossible to effect a change in the ownership structure of the Group, so Royal Dutch would continue to hold 60 per cent of both Anglo-Saxon and Asiatic, plus 470,000 shares in Shell Transport, the largest block immediately after the Samuel family, and the right to nominate directors.[212] The set of contracts binding the various companies together would not be changed either. So Royal Dutch did not surrender anything substantial. The whole exercise amounted to little more than an intricate charade designed to placate excited patriotism in return for serious commercial opportunities, the pressures of war having taught the Group how to perfect this particular art. An Anglo-French memorandum for coordinating oil policy, signed in April 1919, had realigned the TPC shareholdings into 70 per cent British, 20 per cent French, and ten per cent local interests. The Deterding–Harcourt agreement gave Anglo-Saxon half the British share, i.e. 34 per cent, of TPC, which company was to be managed by Shell Transport for the first seven years.[213]

Meanwhile Deterding's wartime charm offensive in France, which had included preferential supplies and a small Dutch hospital largely funded by him, paid very handsome returns. In March Senator Henry Bérenger, the French Government's General Commissioner for Gasoline and Fuel, agreed to have Royal Dutch exploit any oil interests assigned to the French Government by the peace treaties under negotiation. The French TPC share would be assigned to the newly minted Société pour l'Exploitation des Pétroles, and Royal Dutch obliged by having the company set up over a weekend in July 1919, thus underlining once more the importance of dual nationality.[214] With Gulbenkian's famous 5 per cent, the Group now controlled half of TPC, rendering any agreement with the British Government quite redundant. No surprise that Prime Minister Lloyd George complained so bitterly about oil trusts in March 1920. Deterding's deft manoeuvring had defeated his Government's earnest endeavours to bring the Group under control, and with the acquisition of Mexican Eagle the Group dominated British oil supplies, notably for liquid fuel, more than ever. Negotiations about the Deterding-Harcourt memorandum continued in a desultory fashion before petering out in 1920. Once again the Group's articulated structure had shown its curious strength.[215] In one respect these talks did signify a marked change, however. The war had made both the Group and governments around the world fully aware of oil's strategic importance: relations would never be the same again.

Hiding the spoils The Group was already very profitable before the war. Royal Dutch paid 48-49 per cent dividends in 1912-14, while Shell Transport awarded its shareholders a respectable 30-35 per cent. Such figures had already attracted sufficient criticism about the iniquitous market power of oil trusts to make the Group's directors acutely conscious of the need to show restraint if they were to avoid accusations of profiteering. Moderation was not altogether easy, however. Oil prices shot up in the belligerent countries and continued rising, though the US and Asia lagged behind during the first two years of the war. Agents received instructions not to push prices, which Lane termed 'magnificent' anyway, but of course the Group had to follow the market, even if keeping a distance.[216] Directors worsened their own predicament by radical efficiency measures during the financial crunch in the early stages of the war. Bataafsche saved 1.5 million guilders of costs during the three months of August to October 1914 alone.[217] Asiatic made extra profits of £ 275,000 over 1914. Higher prices generated £ 100,000, cost-cutting another £ 175,000.[218]

Consequently, profits rose to embarrassing proportions, defying the burdens of wartime taxation. Some of the profits existed on paper alone, exchange rates and currency controls preventing them from being transferred. Even so giving shareholders their full due would have been a public relations disaster, so the directors kept dividends more or less stable at around 40-45 per cent for Royal Dutch, and a straight 35 per cent for Shell Transport. However, this forced them to find ways of hiding the involuntarily retained earnings. Here as elsewhere, the lack of data hampers attempts to get a clear insight into what happened, but the policy trends are clear enough: the ballooning revenues were absorbed by large investments, more than usually generous depreciations, sums carried forward and other items on the balance sheet. During the first years of the war, the Group could still buy capital goods such as ships or Trumble installations, and directors did so on a large scale. By 1916 however, raw materials and equipment became short even in the US. Just then the Group's companies there entered a very profitable phase. Waley Cohen warned Sir Marcus not to let Shell Transport's profits rise above those of 1914 and 1915 at any cost, because 'being as I am in close touch with the way in which our business is looked upon by the various Government departments at the present time, I feel it my duty to tell you that in my opinion such action would be simply suicidal'.[219] Lane reported to Paris that rising revenues had 'brought about a considerable anxiety on the part of Mr Deterding, because he thinks that for the year 1916 the profits will be so great, it will be impossible to hide them, and the Company will be compelled to pay a very heavy dividend – probably 70 to 75 per cent'.[220] For 1915 the danger was avoided by introducing a novelty on the Dutch money market, dividends paid in bonus shares. After a long debate the Royal Dutch board accepted Deterding's proposal to declare a dividend of 38 per cent, 15 per cent in cash, the rest in bonus shares issued at par. With Royal Dutch quoted at about 530 per cent on the Amsterdam exchange, this meant that shareholders really received a windfall of 108 per cent. At the request of the Paris Rothschilds, whose shares lay in occupied Brussels, special facilities were created for shareholders who could not present their coupons because of the war.[221]

The stratagem of bonus shares could not be repeated with Royal Dutch if the Group did not want to play into the hands of the Dutch politicians calling for oil revenue taxes in the Dutch East Indies. Shell Transport awarded its shareholders free bonus shares in 1918, when they were quoted at around six pounds for a one pound share, and shares at par in 1919, when they sold around nine pounds, i.e. giving them a bonus of respectively 600 and 800 per cent.[222] The balance sheets also had considerable room for manoeuvre. From 1917 both Royal Dutch and Shell Transport created large carry-forward items. At Royal Dutch, the amount rose from just over 100,000 guilders in 1916 to 1.1 million in 1918, with Shell Transport showing an increase from £ 450,000 to £ 1.1 mil-

On Armistice Day, 11 November 1918, French flag-bearers march through the Arc de Triomphe in Paris.

lion. The operating companies, safe from public scrutiny with the exception of some vague and very general data about Bataafsche, probably absorbed most of the revenues in the form of investments, depreciations, and large cash reserves. The Group had always had a very liberal depreciation policy. Exploration expeditions, geological surveys, and initial drilling campaigns in virgin territories were entirely funded from current expenses. Installations and equipment were written off as soon as cash flow allowed, years or even decades before their economic use would end.[223] Rising revenues now served to write down anything in sight, increasing the Group's hidden reserves, already huge before the war. From 1914 to 1918, Bataafsche total assets rose by 120 per cent, but depreciations increased by almost 300 per cent. The 1917 profits were cut down to size by writing off a very large amount on the Russian investments, the political developments there providing a convenient excuse. By 1918 there was not much left to write off.[224] New equipment and rigorous depreciation turned the company into one of the most efficient in the world. There are no cost data for 1918 or 1919, but from 1913 to 1920 the company's total assets rose by 265 per cent and gross revenues by 165 per cent, yet costs rose by a negligible 2.4 per cent to only 4 per cent of assets. Finally, the Group also let the staff share in the prosperity. Notably Deterding grumbled about high wages at every opportunity, but companies compensated wartime inflation with wage rises, and everyone received a peace bonus in November 1918.[225] Moreover, the Group paid an extra 25 per cent per year of every staff member's wages into the Provident Fund. After the war the hidden revenues surfaced at the holdings in the form of assets. From 1918 to 1920, Shell Transport's total assets jumped by 127 per cent, Royal Dutch's by 80 per cent.

[66]

[67]

03|220

The defeated German army struggles homeward.

Conclusion

By breaking up the world economy along national boundaries, the war ended the first era of international business. From August 1914, exchange controls, trade barriers, and chauvinist legislation forced corporations from the highways of global trade into the narrow lanes of national economies. One would have expected the Group to suffer disproportionally from this huge shift, because of its dual nationality, its wide spread around the world, and its global supply chains. However, its structure proved not just resilient, but perfectly suited to cope with the challenges of circumstances totally different from the ones in which the Group arose. The directors largely followed the imperatives of national allegiances, but retained their international outlook; the spread of subsidiaries ensured a continuing supply despite the sometimes violent regional interruptions; the supply chains turned out to be sufficiently elastic to cope with the shocks. Two factors which explain this success may be singled out. First, the flexibility of the organization. The Group succeeded in adapting quickly to circumstances and defending its market share despite losing a very large share of its production. Moreover, the dual nationality proved to be a blessing, both in holding disparate parts of the organization together, and in negotiating for a stake in TPC. However, this flexibility strained the managerial structure to breaking point, with serious consequences during the early 1920s. Second, the operations in the United States which generated the supplies to compensate some of the production lost. From a strategic point of view, the American expansion could not have started at a more propitious time; it would continue to be a vital asset.

Throughout the war, the Group acted as a rational and pragmatic economic actor, not as the 'selfless Government agency' which some claimed it to have been.[226] The Group did, in fact, give crucial support to the Allied war effort, it did exercise restraint in pricing, and did show a keen appreciation of commitments wider than the oil business alone. However, the Group's interests always came first, its long-term survival and prosperity remained uppermost in the minds of directors. Nor would one have expected otherwise.

[1]

Chapter 4

At the peak of its power, 1919-1928

During the 1920s, the Group stood at the peak of its power and had become the almost undisputed leader of the international oil industry. Its strong position was to a large extent based on the global spread of its activities, especially in exploration and production. It continued the expansion worldwide, begun during the pre-war period, and was particularly successful in exploration and production in the Western hemisphere, in the United States, in Venezuela, and in Mexico. The Group's dominance also met with increased resistance, however. Government pressure on the industry increased sharply, with the nationalization of oil industry in the Soviet Union as the most extreme example. Elsewhere governments started to claim a larger slice of the profits, forcing the oil companies to find a new modus vivendi, even in the Dutch East Indies. The oil companies attempted to develop common strategies to counteract growing state power. Building on his rapport with Walter Teagle, Jersey Standard's CEO, Deterding forged a closer relationship with this former archenemy; relations with Anglo-Persian Oil Company also improved. The closer cooperation between the oil majors had important consequences for the strategic game over access to oil. Cooperation, and the development of the Turkish Petroleum Company, which began to develop its concessions in Iraq during the 1920s, helped to cope with the challenge that the Soviet Union offered. These moves culminated in the famous cartel agreement of Achnacarry in 1928, when the major oil companies agreed to freeze market shares and stabilize prices.

Signal Hill, California. Early 1920s.

Dominating the industry Despite suffering serious setbacks during the 1920s, such as the nationalization of its Russian properties and the collapsing production of its most recent acquisition, Mexican Eagle, the Group's rapid expansion continued. Crude output rose by a factor of five between 1919 and 1929, from 5 million to 25 million tons, and the Group's share in global output peaked at more than 12 per cent in 1921 and 1923. The Group now became the world's biggest oil company in terms of crude production, overtaking its old rival, Standard Oil New Jersey. In the annual report of Royal Dutch of 1927 Deterding proudly printed a table showing that the Group's crude oil production was more than 50 per cent higher than that of Jersey Standard and of Gulf, its closest rivals, and that the gap in terms of shipping capacity was even bigger. Only Jersey Standard's refining capacity came close to that of the Group, but still a margin of 20 per cent in favour of the latter remained (Table 4.1).[1]

In the early 1920s Signal Hill, at Long Beach in California, became an industry byword for profligate over-production, with derricks on every available scrap of land, all being allowed by the 'rule of capture' to compete for oil from the same source. In this dramatic picture, dominated by a gusher, the curve of the anticline – the hill itself – is clearly visible. Gushers occurred following an uncontrolled breakage of the crust above a reservoir, and in the earliest days of Signal Hill, Shell's men were often called upon by less experienced competitors to fight fires and bring gushers under control, such as this one, which was brought under control without undue mishap in March 1923.

[2]

	Average crude output, barrels per day	Refinery capacity barrels per day	Number of tankers
Royal Dutch/Shell	344,200	560,500	156
Jersey Standard	214,700	475,200	96
Gulf	212,500	170,000	27
Standard California (Socal)	150,000	135,000	22
Standard Indiana	118,000	340,000	32
Standard New York (Socony)	100,000	179,000	42
Texaco	107,500	209,000	23
Anglo-Persian	102,600	138,000	83

Table 4.1
A comparison of the eight largest oil companies in 1928.

However, the outcome of such comparisons depends rather heavily on the yardsticks used. During most of the 1920s, Jersey continued to outperform the Group by a clear margin in terms of capitalized share value (Figure 4.1), apparently as a consequence of the dynamic demand for its shares on the US market.[2] Dutch investors, and also the Royal Dutch board, commonly complained about the Royal Dutch shares being grossly undervalued on the Amsterdam stock exchange, too small a pool for such a huge fish, and this was to provide a powerful motive for seeking listings elsewhere, culminating in the introduction of the shares on Wall Street in 1954. By contrast, the Group paid out more money in dividends than Jersey (Figure 4.2), resulting in a ratio of share prices to dividends of only about 18 for Royal Dutch Shell, as against 30 to 40 for Jersey Standard.[3]

Only during a brief period in the early 1920s, with oil prices at a post-war peak and the Group very active in acquiring new concessions and oil companies, did its shares capitalization rise above that of Jersey (Figure 4.1). The positions reversed again when the Group entered the difficult period of its grave mistakes in Mexico and Russia. Dividends dropped much more sharply than Jersey's (Figure 4.2), and the Group's share capitalization followed suit.

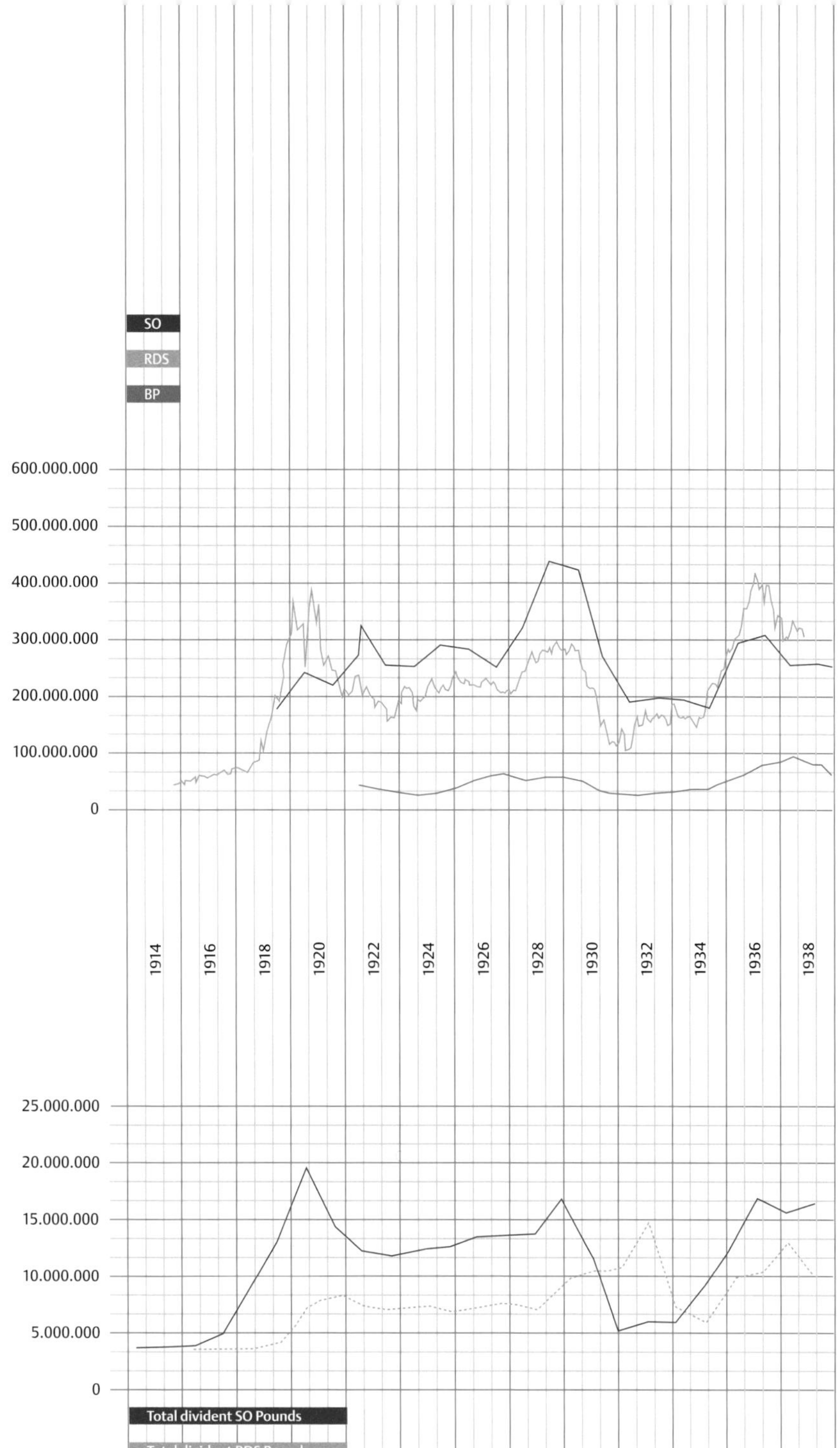

Figure 4.1
Total Share Capitalization of Royal Dutch/Shell, Jersey Standard and Anglo-Persian, 1914-1939 (in pounds).

Figure 4.2
Total dividends paid out by the Group and by Jersey Standard, 1914-1938 (in pounds).

The Mexican adventure was probably decisive here. Mexican Eagle, acquired in 1919 with oil prices at record levels, had been a major move, doubling the Group's crude oil production at a stroke (Figure 4.3).[4] However, despite being situated on the so-called 'Golden Lane', Eagle's properties had begun to suffer from declining yields, inspiring Deterding to link his bid for the company to future production, with a down payment of £ 4 per share topped up with 18 shillings conditional on daily production of crude oil of at least 50,000 barrels a day sustained during five years.[5] This precaution proved insufficient compensation when production from the Eagle fields suddenly dwindled. The Group's Mexican production remained substantial since its other subsidiary there, La Corona, managed to raise its output considerably, but the Eagle mistake encountered open criticism at the Shell Transport AGM, a rare phenomenon indeed.[6]

The Group's strong performance immediately after the war is all the more remarkable given the Russian confiscation and the collapse of production in Romania, which together had supplied more than 40 per cent of the Group's output in 1914.

Consequently, the Group's share in world output declined from 9 per cent in 1914 to less than 6 per cent in 1920, but just then the acquisition of Mexican Eagle plus growing production in the US and in Venezuela more than compensated for the losses sustained (Figure 4.3). As a result the Group's share in world oil output fluctuated between 10 and 12 per cent during the 1920s and 1930s. Jersey Standard's share hovered around 6 per cent during the 1920s, but the company drew level with the Group's 12 per cent between 1931 and 1934. Figure 4.3 also shows the gradually declining importance of supplies from the Dutch East Indies in Group supplies, from a third of total production in 1914 to a fifth in 1930. However, the colony continued to generate proportionally far

Figure 4.3
Crude oil production by the Group and by Jersey Standard as a percentage of world output (including regional breakdown of Group production), 1914-1938.

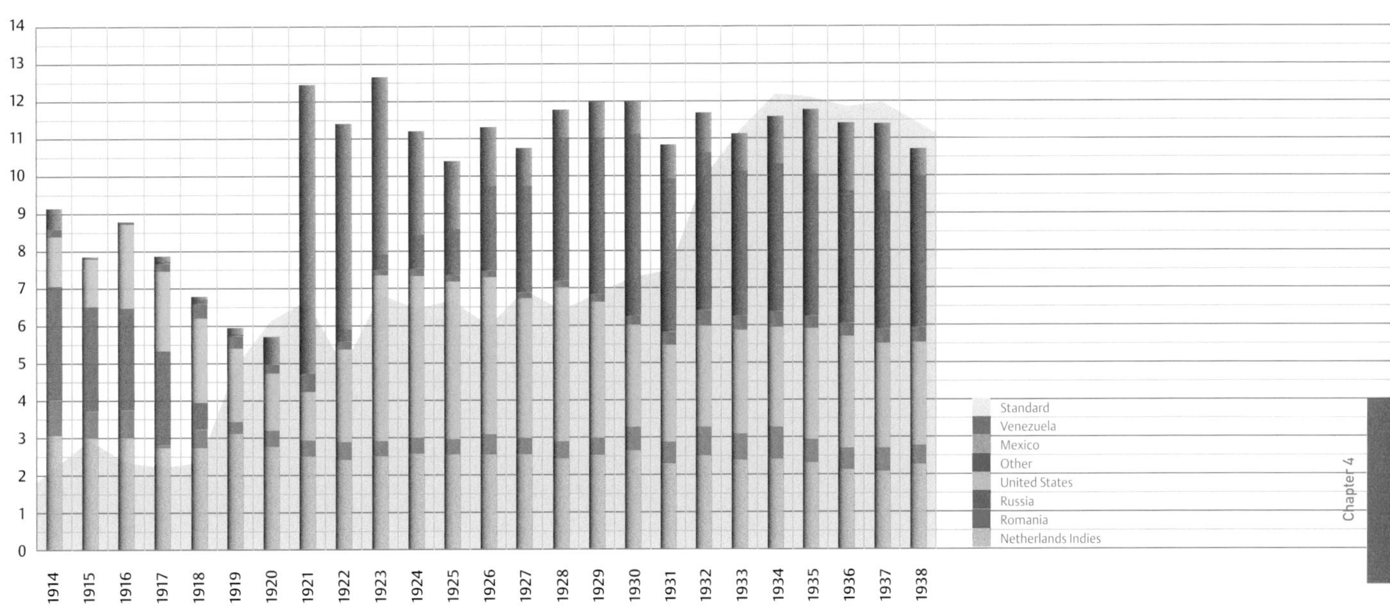

more profits than its share in production, mainly because of the relatively low production costs there. Acquiring production elsewhere, for instance in Mexico or in the US, cost substantially more. Moreover, the Group's tight grip on markets in the East also generated large profits, so in assessing the importance of the Dutch East Indies for the Group one has to bear in mind that Asiatic made substantial profits there as well.[7]

Unfortunately the rudimentary nature of the accounting procedures prevents us from analysing the Group's business in any detail or from pinpointing the profit sources. Consolidated accounts were only produced after the Second World War, and no figures for costs and revenues of individual business functions survive, so we simply cannot tell where the profits actually came from. However, Bataafsche clearly generated the most by far. Before 1914, Bataafsche contributed about 61 per cent of the Group's net profits, and during the war this share fluctuated between 59 and 68 per cent; but after 1920 low oil prices in particular hurt the profitability of the upstream business.

Consequently, the share of Bataafsche profits fell to 40 per cent during the first half of the 1920s, only returning to almost 48 per cent during the second half of the decade, when the Dutch East Indies' share in Group oil production dropped to a fifth.[8]

As the largest oil company in the world, the Group naturally succumbed to the temptation to try and impose its convictions about the oil industry's organization on its competitors, just as Standard Oil had done in the period before 1914, but with a crucial difference. Jersey had used price wars to cow competitors into submission and, given the enormous size of the company before its dissolution in 1911, no other business had withstood for long the power wielded by Rockefeller and his successors. Deterding, however, nursed a strong dislike for the unregulated competition and devastating price wars characteristic of Jersey's regime. His approach was radically different: he preferred to co-operate with his competitors in order to integrate them into rationally organized cartels or combines with stable market shares and carefully managed prices.[9] This approach had been the key to his success in uniting the Dutch East Indies' producers, followed by the formation of Asiatic in 1902 and the Group itself in 1907. He had repeatedly attempted to convince Standard of the benefits of co-operation, only to find his intermittent successes suffering setbacks such as the savage price war of 1910-11.

Since then, dynamic demand and sharply rising prices had made any co-operation between the major oil companies quite redundant, but the situation changed radically after 1920, when fast growing supplies sent prices into a long-term decline that continued until the mid-1930s. Since 1914, production in the American Midwest and in California had expanded strongly, an expansion that continued during the next two decades. During these same years Mexico also became a large oil producer, overtaking the Soviet Union as the world's second producer in 1918. Venezuela rapidly followed suit, in its turn replacing Mexico as the runner-up in 1927. The Group had or acquired large producing fields in all these new producing areas, further strengthening its international position.

Thus by the mid-1920s the situation of strained supplies and high crude prices inherited from the war turned into its complete opposite, and overproduction became the oil industry's main concern. Technological evolution aggravated the problem by offering new methods to increase crude yields. Cracking, for example, which became standard practice during the 1920s, increased the output of petrol by at least 50 per cent (see Chapter 6). Overproduction led to depressed prices and low profits. The huge and unprecedented profits of the 1910s made way for stagnating profitability and declining share prices during the 1920s. When, during the second half of the 1920s, share prices of most industrial firms rose spectacularly, particularly on the New York Stock Exchange, Shell Transport and Royal Dutch lagged behind, partly due to the depressed state of the international oil markets. The consequences of overproduction and keen competition made some US commentators even wish for a re-establishment of the old Standard Oil trust to return the US oil industry to its orderly pre-1911 state. With his preference for order over competition, Deterding shared this view.[10]

Deterding's strategy of cooperation between the major oil companies found its first application in two manoeuvres, an offensive stroke directed at the joint exploitation of new concessions in the Middle East through participation in the Turkish Petroleum Company, and a defensive stroke aiming to boycott the oil exports from the Soviet Union. Reinforced by the discussions at the legendary Achnacarry meeting of August 1928, these tentative moves created the basis for more intensive forms of collaboration during the 1930s.

Managing expansion in the US As related earlier, the Group had first entered the US market in 1910, selling surplus Sumatra gasoline after Standard had terminated a supply agreement. This move had been inspired by Deterding's determination to enter the US, based on his strategic vision that the Group could never draw level with Standard if it did not have a strong position there.[11] In February 1911, Deterding wrote to Loudon that Standard's policy of ruining its competitors by dumping cheap oil products 'has as its basis the sale of products at high prices in America', which made it possible 'to sell abroad at any price, as a result of which competition there is kept in check and, if possible, destroyed'.[12] Consequently, the Group needed to have a foothold in the US, too.

Shell California quickly became a success on the strength of the Californian Oilfields acquisition. In the Midwest Roxana had had some difficulties in getting established, but by 1916 these appeared to have been surmounted, just in time to enable the company to generate crucial supplies and cash for the Group during the war. Expanding the American operations was not very lucrative or always easy, as we will see, but the long-term benefits were very great and would, perhaps more than any other of the post-1907 developments, fundamentally change the Group. From 1922 to 1928 the US was the Group's most important producing area, ceding this position to Venezuela in 1929. Moreover, with the US oil industry generating so many technological and marketing

[3]

Marked in black, the oil fields of California (such as Signal Hill) often lay in already heavily populated residential areas.

innovations, the Group would have been seriously handicapped without its operations there.

The Group's difficulties in the US centred on Roxana, which failed to sustain the sudden production spurt during the war largely due to an unsuccessful E&P policy. The Midwest was a highly competitive environment characterized by many relatively small independents, moving as wildcatters from field to field. In order to keep Standard out, states had often imposed restrictions on the size of properties. This aggravated the fragmentation of the industry and handicapped efforts to reap economies of scale by linking fields with pipelines to refineries. Fragmentation had not prevented Standard from dominating the market both of crude oil and of oil products either. Moreover, many fields had a short life cycle, production peaking early and dropping significantly after a few years, or even after a few months. The vast properties the Group bought in 1913 were estimated to produce about a million barrels in that year, but production declined to half that level within a year.[13]

These problems were well known; Erb in particular remained very cautious, stressing time and again the unpredictable nature of the fields in the Midwest. However, confidence in the Group's superior, methodical approach to E&P won out. As Lane wrote to Weill in 1916: 'our geological work – which we have been conducting for the last two or three years – has put us in possession of such knowledge that we think we shall be able to secure the most valuable part of the new territories available (...) The exact geological knowledge of petroleum territories does not appear to have been practiced very much in the United States, where all prospecting has been done on a wildcatting basis, whereas we have made careful study beforehand in a scientific fashion.'[14]

However, despite hiring an expert geologist, Van der Gracht, for the Midwest, Roxana's field acquisition policy retained a definite element of hit and miss.[15] As President of Roxana from 1917, Van der Gracht conducted a vigorous E&P campaign, in Texas, in Louisiana, in the Rocky Mountains area, employing a total of some sixty geologists and spending a lot of money, but to little avail.[16]

[4]

ELL COMPANY of CALIFORNIA
LONG BEACH, CAL.,
MAY 21-1923

04|230

By 1919, Deterding was getting impatient and he started a discussion about Roxana's performance. He wrote to Van der Gracht criticizing his management, pointing to a number of recent mistakes.[17] In a parallel letter to Colijn he summarized his view on the problems: 'It seems to us that they have permitted the American spirit to influence them too much, and the course of affairs has been quite contrary to good business methods. They have spent thousands of dollars in taking undeveloped properties and testing them, with no success, their own production is gradually declining, and instead of their curtailing their expenditure to offset this decline the expenditure is going up in leaps and bounds, and it appears to us that they are running the whole concern on a much too ambitious basis, and instead of being content to gradually extend their facilities they have from the beginning pressed for them to be much larger than their actual producing capacity, to which I am sorry we agreed at the time, with the result that they now blame us for the excess capacity and use it as an argument that almost at all costs their producing properties must be extended.'[18]

In his letter to Van der Gracht Deterding also sketched the special position the US subsidiaries had had: 'If we had not made an exception for your business, instead of, as we generally do, maintaining our directions here (but we had so much trust in you that we thought you very likely knew better than we and as we had partners we were most anxious to make it a pure American Company in which we would only be advisors) we would be today many millions of dollars better off.'[19]

Van der Gracht replied in a series of memoranda with an in-depth analysis of the problems which the Group faced in the Midwest. He described the difficult geological conditions, and emphasized the '*undue haste in which everything has to be done*, exploration as well as purchases', highlighted the 'methods of our competitor, *the Standard Oil Co.*', and singled out the only strategy with which, as he thought Deterding would agree, these conditions could be overcome: 'we have either to play the game in full (...) or to desist entirely'.[20] According to Van der Gracht, the 'principal sore point with the Midcontinent companies is their shortage of crude', because Roxana had been very unlucky in the exploration of new fields. Because buying proven or semi-proven acreage was

Impressive numbers of personnel and delivery trucks were gathered outside the premises of the Shell Company of California for this photograph, taken in 1923.

[5]

Exploring the Mid-West in 1918 are (from left to right) Dr Josef Erb, head of the Group's geological department; Louis Roark, a geologist working for Roxana; in the car, the company's chief geologist R. A. Conklin; and (to give him his full name) Roxana's president at the time, W. A. J. M. van Waterschoot van der Gracht. The team stopped for the photograph part-way between Healdton and Waurika, Oklahoma.

very expensive, the company had preferred to concentrate on acquiring unexplored fields and on wildcatting. The lack of success meant that pipelines and refineries were under-utilized, pushing up operational costs. Buying crude was an option, but this was rather expensive, since 'S.O.C. [was] making the "market" of both crude and products, and the possibility of high crude versus low products policy'.[21] Finally, Roxana did not have its own sales organization, and had to sell its products to jobbers, thus missing the stability of direct market access. Roxana Vice-President Richard Airey also stressed the latter point, but considered the main problem to be that Van der Gracht was 'not a commercial man'.[22]

Amongst the remedies proposed by Van der Gracht were cost cuts by laying off part of the staff, moving the head office from Tulsa to St Louis, and suggestions 'to obtain American sympathy' by recruiting a US top manager on the board. This initially satisfied Deterding, but the discussion about the quality of management and the degree of independence of the US operations continued. Van der Gracht asked for more independence, but Deterding was unwilling to allow it because, as he wrote in one of his memos, 'you lack the confidence that we can keep sufficiently free from the "American" spirit of boosting optimism, and in fact believe that almost everyone you send there would catch it'.[23]

The reorganization of 1919-20 and the move to St Louis did not satisfy Van der Gracht completely, however. In August 1921 he wrote a long letter to Erb, one of the few who, in his opinion, understood the Mid-Continent problems, and detailed his criticism of the current structure of the Group. He complained about the interference from London in American affairs, often characterized by a lack of detailed knowledge of circumstances on the ground which only a few managers possessed. He suggested that, whereas the marketing of the Group was coordinated by London, The Hague should play a similar role in technical and geological matters, coordinating the policies of the different operating companies in exploration and refining.

'Thus it appears as if a rather liberal policy as to reserves and equipment were sometimes followed for one enterprise, whilst the future of another important investment is jeopardized by exaggerated economy, and important decisions are taken more under the influence of a momentary impression rather than as the result of careful considerations. It is evident that the resources of the Group available for the acquisition of reserves cannot, and should not be distributed with a mere view to existing investment, but solely with a view to the probable profits they can yield to the Group as a whole (...) I strongly feel that the entire mining and engineering of the Group's business should be concentrated in The Hague.'

Moreover, in the management of The Hague 'the various subsidiaries should as much as possible be represented by men sufficiently conversant with local conditions and possibilities of enterprises abroad, and in constant personal touch with them through repeated visits'.[24] De facto, his proposal would imply that The Hague should strongly increase its coordinating function, at the expense of the central control by Deterding in London. These proposals were very similar to the ones suggested by Colijn and Philips in the same period, but they did not leave a mark in the

American staff under training, gathered together at the start of a fire-fighting exercise.

debate conducted in The Hague between Deterding and other directors during these years (see Chapter 5).

In 1920 the Group's American venture was still only a modest success. Two integrated companies had been formed, but in the Midwest Roxana remained handicapped by its erratic and expensive supply of crude. Production in California stagnated after 1918, and even declined sharply in 1920-21. The two companies together produced about 2.4 per cent of the crude oil of the US in 1920, two-thirds of which came from California. Both companies did make quite handsome profits, but this was entirely due to the very high prices of crude and oil products. The years following the First World War have been described as 'the first oil crisis': demand rose spectacularly, notably for gasoline to fuel the rapidly increasing number of cars. Since supplies could not keep up, prices soared, sparking a debate about the long-term availability of crude and about ways to conserve the available supplies. Facing growing concerns about the access of US oil companies to foreign sources of crude, Government agencies such as the State Department assumed a more active role in promoting US oil interests abroad, resulting in demands for participation in the Turkish Petroleum Company, and in pressure on countries, or on oil companies from those countries, refusing American companies access to their territory. As we saw in Chapter 3, this led to ownership of the US Group subsidiaries being transferred to Bataafsche, because US companies were allowed to obtain some concessions in the Dutch East Indies, but did not get access to the British Empire.

The spectacular expansion of crude production rendered the debate about conservation and US foreign policy during the 1920s increasingly irrelevant, however. Between 1919 and 1924 industry output increased by an amazing 100 per cent, as a consequence of the enormous growth of production in Texas and in the Los Angeles basin. The Group operating companies profited disproportionally from these developments. Whereas in the 1910s they had been unlucky in their E&P, now they had good fortune on

[7]

their side. On 25 June 1921, Shell of California brought in Alamitos No. 1, the first producing well in what was perhaps the most spectacular find of the early 1920s, Signal Hill near Los Angeles. Shell had acquired a large share of the concessions there, and even increased its rights soon after 25 June. Signal Hill marked the beginning of the exploitation of a series of large oil fields in the Los Angeles basin, in which Shell of California participated fully. In many ways conditions here were very similar to those in the Midwest during the 1910s and 1920s, characterized by small concessions; Signal Hill, for example, had been developed as a residential area, resulting in land rights being very fragmented. In combination with the American 'rule of capture', which meant that all the crude that could be lifted from a particular piece of land was the property of the landowner, this situation put a premium on exploiting such concessions as fast as possible. This in turn stimulated extreme forms of local overproduction, followed by sharply declining prices. For the Group, however, things were

In overcoming the challenges of undeveloped Nature, the muscle-power of animals – especially mules and horses – was often far more helpful than any motor. Below left, mules are about to drag a wagon ashore, and below right, a large team pulls two wagons carrying a 15-ton oil heating unit from the nearest railroad station to one of the 11 pumping stations being built as part of the Valley Pipe Line in California. Without the heating units the oil would be too thick to pump. Valley Pipe Line Company was organized on 16 April 1914. Construction began in October 1914 and was completed in the summer of 1915.

[8]

In Mexico in 1920, mules are seen (left) levelling the top of the dyke surrounding an open-air oil reservoir. Five years earlier, horses were used every day (right) in the construction of the refinery at Martinez, California.

[9]

[10]

[11]

[12]

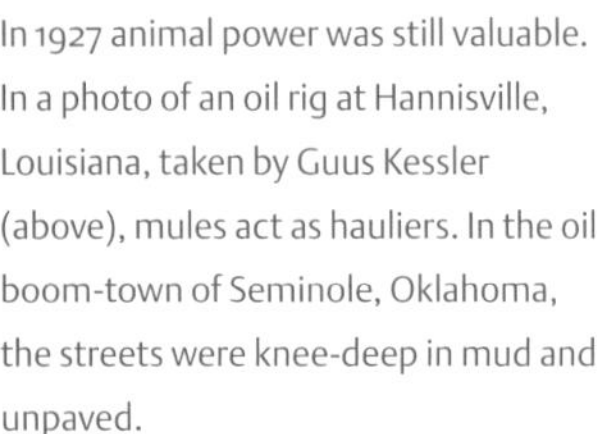

In 1927 animal power was still valuable. In a photo of an oil rig at Hannisville, Louisiana, taken by Guus Kessler (above), mules act as hauliers. In the oil boom-town of Seminole, Oklahoma, the streets were knee-deep in mud and unpaved.

very different now, since it was the first to strike, and not some outsider wildcatting near fields already brought in. At almost the same time, from 1922 onwards, production from the Ventura field, which until then had not given the yields expected on its acquisition by Shell in 1916, began to rise as a consequence of new techniques, such as deeper drilling with heavier rotary rigs. Total production of Shell California, which had dropped from its 1918 peak at 6.9 million barrels to 4.9 million in 1921, rebounded to 21.7 million in 1923, to stabilize at a about 20-21 million barrels thereafter. This in spite of the fact that the Coalinga field was shut in during those frantic years, Shell California preferring to use its

manpower in other fields and save the crude for leaner years.[25]

The fate of Roxana also underwent a marked changed for the better between 1920 and 1925. Early in 1922, Van der Gracht resigned, disappointed by the lack of response to his ideas about the Group organization, and Godber, sent over from the London office to the US in 1919, became president of Roxana. He took action to establish closer connections between the Group's US business and American companies in the same area.[26] As related in Chapter 2, getting access to the United Stated by buying a local company had already been considered in 1913, when the Group negotiated to take over Gulf Oil; the following year another attempt to find an American business partner, Sinclair, also failed.[27] Deterding had reiterated the need to find local connections in 1916 and now Roxana teamed up with a famous Midwest wildcatter, Ernest W. Marland, who was in constant need for cash to finance his many, often very successful, ventures. By becoming joint owner of his companies, Roxana profited from his skills in finding and developing new fields, and in this way began to participate in the rapid expansion of the oil industry in these years, now at a fair price. Investing $ 2 million to acquire a 50 per cent share in one of Marland's companies, Roxana earned $ 30.5 million in dividends, of which a staggering $ 19 million at the peak in 1926.[28] Since Marland was mainly interested in crude production and did not own refineries or a marketing network, Roxana also bought a large part of his oil, in that way also obtaining better use of its assets. Crude production of Roxana and its successors increased from 3 million barrels in 1920 to 7.9 in 1925 and 21 million in 1929; purchases of crude increased similarly from 0.4 million barrels in 1920 to 3 million in 1925 and 11.8 million in 1929. From 1922 Roxana set up a marketing organization in the Midwest so as to gain entry to markets for oil products still dominated by Standard.[29]

With this great expansion, Roxana ventured from Oklahoma into other states, i.e. Kansas, Texas, and Louisana. The success of the joint ventures with Marland also inspired further strategic plans, which eventually resulted in a formal merger of the Group's US business in 1922. Deterding pushed for a consolidation of the American operations and wanted to reinforce them by taking over the rest of Marland's properties.[30] When this idea foundered on Marland's asking price, Roxana and Shell of California merged instead with Union of Delaware, a holding company with as its main property a 26 per cent share in Union Oil Company of California. This latter business was thought to offer good prospects for a turnaround using the Group's experience, but it failed to materialize since Union Oil refused to be integrated, forcing the Group to sell its minority share again in 1924.[31] Meanwhile the merger with Union of Delaware had gone ahead. The Group received an initial 72 per cent share in the new Shell Union Oil

In the early 1930s, an oil boom hit the town of Kilgore, Texas, and – as at Signal Hill in California – oil derricks were erected wherever there was space.

[13]

Corporation, which was lowered to 65.7 per cent to accommodate minority shareholders in Roxana.

Deterding of course became President of Shell Union, underlining his keen interest in the US operations, but the Group wanted a senior American businesman to head the organization because, as Deterding had argued as early as 1916, 'It is, of course, always galling (apart from political considerations) in any country to see an enterprise doing well without local people being interested. It is contrary to human nature, however well a concern like that may be directed, or however much it may have the interest of the people at heart, not to anticipate there will be a kind of jealous feeling against such a company.'[32]

From a different perspective, Erb agreed. Analysing Van der Gracht's resignation, he emphasized the different business culture in the United States, writing to Godber 'I have the impression that Americans can only be fully understood and beaten in the struggle for securing production by their fellow-citizens (...) Then we may be sure that chiefs of American nationality as a rule appeal more to the staff with whom they are in constant touch than foreigners do.' According to Erb, staff trained in the Dutch East Indies were less qualified for top management jobs in the US, because in the colonies, 'where there is no competition, to be a good organizer and administrator are the first qualities for an official of the Bataafsche Petroleum Mij., whereas commercial abilities play a smaller part in a man's qualifications than they do even of a technical employee in the States'. Since the newly consolidated holding company would have almost a third of outside capital, it would also be desirable to have Americans in leading positions. Finally, unmentioned by Erb, only US citizens would be able to take care of any political missions needed.[33] The Group found just the right Chairman in General Avery D. Andrews, who had been active in the Venezuelan oil industry, notably in General Asphalt, one of the Group partners there.[34] His suggestion to let Shell Union follow American business custom by contributing to the Republican Party funds was rejected, however.[35]

Erb's comment formed part of the debate on control and management of the Group's huge US interests which Van der Gracht had started, and which continued after his resignation. He and De Kok, who had become directors of Bataafsche in 1921, liked Van der Gracht's idea that Bataafsche should have responsibility for upstream operations in the US, but Deterding's position was more complex. On the one hand he defended the position of the American subsidiaries against what he considered to be too much interference from The Hague, while on the other he himself tried to exert full control over the American business. In December 1923, after a long trip to the US, he proposed a curious balance. The board of Bataafsche should confine itself to supervision, while instituting a regular staff exchange between The Hague and the US: 'The best people are detained in the Hague to gather more and up-to-date knowledge, but are then sent out again for at least three years, and so on. Apart from the fact that we have thus practical technical control, we also gain this advantage: that anybody desirous of interfering too much, will know that for the greater part of his life he will be subjected to the same interference.'[36]

[14]

[15]

American oilfield workers, Oklahoma, 1939 (left) and Texas, 1937 (right). Those on the left are adding a length of drilling pipe to an oil well in the Seminole field.

When the Bataafsche board discussed Deterding's report in January 1924, he again emphasized the overbearing interference from The Hague departments in matters of technology and geological research, advising De Kok and Erb to travel to America and listen to the complaints of the managers there. Erb disagreed with him; he wanted to centralize knowledge and policy concerning geological research in The Hague, because 'in general the most important successes in the field of geology in different parts of the world have not resulted from local initiative, but from the management of the The Hague office, which often had to surmount fierce resistance of local geologists'.[37] De Kok and Philips agreed with Erb, the latter arguing that the Bataafsche office could not function well without the technical expertise of Erb or De Kok, meaning that they could not be abroad at the same time. Philips also thought that it would be a better idea if Deterding would stay in The Hague for a couple of months, in order to find out if the complaints against the head office were justified.[38] The meeting ended on a conciliatory note: Erb would first travel to the US, followed by De Kok, and Deterding would come to The Hague to study the Dutch side of the matter.

This discussion reaffirmed, as far as the US companies were concerned, the general principle of decentralization, balanced by a regular exchange of staff to keep relations between central offices and operating companies on an even keel. Such exchanges had been quite common in Royal Dutch from its beginning. The first generation of managers such as the elder Kessler and Loudon had started their career in the Dutch East Indies or at least spent a few years there. Now for the new generation of managers a training period of at least a few months in the US became usual. The main instrument for managerial control of the operating companies continued to be the investment budgets. In 1921, probably with the recent problems with Roxana in mind, Deterding called attention to the fact that 'the interests of the parent companies – the Shell and Royal Dutch – or rather wishes of these two concerns, do not quite coincide with the views as to capital expenditure by Bataafsche, the Anglo Saxon, the Shell of California, etc'. The operating companies had a clear incentive to invest as much as possible 'in extensions of the business', whereas 'the parent companies rely on a certain return from the money invested in those companies'. This conflict of interest could only be resolved by restraining the operating companies. Deterding sincerely regretted the need for this limitation of the decentralized model of governance: 'It may be that an instruction of this kind might, to a certain extent, hamper the free development, but it seems to me absolutely necessary that some standard instruction is given. It is exceedingly difficult for me, attending and having responsibilities not only in the affiliated companies but also in the parent companies, that I have to fight for the interest of the parent companies in the affiliated companies, and a very severe instruction of the parent companies will prevent my wasting energy which could be so much better applied.'[39]

A revised procedure for submitting and monitoring all capital expenditure proposals of operating companies to the boards of the parent companies was therefore introduced, and directors spent many an hour discussing these budgets.

Neither staff circulation nor budget procedure terminated the discussion about centralization versus decentralization for, as Van der Gracht had correctly pointed out, this was really a debate about the division of power between The Hague and London. To Deterding, decentralization meant that, even though the actual ownership of the US subsidiaries had been transferred to Bataafsche, London would oversee them; The Hague simply had to keep out. Consequently the Bataafsche board received only piecemeal information about the development of Shell Union, inspiring regular complaints from directors. In September 1925, for example, director IJzerman stated his regret that 'the Board knows so little about the enormous investments of the Group in the United States' and he asked to be informed about it in order to be able to form his own opinion.[40] A month later Kessler gave a detailed response, concluding that 'Although some fields have been better business than others, we all together have been quite fortunate with our investments in the United States, not only because of the direct results, which have to be assessed on the basis of the present value or our assets, and the declared dividends which resulted after generous depreciations, but also because of the position we have acquired in America which brings us enormous indirect advantages, because we can participate, in the country which dominates the oil business, in the market fluctuations and the newest developments in technology, which will bring us advantages elsewhere as well.'[41]

Kessler's comments neatly sum up the importance of the Group's US operations as lying more in the indirect advantages of access to market and technology than in the direct one of return on investment. Dividends had indeed been fairly meagre; during the 1920s the dividend on Shell Union common stock fluctuated between 4 and 10 per cent, with an average of 7 per cent, much below the huge profits the Group made elsewhere. In 1927, when the company's profits fell sharply, dividends could only be maintained at 7 per cent by sacrificing part of the reserves, a policy totally out of character with the Group's normal practice. This modest profitability was the result of intense competition and of the fact that Shell Union had been capitalized early in 1922, just after the peak in oil prices, so most of its assets had been valued at inflated prices. Profits were insufficient to finance the growth of operations, the more so because American investors demanded stable dividends to show that the company was doing well. To sustain its expansion, Shell Union and its subsidiaries borrowed a total of $ 265 million on the New York capital market during the years 1922-29, whereas retained earnings were only $ 57 million.[42] It almost seems as if even Henri Deterding, who had always been so proud of his very conservative financial management, had been affected by the American spirit.

If accidents happened, the results could be spectacular, as in June 1925 when fire raged through the La Rosa oilfield in Venezuela.

Then again, the main motive for entering the United States had not been to make profits, but to get a foothold in the world's biggest oil market and biggest oil producer. Nor can the effects of the American operations on the Group be easily measured in terms of money. First of all, as the US companies grew in size, they altered the balance within the Group, its dual structure with two central offices getting what almost amounted to a third leg on the organization. Secondly, though closely managed from Europe and having a top management of trusted Europeans, the American companies acquired more policy leeway than other operating companies. Finally, as we will see in more detail in Chapter 6, the Group derived great benefits in technology and marketing from its American operations. The US generated many innovations and served as a good testing ground for many imported innovations, thus strengthening the Group's emerging network of worldwide knowledge exchange. The nodes for this network were created by making a training stage with the US companies a regular fixture for aspiring managers and engineers.

In 1929 Shell Union rounded off its expansion with the takeover of New England Oil Refining Company, which provided a core organization for marketing and distribution on the East Coast and outlets for the increasing supplies of Venezuelan oil. Shell Union now had a coast to coast presence, but there was a hidden cost. Oil prices had started to fall, and established marketers on the Atlantic Seaboard were annoyed with the start of Shell Union's aggressive marketing there in 1927; Standard Oil New York (Socony) began a counter-offensive in the Dutch East Indies.[43] With the onset of the Depression from October 1929, Shell Union would be forced to retrench, but that story belongs to Chapter 7.

Venezuela: the embarrassment of riches The development of important producing regions outside the US also changed the strategic balance in the world oil industry in the Group's favour, relying as it did on having supplies close to markets, whereas Standard fed its overseas companies largely with American imports. The 1911 dissolution of the Standard trust increased this advantage by creating successor companies with a lopsided business structure. Jersey Standard, for example, inherited a large part of the trust's refining and transport capacity, but without producing fields to match. Such an imbalance would not have mattered so much before, given the huge power of the trust, but now the competition from Royal Dutch/Shell showed Jersey the superiority of the integrated company.

In response Jersey Standard CEO Walter Teagle developed a strategy of upstream expansion, the implementation of which was accelerated during the 1919-21 American oil crisis. By the mid-1920s Jersey had already increased its share in world oil production to about 5 per cent, up from less than 2 per cent in 1919, by acquiring producing properties in Mexico, Peru, and large new acreage in the US itself. Jersey had also gone into Venezuela, but without much success; according to the official history 'the company had to show for its investment of 27,000,000 dollars over nearly a decade of time only unexplored concessions, inland water craft, drilling rigs, much disappointing experience, miscellaneous camp facilities and forty-two dry holes'.[44]

[16]

[17]

By contrast, the Group had made rapid progress in Venezuela. It had been the first to arrive there, acquiring already before 1914 controlling shares in three companies which together possessed a very large part of the Venezuelan oil concessions: the US-based Caribbean Petroleum Company (CPC), and the British-based Venezuelan Oil Concessions (VOC) and Colon Development Company (CDC). All companies had convoluted ownership structures with minority shareholders and holding companies. The Paris Rothschilds, the General Asphalt Company, an American company, and the Burlington Investment Company all participated in the Caribbean, for instance. The Group owned half of the Burlington Investment Company, again with participation from the Rothschilds and from General Asphalt. Part of the share capital of the VOC was traded at the London stock exchange, and

[18]

[19]

Venezuela in the 1920s was an oilman's dream as well as sometimes being a nightmare. In 1922 (left), a remarkably relaxed and neat group stand in front of an uncapped gushing well in the La Rosa oilfield, north-east of Lake Maracaibo. In 1926 more oil spouts from well T11 in the El Cubo oilfield, district of Colon, south-west of the lake (above left); nearby, workers pose by well T13, Los Cruces, Colon (above centre); and surplus gas is flared at Casigua, Colon (above right).

[20]

[21]

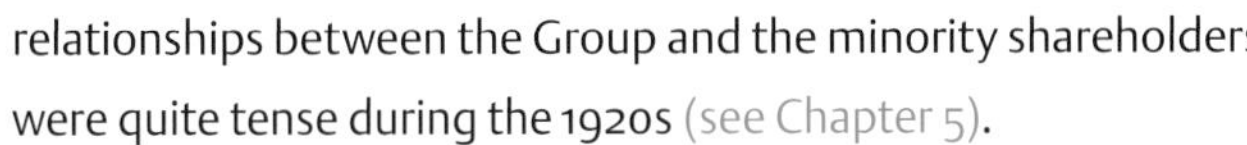

relationships between the Group and the minority shareholders were quite tense during the 1920s (see Chapter 5).

The combined concessions were huge. The Group really suffered from imperial overstretch: it simply had too many concessions for its resources. To some extent managers considered the Venezuelan concessions as strategic reserves, with Mexico receiving priority. Development was also retarded by particular obstacles encountered on the concessions, such as difficult soil conditions and, in some parts of the country, hostility from the local Indian population who regarded the oil companies as intruders. The shortages of materials and staff during the First World War then imposed further delays on active exploration, which was largely performed by US geologists supervised by Erb. Finally, conditions in Venezuela were very poor, far worse of course than in the US where operations had to compete for staff, generating a constant flow of complaints from the technicians and managers who were dispatched to the country.[45]

The slow initial progress contributed to the problems that the Group encountered during the early 1920s. When American oil companies such as Jersey and Texaco wanted to establish themselves in Venezuela, the government of Juan Vicente Gómez had good arguments to reduce the Group's dominant position. Gómez introduced legislation forcing the Group to concentrate on those parts of its gigantic concessions under active exploration and return the rest to the state and also started challenging the validity of the VOC and CDC concessions, as part of an overall review of the country's oil policy. American pressure played an important part in the debate on the validity of these concessions, with Jersey Standard in particular considered likely to make 'determined efforts to turn out the British companies' in order to acquire the concessions themselves.[46] Gómez did not want to free himself from the British to deliver the oil industry to Americans, however. In the long run he would profit far more from being able to play the Americans against the British and vice versa, so he did not want to reduce the Group's presence entirely.

The first round of the conflict over the concessions ended in 1922 with a new oil law which meant that VOC and CDC had to concentrate on much smaller fields in the future. In weighing the available options, developing the CDC concessions was considered very expensive, whilst the two other companies promised quicker returns on investment. Moreover, the CDC's articles of association contained a clause granting the minority shareholder, the US-based Carib Syndicate which had originally acquired the Vigas

RADIOGRAM

Via RCA WORLD WIDE WIRELESS

R.C.A. COMMUNICA

RECEIVED AT 64 BROAD STREET, NEW YORK, AT

PPCGRA HL320 Y 169

DENHAAG 9 28 1756

DETERDING ASPETCO NEWYORK

HWEYIFDEMG ODNTUNUKMI HARSUNUXO

FRYJIMZUSK

I am ~~we are~~ pleased to inform you

gas 27.000 ~~~~ no water

TELEPHONE: HANOVER 1811 To secure prompt action on inqui R.C.A. COMMUNICATIONS, Inc.

FORM No. 112

Commercial Cables

CLARENCE H. MACKAY, PRESIDENT

ALL AMERICA CABLES — THE COMMERCIAL CABLE COMPANY — MACKAY SYSTEM — POSTAL TELEGRAPH

PACIFIC — ATLANTIC

TELEPHONE HANOVER 1140 — "CABLES TO ALL THE WORLD" — ADDRESS 20 BROAD STREET

FORM CCC 3

APR 23 1930

ER5 XP2646

SGRAVENHAGE 47

NLT DETERDING CARE VANECK NEWYORK

YOUR PARTICULAR OUTSTEPPING WELL 168 SOUTHERN LAGUNILLAS
CAME IN WITH 166 TONS THEREFORE NO CHAMPAGNE FULLSTOP ON OTHER
HAND WELL ABOUT A MILLE FURTHER SOUTH BUT ABOUT HALF WAY BETWEEN
SHORE AND 168 CAME IN LAST WEEK WITH ABOUT 350 TONS
DEKOK

The brackish waters of Lake Maracaibo cover 5,100 square miles or 13,210 square kilometres – equivalent to nearly one-third of the Netherlands. From there and other parts of the world telegrams, often encoded (far left), flowed in to London and The Hague, informing Deterding of progress or otherwise. De Kok's telegram (near left) from Lagunillas, on the lake's north-eastern shore, shows that not every discovery was worth a celebration.

concession, the right to claim a quarter of any increases in the company's capital which were needed to develop the concession, without having to supply additional funds. In other words, they could claim a quarter of any capital supplied by the majority shareholders.[47] Consequently, the Group put CDC up for sale with an asking price of $ 10 million, with $ 7 million considered an acceptable bid.[48] Neither Texaco nor Jersey Standard showed interest, presumably put off by the strong position of the minority interests in CDC.

In 1924 the Gómez government again challenged the validity of the VOC concessions. The Group reacted immediately, after being informed about the renewed discussion of the legality of their titles in early November, 'on the 12th instant, two telegrams were dispatched to General Gómez through the Foreign Office, which would be presented by the British Minister in Caracas. One of these telegrams (...) pointing out the effect of any contest as to the V.O.C.'s rights on the credit of Venezuela, was signed by Messrs Rothschild, Baring and Schroder'.[49] By March 1925 the situation had not improved, and the discussion 'to legalise deeds and titles which are invalid and which may give rise to actions by third parties', as it was formulated by Gómez, continued. The VOC board replied that 'regarding compensation we still feel that this should be handled direct with Gómez through his private secretary or other acceptable person and we are still agreeable to payment of not exceeding £ 100,000 (...) after everything has been settled to our satisfaction.'[50] The situation had meanwhile deteriorated, however, because General Aranguren had leaked via an intermediary that the Group might be willing to pay as much as a million pounds to get their claims confirmed. The VOC board refused to consider such an enormous sum. One month later, after the resignation of Aranguren from the board, the sum requested had dropped to £ 350,000. Agnew travelled to Venezuela to negotiate an agreement on that basis and, thanks to 'most valuable assistance (...) rendered by the British and Dutch Foreign Offices', succeeded in settling on a total of almost 10 million bolivars, the equivalent of nearly 360,000 pounds, plus another 25,000 pounds for 'extra expenses incurred over and above the official payment to the Venezuela Treasury'.[51]

The 1922 oil law and its subsequent rather difficult implementation both stabilized the situation and created incentives and opportunities for the American oil companies to acquire concessions. The Group's share of production in Venezuela dropped rapidly, from 99 per cent in 1922 to 55 per cent four years later and 45 per cent in 1930.[52] This was a relative decline in an extremely dynamic industry, however. From 1926 Venezuela became the third largest oil-producing country in the world, after the US and the Soviet Union. Group production also rose very rapidly, from 240,000 tons in 1921 to 8.8 million tons in 1929. Because there was

04|250

deemed to be no suitable location on the Venezuelan coast for siting a refinery, the crude was transported by pipeline to the shore of Lake Maracaibo and from there shipped by barge to the refineries on the Dutch West Indies islands of Aruba and Curaçao, locations which also offered the advantage of political stability and attractive tax benefits obtained from the Dutch Colonial Office.[53] The Curaçao refinery, established during the First World War, was by far the bigger of the two. With Martinez, Wilmington, and Houston, it became the Group's main processing centre in the Western Hemisphere.[54] Until the early 1930s, when the American government banned foreign oil imports, most of the fuel oil and gasoline produced at Curaçao was exported to the US; in this respect Venezuela really functioned as an extension of the oil industry in the States.[55]

One of the companies that immediately took the opportunities opened in 1922 was Jersey Standard. When its own venture Standard of Venezuela failed to make much headway, Jersey made two acquisitions, buying Creole Petroleum Company in 1928 and Lago Petroleum Company four years later, both companies holding large concessions in and near Lake Maracaibo.[56] With these two crucial moves, Jersey Standard became at a stroke the most important producer in the country and at the same time became the largest crude oil producer in the world, overtaking the Group in that position (see Figure 4.3).

Cooperation and conflict over Iraq: The Turkish Petroleum Company The Group's involvement with the Turkish Petroleum Company (TPC) dated originally from 1912, when Anglo-Saxon had committed itself to a 25 per cent participation in the concern, which had been turned into actual shares at its final formation two years later. Other shareholders were the D'Arcy Group, i.e. Anglo-Persian, with 50 per cent, and Deutsche Bank with the other 25 per cent. TPC had been one of Gulbenkian's projects, the company's ultimate success largely due to his skilful mediation between the various groups involved, pasting over their sometimes quite antagonistic positions. He possessed a wide network of business relations in Turkey and at one point also acted as the official financial adviser of the government. In return for his efforts Gulbenkian received his famous 5 per cent in TPC, an interest that according to the 1914 agreement would be bought back by the D'Arcy group and Anglo-Saxon 'in the event of his death'.[57]

Chapter 3 showed how, at the end of the war, the lure of Mesopotamian oil induced the Group first to consider a full merger with Anglo-Persian and form a new and British company to gain access to concessions there, then to the 1919 Harcourt–Deterding memorandum aiming to bring the Group's main operating companies under British control.[58] Deterding's move to secure the management over the French share in TPC had rendered the memorandum quite redundant for the Group, but during the summer of 1919, Cohen Stuart found that it would also be very unwise to carry out the arrangements made. In a detailed memo dated 15 August 1919, he raised the issue of anti-British sentiments in various parts of the world. He argued that the Group owed its ready acceptance overseas at least partly on it being controlled by 'the subjects of a small, secondary, power'.[59] More importantly, Stuart questioned whether the governments of Mexico, Venezuela, Romania, and the United States would tolerate a company controlled by the British Government and possessing large concessions and other facilities in their respective countries. Indeed, the Mexican government had already indicated that it might declare the concessions of the Group's newly acquired Mexican Eagle forfeited 'in the event of

Chapter 4

[26]

Despite its tropical location, Lake Maracaibo (top) was not very close to being a paradise; but when oil was to be found, people swept in (below), coping with the climate and introducing all the essential aspects of their time and culture.

[27]

[28]

A panorama showing the vast extent of La Rosa oilfield, Venezuela, 1924. Derricks stand in the lake and on shore, along with storage tanks, warehouses, offices, an aircraft hangar and an entire town in what had been virgin jungle.

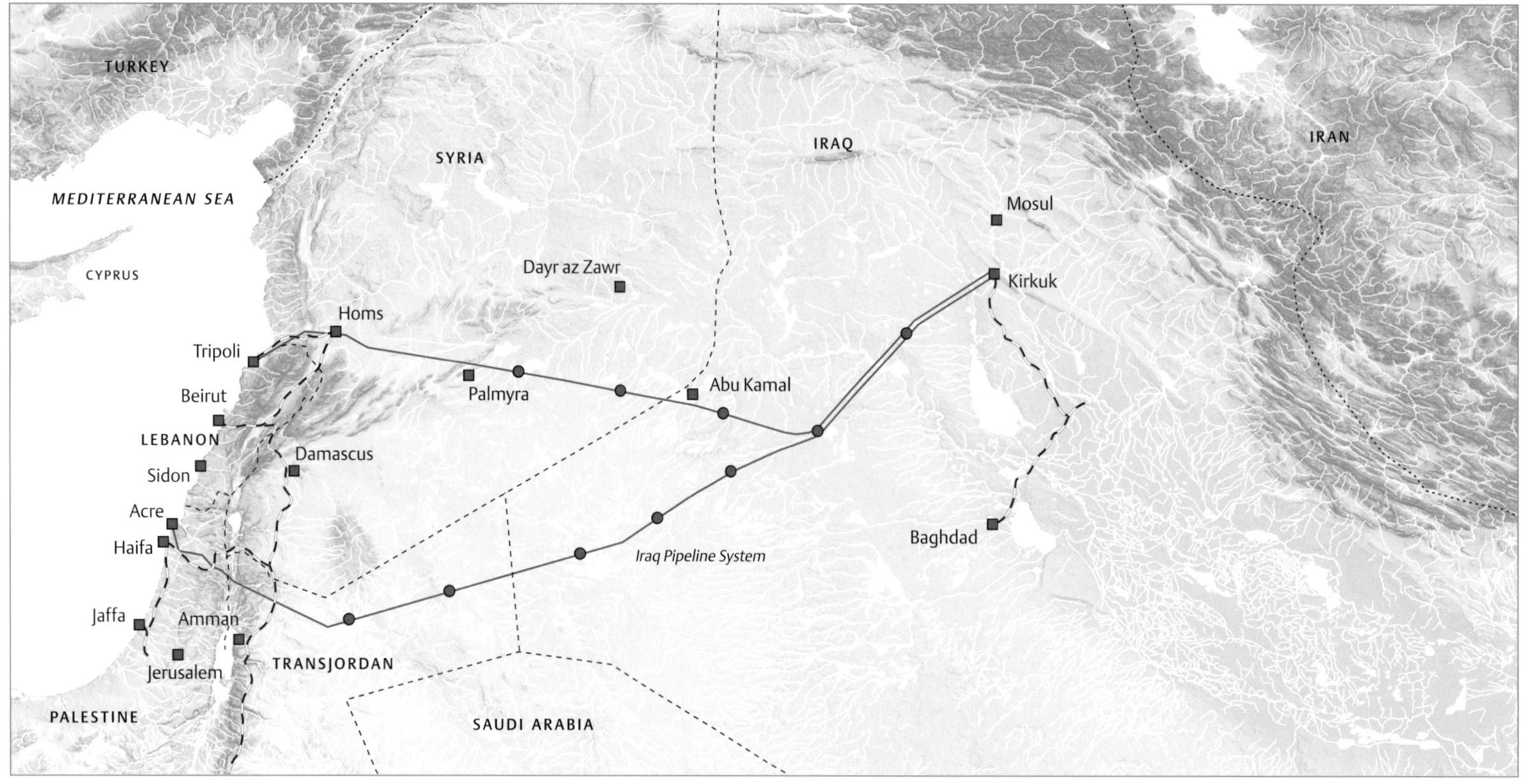

[29]

them being transferred to a foreign Government, or in case any foreign Government becomes partner in said concessions'.[60] Romania was considered to present a problem of similar urgency. Consequently Group managers decided to lift the operating companies in these two countries from the Harcourt–Deterding agreement for the time being which, of course, complicated the whole operation still further.

Stuart's premonition about imminent objections in the US to British control over oil concessions became a reality in 1920, necessitating the transfer of the Group's substantial and growing American assets from Anglo-Saxon to Bataafsche. In a manoeuvre of sublime ironic beauty, the susceptibility of British companies to chauvinism overseas defeated the government's patriotic endeavour to secure Britain's strategic oil position through control over the Group's main operating companies. When in 1922 the Harcourt committee reconvened and interviewed Walter Samuel about a merger with Anglo-Persian, the latter demanded as a precondition the sale of the Government's shares in that company.[61] Since surrendering the Admiralty's cherished supplier was unacceptable, the Harcourt committee concluded that an arrangement with the Group had finally become impossible.

Meanwhile American oil companies had turned their attention to Mesopotamia, especially Jersey Standard, which as we have seen had inherited large refining plants and global distribution networks from the 1911 dissolution, but only relatively small oil production, and actively sought concessions outside the Western Hemisphere.[62] The US companies, *de facto* excluded from the Middle East by the Franco-British agreement about cooperation there, began to argue for an 'open door' policy, i.e. the admission of American oil companies to TPC.[63] Towards that end Jersey Standard formed a consortium of seven companies interested in Mesopotamian oil. Strongly supported by the US State Department, this consortium gained access to the negotiations about a restructuring of TPC in 1922 by threatening to challenge the shaky concession that the company claimed to have in Iraq, concerning the Mosul area in particular.[64] By that time the position

Map of the Iraq pipelinesystem in the 1930s, from Kirkuk to the eastern Mediterranean.

— - Pipeline

● - Pumping stations

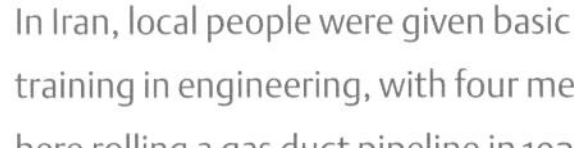

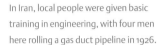

In Iran, local people were given basic training in engineering, with four men here rolling a gas duct pipeline in 1926.

[30]

of TPC and the recognition of its concessions had become a very complicated and delicate issue indeed, so it took years to arrive at a settlement. The Mosul region was disputed territory between Iraq and Turkey, a matter only settled in 1926. Britain governed Iraq under a League of Nations mandate. The Iraqi government itself claimed a right to oil concessions and a share of TPC, which had been promised to it by the French and British governments in 1920. To accommodate the Americans, France and Britain now reneged on their promise and exerted pressure on the Iraqi government to recognize the TPC concessions on new terms, which it did, reluctantly, in 1925. This appeared to clear the road for reconstituting TPC with the Compagnie Française des Pétroles, the US consortium, the Group, and Anglo-Persian obtaining 25 per cent each. Following a *modus operandi* agreement reached in July 1924, TPC became a joint venture, not a profit-oriented company in its own right, producing crude oil and oil products supplied at cost price to the shareholders in equal proportion to their participation. Two pipelines were to be built for transporting oil to the Mediterranean.

However, one crucial obstacle remained, Gulbenkian's 5 per cent claim. An opening bid to give him a 4 per cent 'lifelong' interest was fiercely rejected by him. Gulbenkian strenuously maintained that in confidential negotiations with Deterding and Nichols, the managing director of TPC on behalf of Anglo-Persian, the lifelong interest mentioned in the 1914 agreement had been transformed into a permanent claim on 5 percent of the TPC shares, in other words, the other shareholders could not reclaim his shares after his death. Deterding failed to remember this, but did not really deny it either, which Gulbenkian took as final proof of Deterding's unreliability. Then the July 1924 agreement gave him title to 5 percent of the oil produced, but Gulbenkian again swept the offer from the table with the argument 'that he was not an oil trader and that he might find himself in the position that he could not get rid of his oil'.[65] He maintained that the agreement encroached on his rights, to which he strongly objected.[66] In September 1924 Gulbenkian threatened litigation, and both sides instructed solicitors to prepare their cases.[67] With these threats and counter

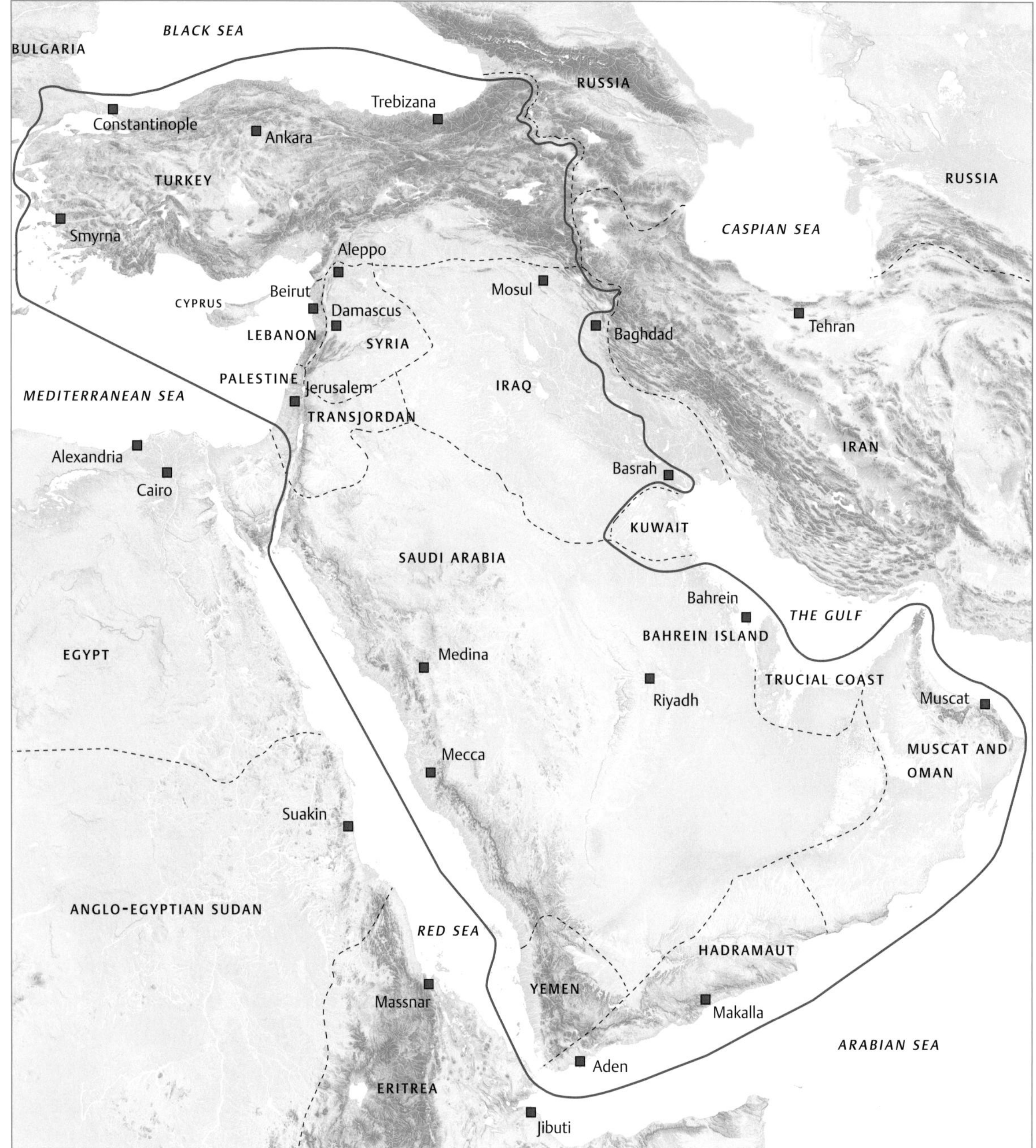

[31]

The original Red Line Area, 1928

threats from Deterding poisoning the atmosphere, the TPC conflict contributed to the definitive breach between the two men, to be analysed in greater detail in Chapter 5. The Americans added further pressure by threatening to withdraw altogether from TPC and strive to obtain their own concessions from the Iraqi government.

When in 1927 the first well of a spectacular new field came in at Baba Gurgur, north-west of Kirkuk, the negotiating sides finally made haste. Gulbenkian received his full due, including a royalty on all crude oil produced by TPC, and the company was finally established. The Compagnie Française des Pétroles, the US consortium, the Group, and Anglo-Persian acquired 23.75 per cent, Gulbenkian his five. In a crucial clause, the TPC shareholders agreed that they would only work on a joint basis in the region assigned to TPC, i.e. the old Ottoman Empire as defined by a Red Line drawn by Gulbenkian on a map, another one of his ingenious ideas which was to have profound consequences.[68] And so the Red-Line agreement was born.

The rising tide of nationalism

The new state which emerged in Iraq during the 1920s was unable to wield any influence over its territory at all, but other states could, and did. The fundamental shift in the relations between governments and the oil industry during the 1910s and 1920s was perhaps nowhere so dramatic as in the producing countries, which developed major policy offensives targeted at the oil companies in an attempt to raise more revenue from this highly profitable industry and to obtain a better grip on this strategic economic sector. In two extreme cases, Soviet Russia in 1920 and Mexico in 1938, oil companies were nationalized, but elsewhere governments acknowledged their dependence on the know-how and marketing networks of the oil majors and sought alternative ways to increase their slice of the cake. Political instability in many oil producing countries contributed to the industry predicament, if only because for obvious reasons political conflicts tended to focus on oil and on ways to deal with the foreign oil companies.

Mexico is a good case in point. The Group had acquired its first possessions there as early as 1912, building up a relatively small company, La Corona. Its production fluctuated sharply because of the curious geological circumstances in the region.[69] As noted above, the Group's 1919 acquisition of Mexican Eagle, at the time the largest producer in Mexico, turned into a fiasco when the com-

[32]

[33]

Portraits of dictators: Juan Vicente Gomez (right) and Porfirio Díaz (far right). The guerrilla leader Gomez (1857-1935) seized power as president of Venezuela in 1908 and ruled as an absolute tyrant until his death. Díaz (1830-1915) was twice president of Mexico, in 1876-1880 and again in 1884-1911, encouraging foreign investment in the nation but ruthlessly ruling in favour of the few and at the expense of the peasantry.

[34]

The regimes of Gomez in Venezuela and Díaz in Mexico were distinctly unsavoury but did provide the oil industry with essential long-term stability. Popular resentment of the regimes meant that could not last forever, and in Mexico (above left and right), the revolutionaries Pancho Villa (c. 1877-1923) and Emiliano Zapata (c. 1879-1919) joined forces in a path leading to the eventual nationalization of the oil industry.

pany's production suddenly plummeted in 1921-2. Eagle's three refineries proved vital assets in coping with a sudden surge of production at La Corona, from 4 million barrels in 1920 to 14.5 million barrels in 1922, only to find themselves threatened by renewed idleness when output plunged again within an even shorter time.[70] To stabilize the business in Mexico, the Group needed new producing properties, but political instability made exploration unattractive. When the conservative Díaz regime collapsed in 1911, inaugurating a period of protracted revolutionary upheaval, the American oil companies reacted by forming in 1912 an Association of Petroleum Producers in Mexico to protect their interests. The Association wielded considerable political force in the conflicts which followed, not least because of the incidental support from the State Department.[71] As a result the oilfields were protected by, successively, foreign war ships, including a Dutch gunboat sent to Tampico in 1914; an expeditionary force from the US; and a rebellious general seeking funds for his war chest.[72]

With the return of more orderly circumstances in 1917 a new threat to the oil business emerged. The Díaz administration had followed the American example and treated subsoil resources as belonging to the owner of the land. By contrast, Article 27 of the new constitution of 1917 proclaimed such resources to be state property. The constitution thus formally declared all existing oil concessions legally void, but opinions as to the consequences of this annulment changed with the political orientation of governments, which succeeded each other with some speed, tying the industry's fate to the ever shifting balance of forces within the administration and in Mexican society at large. As a first step towards asserting its rights, the Mexican government increased taxation on production and exports.

In 1925 a new oil law raised the stakes by requiring the reconfirmation of all concessions. The US oil companies threatened to destabilize Mexico by withdrawing their capital and halving production, inspiring rumours about military intervention by the big neighbour. The Group was slightly less concerned about the law and did not object to the constitutional clause itself, because in the Dutch East Indies and in most other producing countries the state also claimed the ownership of subsoil resources. Perhaps the Group also hoped to profit from the anti-American sentiments that were also part of the Mexican political scene. During the following ten years the Group would even enlarge its possessions in Mexico until, in 1936, it controlled almost two-thirds of output. By then output had greatly diminished, however. Having peaked during the early 1920s, when Mexican oil production supplied almost a quarter of global output, ranking second after the United States, the industry declined to relative insignificance.[73]

To a large extent Romania offered a European parallel to Mexico. As in so many countries, the First World War and its aftermath brought radical political change. A new regime dominated by the Liberal Party took power, determined to effect a rapid

By the 1920s Romania was also an important oil producer – indeed, the most important in Europe. Drilling at Ochiuri in the summer of 1929 produced a very messy result when a well blew out and covered workers in oil.

[35]

Above (left to right), in Astra Romana's construction projects, women were just as likely to be employed as men; and in daily operations cattle were just as likely as motor vehicles to be used in transport – even if they might sometimes block the road. Similarly on payday, workers would block the office entrance when crowding in to collect their cash.

economic and social modernization of the country. The government was also hostile to the oil industry and its domination by foreign companies. In 1919-20 bankers and businessmen linked to the new regime set up two national oil companies with the explicit aims of developing the oil reserves on state-owned land, achieving a monopoly on the internal distribution of oil products, and acquiring the German and Austrian interests in the industry, which had been sequestered as enemy property. This latter acquisition proved to be quite difficult to effect, since the ownership of these companies had already been transferred to citizens of neutral countries. Even so the government began push the national companies at the expense of the foreign oil companies, in particular the two leading ones, the Group's Astra Romana and Romano-Americana, the Jersey Standard subsidiary, which together dominated the industry. In 1920, the public debate in Romania began to call for the outright and complete nationalization of the industry. Astra Romana and Romana-Americana faced demands that they nationalize their companies by transferring the majority shares to Romanian holders, in exchange for concessions on state lands. The companies withstood the pressure, however, and tried to continue their operations as best as circumstances permitted.[74]

Both Standard and Royal Dutch felt rather defenceless against the adverse climate in Romania. As Kessler put it in a letter from August 1920, 'It is worse than useless and entirely wrong to fight anything made by the Roumanian Government which they consider in the interests of their country. If their decision is wise and sound we should support them instead of fighting them and if their decision is unsound it is no use fighting it because such a decision will never stand and anything which is economically wrong will disappear after a short while. We do not wish to mix in politics in any way and we want to look after our own business and do all we can to try to develop our company into a prosperous and sound one for the benefit of the country in which it is working, its shareholders, employees and workmen.'[75]

[38]

[39]

During 1923-24, the Romanian government took further steps. The oil distribution networks, notably the pipelines, were brought under state control and new legislation provided a firm basis for the policies which the Liberal Party had implemented since 1918-19. The constitution acquired a clause giving the ownership of all subsoil resources to the state and a mining law reserved new concessions for Romanian companies. Such companies needed to have at least 60 per cent of their shares held by Romanian citizens, although for existing businesses the threshold could be lowered to 55 per cent. Moreover, the president and two-thirds of directors of these companies had to be Romanian as well. Gradually the foreign concerns were forced into widening the participation of Romanian citizens and into collaborating with Romanian companies for getting new concessions. The companies refused to give in, however, because they trusted in their superior technological expertise, access to markets, and finance to give them a clear edge over any local rivals. And so it proved. By the mid-1920s four Romanian companies together controlled 26 per cent of the oil industry, against 4-5 per cent in 1914, but they had a hard time competing with the foreign firms. The companies were boycotted on foreign capital markets, for example, as a result of negative publicity about the dubious nature of Romanian legislation.[76] However, the Liberal Party encountered increasing opposition within Romania itself and in 1928 the National Peasant Party took over. The new government was much more friendly towards foreign capital and soon modified the mining law accordingly.

These vicissitudes did have a serious impact on the Romanian oil industry, of course. Romania had been an important producer of petrol before the war, but the country recovered its position slowly during the 1920s. Investment in the industry remained relatively low and the Romanian share in world output declined markedly. The Romanian share in the Group's output (see Figure 4.3) followed a similar curve; Astra Romana's production only surpassed its pre-war level after 1927, with the return of the more favourable investment climate.

The Jambi saga The Group also struggled with increased demands by a state bent on defending national interests in what it considered to be its own backyard. The early 1920s saw confrontation and conflict between Royal Dutch and the governments of both the Netherlands and the Dutch East Indies over the export tax of 1921 (which will be dealt with in Chapter 5), but before that the long-running Jambi saga had already helped to sour the complex relationship between the Group and its most profitable host country.

Jambi referred to an area with ostensibly very promising oil-fields on Sumatra, considered to be as rich in terms of light oil fractions as the best fields in the Palembang area on the same island. Royal Dutch had trained its beady eye on these fields since 1902, but its application for oil concessions had become stuck in a political quagmire, partly of its own making. The company's towering profitability and strength in the Dutch East Indies had made politicians and civil servants alike wary of granting further concessions to Royal Dutch as a matter of course and on the same conditions as before. For Jambi, the modest fixed royalty terms were replaced by a scheme guaranteeing the colonial government a substantial profit share and the concession rights were to be put up for public auction so as to give outsiders a chance to break into Royal Dutch's monopoly on oil production. Jersey Standard, annoyed by Deterding's spectacularly successful entry into the US, was keen to do just that with a Dutch-registered subsidiary set up for the purpose, the Nederlandsche Koloniale Petroleum Maatschappij, known colloquially as the 'Koloniale' or NKPM, which had bought a few small concessions and spent large sums of money on exploration without getting much production.[77] A third contender emerged in the form of the Zuid-Perlak Maatschappij organized by

[40]

[41]

Exploitation of the East Indies involved the typical colonial pyramid of government, business, and workers. Clockwise from top left, A. W. F. Idenburg (1861-1935), governor-general of Surinam from 1905 to 1916, is seen during his 1915 trip to the 'Vorstenlanden' – the cultivated regions of Java – with Paku Buwono X (1886-1939), hereditary sultan of Surakarta, on his right. Business is represented by the Group's 1921 coded messages about high export taxes; and below, workers struggle through the jungle carrying components of a new installation.

KOPIE VAN TELEGRAM

No: 4326

dd: 21.7.'21

AFGEZONDEN DOOR: BRANDAN HAAG

GEADRESSEERD AAN: JUDEX WELTEVREDEN

JUDEX WELTEVREDEN

251) lobuk Met oog op onze groote voorraden kerosene benzine en het feit dat wy in naaste toekomst geen verbetering in huidige ongunstige toestanden verwachten en vooral in verband met de sinds 18 Mei in Indie geheven uitvoerrechten moet worden overgegaan tot inkrimping der productie en van het boorbedryf U gelieve daartoe de volgende maatregelen te nemen

tunyhvgitu	firstly on all
yfalgtsope	small fields
vgiriethga	on Palembang
oxuzhumwry	and Java
ziizjscaco	where costs of
winning	winning
vebokenurm	of oil
pjiizudtze	are high
yvbigmgonu	to reduce
wilubyvbig	production to
ykpfutorvi	such extent
ytcigsesoz	that damage to
tsopcyfbfa	field as small as possible
tyobnpwucy	fullstop If but
vytdemgoov	possible stop
wilubtifda	production entirely
undabugjax	keeping in
uwmpuvboep	mind obligations
mhajitylab	contracts fullstop
owynwtomaw	All exploitation
tcagyvgoti	drillings on these
tsopemuowb	fields to be stopped
xuxluvgose	secondly on the
unihgtsope	large fields
ugjaxethga	in Palembang
oxuzhetrax	and Brandan
vaffuvabbu	no new
poolpziefm	big wells
msukatcafu	must be drilled

Model No. 750 200.000 Juli 20a.

[43]

[42]

50 fullstop it is

51 self evident
52 that where
53 on the one hand a
54 competitor is
55 so favourably
56 placed and
57 on the other hand
we are
58 continually
59 harassed
60 by the Government by
more
61 and more unreasonable
62 taxes we have no
63 choice but to
64 limiting production
to
65 most economical
wells
66 and reducing
67 personnel
68 to a minimum
69 in order to
prevent that
70 at a future
71 date we would
have to
72 cease all our
73 operations in the
74 East and
75 discharge whole
76 staff fullstop
77 You may communicate
contents
78 this telegram to
79 Personeelsvereeni-
ging

Behandeld door: PS(DI).

the Deen brothers, who had not lost their nose for a speculative opportunity in oil concessions; four also-rans made up the pack.

The procedure got underway in 1912 with the contenders putting in bids for the government's share in net profits on the oil produced. NKPM offered 40 per cent to the government; Bataafsche, determined to block the Americans' advance in the Dutch East Indies at almost any price, bid 50 per cent; Zuid-Perlak came out on top with 62.5 per cent, a figure which according to

As the transcript (below left) of a June 1921 newspaper article shows, the Netherlands East Indies' increase in export tax was seen in France as an indirect tax on non-Dutch shareholders and brought a drop in Royal Dutch's share price. Part of Bataafsche's reaction, in August 1922, was to attempt a cheese-paring economy (below) with an order to save paper by reducing the spacing in outgoing typed letters. This rule produced such unsatisfactory results that it was reversed in December.

[44]

P.

Paris, 28 Juin 1921

BAISSE des ACTIONS "R O Y A L   D U T C H"
sur le Marché de PARIS.
-:-:-:-:-:-:-:-:-:-:-

Depuis plus d'un mois, le marché de cette valeur a été très agité, et les cours ont marqué une étape sensible vers la baisse.

Le public français a, en effet, été assez mal impressionné par la récente augmentation des droits d'exportation des pétroles dans les Indes Néerlandaises, augmentation qui a d'ailleurs coïncidé avec la baisse des prix.

Le public français a estimé que cette augmentation des droits était difficilement justifiable, étant donné que sa répercussion devait atteindre non seulement les sujets Hollandais, mais encore tous les Français, (et ils sont très nombreux) porteurs d'actions "ROYAL DUTCH".

L'augmentation représente un véritable impôt indirect hollandais qui frappe les porteurs étrangers.

-:-:-:-:-:-

[45]

AG(DE) - **Correspondentie.**

's-Gravenhage, 9 December 4.

33893 B
orig. in 651.3 B

Aan het Laboratorium van
DE BATAAFSCHE PETROLEUM MAATSCHAPPIJ.
Overzijde v/h Y - Badhuisweg 3-5,
A m s t e r d a m.

Mijne Heeren,

Onder verwijzing naar ons schrijven van 30 Aug. '22, deelen wij U mede, dat het typen van brieven zonder tusschenruimte tusschen de regels geen voldoening heeft gegeven en dat derhalve besloten is, al onze brieven voortaan weder met een regel tusschenruimte te typen op de wijze als van dit schrijven.

Wij verzoeken U, dienovereenkomstig te willen handelen en teekenen.

hoogachtend,
DE BATAAFSCHE PETROLEUM MAATSCHAPPIJ,
W.g. W. C. KNOOPS

22.

Bataafsche's calculations rendered a profitable exploitation impossible.[78] Unfortunately the winner lacked the necessary capital to work the concessions and had to back out. Bataafsche claimed them as the runner-up, presenting itself as the obvious candidate, because of its unbeatable experience in exploration and production and its role as main benefactor in terms of employment and income of home country and colony alike. Granting the concessions to Bataafsche was not so easy and straightforward, however. Members of Parliament objected that it would bolster the company's position still further and voted to have the fields exploited by a state-owned company.

Talks resumed in 1919, Colijn opening with a memo which tried to convince the Minister of Colonial Affairs, A. W. F. Idenburg, a close political ally of Colijn and co-leader of the orthodox-protestant party, that the Dutch East Indies could only successfully develop Jambi if it cooperated with either Jersey or Royal Dutch. Given Royal Dutch's much stronger position in the archipelago, with an 86 per cent market share, control over the East coast of Africa, and an influence in China equal to that of Jersey, the Group was the logical partner for the colonial government. Colijn also argued 'that the Dutch character of the largest Dutch enterprise has to be maintained as much as possible'.[79] Relations with Idenburg's successor, S. de Graaff, were initially even better. In 1920 Colijn and De Graaff worked out the details of a joint venture between Bataafsche and the colonial government, presented to Parliament in early 1921. Bataafsche agreed in principle to manage the new venture as the junior partner, on condition that the company would get a preference right to any new concessions given out in the Dutch East Indies. De Graaff even consulted Colijn and his fellow Bataafsche director De Jonge, another colonial specialist, before answering the questions raised by Members of Parliament.[80]

Interlude: the colonial export tax The Jambi saga took yet another turn when Royal Dutch and the Government came into conflict over the export levy on oil produced in the Dutch East Indies. During 1919-20, the colonial government introduced export levies to tax excess profits made in the major commodity sectors because of high world prices. The Group had also made stupendous profits during and immediately after the war. Net profits of Royal Dutch and Shell Transport soared from almost £ 4.4 million in 1913 to more than £ 12 million in 1918 and £ 20.5 million in 1920. What had been an acute embarrassment in wartime turned into a serious political liability in peacetime. Civil servants and Members of Parliament in the Netherlands, convinced by these riches that extracting crude from the Dutch East Indies was unduly cheap, argued for the introduction of special taxes to keep some of the profits in the colony. Just around that time Marcus Samuel stated in public 'that 95 per cent of the profits of our group are made in the Dutch Indies', a typical gaffe for which the board of the Bataafsche rebuked him, but the damage had been done.[81]

The Group redoubled its efforts to keep published profits within bounds. In January 1920, Lane explained to his Rothschild friends: 'we are up against the same old problem, what to do with the profits.' The solution was 'to somehow or other hide the profits' in order to keep the dividend to a 'reasonable limit'.[82] This meant that the Group began to erect a screen around its finances for, wrote Lane, 'the liability of taxation of the Shell and Royal Dutch is becoming so complicated that every care has to be taken to publish as little information as possible'.[83] In spring 1921 the problem had become almost unmanageable. In May 1921, Lane talked with Deterding about 'the manner in which the profits of 1920 should be dealt with: net profits, after depreciation has been made, shall be dealt with in the following manner: 60 per cent of these profits shall be retained by way of further depreciation, and 40 per cent allocated to dividend'. In order to keep the dividend between 35 and 40 per cent the issue of bonus shares was necessary, because 'it is absolutely unavoidable if trouble with the Dutch Government is to be avoided (...) the Company dare not pay more than about 40 per cent dividend, as it would certainly lead to disadvantageous legislation'.[84]

As so often, Lane's foresight proved dead right. At the end of 1920 the colonial government began to scrap some of the export levies, but not the one on oil, and in May 1921, with world oil prices falling, the government even proposed to raise it substantially to ease the colonial budget deficit. Deterding was furious and ranted that the decision heralded 'the beginning of Bolshevism in the Dutch Indies'.[85] He started a full-scale political campaign against the export tax, drumming up support both within and outside the Netherlands and devoting four pages of the 1921 Royal Dutch annual report to the issue. This was the first instance of Deterding using the report, required reading for Dutch investors, to air his political views. In August 1921 the Royal Dutch board issued a sharply worded request to De Graaff warning of the imminent ruin of the oil industry in the Dutch East Indies if this policy continued; the company also announced cutbacks in production there because operations were becoming too expensive.[86] To dramatize the impact of the taxes on the competitiveness of the oil industry in the Netherlands Indies, the free Saturday afternoon, only recently introduced after long lobbying by employees, was abolished again. Bataafsche's newly set up association of European employees was induced to send a telegram to the Minister of Colonial Affairs arguing that the employment of 2,700 Europeans and 54,000 Asians was endangered.[87]

Colijn and Philips coordinated a more subtle lobbying of members of Parliament, whilst Deterding used all his influence to gather support. He asked for an audience with Queen Wilhelmina and announced that he would use the occasion to discuss the export tax.[88] He also began mobilizing his relations outside the country, trying to convince the Rothschilds and Deutsch, a French oil company closely associated both with the Rothschilds and with the Group, to ask the French government for a *démarche* with the Dutch government. The request greatly annoyed the Rothschilds. Deterding had wanted Baron Edmond to 'explain to our ambassador in Paris the injustice' of the export tax, but Rothschild flatly refused to act as messenger boy for Deterding by lobbying for Royal Dutch.[89] Deutsch proved more forthcoming; his request for intervention to the French minister of commerce inspired a formal 'friendly démarche' by the French envoy in The Hague.[90]

As in the Jambi case, Royal Dutch took the chauvinist line in its attack on the export levy, arguing improbably that cutting the profitability of the Dutch East Indies operations undermined the Dutch interests within the Group during the revived debate in Britain about a possible merger with Anglo-Persian and potential changes to the ownership structure of the Group. After a long interview with De Graaff in January 1922, Philips for example wrote to the Minister that 'negotiations about a new relationship [between Royal Dutch Shell and Anglo-Persian] had become necessary, in which also the English government is very much interested, and that it is not impossible that these negotiations will eventually result in a Combine, in which Dutch interests will no longer hold a majority'. And he added that 'the position of the Dutch group (...) is greatly weakened by the drastic increase in export duties on Netherlands-Indies kerosene'.[91] Philips and Deterding also agreed to cancel the Royal Dutch interim dividend for 1922 in protest if the export tax remained.[92]

The Royal Dutch offensive was totally out of proportion to the impact of the tax, as a few figures demonstrate. The total yield of all export taxes of the Dutch East Indies in 1921 amounted to slightly less than 8 million guilders, against a total dividend paid out by the Group of 163 million guilders. If we accept Lane's statement cited earlier that only 40 per cent of profits were paid out, total profits were really two-and-a-half times that amount, or more than 400 million guilders. Even if we assume that oil generated all export tax revenues, which was definitely not the case, then the export tax cannot have lowered the Group's total profits by more than 2 per cent.[93]

The export levy so incensed Deterding that he threatened to veto the NIAM, the proposed joint venture with the government in the Netherlands Indies. In a draft letter to De Graaff read out during a meeting with the Minister on 29 November 1921, the Board of Bataafsche expressed 'serious doubts if they would be capable of implementing the plan to establish (...) the Nederlandsch-Indische Aardolie Maatschappij', because the export tax had fundamentally changed the basis on which the proposal was based.[94] This threat was very effective; in that same meeting the Minister promised to either abolish or substantially lower the export tax, after which the agreement to form the NIAM could be signed. The Board of Bataafsche added that this was based on the understanding that the changes proposed by the Minister would be approved by Parliament.[95] The export tax was abolished in 1923 and that same year NIAM began operations in Jambi.[96]

Meanwhile Koloniale's protests at being excluded had made Jersey Standard complain to the US State Department over being denied access to concessions in the Dutch East Indies, whereas Royal Dutch enjoyed free access in the USA. At that juncture the tide of nationalism, growing concerns about declining oil reserves in the US, and the need to gain access to oilfields overseas, made Congress aware of the need for better regulation of the granting of concession rights in the USA. Towards that end the 'Oil Leasing Act', passed in February 1920, had a special reciprocity clause, stipulating that 'Citizens of another country, the laws, customs or regulations of which deny similar or like privileges to citizens or corporations in this country, shall not by stock ownership, stock holding, or stock control, own any interest in any lease acquired under the provisions of the Act'.[97]

This article created a serious complication for the Group. Its American companies were owned by Anglo Saxon, and many formal and informal barriers prevented American companies from getting concessions in the British Empire. Knowing that changing the rules of the Empire was well beyond its powers, the Group swiftly shifted the ownership of its American subsidiaries to Bataafsche, but that strategem merely helped to concentrate the attention of the State Department on access to the Jambi concessions. The US envoy in The Hague began to exert pressure on the Dutch government to grant Jersey Standard a share in the Jambi fields, threatening that 'temporarily, until the American Government has been satisfied, no new concessions will be given to companies of which the majority of shares is in Dutch hands'.[98] The Dutch government deftly sidestepped the issue by pointing out that the 1912 auction had ended inconclusively; since then, Parliament had voted to reserve oil exploitation in Jambi to a state-owned company, so there was no question of Koloniale being unfairly shut out. Since by then Jersey Standard had largely lost interest, the US government took no further action.

Royal Dutch won the battle for Jambi, but the struggle made the company reconsider its resistance to the entry of American oil companies in the Dutch East Indies. Though ultimately unsuccessful, the State Department's pressure had given a clear signal of intentions which the Group, considering its extensive and growing interests in the USA, could simply not afford to ignore. In 1919 Bataafsche had already sold the Talang Akar field in the Palembang area, which it considered worthless, to Koloniale. With deep drilling NKPM was almost immediately successful and its field became one of the most productive in the archipelago, more so in fact than the hotly contested Jambi fields. After the usual lengthy negotiations, NKPM acquired new large concessions in the same region, as well as on Java and Madura, in 1925, rather more than it had hoped for, enabling the US envoy in The Hague Richard Tobin to conclude that 'our legislative reprisals have been more effective than had been anticipated'.[99] With the 1925 agreement, ratified by the Dutch Parliament in 1928, the Group and Jersey Standard finally eliminated an old source of friction.

REVOLUTION IN RUSSIA; CZAR ABDICATES; MICHAEL MADE REGENT; EMPRESS IN HIDING; PRO-GERMAN MINISTERS REPORTED SLAIN

The New York Times - 16th March 1917

[46]

[47]

In a propaganda poster (above), the rising sun silhouettes Russian soldiers, sailors and citizens marching under a banner extolling the values of freedom and industry in the Russian revolution. To Westerners the reality was no less disconcerting, with public demonstrations in 1917 – not only civilians in Moscow's Palace Square (right), but also soldiers (far right) marching in Petrograd (St Petersburg).

Coming to terms with the Soviet Union The Russian Revolution of 1917 and the outright nationalization of foreign and domestic companies which followed it in 1918-20 presented the Group with a fresh challenge. The expropriation of voluminous assets such as the Group owned in Russia was unprecedented; groping one's way towards dealing with the new Socialist state equally demanded a substantial capacity to improvise and experiment, if only because the Soviets themselves did not follow a clear-cut course and changed their policies sometimes quite radically. Moreover, the Group did not pursue a coherent strategy either. Of course, the return of its properties or, failing that, obtaining adequate compensation received top priority, but managers allowed themselves to stray from this objective by opportunistic attempts to profit from the new circumstances, both by supporting the White Russian forces and by trying to find some common ground with the new regime.

During 1918-20, a period of great turmoil, foreign intervention and civil war in the Caucasus, Gulbenkian devised plans to regain control of the oil industry and start reconstruction.[100] One scheme envisaged producers in Baku and Grozny pledging their supplies to a number of oil companies, such as the Group, Anglo-Persian, Nobel, and Jersey Standard, helping 'to raise funds for General Wrangel (...) to relieve the Caucasus of the presence of the Bolsheviks'.[101] At the same time the Group also acquired, at the considerable expense of more than £ 1.3 million, stakes in the Grozny oil companies Anoto and Benzonaft. While of course aiming to strengthen the Group's overall position as a producer in Russia, this move also had the more speculative motive of profiting from the buoyant demand for oil shares on the stock market by reorganizing the companies into Dutch holdings and then floating the shares in France and in the UK.[102]

[49]

[50]

Leonid Krassin (1870-1926), a successful businessman before the revolution, became People's Commissar for Foreign Trade in 1920 and is seen (above) returning from Britain that year, after discussing the nationalization of Russia's oil industry. Left, General Baron Peter Wrangel (1878-1928) was badly defeated as commander-in-chief of White Russian forces in the Crimea, and died in exile in Brussels.

This particular aim became impossible before it could be realized. On 28 October 1920 Gulbenkian wrote a letter to Philips urging him to continue with the preparations for the intended resale, but three days later the White Russian General Wrangel was defeated by the Soviets and the Group's attitude suddenly changed. On 2 November Philips wrote to Colijn, together with Deterding an outspoken defender of the project, once again warning about the risks involved. The property titles of the companies concerned were not recognized by the new regime, and in his view Royal Dutch ought not be associated with highly speculative ventures which might tarnish its reputation. Moreover, many details of the agreements still needed to be worked out, adding to the general uncertainty of the situation.[103] The two opinions clashed at the next meeting of the Bataafsche directors, Colijn defending the original plan, Philips emphasizing his concerns. Gulbenkian's role and his apparent claim to a 50 per cent stake in the companies to be created also came under scrutiny. The rest of the board shared Philips's objections and de facto put an end to the project.[104]

At that very time Colijn and his secretary Gerretson were already deep into secret negotiations with the head of the Russian Trade Delegation in London, Krassin, about ways in which the Group might acquire, or in fact, re-acquire, and exploit concessions in the Soviet Union.[105] The first overtures took place somewhere around the middle of October 1920 and on 6 December Krassin sent a detailed memorandum entitled 'Fundamental principles for the granting of concessions in Soviet Russia to the foreign states, workmen's associations, co-operative and private capitalist undertakings and corporations' to Colijn, referring to a conversation the two had had in London. The documents do not reveal how Colijn and Krassin considered circumventing the Group's formal condition for entering into official negotiations with the Soviets, i.e. full recognition of its rights prior to 1920. For his part, Krassin attempted to put pressure on the negotiations by threatening to conclude an agreement with Anglo-Persian, which had not been affected by the Soviet nationalizations, about exploiting the Grozny wells. Colijn immediately wrote to Sir Robert Horne, President of the Board of Trade, to stop Anglo-Persian, arguing that this would be tantamount to recognizing the expropriations: 'I do not for a moment believe that a British Company, in which the British Government is a prominent shareholder, will be allowed to depart from the attitude which the British Government has advised us to take in connection with former rights'.[106] The intervention succeeded in scuppering the Anglo-Persian threat, so Colijn and Gerretson on the one hand and Krassin on the other started to work out detailed plans for 'building oil pipe lines for the Russian socialist federal soviet Republic'.[107]

Now the Soviets had changed their mind, however. Another agreement concluded by Krassin, concerning the exploitation of new oil properties by an American company, was eventually

An atmospheric photograph from 1919 shows four leaders of the revolution – left to right, Josef Stalin, Alexei Rykov, Grigori Zinoviev and Lev Kamenev.

After show trails, Stalin ordered the executions of Zinoviev and Kamenev in 1936, and Rykov in 1938.

[51]

[52]

In poses perhaps symbolizing rising productivity, Russian students in the late 1920s carry posters depicting Soviet achievements in the metals, oil, coal, and textile industries.

vetoed by Lenin himself, ending the brief period of the New Economic Policy during which a return of the foreign oil companies figured as a serious option. Now Group policy also took a radical new turn, inspired by expectations about the imminent collapse of the new regime. Deterding firmly believed that the Soviets were unable to run their state, let alone their economy, and expected them to 'be cleared, not only out of the Caucasus, but out of the whole of Russia in about six months',[108] an opinion repeated indefatigably during the next ten years or so. His optimistic expectations were shared by Jersey Standard, which bought a large part of the Nobel company in 1920.[109]

Meanwhile the rapprochement between Britain and the Soviet Union had resulted in a trade agreement which included a recognition by the latter country of the principle of compensation for nationalized assets. Since this agreement put the British government in a position to press claims, the Group transferred the ownership of the Russian companies from Bataafsche to Anglo-Saxon, for the Dutch government adamantly refused even to consider talks with the Soviet Union and would obviously have insufficient leverage anyway.[110] This manoeuvre turned the Group's lost Russian assets, estimated at over £ 27 million in 1927, into a loss suffered by a British company.[111]

The Genoa Conference of May 1922 failed to improve the relations between the Soviet regime and the Western democracies, however, bringing the compensation issue no nearer to resolution. Deterding could not decide what to do about the matter. In June 1922, when the Syndicat Franco-Belge, an organization of French and Belgian oil companies with interests in Russia set up directly after the failure of the Genoa Conference, asked Royal Dutch to join their boycott of the Soviets and to stop negotiating with the Russians, the prospects for an agreement with Krassin still looked too good to do so. As an alternative, Deterding approached Jersey Standard to establish some common position against the Soviets. In July, however, he appeared to have changed his mind.

Lydia Koudoyaroff, Deterding's second wife, came from an aristocratic Tsarist family dispossessed by the Soviets and naturally felt deep opposition to the regime. Marriage to her probably helped to cement Deterding's fervent anti-Communism.

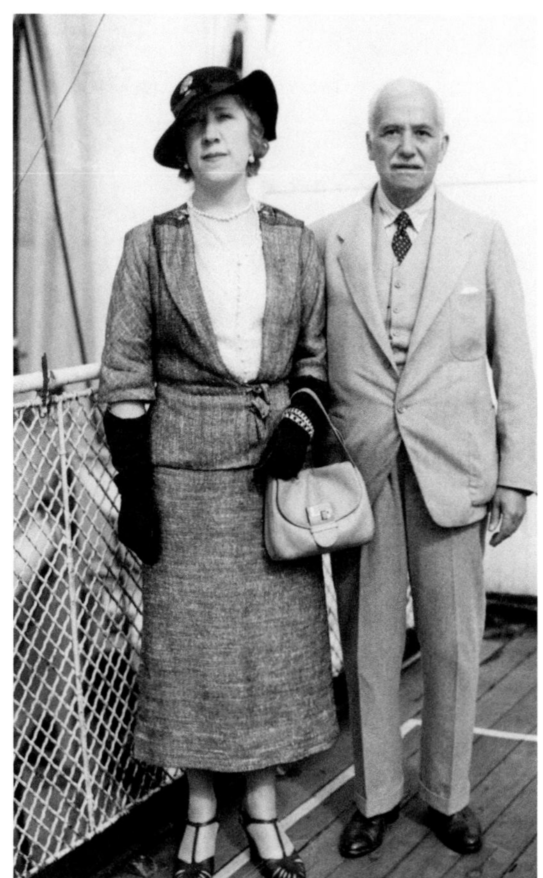

[53]

A meeting with Teagle, Riedemann, and Nobel produced the London Memo, in which the three companies agreed to coordinate their actions with obtaining compensation for their nationalized properties as the main object.[112] The London Memo provided the basis for the Front Uni, formed by the companies which had signed joining the Syndicat Franco-Belge; Deterding chaired its meetings, and in his absence Philips did so. Even at this stage Deterding still had no clear ideas about how the organization was to achieve its objective. He supported a boycott, but at some meetings he took the completely opposite position, suggesting use of the Front Uni as a syndicate for major oil purchases from the Soviets.[113] The idea made some sense, especially given Deterding's old and enduring advocacy for stabilizing markets through alliances. Oil prices had dropped sharply from their 1920 peak, following the onset of an international depression; Russian oil exports had returned to the world markets, pushing down prices even further. Unless the market recovered, it would be very hard to keep the Front united, and buying up excess oil might achieve some stabilization, as it had in the past. Deterding's commercial opportunism, another lifelong habit, got the better of him before the Front had formulated clear policies anyway. In March 1923, Deterding allowed Debenham to buy large quantities of Batum kerosene at very keen prices for Asiatic.[114] This 'treason' was immediately made public by other members of the Front Uni; Deterding's offer to share the profits of the transaction did little to placate them. With the basis for a common policy shattered by the Group, Socony and Vacuum, which had no lost assets to mourn, opted out and began making their own purchases of Soviet oil.[115]

After the Front Uni's collapse Deterding persuaded Teagle to set up a joint venture for buying Russian oil. A very detailed contract, which carefully kept the door open for Front Uni members to come in, was duly drafted and signed for the company, which received as its official aim finding compensation for former owners of oil properties. The business remained dead on the page. Negotiations with the Russians proved quite troublesome and stranded on irreconcilable differences. Deterding and Teagle preferred a long-term contract, whereas the Russians only wanted to sell whatever they had in surplus. Deterding's marriage, in 1924, to Lydia Koudoyaroff, a Russian refugee bitterly opposed to the new regime, finally appears to have turned him against the idea of doing business with the Soviets and into a staunch advocate of the boycott which he himself had helped earlier to frustrate.[116]

During these negotiations Jersey Standard had succeeded in restraining its American colleagues, in particular Socony and Vacuum, from spoiling the game, but once it had become clear, in November 1925, that the talks would fail, these two competitors

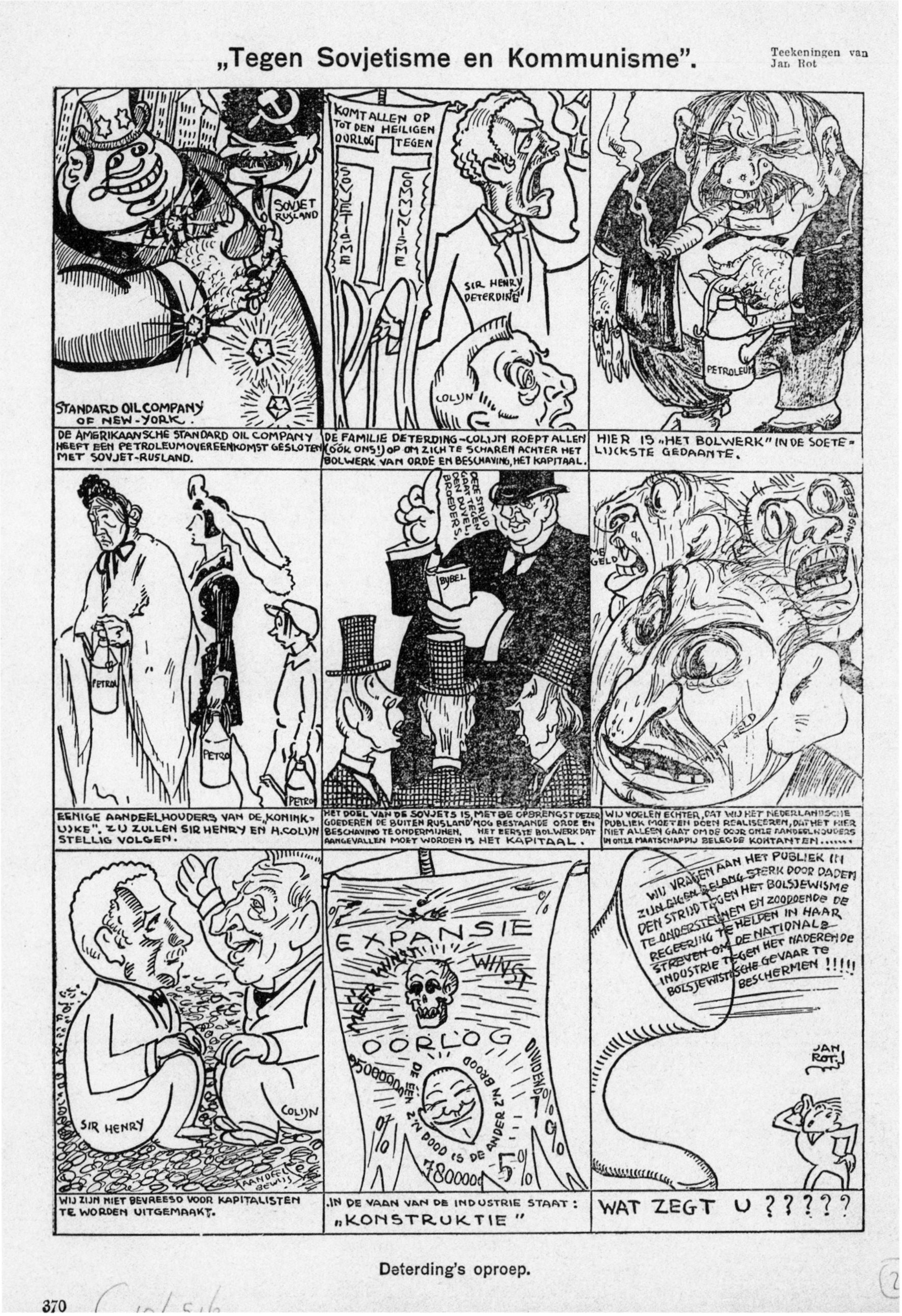

„Tegen Sovjetisme en Kommunisme".

Teekeningen van Jan Rot

STANDARD OIL COMPANY OF NEW-YORK.
SOVJET RUSLAND

DE AMERIKAANSCHE STANDARD OIL COMPANY HEEFT EEN PETROLEUMOVEREENKOMST GESLOTEN MET SOVJET-RUSLAND.

KOMT ALLEN OP TOT DEN HEILIGEN OORLOG TEGEN
SOVJETISME
COMMUNISME
SIR HENRY DETERDING
COLIJN

DE FAMILIE DETERDING–COLIJN ROEPT ALLEN (OOK ONS!) OP OM ZICH TE SCHAREN ACHTER HET BOLWERK VAN ORDE EN BESCHAVING, HET KAPITAAL.

PETROLEUM

HIER IS „HET BOLWERK" IN DE SOETE-LIJCKSTE GEDAANTE.

PETRO

EENIGE AANDEELHOUDERS VAN DE „KONINKLIJKE". ZIJ ZULLEN SIR HENRY EN H. COLIJN STELLIG VOLGEN.

DEZE STRIJD GAAT TEGEN DEN DUVEL, BROEDERS.
BIJBEL

HET DOEL VAN DE SOVJETS IS, MET DE OPBRENGST DEZER GOEDEREN DE BUITEN RUSLAND NOG BESTAANDE ORDE EN BESCHAVING TE ONDERMIJNEN. HET EERSTE BOLWERK DAT AANGEVALLEN MOET WORDEN IS HET KAPITAAL.

WIJ VOELEN ECHTER, DAT WIJ HET NEDERLANDSCHE PUBLIEK MOETEN DOEN REALISEEREN, DAT HET HIER NIET ALLEEN GAAT OM DE DOOR ONZE AANDEELHOUDERS IN ONZE MAATSCHAPPIJ BELEGDE KONTANTEN......

SIR HENRY
COLIJN
AANDEELBEWIJS

WIJ ZIJN NIET BEVREESD VOOR KAPITALISTEN TE WORDEN UITGEMAAKT.

EXPANSIE
MEER WINST
WINST
OORLOG
DIVIDEND
DE EEN Z'N DOOD IS DE ANDER Z'N BROOD

„IN DE VAAN VAN DE INDUSTRIE STAAT: „KONSTRUKTIE"

WIJ VRAGEN AAN HET PUBLIEK IN ZIJN EIGEN BELANG STERK DOOR DADEN DEN STRIJD TEGEN HET BOLSJEWISME TE ONDERSTEUNEN EN ZOODOENDE DE REGEERING TE HELPEN IN HAAR STREVEN OM DE NATIONALE INDUSTRIE TEGEN HET NADEREN DE BOLSJEWISTISCHE GEVAAR TE BESCHERMEN !!!!!

JAN ROT.

WAT ZEGT U ?????

Deterding's oproep.

370

[54]

Cartoons often capture the mood of a period better than official photographs. *De Notenkraker* (The Nutcracker), a Sunday supplement to the Dutch socialist newspaper *Het Volk*, carried many cartoons – often brilliant, often vicious – and (left) in 1927 displayed its opposition to the oil industry, with obese top-hatted capitalists encrusted with diamonds, and grotesque Jewish parodies under the heavily sarcastic title 'Against Sovietism and Communism'. This was in response to Deterding's call to boycott the Soviet Union.

Other late 1920s cartoons in *De Groene Amsterdammer* (below and right), focussing on price increases and the race between Royal Dutch and Jersey Standard, were less vitriolic but just as telling.

[55]

became restless and started their own negotiations with the Soviets. In August 1926 Deterding, who had reinvented himself as the main spokesman of the pro-boycott party, travelled to New York to convince the presidents of Socony and Jersey Standard not to buy Russian oil, which resulted in a boycott agreement for eight months, the maximum period Socony was prepared to consider.[117] In September 1927, however, Socony was buying Russian oil, and intended to build a kerosene plant in Batum for upgrading the low-quality Russian export products. This enraged Deterding, who accused them of breaking their promises.[118] Deterding similarly tried to ward off another danger. During 1925-26 there was a brief thaw in the relations between France and the Soviet Union, which opened the possibility of partial compensation for the owners of Russian stocks. One of the options considered was the formation of a French company for exploiting oil concessions in Soviet Union, the proceeds of which would be used to pay off French claims. Deterding asked the Rothschilds to arrange an interview with the French President, Raymond Poincaré, for him so that he could dissuade him from this idea, but they refused, fearing that 'a conversation between these two Napoleons (...) will turn sour'.[119]

Thus Deterding's attitude towards the Soviet Union changed markedly over the years; he became a staunch champion of a total boycott only in 1925/6. He used the Royal Dutch annual reports to trumpet his cause. During the early 1920s the annual reports hardly paid any attention to the situation of the Russian oil industry, apart from noting that no change had occurred, but in 1926 and 1927 he wrote extensively about the damage done by the dumping of Russian oil on world markets and about the cooperation of some oil companies with the Soviets. Deterding also started an orchestrated press campaign, which included Group donations to an anti-Soviet newspaper, accusing Socony and Vacuum of buying 'communist' oil.[120] The boycott clearly became a matter of great personal importance to Deterding, perhaps because he believed that an effective boycott would lead to the total defeat of the Soviets, for which he was prepared to mobilize substantial forces.[121] In 1927, he began a price war against Socony and Vacuum, attempting to spoil their markets for Russian kerosene and letting Shell Union conduct an aggressive campaign on the US East Coast. India, where Socony sold kerosene manufactured in Batum,

[56]

[57]

became the principal battlefield.[122] Jersey Standard remained neutral, but Socony struck back in March 1928, taking the offensive to normally quite profitable markets for the Group, such as Belgium, Italy, the United Kingdom, and even the Dutch East Indies.[123] This contest, while less disruptive and global than the previous large-scale conflict in 1910-11, still caused considerable and needless waste. Moreover, the publicity emanating from the press campaign and the high drama created by Deterding's sudden, violent eruptions at dinner parties and at receptions meant that the private tussles between the oil companies now spilled into the open, helping to raise public awareness of these conflicts.

Achnacarry

The 1927 price war was not really a coincidence. Following the brief period of scarcity during 1918-20 production rose everywhere. The American oil industry raised output particularly rapidly. From the mid-1920s excess production from the Midwest and from California began overflowing first the home markets and subsequently those of the rest of the world. Rising US exports coincided with increasing exports from the Soviet Union and from Venezuela to create a worldwide glut of oil.

As on previous occasions, Deterding responded to the growing crisis by forging alliances with those of the Group's competitors over which he had some influence. Since 1905 the Group had collaborated very closely with Burmah Oil in marketing oil in India. When the original agreement lapsed in 1927, both parties chose to intensify their cooperation by merging the two distribution networks into the Burmah-Shell Oil Storage and Distribution Company. Burmah Oil also acquired a substantial share in Shell Transport, partly from the sale of shares by Royal Dutch, purchased from Sir Marcus as part of the 1907 amalgamation. To cement this alliance R. J. Watson, the Burmah chairman, joined the Shell

[58]

Henri Deterding regarded the Achnacarry agreement as one of the most notable achievements of his life, and after the conference Lady Deterding received this attractive souvenir. Its design may have been inspired by the traditional Scottish 'quaich', a cup used to offer guests a drink of welcome and departure.

[59]

Walter Clark Teagle (1878-1962) was president of Standard Oil Company (New Jersey) for very nearly 20 years, from 15 November 1917 to 1 June 1937. From that date he was chairman of the board until 30 November 1942.

Transport board in 1929. Deterding also established closer relations with Anglo-Persian, helped by the fact that Greenway had meanwhile been succeeded as chairman by John Cadman, who did not nurse the same keen antipathy to Shell which had marked his predecessor. In 1928 Anglo-Persian joined the Burmah-Shell arrangement in India and the Group and Anglo-Persian concluded a similar agreement for West and South Africa, forming the Consolidated Oil Company.[124] Three years later the UK marketing companies, Shell-Mex, established in 1921 after the acquisition of Mexican Eagle, and BP, which Anglo-Persian had taken over in 1917, merged as well.[125]

This partial consolidation of the Indian and West African markets provided a good basis to enter into negotiations with Socony over its sales of Russian kerosene. In June 1928, Deterding and Socony CEO Herbert L. Pratt resolved their differences, which included making Socony part of the arrangements with Anglo-Persian and Burmah. Socony's Russian kerosene in India would henceforth be distributed by Burmah-Shell.[126]

The entente between the Group and the other British oil companies, the peace accord with Socony, and the continuing good relations between Deterding and Teagle, together formed the basis for the famous talks during August and September 1928 at Achnacarry, a Scottish castle rented by Deterding for the shooting season. Teagle had already decided in spring of that year to travel to Britain for discussions with Deterding and Cadman, with Deterding's invitation to go shooting grouse together providing the official reason for his trip. Representatives of the Standard of Indiana, Sinclair, Texas Company and Socony boards were invited to join them. Conditions at Achnacarry were rather Spartan for the summit, the castle lacking such basic business amenities as a telephone, and the men spent most of the time in talks with each other. Deterding wrote to Agnew: 'Since Friday morning, when Mr Teagle and his party arrived, we have been talking a good deal, and I was able on Saturday to put something in writing, just so as to start a foundation on which to build something.'[127] The final draft agreement was probably the work of Deterding, Teagle, and Cadman together.[128]

The aim of the 'As Is' agreement, as it became known, was to stabilize oil markets by having the participating companies respect each other's market share, limiting exports and/or production to keep the respective shares at the 1928 level. Since regulating US markets would be in conflict with anti-trust regulation, the agreement targeted markets outside America. A formal association of oil companies would allocate to each producer quotas in export markets equal to this market share.[129] Price levels in different markets would be based on the US Gulf price plus transport costs. Such a cartel agreement was not unusual during the Interwar Period.

Pictured below, Achnacarry Castle (ancestral home of the Cameron of Lochiel, hereditary chieftain of the Cameron clan) is not really a castle, but a quite large, elegant castellated house, tucked away very privately in its beautiful estate: the nearest public highway is more than a mile away. Its rivers are well stocked with fish, its woods and moors with birds and deer, and the family's gun logs and fishing logs record all the results of their sporting activities over many decades. In the past, the castle would be leased in summer, often to one of the great shipping companies, which would then sublet it to a distinguished customer. The Cunard Steam Shipping Company sublet it to Deterding for the whole of August 1928, with shooting and fishing rights; but the logs show that he gave little time to such activities.

At his invitation a large group assembled, of whom in historical terms the most important others were Walter Teagle of Jersey Standard, and John Cadman of Anglo-Persian. All were accompanied by members of their families and staff, including Teagle's close colleague Heinrich Riedemann.

It was impossible to keep such a congregation of oil magnates secret from the Press, but journalists could only guess what they might be talking

04|280

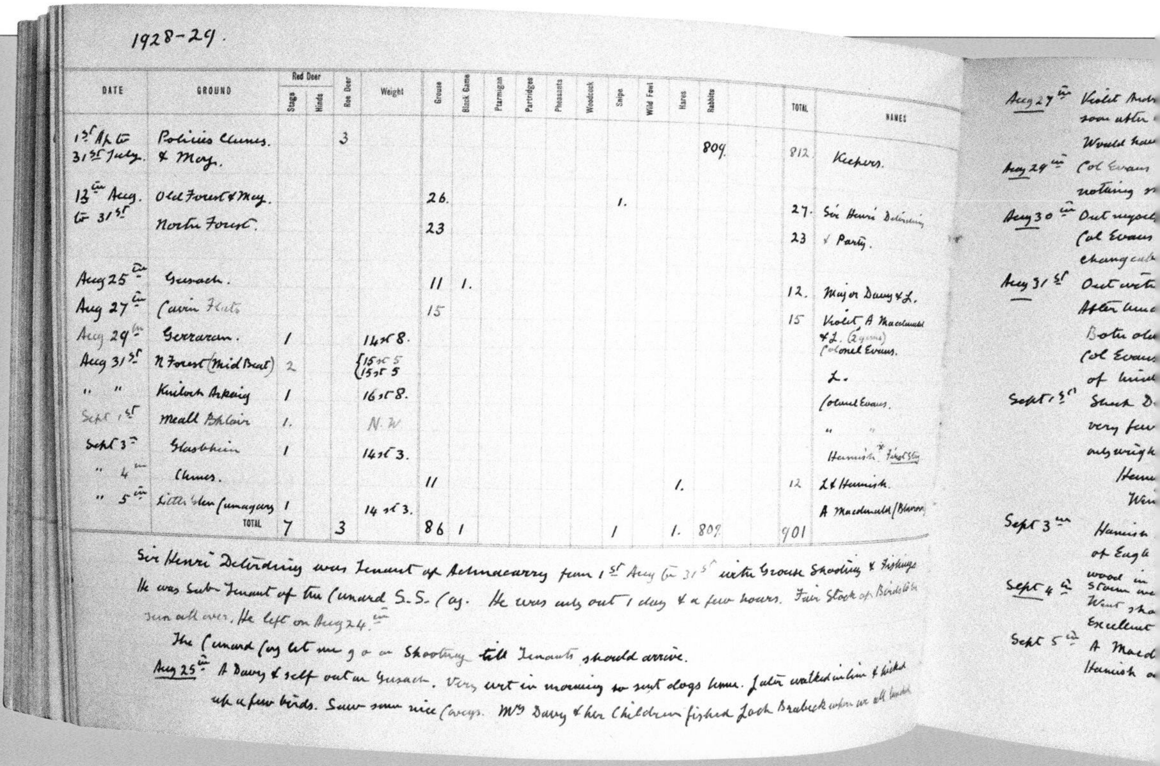

about; during the conference there was no leak as to its subject. Deterding would not have been worried by the facts that the castle's telephone was not working and that the nearest hotel had no telephone, but in a letter to his colleague Andrew Agnew he did complain that the party had been approached by Mr Milligan, special correspondent of the Daily Express, who was 'devoid of any good manners, and a rather silly fool'. Teagle likewise sent a bland disclaimer to Time magazine, saying 'There is no mystery about my being here' – it was merely a grouse-shooting party. He added that their shared interest in oil simply meant that the party would always have 'a wide field for conversation'. Teagle arrived on Friday 10 August and the business began at once. Draft agreements were prepared over the weekend: on the Monday Deterding wrote to Agnew that 'I was able on Saturday to put something in writing, just so as to start a foundation on which to build something'. Cadman's colleague William Fraser also worked on a draft (probably the same one, although this is uncertain) which he dated 12 August, the Sunday.

Ostensibly a social gathering, it seems the group actually went out to shoot only once – just enough to keep up the pretence – and not at all to fish. If they did fish, they were unsuccessful; Teagle later remarked that Cadman was a poor fly fisherman, and the log shows no record of any excursion. Monday 13 August brought their sole recorded shooting episode, which lasted only a few hours; and although the gun log (above) shows that in their hunting ground, the Old Forest of Moy, there was a 'Fair stock of birds to be seen all over', the party bagged just 26 grouse and one snipe. The hunting may have been 'lousy', as Teagle recollected, but on Saturday 18 August the oilmen approved (but did not sign) the 'Achnacarry Agreement'. Thereafter, they dispersed quickly, and although his sub-tenancy ran until the end of the month, Deterding left on 24 August. The conference had achieved its purpose, and there was no point in staying on.

Many other industries, such as chemicals, electronics, steel, fertilizer and shipping, had set up similar and often much more detailed cartels to regulate competition.

By October 1928 the boards of all oil companies with representatives at Achnacarry had ratified the draft. Only Texaco had not and the company was unlikely to agree anyway, forming an important exception. Deterding thought that it would take some more years to formalize the agreement, but he was very optimistic about its prospects.[130] Kessler had his doubts, arguing in a letter to Teagle that 'a large part [of oil production] is controlled by companies which are not controlled by you or us or by any of the few other large oil companies. From this followed that the present balance in the world's oil production cannot be maintained by you and we only'.[131] One of the things left undecided was whether or not to include lube oil, wax, and other by-products in the agreement. Numerous conferences discussing the implementation of the ideas adopted at Achnacarry followed. In January 1929 the US oil companies took an important step forwards by establishing the Export Petroleum Association to regulate American exports. Attempts to control production in Venezuela, where both Jersey Standard, Gulf, and the Group had very extensive fields with a rapidly expanding production, led to an agreement in November 1929.[132]

However, the efforts to stabilize production and market shares were continuously handicapped by a fundamental disagreement between the participating companies. The two global players, Jersey Standard and Royal Dutch/Shell, strove to uphold the Achnacarry principles, whereas the other companies, the Texas independents in particular, remained uncommitted, advocating changes, attempting to increase their market share, often acting as free riders. Consequently, it proved hard indeed to implement the Achnacarry agreement. The opening of large new oilfields, notably in Venezuela and the American Midwest, added ever more oil to the existing surpluses, and the onset of the Great Depression in 1929 reinforced the independents' marked preference to fend for themselves and forget about the rest. Where Standard and Royal Dutch/Shell dominated markets, their stabilization efforts met with some success though, as we will see in Chapter 6.

Even so Achnacarry did have an important result in the form of a much closer relationship between the Group and Jersey Standard. From now on until well into the 1950s and in some cases even the 1960s, Royal Dutch/Shell and Jersey collaborated in numerous markets and joint ventures. What had started in 1907 when Teagle and Deterding discovered a mutual rapport, now turned into a regular understanding between the two multinationals, which continued long after the two founding fathers of the entente had disappeared from the scene. The Group made, in 1932, an agreement with Jersey Standard about joint exploration in Cuba and in the Netherlands, with Deterding suggesting to Teagle a more comprehensive arrangement covering other areas around the world.[133] In 1938 the Group concluded a comprehensive agreement with Anglo-Iranian, the name adopted by Anglo-Persian in 1935, and Burmah for joint exploration in India and Burma.[134] Jointly they decided to boycott Spain after the nationalization of the distribution networks there,[135] and each took a 25 per cent each stake in the Hugo Stinnes Ölwerke, a company set up by IG Farben to manufacture synthetic oil from coal.[136]

Missing out on the Middle East The place where the Group really missed several chances of access to crude oil was the Middle East, a failure which was to have long-lasting consequences for the period after the Second World War. In 1925-26 the company was offered a number of options on oil concessions in Bahrain, in the al-Hasa region of Saudi Arabia, and in the neutral zone between Kuwait and Saudi Arabia. Major Frank Holmes had developed these options on behalf of the Eastern and General Syndicate, a group of London mining engineers. The Group was, however, not interested in these concessions; nor, for that matter, were Anglo-Persian or Jersey Standard. An outsider in the area, Gulf, took up Holmes's offer in 1927, a tactical decision which would turn out to have great strategic value.[137]

From 1928 onwards the Red-Line agreement underlying TPC stymied expansion in the Middle East. Attempts by the American partners in TPC to broaden the company's strategy, for example by buying the Bahrain concession, met with stiff resistance from the British partners, from the French participant TPC and, of course, from Gulbenkian who wanted compensation for any watering down of the original agreement. As a consequence Gulf Oil sold its Bahrain option to Standard of California (Socal), which successfully developed it during the early 1930s. With the pessimists proven wrong and the Holmes concessions shown to be very valuable, other companies became increasingly interested in potential oil fields on the eastern shores of the Red Sea. Gulf Oil began to develop its Kuwait option, and Socal opened negotiations with Saudi Arabia and soon acquired the al-Hasa concession.[138] Because Socal had difficulties in developing this field and in selling the crude, the company made another attempt to open up a Red-Line agreement and invited TPC shareholders, including the Group, to participate in the al-Hasa concession. In early 1934 the TPC was directly negotiating with Ibn Saud, and it was proposed to make a 'definite offer (...) to Ibn Saud which will be more attractive to him than one likely to be made by the Standard Oil of California'.[139] The Group remained unconvinced about the value of the concession, however; when consulted by London in 1934, the geologists in The Hague judged that there was no oil in Saudi Arabia.[140] The following year Anglo-Saxon accepted a proposal for a concession in Qatar, though for defensive reasons, to keep out the competition.[141]

As a result, the Group missed the opportunity to participate in the most productive concessions in the history of the oil industry. Its failure stemmed from a combination of factors. Falling oil prices, the imperial overstretch evident in the lack of resources to develop the concessions in Venezuela, and the negative expert opinion about the likely prospects dampened enthusiasm for new and extensive E&P programmes and eliminated the arguments needed to overcome stiff resistance from fellow TPC shareholders.

Moreover, the Group also thought that oil from the shores of the Persian Gulf would be a problem for some of its competitors, in particular for Anglo-Persian and for Burmah Oil, because it would spoil their markets in South Asia, but less so for Royal Dutch/Shell. After all, via Turkish Petroleum Company the Group controlled the main source of oil in the region which could reach European markets at an unbeatable price through pipelines to the Mediterranean and from there by tanker.

Finally, another, less obvious factor may have played a role – namely, the Group's consolidation strategy. From the mid-1920s the Group aimed to stabilize relations in the oil industry, rather than pushing to change them for its own advantage. The Red-Line agreement consolidating the oil industry within the borders of the former Ottoman Empire fitted perfectly with this conservatism. After 1945 the Group would pay a substantial price for this policy choice.

[62]

The interior of an oil derrick in the 1930s shows everything in place for deep drilling, with the opening drill pipes ready to go and many more stacked in position.

Conclusion

For the Group, the Roaring Twenties ended on a high note. Royal Dutch and Shell Transport announced bumper profits totalling £ 17.2 million for 1929, second only to 1920. The latter record was a somewhat artificial one, however. Inflation, pushing US crude prices to their highest point since 1872, and the channelling of war profits from hidden reserves into revenues had boosted the figure. Group production peaked that year at 25 million metric tons or 175 million barrels, constituting 12 per cent of world production amounting to about 211 million tons, nearly 1.5 billion barrels. The Americas now supplied two-thirds of the Group's oil, Venezuela producing 35 per cent and the United States 30 per cent, pushing Indonesia into third place with only 18 per cent. Mexico, Romania, and Brunei and Sarawak produced substantial volumes of crude as well (Figure 4.3). New areas that began contributing to Group production during the late 1920s and 1930s were Argentina, Trinidad, Iraq and, from 1938, Germany. Because of the prevailing oil surplus, the Group entered few new areas for exploration and missed out the development of Saudi Arabia. Initial explorations in Colombia yielded few results.[142]

As for the Group's competitive position, its eternal rival, Jersey Standard, had made great strides under Walter Teagle to acquire more overseas production to supply its markets. With a refinery run of 28 million tons or 197 million barrels Jersey Standard appears to have had a slightly higher throughput, but then the Group bought some 3 million tons on the market, including gasoline and fuel oil from Anglo-Persian.[143] Consequently, in distributed profits per unit Jersey probably had an edge. Royal Dutch and Shell Transport paid a total of £ 16.3 million in dividends, Jersey Standard $ 46.5 million, or £ 16.6 million.

Consistent with the Group's prominent position in the international oil industry, Deterding fashioned a new role for himself as captain of the international oil industry. Achnacarry was, in many respect, his finest hour, and at the same time a clear marker of the sea change in the Group's business, from the dynamic expansion of its first fifteen years to consolidation. After the disappointments in Mexico and Russia, the Group largely restricted its acquisitions of companies and concessions to the US; the Brunei concession taken in 1922 was probably the only one elsewhere. This is in marked contrast to Jersey's rapid build-up of crude production outside the US, formerly the weakest part of that company's operations. The Group also remained overly dependent on profits from the Dutch East Indies.

On the other hand, the Group stretched its resources to expand its operations in Venezuela and the US as fast as it could, and still had to let go of promising opportunities. Managers thus had good reasons not to venture further afield. Concentrating on a few known areas was also a logical response to the growing nationalism which ran counter to the Group's rationale of a world-spanning company, acquired during the first era of globalization. The 1920s were a learning experience in that respect, exemplified by the erratic response to the Russian nationalization, and the lessons learned then would come in good stead during the next decade of still greater challenges. Finally, the Group needed a breathing space after the hectic expansion of the 1910s, time to adapt the management structure to the increasingly complex usiness organization, time also to transform research and marketing.

[1]

Chapter 5

The contours of a mature corporation: management and staff relations

Henri Deterding in Italy, 25 June 1922, with members of Shell's company Nafta Societa Italiana.

The early 1920s exposed the strains created by the Group's expansion and the aftermath of the First World War. Deterding became caught between his immense workload and his inability to delegate decisions to others. Proposals for fundamental management reforms from Colijn and Philips were blocked by Deterding, probably because he feared a check on his powers. His increasingly temperamental attitude alienated him from the colleagues with whom he had managed the company until then, resulting in the premature departure of some of them. Deterding also came into sharp conflict with his close ally and friend Gulbenkian, which ended with the latter's withdrawal from his very successful association with the Group. Once Deterding had consolidated his grip on the company, however, a new generation of managers accommodated to his style of dominating leadership, helped to restore the managerial balance, and limit the damage that his erratic style did to the Group. The 1920s also witnessed the emergence of a corporate culture in staff relations and recruitment and training patterns.

[2]

A 1932 board meeting in Royal Dutch's head office, 30 Carel van Bylandtlaan, in The Hague, captured by the celebrated photographer Erich Salomon. From left to right, Andrew Agnew is just in the picture. Jonkheer Hugo Loudon gazes attentively at Deterding. Sir Robert Waley Cohen's distinctive profile is clearly visible, but next to him G.C.D. Dunlop is almost completely obscured, as is August Philips. J.E.F. de Kok can be seen at the end of the side, hand to chin. Sir Henri Deterding is naturally at the head of the table and centre of the picture, with J.B.A. Kessler next to him at the corner, followed by J. Th. Erb, C.J.K. van Aalst, J. Luden and N. van Wijk. Gaining Deterding's permission to take the photograph took three months, and his consent was only granted after Salomon had provided him with a photograph of the Dutch Cabinet in session.

Managing a sprawling empire As the Group expanded, the structure of two holding companies controlling three operating companies held together by interlocking directorships became ever more elaborate. Operating companies in Russia, Romania, Sarawak, the United States, Trinidad, Mexico, Venezuela, and Egypt were more or less randomly inserted in the structure. Several factors added to the increasing complexity. Political or fiscal expediency, or the need to accommodate outside shareholders, sometimes necessitated setting up a two-tier structure, with a separate holding company managing a particular operating company. Such companies could have different legal regimes, say a holding company under Dutch or British law controlling and an operating company under local, say US, law. Sometimes, notably in the US and in Venezuela, the Group had separate companies in the same country, exploiting particular concessions, working in particular areas, or performing particular functions, such as shipping or marketing. These companies could formally belong to a company managed from The Hague, but report to another located in London for some or all of their operations. All companies were firmly tied to the Group fold by funding arrangements, management agreements, and service contracts, except in the US, where the Group followed a fairly deliberate policy of developing strong local roots and direct links to the American capital market. Finally, some companies had the interests of outside shareholders to consider, such as the original concession holders, other oil companies, participants like the Rothschilds or Dutch banks, or business associates such as Gulbenkian. These participants shared in the risks of developing the operations concerned, but they did not necessarily have a proportionate say in the actual management of them.

Marketing subsidiaries proliferated primarily for fiscal reasons. The spread of income and profit taxation around the world meant that in order to avoid dual taxation the Group had to set up separate companies nearly everywhere. In the Dutch East Indies, for example, Asiatic dictated the local market and made huge profits. The company traded through a local branch office which did not have a separate legal identity from the UK head office. The Dutch East Indies government therefore threatened to tax Asiatic itself, on the basis of its worldwide profits. Consequently the Group's marketing operations in the Indonesian archipelago were reorganized into a separate company by resurrecting the Handelszaken department of the dormant Dordtsche Petroleum Company.[1] In this way Asiatic gradually acquired dozens of subsidiaries to manage local marketing operations, and sometimes refining as well.

In the years leading up to the First World War, uncontrolled growth and increasing complexity had already created serious strains within the Group's original management team. The management changes of December 1913 had removed some of these, but had failed to resolve the issues. The individual workload of the The Hague managers was reduced by the appointment of Colijn and Pleyte as managing directors of Bataafsche, leaving Loudon and Cohen Stuart to manage Royal Dutch affairs. Cohen Stuart had also resigned as managing director of Anglo-Saxon, Waley Cohen succeeding him there. The root of the problem, the feeble management structures at the top and Deterding's wilful leadership style, remained untouched, however. The First World War accelerated the fragmentation of the Group's operations. The global markets for products, capital and labour that had existed before 1914 disintegrated, raising no end of complications for the Group, built as it was on global production, transport, and distribution like no other company. Moreover, the increasing government interference with the oil industry, ranging from the imposition of higher royalties and higher taxation, via the setting up of national oil companies, to the threat of state monopolies and de facto nationalization of private enterprise, placed the oil companies in an entirely new and unaccustomed position. With these fresh challenges a thorough overhaul of the top management structure became a vital necessity. On the return of peace, the prewar tensions within the managerial team returned with a vengeance in the form of a series of bitter conflicts between Deterding and his colleagues.

Deterding's long holiday On 30 September 1919 Cohen Stuart informed the Board of Royal Dutch that Deterding wanted to take leave from his position and rest for a while. Colijn was to replace him as managing director of Asiatic and Anglo-Saxon in London.[2] B. C. de Jonge, like Colijn a former politician with close ties to the Dutch East Indies, became acting manager of Bataafsche, Pleyte having resigned in 1919 for health reasons.

At the moment of Stuart's announcement Deterding was already on leave, having stopped attending meetings of the major companies sometime around the middle of September 1919 and withdrawn to a hunting lodge in the Midlands. He intended to have a long holiday, perhaps even for a year, but in November Gulbenkian advised him against that idea, writing that 'I think your best course would be to stay more often in the country than you do now and to take weeks off at a time, get in more sports and country life, rather than to take such an exaggerated measure as a holiday of months' duration.' He summarized his diagnosis of Deterding's situation as 'I feel that we are all suffering from a state of over-excitement'.[3] Deterding did not take his advice. In early December he officially notified the boards of Anglo-Saxon and Asiatic that he had resigned, nominating Colijn to be his successor as managing director of both companies. His absence lasted for about six months, until March 1920. On 16 March he attended a Shell Transport board meeting; on 24 March he returned to the board of Asiatic, chairing a meeting on 14 April in the absence of Marcus Samuel. On 20 April Deterding's long holiday formally ended with his reinstatement as managing director of Asiatic and Anglo-Saxon. If anything, Deterding's six months' leave showed that the Group's brilliant leader was human after all, that there were limits to his ability to cope with the strains of work, multiplied by the war and exacerbated by a grievous personal loss, his first wife having died in 1916. His absence had opened a window of opportunity for a debate about the Group's management structure. The main problem concerned the sheer size and complexity of the organization, generating a volume and variety of dossiers too large for a single manager to oversee, let alone handle. Moreover, the top managers did not have clearly defined tasks and their

[3]

Jonkheer B. C de Jonge (right), governor-general of the Netherlands East Indies and former director of Bataafsche, hands over to his successor A.W. L. Tjarda van Starkenborgh Stachouwer in Batavia, 1936.

vague responsibilities often overlapped, so most boards left key issues for Deterding to decide.

As discussed in Chapter 3, Colijn had clearly identified these bottlenecks as early as 1915, and he had reorganized Bataafsche accordingly.[4] Observers inside and outside the company appear to have agreed that the Group needed a reorganization along the same lines. In January 1920, Fred Lane once again showed his peerless insight into the Group's problems, writing to the Rothschild manager Weill: 'Deterding's great strength was that he followed and discussed the details of every department in the minutest fashion. (...) I have felt for a long time past that if Deterding sought to maintain the excessive labours he has hitherto undertaken, then there would be an end of it. It is impossible for any man, however strong, to sustain such a strain after such years of laborious work.' Lane considered Colijn a good successor, commenting that 'while he is not a Deterding, he is a highly capable man (...) a man of method. The lines upon which he is working seem to be more methodical than those hitherto pursued by Mr. Deterding, throwing greater responsibility upon subordinates, and regularizing the relations between subordinates and the chief, so as to relieve the chief of the detailed work of each Department'. Moreover, Lane expected the Group to emerge stronger from a reorganization. Deterding would still monitor the business, taking care 'that no arrangements are made in the executive which do not meet with his approval'; however, he would now be able to concentrate on the main issues. 'I think the present organization will be very successful, and the relief from the excessive labour of the past should make him [Deterding] more powerful in the general control'.[5] Another colleague and friend, Calouste Gulbenkian, wrote a personal letter to Deterding in the same vein.[6]

Personal vs departmental responsibility

However, nothing definite appears to have been resolved during or after Deterding's holiday, perhaps because he returned early to reassert his control and thwart fundamental change. As before, the Group sought a remedy in appointments rather than in confronting the issues at hand. In August 1920, Royal Dutch chairman Capadose informed the board about the outcome of a meeting in London to discuss the company's organization with the three managing directors there, Deterding, Colijn, and Waley Cohen. Four months later the board discussed the issue again, this time requesting Deterding's presence at the next meeting in January to answer questions raised. Finally a reorganization plan emerged in February 1921. The number of Royal Dutch directors was increased from three to five. Loudon and Cohen Stuart stepped down as managers to ascend to the supervisory board, raising the number of supervisory directors to nine.[7] Four new directors were elected: Erb, De Kok, Colijn, and De Jonge. The former two succeeded the latter two as managers of Bataafsche, De Jonge following Colijn to London to take Cohen Stuart's place there. No details were given about the preceding discussion 'because secrecy is necessary', as Loudon stated.[8]

These changes helped to reinforce the top management of Royal Dutch and Bataafsche, but failed to solve the fundamental problem of vague functions and overlapping responsibilities. During the ensuing debate about corporate organization, two main issues emerged, i.e. the formalization of decision-making procedures and the internal accounting practices. August Philips, Royal Dutch and Bataafsche non-executive director and legal counsel of those companies, drew up the charge sheet on the former issue in a long letter to Capadose, written in September 1921. He analyzed the organizational malfunctions in detail and pointed to a number of cases where the management had failed, notably in Mexico and in Russia. His indictment sounds very similar to the one presented by Lane, excoriating the Group for its inefficient organization of work, the insufficient definition of managers' functions, and the duplication of functions between London and The Hague often resulting in issues being dealt with twice, each office con-

-2-

het beheer niet zoo goed plaats heeft als redelykerwyze mocht worden verwacht met het oog op de talryke eminete krachten,welke,van den hoogstgeplaatsten leider onzer groep af,door alle rangen heen en in alle departementen van ons bedryf worden aangetroffen, de zeer aanzienlyke geldmiddelen,waarover wy nog altyd beschikken,den velen tyd en de groote moeite die er aan de zaken wordt besteed. Het schynt my niet twyfelachtig,om slechts een paar grepen te doen,dat noch in Mexico,noch by het sluiten en uitvoeren onzer jongste contracten betreffende Rusland, noch in ons optreden in de landen der voormalige Oostenryk-Hongaarsche monarchie,door ons wordt gehandeld op zoodanige doeltreffende wyze als noodig en mogelyk zou zyn. Toch zyn by de twee eerste der bovengenoemde onderwerpen zulke enorme bedragen betrokken,dat misslagen daar zoowel voor de financieele positie onzer groep als voor hare reputatie en crediet de bedenkelykste gevolgen zouden kunnen hebben. En ik ben overtuigd dat het op tal van andere gewichtige gebieden,waarop myn aandacht toevallig niet zoozeer gevallen is,niet beter is gesteld.

Wat is de oorzaak van dien onbevredigenden toestand?

Naar myne vaste overtuiging gebrek aan een goede organizatie der werkzaamheden,onvoldoende afbakening van ieders taak,geen praktische samenwerking tusschen de verschillende krachten waarover wy beschikken, daardoor gebrek aan consequentie in de behandeling der verschillende détails,vermindering van verantwoordelykheidsgevoel by sommige leiders,usurpatie

-4-

vaak behoefte hebben.- Ik geloof,om my tot deze twee reeds genoemde punten te bepalen,dat de zaken in Mexico er anders zouden voorstaan,indien de aanwyzing der voornaamste bedryfsleiders aldaar minder eigenmachtig en abrupt ware geschied,de algemeene organizatie te Tampico en Mexico City kalm ware overwogen en bediscuteerd,en dat de Russische contracten misschien nooit gesloten waren,of er althans heel anders zouden hebben uitgezien,indien ook daarby tydig het inzicht van anderen ware gevraagd.

Dat deze stand van zaken allengs is ontstaan is niet onverklaarbaar.- Ons bedryf is in een buitengewoon kort tydsbestek reusachtig gegroeid. Door de samenwerking met de andere Oostersche producenten van 1903 en de fusie van 1907 moest het beheer deels van uit Londen,deels van uit den Haag geschieden. De afbakening van de werkkringen der drie groote zustermaatschappyen,zooals die oorspronkelyk was ontworpen,bleek moeielyk vol te houden; veel van wat aanvankelyk was gedacht als tot de taak der Bataafsche te behooren,kwam ten slotte terecht by de Anglo-Saxon,waardoor de sferen van de bemoeiingen dier beiden door elkaar gingen loopen. Dit werd nog bevorderd doordat onze krachtige directeur-generaal,in wien door iedereen terecht de alles overheerschende persoonlykheid in onze groep werd gezien,te Londen was gevestigd en daar

[4]

vinced that they were in charge. According to Philips some top managers took their responsibilities too lightly, allowing middle managers to usurp power and influence. In a direct attack on Deterding's leadership, Philips laid the blame for this unfortunate situation first of all on the Group's top management and its lack of systematic and effective organization evident in the confused distribution of work between the various boards. Secondly, he pointed to the rapid growth of operations and the difficult communication between the boards in London and The Hague as important causes. Finally he singled out the very autocratic way in which the business was run, more so than previously, Philips argued, 'colleagues of our director-general had exerted considerable influence on him; they were consulted on almost all important matters (...) Once these colleagues had retired from daily management, and with the management in The Hague also becoming rather autocratic, the dangers of this approach had increased considerably'.[9] Philips proved better at analysis than devising remedies, which remained rather sketchy in his letter. Echoing earlier suggestions raised by Marcus Samuel, he proposed merging the supervisory board of Royal Dutch and the board of Shell Transport into a single board, and to form subcommittees of directors to monitor specific policy areas, such as Mexico, Russia, or the tanker fleet.[10] Those subcommittees would then coordinate operations and monitor the managers concerned.

Scrawling in thick red pencil across the text of a letter written by Philips on 27 September 1921, Deterding gives vent to his fury.

Philips' radical and highly critical statement found a warm welcome amongst the members of the Royal Dutch supervisory board, who appear to have agreed with his analysis.[11] Some of them wrote to Capadose adding their own worries. One such letter, unfortunately not signed, mentioned as a key problem 'not that Colijn is a politician and that Deterding is often away during the winter', but that the organization failed cope with this, because 'we are much too centralized', and therefore do not 'breed new leaders'. To make his point, the author listed a number of managers who had recently failed in their assignments.[12] Colijn also responded immediately: he agreed to some extent with Philips's views, but not with his solution.

However, Deterding's reaction was most significant of all. His objections written in red on his copy of Philips' letter show that he felt attacked and offended. And he disagreed fundamentally. His comment on the Group's alleged lack of management transparency went directly against the main thrust of the reforms proposed by Philips and implemented by Colijn at Bataafsche, arguing that 'the key issue is the recent trend towards departmental responsibility over personal responsibility, which in my view is completely wrong'.[13] In another comment he blamed the frictions on a supposed Dutch trait of jostling for the limelight rather than remaining in the shadows and letting the business take the credit.[14] As far as we know Deterding never reacted formally to Philips' letter, so we can only speculate about his motives for rejecting these and similar proposals, the sensible arguments put forward by old friends like Lane and Gulbenkian, and the strong recommendations from Colijn and Philips.

His reasons for doing so appear to have been rooted in his inability to accept the impact which the Group's expansion had had on his own position as chief executive. For already quite some time, Deterding had found it difficult to cope with the stress of running the huge organization, but he refused to accept the obvious solution, delegating some of his powers. With his own intuition and vision, he had almost single handedly built a formidable enterprise, and arguably had become the most successful businessman of his generation. If the business now suffered from friction, the people running it were incompetent: so why compromise and delegate power to people whose competence he could not trust? Formal procedures and clearly defined functions within the top management would restrict Deterding's control. Such a reorganization would also definitely have conflicted with his very personal style of management. He switched from one dossier to another, more or less following his instinct and without a clear plan, giving very detailed attention to matters which interested him, while leaving most other issues to be dealt with by his fellow directors. Deterding did not, however, give his colleagues full responsibility, for he reserved the right to interfere in any topic at any time. This erratic style made him rather unpredictable, a constant source of annoyance to his colleagues. Since he could clearly no longer control everything, however, managers who ran operations which happened to fall outside his gaze gained considerable leeway. Finally, Deterding's continuing temperamental behaviour frayed relationships. He took to violent displays of public anger, as often as not out of all proportion to the importance of the issue which aroused them, alienating colleagues and business relations alike. His intemperate language and canvassing for support over the Dutch East Indies export levy (see Chapter 4) greatly puzzled the Rothschilds; as Gulbenkian wrote to Deterding, they and other friends of Royal Dutch were 'amazed at the importance you attached to the matter and at the severity of your attacks on all fronts'.[15] From the summit of a career sculpted by sheer willpower, Deterding could understand life only as an expression of the will, and consequently any adversity as ill will directed at him personally. Consequently, as disappointments mounted with the advancing years, his outbursts became more frequent, making those around him want to keep their distance.

The profits maze

Consequently, the push towards more formal management procedures and a clearer demarcation of managerial responsibility foundered on Deterding's wish to retain full control. Simultaneously the initiative to reform internal accounting practices ended in defeat for the advocates of change. Vehement debates about pricing and accounting practices had dogged Asiatic from the start, resulting in almost continuous tensions between the partners in the joint venture. The Group's tiered structure of holding companies and operating companies billing each other for products and services had created a complex internal accounting system based not on actual cost, but on fictitious prices. Rising taxation and diverging tax regimes subsequently put a premium on allocating profits not to the operating company which produced them, but to the company most suited to receive them from a fiscal point of view, resulting in a veritable accounting maze which obscured the objective assessment of costs, profits, and performance. As a consequence, managers had a weak case when confronting governments basing tax claims on profit figures, as had happened to Bataafsche in its dispute with the Dutch colonial government over the export levy (see Chapter 4). There was also no systematic data collection at Group level. Since Colijn's reorganization, Bataafsche had a statistical department producing detailed figures of almost all aspects of its organization, such as cost prices, production levels, number of employees, their remuneration, etc. Such data were not systematically compiled for the rest of the Group, presumably because Deterding considered such systematic information quite redundant to his needs as a manager.

In a detailed memo dated 6 September 1921, Colijn's personal assistant Gerretson focused on this problem and argued that it resulted 'from endeavour to let the Bataafsche make all profits, which again was a means to escape excess profits duty in England'. He pleaded for a reorganization that would result in all companies making 'such profits as they would make in the natural course of event', meaning 'profits that would be made by the various companies if they did not form part of a vertical trust'. 'Nothing has created so much diffidence against us as the continuous passing on of natural profits between the Anglo-Saxon, Asiatic and Bataafsche.' As a solution, Gerretson proposed replacing the fictitious internal prices by market prices, and scrapping the special arrangements channeling profits from one part of the Group to another. This would, he argued, also solve debates with foreign governments, by removing 'the incentive for political attempts either with the object of nationalizing the fields or of compulsory production'.[16]

By the time Gerretson's sensible and far-sighted proposals were tabled, Colijn's position had already become too weak to effect such fundamental changes. Moving him to London to replace Deterding may have been a mistake in the first place, for his English never became sufficiently fluent and he felt increasingly isolated from his roots and from his political constituency. His frequent visits to the Netherlands to attend his interests did little to improve his position in London. Moreover, Colijn's long-term ambition was to be a Dutch statesman, not an oil baron. On the death of Abraham Kuyper in 1920, he had become leader of the orthodox-protestant political party Anti-Revolutionaire Partij (ARP), and his intensive contacts with the party, its newspaper, and its supporters irritated some of his colleagues who, unlike him, were one hundred per cent committed to Royal Dutch. Colijn was also rather embarrassed by the anti-Dutch sentiments at the London office, which Deterding fanned with vociferous condemnations of the export levy. He had also lost the struggle over rationalizing the Group's organization. Finally, Colijn had fallen short of Deterding's expectations in failing to secure the Jambi concessions, and he had, in the latter's view, committed a cardinal sin by showing Bataafsche's cost prices to the Colonial Ministry during the Jambi negotiations, presumably to convince them of the market-oriented reforms which he had in mind.[17]

On his last day at the London office, Colijn summarized his views on the weaknesses of the Group in a valedictory message, emphasizing the antagonism between British and Dutch, and particularly Deterding's anti-Dutch feelings, which he blamed for

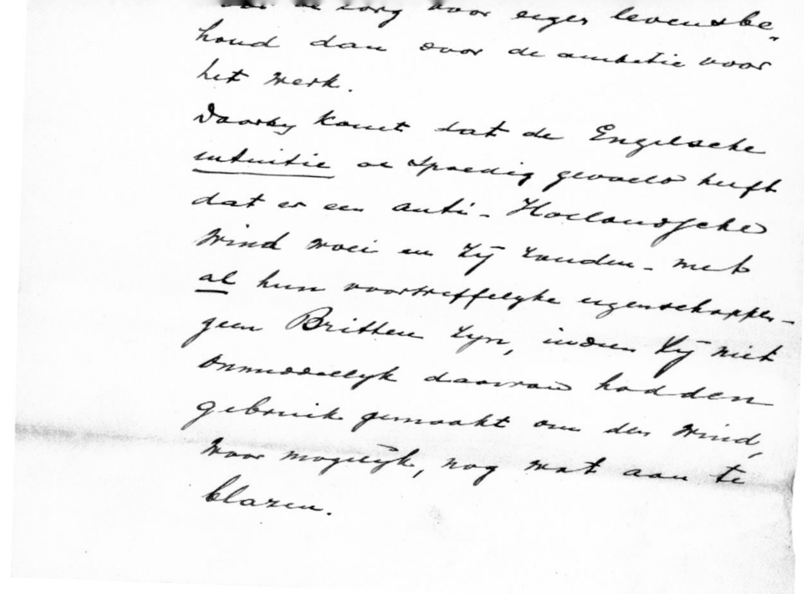
... de zorg voor eigen levensbe-
houd dan voor de ambitie voor
het werk.
Daarbij komt dat de Engelsche
intuitie al spoedig gevoeld heeft
dat er een anti-Hollandsche
wind woei en zij zouden - met
al hun voortreffelijke eigenschappen -
geen Britten zijn, indien zij niet
onmiddellijk daarvan hadden
gebruik gemaakt om den wind,
waar mogelijk, nog wat aan te
blazen.

Colijn's parting shot: on his last day in office, Colijn summarised his views on the Group's organization.

[5]

undermining his position.[18] He reiterated his plea for a more rational organization of the Group and notably a clear separation of tasks between the various managers. Colijn ended his letter on a slightly dramatic note, saying that he would have preferred to discuss these matters face-to-face with Deterding, but the latter had been too busy to meet him. Nor did Deterding attend Colijn's farewell lunch given by Bataafsche in The Hague on 6 April.[19] With Colijn the proposals to introduce more market oriented internal prices disappeared from the agenda as well. A few months after Colijn, De Jonge resigned as well, finding the London office work uncongenial.[20]

After considerable discussion, the Royal Dutch board appointed Philips to succeed Colijn as managing director of Royal Dutch, and as director of Shell Transport, Anglo-Saxon, and Asiatic. This choice was a clear sign of both the board's critical attitude towards Deterding and of its support for the rationalizations proposed by Philips and by Colijn. The board failed to take the necessary steps to effect their recommendations, however. Consequently the more systematic and bureaucratic management model which Colijn had introduced at Bataafsche was not applied to the top of the business.

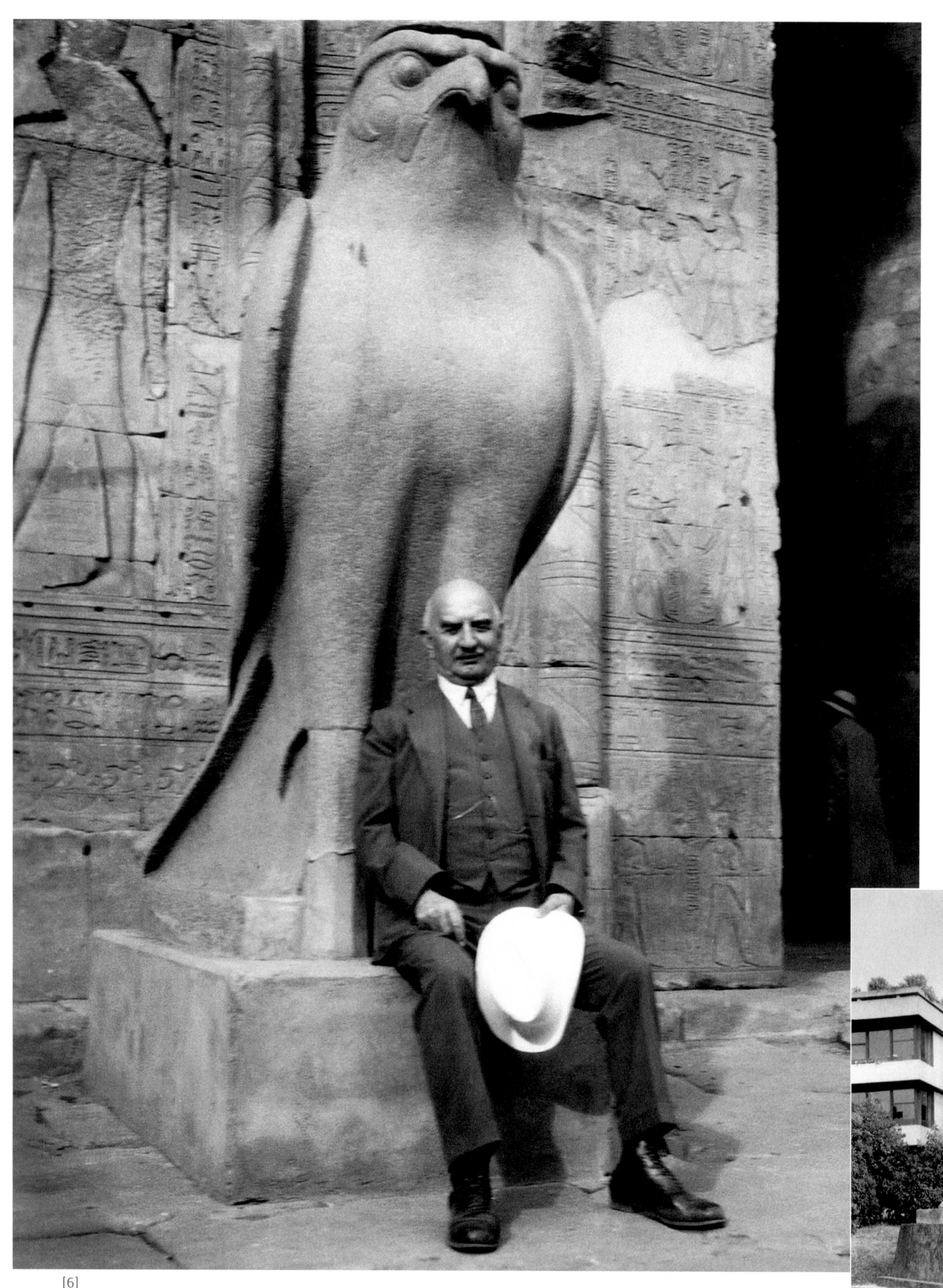

[6]

The Armenian oil magnate Calouste Gulbenkian (1869-1955), seated on a statue of Horus, god of light and wisdom, at Edfu in Egypt, 1934. The striking likeness between man and statue later inspired the sculptor Leopoldo de Almeida (1898-1975) to create a new statue based on the photograph and now on display in the Calouste Gulbenkian Park, Lisbon, Portugal.

[7]

Accounting practices and minority interests: the Venezuelan Oil Concessions Ltd The opaque accounting and circulation of profits criticized in the Colijn-Gerretson memo did not just occur between the main operating companies, but also between Group companies and subsidiaries in which outsiders had minority interests. These outsiders could be former concession holders, shareholders of companies taken over, or financial institutions participating in a particular venture. There were quite a number of such companies with minority interests, notably because, during its expansionary phase, the Group could not always buy out concessionaires or shareholders straight away and had also regularly needed to draw finance from commercial banks or professional investors. The expansion into the American Midwest, for instance, had been financed through the 's-Gravenhage Association, set up in 1912 with a capital of £ 1 million, partly subscribed by the Paris Rothschilds and Dutch banks. The Association controlled the Roxana Petroleum Company, established in the same year, which in turn owned and managed the producing properties from its seat in Tulsa, then the oil capital of the Midwestern world. Another such vehicle was the Burlington Investment Company which participated in the Group's Venezuelan operations.

It was Deterding's firm policy never to invest in a company that he could not control, so in almost all cases the Group had a clear majority interest in any subsidiaries, bolstered by service contracts with Group operating companies covering the management and the actual exploration, drilling, exploitation, and marketing. This complete control could render it quite difficult for minority shareholders to protect their position. Long-running, capital-intensive projects are likely to generate conflicts of interest between the participants anyway, but the Group's joint ventures set up during 1910s and 1920s were unusually prone to such clashes, because most had outside shareholders who were not themselves active in the oil industry and thus could not always see through the web of conditions attached to a given venture. This changed only in the wake of Achnacarry, which laid the basis for a closer collaboration between the oil companies.

One such conflict centred on the Group's revenue accounting methods and led to a permanent break with one of its staunch supporters. Gulbenkian was a minority shareholder in a number of the Group's ventures, and he had, among many other activities, also helped to get the Venezuelan expansion off the ground. The Group had acquired interests in three companies there, which owned huge concessions in various parts of the country. These were the Caribbean Petroleum Company, the first to successfully develop an oil field in the country; the Colon Development Company; and finally the Venezuela Oil Concessions Ltd (VOC), acquired in 1915 through the intermediation of Gulbenkian. VOC owned the Aranguren concession near Lake Maracaibo, named after General Aranguren who had owned the concession since 1907. To complicate matters further, the three different production companies were represented by a single organization in Venezuela run from The Hague, but marketing remained separate from the E&P and was directed by London.

The first attempts to develop the VOC concession failed however, and the Group subsequently concentrated its efforts on the Caribbean Petroleum Company. Because VOC lacked sufficient working capital and only made losses, the company was restructured in 1921 by the formation of a separate holding company to attract fresh outside capital. At about the same time, a series of contracts between VOC and various Group operating companies was concluded, notably a sales contract with the Caribbean Petroleum Company.

Already in 1921 and 1922 complaints surfaced about the slow progress of work on the VOC concessions, but the Group largely ignored them. The situation changed completely in December 1922, when the Barroso well in the La Rosa field came in, a spectacular gusher producing an estimated 100,000 barrels daily. VOC now became one of the most important producers within the Group, raising the stakes on the question of who would profit from these riches. The set of agreements which VOC had concluded in 1919-21 with several Group subsidiaries now turned out to be rather unfavourable. Gulbenkian became increasingly worried about the implications. He maintained that he had 'protested against it at the time' and, in spite of some changes, still did not consider it 'a fair document'.[21] Deterding had not been involved with the drafting and conclusion of the agreements, as he was then not very interested in developments in Venezuela, so initially Gulbenkian thought that after proper study Deterding would see the need for amending the sales agreement and resolve all disagreements. The two men's relationship had already deteriorated somewhat, however. Differences of opinion over Gulbenkian's share in the Turkish Petroleum Company soured them; probably Gulbenkian also nursed an aspiration to become managing director, which Deterding refused to consider.[22] The tone of their frequent corres-pondence, which had been very cordial, changed noticeably. 'So it is very, very depressing for me, and somewhat heartbreaking when my friend Deterding forgets the past and writes as in his last letter', Gulbenkian lamented in January 1923.[23]

In May 1923 Andrew Agnew, managing director of Anglo-Saxon and president of the VOC board, suggested asking H. van Blommestein, a legal counsel of Royal Dutch, for a review of all contracts between VOC and the Group. In June Van Blommestein submitted his report to Agnew and De Kok, who handled this issue at Bataafsche. It was very favourable for Gulbenkian. Van Blommestein wrote with the dry understatement of a legal expert that 'the complaints in fact are not altogether unfounded on some points'. In particular the conditions at which the crude was sold to Caribbean were very unfavourable for VOC and needed to be amended, but Van Blommestein also considered the other agreements very one-sided. A recent £ 100,000 loan from Bataafsche to VOC, for example, carried not only the fairly high interest rate of 8 per cent, but also included a clause imposing a 5 per cent royalty on all oil produced by VOC. Another odd clause in the sales contract gave Caribbean the right to limit the production of VOC to as little as 1,000 barrels per day. Van Blommestein's report was suppressed, however. De Kok wrote to Agnew that 'of course Mr. Gulbenkian never had any right to the report, as Van Blommestein's investigations were being carried out as a matter of ordinary staff routine', and, of course, 'the report was prepared from quite a wrong angle'.[24] Time and again Gulbenkian requested a copy, but he never got one.[25]

[8]

Sir Henri Deterding, K.B.E. -2- 30/9/25

justice? But for goodness' sake let us have no more correspondence about this matter, but peace.

With kindest regards, I am, my dear Deterding,

Yours very sincerely,

C.S. Gulbenkian

The conclusion of one of Gulbenkian's last letters to Deterding, 30 September 1925. His hope for renewed friendship would not be fulfilled.

Other critics of the VOC policies did not restrict themselves to writing angry letters to Deterding or Agnew. In August 1923 General Aranguren, still an important shareholder in the company and one of its directors, came to London and published a lengthy pamphlet about the mismanagement of VOC, focusing on the sales contract and on the slow progress made by the company in developing its concessions. In December 1923 the pressure was increased by a number of London stockbrokers who, 'having been responsible for having placed a very large number of shares' demanded that 'the whole question of the contracts should be reviewed'. They suggested mediation by Gulbenkian and Deterding together, probably without knowing that their relationship was rapidly falling apart.[26] When the conflict burst into the open at the meeting of VOC shareholders on 30 December 1923, the opposition achieved very little because the Group could outvote them easily. Pressure on the VOC board increased, in particular because two London stockbrokers, A. Chester Beatty and A. E. K. Cull, continued to lobby against the management of VOC. After some delay an extraordinary meeting of the VOC Holding was called for 4 April 1924, which appointed a special committee, counting amongst its members the two stockbrokers Beatty and Cull plus General Aranguren, 'to consider the position of the VOC Ltd and the VOC Holding Ltd with the Royal Dutch Shell Group, and to negotiate the modification and/or the anullment of certain contracts' between VOC Ltd and the Group.[27] The proposals of the committee were implemented quickly; in the sales agreement the price to be paid by Caribbean was linked to the average price on the day of delivery f.o.b. Tampico, i.e. the Mexican export market, and some constraints in the previous contracts were lifted as well.

In an attempt to take opposition against the VOC management on board, Beatty and Cull were also appointed to the board of VOC, but even so a different yet related contentious issue emerged during the next few years. The non-Group directors, including Gulbenkian, wanted to steer VOC on a new course and develop it into a more or less independent producing company, with its own management in Venezuela and its own facilities for exporting crude to the world markets. In November 1924, Beatty proposed 'to consider the advisability of having a manager in Venezuela who should devote his attention exclusively to the interests of this Company', and the following year similar proposals were repeated.[28] Meanwhile, production in Venezuela expanded so rapidly that the revised 1924 contracts were rapidly overtaken by the new developments. The group around Beatty now proposed to find out the conditions for selling oil to outsiders, and the size of investment needed for building an ocean terminal 'to enable the Company to deliver its production at a deep water point'.[29] Such a strategy was unacceptable for the Group members of the board, for whom it was logical that VOC would continue to sell its oil to Caribbean at the prices agreed upon in 1924. The issue became a tough struggle over principles. When in January 1926 the VOC board voted on commissioning a feasibility study for an ocean terminal, the six Group directors voted against, forcing the five other directors representing minority interests, including General Aranguren, Beatty, Cull, and Gulbenkian, to resign.

The Group paid a price for these conflicts in the form of a dent in its stock market reputation plus the growing hostility of Gulbenkian. The close cooperation between Deterding and Gulbenkian had been one of the keys to success before 1920, but the disagreements over Venezuela and over TPC (see Chapter 4), turned their friendship into bitter enmity. In 1925 they still exchanged hostile letters in which, occasionally, Gulbenkian attempted to leave things behind and return to normality, at one point exclaiming 'but for goodness' sake let us have no more correspondence about this matter, but peace'.[30] A few months later he resigned all his directorships within the Group; Deterding ignored his wish for his son Nubar to succeed him. The Group lost a valuable asset in this way, and acquired a powerful enemy.

Alone at the top By the mid-1920s Deterding had become estranged from nearly all his old colleagues. Marcus Samuel had already retired in 1921, his son Walter Samuel succeeding him as Shell Transport chairman. Fred Lane moved out of the inner circle and concentrated on representing the Rothschild interests. In 1921 he confessed that when he met Deterding he 'could make very little of what he said' because of 'his mumbling and my being very deaf'.[31] He died five years later, a director of Bataafsche and Royal Dutch to the end. In April 1925, August Philips resigned as managing director after a series of clashes with Deterding over various issues.[32] Finally, Waley Cohen, perhaps Deterding's most loyal lieutenant, retired as managing director of Anglo-Saxon in 1928 at the still young age of 51 because, as his biographer Henriques writes, 'after more than twenty years of working together, Bob and Deterding no longer trusted each other. The tension between them was often intolerable, and outbursts of temper when both men shouted and raved their recriminations at each other were not uncommon'.[33] Both Philips and Waley Cohen remained closely involved with the Group's business as non-executive directors, Philips at Royal Dutch and Bataafsche, Waley Cohen at Shell Transport, Anglo-Saxon and Bataafsche, clearly preferring, like Loudon earlier, to keep a distance from Deterding.

Having gained undisputed control, Deterding sought further ways to enhance his position. In October 1924, the Royal Dutch board adopted a resolution giving him an appeal to the full board if the managing directors were to outvote him. The resolution was recorded in a secret annex, which also outlined a procedure for dealing with complaints (*grieven*) of directors.[34] His next step was to change the split of the bonuses paid to the Royal Dutch managing directors, defined by the the articles of association as 3 per cent of the dividends paid by the company. In 1907, Deterding himself had proposed a very equitable scheme which gave him one-third of the bonuses paid by Anglo-Saxon and Bataafsche, with Loudon, Cohen Stuart and Waley Cohen sharing the rest. The split of Royal Dutch bonuses between Deterding, Loudon, and Stuart was probably also fairly even. When Colijn became Royal Dutch managing director, Deterding even conceded part of his profit share to him, presumably because the appointment was his personal idea.[35] However, when a new formula for splitting the managing directors' bonus needed to be worked out following Colijn's resignation and the appointments of De Kok, Erb, and Philips, Deterding no longer considered his colleagues worth a fair share. In 1923 Philips wrote to Deterding to sound out his opinion about a new formula, asking him, 'Which share do you think you should have? 97% for you and 1% for De Kok, Erb and me would of course be too much for you; 28% for you and 24% for each of us too little.'[36] Philips suggested a figure in between these two extremes as a fair one, but Deterding disagreed and simply grabbed nearly everything, except for a meagre 10,000 guilders for the other directors.[37] In 1929, for example, he received 2.3 million guilders in Royal Dutch bonus, leaving a pitiful 3,000 guilders for De Kok and 3,700 for Erb.[38]

Finally, Deterding asserted his leadership when the Royal Dutch preference shares were redistributed. These preference shares had been introduced in 1898 to protect the company against a hostile takeover bid from Standard Oil, by vesting the right to appoint Royal Dutch directors in the board of preference shareholders. Since the shares had proved difficult to sell during the company's production crisis, they had been placed with trusted friends. Kessler's untimely death created further complications, for he still owned 250 of the 1,500 preference shares. His widow had sold the shares to Deterding, Loudon, and Capadose on the condition that her sons would be allowed to choose a career in the company and, if suitable, be groomed for a directorship.[39] Two sons chose to join the Group, G.A. (Dolph) Kessler (1884-1945), and J.B. Aug. (Guus) Kessler Jr (1888-1972).[40] In 1915 Dolph resigned, disappointed in his unduly high career expectations; in September 1924 Guus became a managing director of Bataafsche and of Royal

Dutch, fulfilling the company's outstanding obligation on the preference shares.[41] Some board members argued that they could now be simply abolished, but after some discussion the Royal Dutch board decided to keep them and concentrate ownership in the hands of the managers and non-executive directors of Royal Dutch. All outsiders, some of them relatives of Loudon and Kessler, were asked to sell their shares to Deterding, Loudon or another insider. Already in February 1924, those insiders owned all shares except for one lot of fifty. Deterding increased his ownership from 50 to 500 shares and became the largest holder of preference shares with exactly one-third of the total, thus ensuring a controlling vote in any management appointment. Loudon raised his holding to 300 by buying all shares belonging to members of his family, and a few other directors such as Van Aalst and Erb acquired preference shares as well.[42]

With the redistribution of the preference shares Deterding's ruthless quest for power ended in total victory, but at a price. From captain of a team composed of like-minded colleagues, he had become an absolute ruler over subordinated managers with whom relations were tense. Still arguably the best expert on the oil market and the industry, he was now an increasingly poor judge of his colleagues and, perhaps most of all, of himself. Only a few members of the Royal Dutch board, in particular Hugo Loudon and August Philips, could afford to criticize him; new managers learned to keep a safe distance.

Moulding a new managerial team Kessler was the most prominent of the managing directors who ascended to the boards in the later 1920s. His interests really covered the whole of the business, showing as sure a grasp of gasoline and anti-knock dopes as of lube oil, the latter during the 1920s and even 1930s still very much a recondite subject.[43] In addition to his appointment at Royal Dutch and Bataafsche, Kessler also became managing director Asiatic and Anglo-Saxon during 1922-24, and in 1929 he succeeded Erb on the board of Shell Transport. This step made him in effect second only to Deterding at the top of the Group, Waley Cohen having resigned as managing director from Anglo-Saxon the year before. Deterding and Kessler alone sat on the boards of both holdings and all three central operating companies, and thus were party to all information and decisions.[44] Kessler gradually assumed a role as roving ambassador for the Group. He travelled frequently and far, his responsibilities including, in addition to research and chemicals, the management of Group companies, installations and buildings in Continental Europe, and financial relations within the Group.

By contrast, judging by their more frequent attendance at board meetings his fellow directors in London and The Hague appear to have remained somewhat more deskbound. The very sparse information we have about A. S. Debenham, appointed director of Shell Transport in 1921, of Bataafsche in 1924 and of Anglo-Saxon in 1925, suggests that, as pivot of the London office organization, he rarely ventured abroad.[45] Other managing directors must have made regular trips abroad for inspections or board meetings overseas. Andrew Agnew (1882-1955) had worked his way up at the Singapore firm of Syme & Co., the Samuels' agent there, and subsequently at the Asiatic office there. He transferred to the London Anglo-Saxon organization in 1919 and was one of the staff to receive Royal honours for his services to the Allied war effort during the First World War. Agnew became managing director of Anglo-Saxon and Asiatic and a director of Shell Transport and Bataafsche in 1923.[46]

ONZE VLIEGENDE DIRECTEUR-GENERAAL.

Cliché „Deli Courant".

Op 11 Febr. j.l., arriveerde onze Directeur-Generaal Ir. J. E. F. de Kok, vergezeld van den heer H. M. Schmidt Crans, na een schitterende vlucht met zijn sportvliegtuig in Indië.

Hierboven een foto na de aankomst te Medan. Van l.n.r. de heer H. M. Schmidt Crans, Mevrouw de Vries, echtgenoote van den Agent der B.P.M. te Medan, Ir. J. E. F. de Kok en de heer P. M. de Boer, Administrateur van Pangkalan Brandan.

Like Agnew, Fred Godber (1888–1976), appointed to the Anglo-Saxon, Bataafsche, and Shell Transport boards in 1929, the same year as Kessler, came up the long way. He had started as office boy at Asiatic in 1904 and helped to draft documents for the formation of the Group at George Engle's Secretarial Department in 1907, before moving to the staff of Dolph Kessler's nascent American Department. In 1919, Godber was sent to the US as vice-president of Roxana Petroleum, becoming president three years later. During his presidency Roxana's production and sales tripled, but the acquisition policy which drove the expansion was to turn sour with the onset of the 1930s Depression. Godber was probably Agnew's closest colleague and the two men had close family connections, too.[47]

After the experiments with the outsiders Colijn and De Jonge, the Bataafsche management had fallen to Erb and De Kok. Erb resigned from active management in 1929-30 and moved to the supervisory boards of Bataafsche and Royal Dutch. His seats on the boards of Anglo-Saxon and Asiatic were taken by Loudon, recently elected chairman of Royal Dutch and Bataafsche on the death of Capadose.[48] The management in The Hague now came to rest entirely with De Kok, a former army officer who had retrained in Delft as a chemical engineer. Spotted by Loudon as a promising man, he worked in various research functions until Colijn, having reorganized Bataafsche in 1916, promoted De Kok to head of the newly created Technical Department. Colijn relied on him and on Erb to make the new office structure work, which resulted in both becoming managing directors of Royal Dutch and Bataafsche in

[10]

'Our flying director-general' – good-looking and glamorous, J. E. F. de Kok was greatly admired. After gaining his private pilot's licence he travelled widely, even flying to the Netherlands East Indies. Far left, with their light aircraft in the background, stand (from left to right) De Kok's co-pilot H. M. Schmidt Crans; Mrs de Vries, wife of Bataafsche's agent in Medan; De Kok; and P. M. de Boer, Bataafsche's administrator in Pangkalan Brandan. Left, after a private showing at Carel van Bylandtlaan of the film of their flight, Prince Bernhard congratulates the beaming director-general.

1921, and directors of Anglo-Saxon the following year. De Kok made himself very much the soul of the The Hague office, building up great respect and loyalty amongst the staff. Though a bachelor and thus free to travel, he only made occasional trips to the US and to Indonesia; nor did he attend London board meetings more than six or seven times a year, still enough for him to succeed in nursing a close rapport with his colleagues there. De Kok only began travelling extensively after learning to pilot his own aircraft in 1935.[49]

With the appointment of these managers a more collegiate form of governance reappeared, spreading managerial responsibilities over a wider group and thus gradually counterbalancing Deterding's towering position. He retained an overriding influence and his privilege of the last word; he remained difficult to handle, his almost paranoid suspicion of being left out leading to regular temperamental outbursts which, as Lane had diagnosed as early as 1913, resulted in him being increasingly bypassed, thus confining him to the very corner which he detested most.[50] One of Deterding's favourite games was speculating openly about an imminent retirement, or even announcing this. In December 1930, he wrote a long letter to Kessler telling him that he would retire on 1 July 1931, because he was tired of the many battles he had to fight; 'it is high time, especially for myself – if I am not to become completely obsolete – to say goodbye to the business'. Before doing so, he wanted Kessler to consult Debenham, Godber, Agnew and De Kok. Deterding made it clear that he could no longer handle opposition to his leadership from other boards, and in particular the supervisory board of Royal Dutch. He argued that either he alone had to be responsible, which was what the public thought was the case, or an 'ensemble' of four or five directors had to take over, in which case he would resign.[51] The story goes that Kessler did not dare to respond to the letter, because he feared that openly agreeing with Deterding that it was time for him to go would bring his career to a

[11]

05|304

premature end, since the latter would interpret his opinion as a disloyal act.[52]

However, by the late 1920s the team of managing directors around Deterding had learned to cope with these handicaps, keeping him informed, pandering to some of his whims, and generally running their own affairs. Gradually Deterding's position evolved from the determined autocracy which had estranged so many of his former fellow directors, to that of a chairman, overseeing general developments, randomly intervening here and there, less and less actively involved in the daily management. Some sort of balance emerged, managers obtaining freedom of action in general matters in return for allowing Deterding to exercise his hobby horses. The directors in The Hague for instance ran the affairs of Bataafsche and Royal Dutch more or less on their own, consulting Deterding where needed and attending to his requests, but he did not claim or exercise authority over their handling of affairs.[53]

As before, practical considerations rather than functional ones appear to have guided the division of management tasks. There was no clear-cut demarcation on the basis of geographic area or subject matter. Some areas and issues belonged to the competence of one particular director, others were spread over several. Deterding, Agnew, Godber, and Kessler all conducted their own correspondence with Shell Union officials, as often as not the same ones, sending copies to The Hague and so creating an ostensibly needless multiplication of information.[54] When De Jonge succeeded to Cohen Stuart's London directorships in 1921, he did not get a defined portfolio of tasks and had great difficulty in carving out a meaningful position as managing director.[55]

In 1921 Deterding's silver jubilee as director-general of Royal Dutch was marked by this picture of the board, showing (from left to right) Luden, Philips, Erb, Loudon, De Jonge, Capadose, Deterding, IJzerman, Van Aalst, De Kok, and Colijn. Their smooth and formal front masked many undercurrents, rivalries and jealousies.

The corporate balance

During the First World War, London, and more specifically the Anglo-Saxon board, had become the centre determining the outlines of Group policy, but in the 1920s the balance of power shifted back and The Hague regained most of the lost ground. After the acquisition of the Rothschild share in 1918, Asiatic gradually turned from an independent company into Anglo-Saxon's commercial and management department. The boards were largely identical and all key data and policy decisions concerning Asiatic were recorded in the Anglo-Saxon minutes. By 1939 Asiatic no longer ranked with Anglo-Saxon and Bataafsche as one of the three top operating companies, effectively reducing their number to two.[56]

For a time it looked as if Bataafsche might also be relegated to the sidelines as revenue source, supplier of specialist knowledge, and hardware provider, but the resumption of the regular meetings of all directors in The Hague after the war halted this slide. With the increasing importance of science and technology for mapping corporate strategy, Bataafsche regained its position as co-pilot, assessing Group policy, monitoring new technical developments around the world, coordinating research, and managing patents (see Chapter 6). Unfortunately no board minutes survive for most of the 1930s, greatly handicapping our understanding of the company's policies during this period. The London office remained responsible for wider policy decisions involving patents and technology; when in 1927 the Bergius patents graduated from being a Bataafsche concern to a Group matter involving talks with IG Farben and Jersey Standard, De Kok handed the file over to Kessler (see Chapter 6).[57]

From 1927 The Hague also became the venue, in May, for a meeting of directors to discuss the previous year's results, which would then split into separate meetings for the Royal Dutch and the Shell Transport directors to decide on the dividend for their respective companies. During the 1930s the Shell Transport board sent delegates to this annual results meeting, the remaining directors in London voting on the dividend after getting the profit figures from The Hague by telephone.[58] Like the capital expenditure surveys, the coordination of results and dividends, formerly left to the handful of interlocking directors, simplified and reinforced the Group's organization.

As before, the five-man team of Deterding, De Kok, Kessler, Godber, and Agnew had close support in London from the non-executive directors there. The Shell Transport board met once a month, but the directors kept themselves well informed by frequently attending the weekly Anglo-Saxon board meetings, which the Royal Dutch directors did only occasionally. Moreover, the Shell Transport board had a rather more varied composition than the Royal Dutch one, underlining the company's need to cultivate links with other oil companies such as Burmah and Anglo-Persian, the armed forces, and the British government. Bankers and former managers continued to dominate the Royal Dutch board, though the 1930s did witness some change with a sprinkling of scientists and one isolated industrialist coming to the fore. Given the Group's overwhelming importance for the Dutch and Indonesian economy, Royal Dutch needed no ushers to gain entry to the corridors of power. If anything the partnership arrangement in the NIAM had cemented the bonds with the colonial government, enabling De Kok to write directly to the Governor-General if he wanted to. Though a member of the *Ondernemersraad voor Nederlandsch-Indië*, a corporate lobby group for trade and industry in Indonesia, Bataafsche negotiated with the authorities in The Hague and in Batavia on its own.[59] Consequently Royal Dutch had little need to seek board candidates from a wider field, at the penalty, perhaps, of keeping the discussions somewhat inward looking and provincial. The structure of decision making of course contributed to that. Key issues were still often first debated in London, putting the Dutch non-executives at a disadvantage. The Bataafsche board concentrated on technical and production issues, leaving little room for general policy debates. Nor could the executives of Royal Dutch and Bataafsche give up their bad habit of springing surprises on their supervisory directors. Important transactions such as floating a $ 40 million bond loan in New York were rushed through without proper preparation or consideration, always with the argument of necessary haste to take advantage of supposedly favourable conditions.[60]

Shareholders were treated with even less respect. The annual reports of Royal Dutch, fairly transparent during the early years of Deterding's reign, became wholly unsatisfactory as a source of information on the true state of the Group. From 1920 onwards the reports only gave information on Royal Dutch and selected data on Shell Transport, but no longer any data on the three main operating companies. Moreover, the text was increasingly dominated by any political messages Deterding wished to peddle.[61] The Shell Transport report was similarly lacking in data, though mercifully free from Deterding's harangues. As a consequence nobody outside the Group, and precious few people within it, if any, could know exactly how large sales, profits, assets, or debts of the whole Group really were. Only at the end of the 1920s did Bataafsche begin sending the Amsterdam stock exchange the minimum financial information required to have its bonds quoted.

For a long time, Royal Dutch managers chose to ignore calls for changes to the company's financial reporting. During the 1920s and 1930s the Dutch financial press, formerly in thrall to the company, became increasingly critical, drawing attention to the company's failure to give investors proper information, but to no avail.[62] When in 1931 the outside auditor demurred about the way in which he was asked to draw up the balance sheet, he was simply sacked.[63] The secretiveness displayed by the Group was partly due to an aloofness bordering on arrogance, but definitely also to the rudimentary state of the central accounting procedures and the lack of standardized information at central offices. Budgets remained vested in the main operating companies, but some form of centralization became visible from January 1927, when directors received the first Group-wide overview of capital expenditure by operating company during the preceding months, subsequently a monthly return exercise.[64] Until then not all directors of, say, Anglo-Saxon would necessarily have been aware of sums spent by Bataafsche and of the Group's overall obligations, so the introduction of this survey marked a conspicuous advance in the operational management. In April 1927, the London office also sent a statement of Group sales in tons and proceeds in sterling during 1926 to The Hague, judging by the correspondence to which the document gave rise, the first such compilation.[65] The capital spending and sales figures were presumably compiled by a Group Finance Committee which came into existence at the end of the 1920s.[66]

Drafting financial overviews was only a tentative step forward, however, since as yet Group managers do not appear to have made efforts to standardize accounting and reporting procedures. Thus as late as 1937 Godber confessed to being flummoxed by Bataafsche's annual accounts for 1936, writing to Van Wijk in The Hague 'While from the statements and exhibits most of the accounts can fairly well be understood I must admit that the mass of figures is somewhat bewildering, particularly as the information in which I am interested is to be found only by looking at different exhibits at the same time. As a result I am somewhat at a loss to follow what is the exact position in regard to assets and depreciation.'[67]

A few months later, Godber's complaint was echoed by Van Eck in a letter to Van Wijk discussing the variety of depreciation and reserve building policies amongst operating companies.[68] The fact that people like Godber or Van Eck could not fully understand the accounts of operating companies with which they were intimately familiar shows a glaring absence of proper structures of control at the heart of the Group. This was now evidently beginning to matter. In February 1940, W. H. van Leeuwen, a prominent captain of Dutch industry and a Royal Dutch and Bataafsche non-executive director since 1936, wrote to the Bataafsche board asking for proper consolidated accounts, as he had until then been unable to grasp the Group's financial structure, and thus was not in a position to assess financial policy.[69] No reply is recorded; consolidation came only after the war.

Towards a new corporate culture: staff management and labour relations

With the continuing growth of the business, labour relations and staff management became of ever greater significance to Group managers. Here as elsewhere, the decentralized nature of the organization created both considerable differences in actual labour relations, and a noted scarcity of data at central level: there are no figures for total Group employment until the mid-1930s. In his ghosted autobiography *An International Oilman*, Deterding relates the story how, during his audience with Pope Pius XI in the early 1920s, his Holiness had asked him how many people he employed, to which he answered, after thinking a while, 'In England over six thousand, and over the world from thirty to forty thousand.'[70]

In answering the Pope's question, Deterding probably considered only the employees forming the core of the organization, those holding a job with which they qualified for the Provident Fund, established in 1913 as a voluntary savings fund fed by employee contributions and bonus payments by participating Group companies. In 1922, the fund counted 11,890 members, rising to 33,000 at the end of 1929 and 37,000 in December 1930, i.e. more or less the number Deterding mentioned. However, the Group depended on a far larger number of people in more or less regular employment. In the Dutch East Indies alone, Bataafsche employed more people than the entire membership of the Provident Fund, 56,983 Asian and 2,156 European employees in 1929.[71] That same year, the US companies counted 35,000 employees.[72] By 1935 the first reliable figure for total Group employment gives 180,400 people.[73] Since worldwide employment shrank dramatically between 1929 and 1936, total employment in the late 1920s must have comfortably exceeded 230,000 people. The other employees, numbering as many as 200,000 in 1929, had a much looser relationship with the Group, or rather, the Group had a much looser relationship with them.

One of the characteristic features of the evolution of big business during the first half of the twentieth century is the development of a wide range of services for their core employees: housing, recreation, insurance against old age, sickness, and sometimes even unemployment; staff members were cared for almost from the cradle to the grave. In this way companies strove to reinforce ties with their employees, to nurture their scarce, sometimes even unique, human resources. The management of this vital resource increasingly became a focus of attention: how to recruit people with a high potential, how to train them and supervise their career became of strategic importance.

As with other companies, the evolution of staff care began on a fairly small scale. When, for example, in 1914, the London head offices moved to St Helen's Court, the board of Anglo-Saxon discussed a proposal supported by Deterding and Waley Cohen for 'the supplying of lunches in the lower basement on certain terms to the staff'. It was vetoed by the Samuels.[74] During the war, this decision was reversed and a cafeteria opened. At that time, staff management was still part of the accounts department and thus one of the varied tasks of George Engle, the chief cashier and later accountant of Anglo-Saxon.[75]

The Provident Fund was another example of a modest beginning which also encountered resistance, both from the Samuels, again, and from employees who did not like to have their regular bonuses converted into a contribution to the scheme. Regular extra donations from surplus Group profits, notably during and immediately after the war, made the Provident Fund increasingly attractive as a form of voluntary saving towards retirement provision. The fund remained a provision for an elite of Group employees, however, and its existence formed an obstacle to the introduction of a proper pension scheme. Deterding was said to be against, presumably because he wanted staff to make arrangements themselves, with the Group, in the Provident Fund, already offering them an excellent way of doing so. The Fund provided insufficient cover, however, even for staff members who used it. The London boards discussed hardship cases amongst staff and the families of staff with some regularity, usually granting a pension as a favour and 'at the board's discretion'.[76] One such case, in 1932, concerned a 60-year old employee who had entered the service of Samuel & Co. in 1897. He had £ 4,300 in the Provident Fund, still not enough to retire, so the Shell Transport board granted him an additional pension of £ 200.[77] Bataafsche set up a scheme after Deterding's retirement in 1936, just in time to meet a deadline imposed by new legislation taxing contributions to savings funds, while exempting those to pension funds. In 1938 the pension fund was extended to all Group companies.[78]

Salary regulations too were increasingly focused on the needs of employees. During the First World War and its aftermath, salaries were raised to meet spiralling prices, initially in the form of special allowances for high costs of living, but after 1918 in a more structural way. In 1919, Colijn introduced new guidelines for setting salaries at Bataafsche with a decidedly Dutch flavour; he may have copied them from similar guidelines for government employees which he would have known as Minister for War. Important guiding principles were (a) at age 21 unmarried employees had to be able to live independently, so their minimum salary was set at 1,200 guilders; (b) at age 25 the salary had to increase to 2,000 guilders, so as to enable them to start a family; (c) on top of the salary employees would receive a family allowance of 10 per cent for the wife and 2 per cent for each child.[79] In addition to salary rises, the London office developed two new facilities during the early 1920s to help staff cope with the high costs of living. St Helen's Housing Company Ltd. provided mortgages at favourable rates for staff to buy their own homes; the company also developed its own residential housing, the Grove Estate. During its first decade, St Helen's Housing accepted 1,138 loan applications for a total sum of £ 897,594.[80] St Helen's Court Stores (1924) was a cooperative shop for employees.[81]

detected in a misguided and futile attempt to fasten on his collar at a time when he was *in puris naturalibus* from the midriff upwards!

Mr. Willemsen, Mr. Faas, Miss de Vries, Miss Hartogs, and the amiable Mr. Arntzenius (familiarly dubbed "Auntie" by those who were unable to pronounce his name) travelled with us from the Hook to the Hague and imparted much sage counsel and information destined to serve us in good stead during our sojourn in their native land. On reaching the Hague we were duly conveyed by a special service of trams to our respective hotels, where we breakfasted with wolf-like avidity on sausage, cheese, butter, new rolls, coffee and honey. Black honey was evidently a novelty to the stalwart exuberant embraces and other public demonstrations of affection exchanged between the trio added yet another nail to the coffin of the belief once prevalent in England that the Dutch race is notable for impassivity and phlegm.

At 2.30 three motor chars-à-bancs took us, via Scheveningen, to Bosch Hek, an open-air café in an idyllic situation. Here we dallied for a spell under a leafy canopy, and ate, drank, smoked, and sported with Amaryllis in the shade.

After dining at the Riche we attended in full force at the Hague swimming baths, where, together with Sir Henri and many other directors, we enjoyed an ex-

Vol. 6. No. 141 WEDNESDAY, JUNE 2ND, 1926. *Price twopence*

EDITORIAL.

This photograph, for which we are indebted to the "Nafta" Societa Italiana pel Petrolio ed Affini, was taken on the occasion of the visit of the Crown Prince of Italy to the Fair recently held in Milan. H.R.H., accompanied by Gr. Uff. Attilio Pozzo, is descending the steps of the "Nafta" stand.

* * * * *

The following announcement appeared in *The Times* on May 12th:—

On the 11th May, at The Barnes, Radlett, Herts, to Evelyn, wife of C. F. A. Greenslade—a daughter.
Congratulations!

We gladly give publicity to the following letters:—

From Viscount Knutsford, Chairman of the London Hospital, to Mr. R. F. Jenner, St. Helen's Court.

What splendid friends you have all been to "The London"! I have just been shewn the figures for the year's collection at the Asiatic Petroleum Company. They stand at £329 2s. 9d.—a record of help for which it is not easy to find a parallel.

I should like, if I may, to send these few lines of warmest thanks and real congratulations to all concerned.

What appeals to me specially in this gift is the fact that there are so many people concerned in it, and there is nothing more encouraging to me as "The London's" Chairman than such knowledge.

A huge place like this, where there is always something to be done, or someone to be helped, needs just as many friends as it can possibly find, and I do hope, most sincerely, that you will be able to continue this help with the same generosity and the same enthusiasm.

Can you find some way of passing on these thanks of mine? I should like as many as possible to know how truly grateful I am to them.

From Captain Ian Fraser, C.B.E., M.P., Chairman of St. Dunstan's, to Mr. H. Colville, St. Helen's Court.

My attention has been called to the collection made by you in four St. Dunstan's collecting boxes in connection with the recent Flag Day, and to the splendid sum of £34 1s. 7d. which you raised, and for which I now enclose an official receipt.

This is just a line of personal thanks for the work in which this collection must have involved you, and of appreciation for this very practical expression of your helpfulness and loyalty to St. Dunstan's. It is a very pleasant thing that an old St. Dunstaner should put himself to so much trouble to help us, and I can assure you that we are most grateful.

Will you convey my most cordial thanks for their kindness to all who contributed to the collection. It was splendid of them to respond so generously to your suggestion.

* * * * *

Mr. A. R. McVeagh, Superintendent of Shell-Mex, Ltd., South Quay Depôt, Douglas, Isle of Man, in forwarding a local holiday address, adds:—

I was very glad to read recently, in your bright notes of the various doings of Shell-Mex Ltd. in the Divisions, etc., that you had had a number of new subscribers from "Shell-Mex 2," late M.V. "Hera." This vessel discharged here on one of her earliest voyages under her new colours, and I happened to have met the new skipper before. In congratulating him, I brought our PIPE LINE to his notice. I gave him a copy, which had just arrived, and I told him that he should subscribe, as it was the best way to get and keep in touch.

We much appreciate our correspondent's efforts in THE PIPE LINE cause.

pill to swallow. The Lensbury team, despite their youth and physique, were well beaten in two successive pulls.

Shortly after 4 o'clock Sir Henri Deterding presented the medals to the successful competitors, with a few happy words of congratulation to each, and our Hague comrades made the welkin ring (if welkins are indigenous to Holland) with their full-throated and impartial cheers. The way in which the various events were contested, and the obvious feeling of good fellowship that existed between the rival athletes, afforded a convincing proof, if such were needed, of the splendid sporting spirit of the Dutch nation.

The battle over, and the captains and the kings having departed, we hied us to our hotels, donned the "immaculate evening dress" of the lady novelist, and proceeded

curtain, and cheered whole-heartedly when we saw upon the stage not only Sir Henri Deterding but such notabilities as Jhr. H. Loudon, Mr. Colijn, Dr. Capadose, Mr. Frederick Lane, and a host of other directors.

At this point addresses to Sir Henri (in Dutch) were given respectively by Dr. Capadose, Mr. Colijn and Mr. van den Broek.

When Sir Henri prepared to respond the whole audience rose and cheered with such genuine enthusiasm that he was unable to speak for some minutes.

Sir Henri Deterding spoke first in Dutch, then in English. He was very pleased, he said, to see in the audience so many members of the London staffs. He expressed the hope that they were thoroughly enjoying themselves, and that their visit to Holland would be a

IR HENRI DETERDING, MR. COLIJN AND SIR REGINALD McLEOD AT THE SPORTS MEETING.

feast luxuriously and hilariously at the Riche. Some itement was caused by the belated advent of the ni-distraught Meakin, who, on leaving the Sports ound, had artlessly boarded the wrong tram and paid involuntary and premature visit to Scheveningen. Anon, our fleet of chars-à-bancs, by this time a familiar ctacle to the populace, conveyed us to Scheveningen, h its lordly hotels and apparently illimitable ex- nse of golden sand. For the nonce the Ambassadors' eatre was our private property and that of the Dutch ffs, and it was crowded to capacity. Ushered into stalls we expectantly awaited the ascent of the

pleasant memory in the years to come. For the benefit of his English friends he took the opportunity of saying that his great affection for their country, in which so many eventful years of his life had been spent, had decided him to approach the British Government with a view to the foundation of a colony in which farming could be carried on under Dutch methods and thus assure a livelihood to disabled ex-service men. He told the Government that he was prepared to shoulder the responsibility for providing the first £50,000, and if necessary the second £50,000, that would be required for the inauguration of this scheme. The various Ministers

Created during the First World War as a link between staff at home and those in the armed forces, the matured post-war into *The Pipe Line*, 'the link that binds', and was enduringly popular. Today its pages offer a vivid and fascinating link with the past. Thus one front page in 1926 featured a royal visit by the Crown Prince of Italy to the imposing stand of Nafta, the Group's Italian marketing company, at a trade fair in Milan, and sport was a perennial favourite. In the days before Royal Dutch had its own country club, Te Werve, Deterding and Colijn could be seen watching the 'Oilympiad' sports event on the lawn outside the head office in Carel van Bylandtlaan; and when the handsome cup donated by De Kok (overleaf) was won twice in a row by the British, it became their property.

[14]

THE De KOK CHALLENGE CUP COMPETITION

Mr. J. E. F. De KOK

Our good friends from the Hague are unable to utter the Cesarean boast—*Veni, vidi, vici.* The De Kok Challenge Cup, the acquisition of which was the primary object of their descent upon these shores, remains in the Lensbury club-house. Moreover, by winning it twice Lensbury have made it their permanent property. It is gratifying to know that in doing so they have not killed the goose that lays the golden eggs, for Sir Henri Deterding, with characteristic generosity, has announced his intention of offering a similar trophy, and th s will be duly competed for in 1923. As Sir Henri pointed out when he presented the Cup and medals on June 17th, we are in the fortunate position of being able to select our representatives from a membership of over 2,000, as compared with the A.S.V.'s 400, so that the latter, in scoring 26 points, put up quite a creditable show, and incidentally proved themselves excellent sportsmen. The broad hint thrown out by Sir Henri on the same occasion as to the inception in Holland of a club-house on Lensbury lines will serve as an incentive to our Dutch comrades to devote themselves heart and soul to the various branches of athleticism which form the basis of the annual competition.

The Hague party of 52 duly arrived at Harwich in brilliant weather on Friday morning, June 16th, where they were met by Messrs. E. H. Layton, R. F. Jenner and A. J. van 't Hoff, who (together with Miss Boxall) constituted the Entertainment Committee, and who travelled with them to London. In addition to the competitors, the party included four members of the A.S.V. Committee, Messrs. C. Sauer (President), J. W. Faas, W. D. A. Arntzenius ("Auntie") and Miss Hartogs. It was a matter for regret that Mr. J. Willemsen, who took a prominent part in the reception of our representatives at the Hague last year, was unable to make the journey owing to ill health. The tourists reported a rough crossing, but fortunately, in the words of *Ingoldsby*, "nobody seemed one penny the worse."

Piloted by the above-named Lensbury trio to St. Helen's Court, the Dutch visitors assembled in the board-room, where they exchanged greetings with many old friends and Messrs. A. S. Debenham and G. S. Engle extended to them a cordial welcome. Each member of the party having been provided with a Sports Programme and a souvenir boo' let presented by Mr. G. S. Engle, which embodied a series of Lensbury views, they motored to Teddington, accompanied by Mr. Engle and other members of the Sports Committee.

Lunch at the Club-house was followed by an inspection of the grounds, and these, especially the lawns at the rear of what used to be called Weir Bank, seem to have been greatly admired. Next came "the cup that cheers but not inebriates," and then the party motored to the West End, dined at Stewart's, in Bond Street, and spent a cheery evening at the London Pavilion, returning to Lensbury at midnight. And so to bed. Sports Day dawned in somewhat bleak fashion, but providentially the sun came out before breakfast-time and, apart from a dull spell at noon, was in evidence throughout the day. The tennis matches, the first item on the Sports Programme, are described on page 179.

THE CUP.

The Enclosure.

Mr G.S. Engle, Mr J.E.F. de Kok, Sir Henri Deterding.

Fencing Protagonists J.W. Lake Lake, G.A.A. de Voogd (A.S.V.)

The Tug-of-War.

2 Miles Walk. A Dutch competitor.

About with the Foils. Miss Sasberg, Miss Roberts.

Finish of the 2 Miles Walk.

Onlookers.

At the conclusion of the Meeting a general ... made in the direction of the cricket pavilion with a view to Sir Henri Deterding presenting the Cup and medals. Mr. G. S. Engle opened the speech-making with the following remarks:—

Before Sir Henri Deterding presents the medals, and also the Cup which he has so generously given for competition between Lensbury's representatives and the members of the Royal Dutch-Shell Group on the Continent, I should like to express the hope that our visitors will take away with them very pleasant recollections of this—shall I say?—typical English summer day (laughter), and that what little we have been able to do for them will make them want to come here again and accept more of the hospitality which we are so keen on showing to them. You all know that the inception of Lensbury Club is solely and entirely due to Sir Henri Deterding (loud applause), assisted by some of us who know what his wishes are. We know that at any rate in certain other countries the staffs already have sports grounds. The "Te Werve" Club at the Hague is well known to many of us, and I understand that sports grounds are being opened in Sweden, Norway and Denmark (Applause). No doubt those countries which do not yet possess grounds will take a suitable opportunity of putting their views before Sir Henri (Laughter and cheers).

Sir Henri Deterding, who was greeted with loud and prolonged cheers, said:

When I addressed our foreign visitors yesterday I was asked to speak in French, but I hope that next time they come over they will all be able to understand English, which is quite an easy language to learn. (Laughter). When we started these sports meetings we had no idea that they would develop into international events. But last year, when I was in Italy, I was discussing the matter with my friends there. I gave them every encouragement and met with a hearty response, and was assured that they would like to come over to take part in the annual competition. We have now started making these meetings international, and I believe—perhaps it sounds somewhat highflown—that in bringing together people of so many nationalities we shall perhaps do more for the world than the League of Nations. (Hear, hear and cheers). I don't want to enter into politics, but I think that a combination of commercialism and good fellowship, as we see it here to-day, will do more good than a great deal of talk. (Renewed applause). Lensbury have won the Cup this time. They have the advantage of better conditions. The Continental competitors have made a magnificent show, and I am very proud of them. It shows what can be done by training, and I am certain that some day the Cup will go to the Continent. (Cheers).

Sir Henri, amidst loud applause, then presented the Cup to Mr. G. S. Engle on behalf of the Lensbury Club, and the medals to the successful competitors. He also expressed his high appreciation of the work done by the

Some of the Continental Competitors.

At the suggestion of Mr. A. Agnew, three hearty cheers were given for the visitors, and also for Sir Henri Deterding for his kindness in presenting the Cup and medals.

The proceedings concluded with the rendering by the Band of the respective national anthems of the countries represented at the Meeting.

The rest of the evening was devoted to music, dancing and boating. On Sunday, July 8th, the visitors, piloted by the Lensbury Entertainment Committee, spent the morning in sight-seeing around London. Luncheon at Stewart's, Bond Street, was succeeded by afternoon tea at the Club-house, and after dinner a launch trip on the Thames rounded off another highly enjoyable day. On Monday the whole of the party attended a matinée performance at the Coliseum, and at 6.15 p.m. attended a Farewell Dinner at Frascati's, at which Mr. A. S. Debenham took the chair, and at which Mr. and Mrs. G. S. Engle, the members of the Lensbury Sports Committee and the British competitors at the Sports Meeting were present. We give below a brief report of the speeches:

Mr. A. S. Debenham said that it was a great pleasure to meet so many friends from abroad—from Holland, Italy, France, Denmark and Sweden—and also to know that the Sports Competition had been so much more closely contested than it was in the preceding year. He hoped that now their Italian, French and Scandinavian colleagues had seen what they had to compete against they would next year be able to put up an even better show than they had recently done. He wished the Continental visitors a safe return, and looked forward very much to seeing them again.

1. The Start, 880-Yards.
2. The Finish, 2-Mile Walk (B. Veen, Holland).
3. The Finish, 880-Yards (G L. Hankey, Lensbury).
4. The "Deterding" Challenge Cup.
5. Long Jump (G. Arfvidsson, Sweden).
6. High Jump.
7. Fencing Competition (Epée).
8. " " (Ladies' Foils).
9. Presentation of The Cup to Lensbury's Chairman.

[15]

Reinforcing ties with employees also meant creating a sense of corporate identity. Staff magazines emerged during the First World War. The first staff magazine for the Dutch employees was *Maandblad van het Personeel der verbonden Oliemaatschappijen*, launched in 1917 and subsequently renamed *De Bron* (The Well). In 1920 its British counterpart *The Pipeline* took over from the wartime *St Helen's Court Bulletin*, originally intended only to keep staff members in the armed forces in touch with the Group companies. The employees in the Dutch East Indies, who had an *esprit de corps* very much of their own, also set up a magazine, called *Minjak* ('oil' in Malay) in 1921, though the management in the Dutch East Indies needed some persuading before it allowed publication.

The magazines shared one regularly returning feature in reports about the local sports events and, from as early as 1921 onwards, also between different national teams. In 1921 the Lensbury club, one of Engle's creations, hosted the first of an annual fixture between teams from the Netherlands and the UK, later extended by a team drawing on staff from companies in other European countries, competing in fencing, tennis, and athletics. De Kok, a keen sportsman himself, donated a cup, leading to the event being named after him. When the British team, building on a greater enthusiasm for sports on their side of the North Sea plus a larger staff pool, had won the cup twice and rightfully claimed the De Kok cup as theirs, Deterding donated a cup and the event was rechristened accordingly, keeping that name throughout the rest of the interwar period.[82] At the second 'Oilympiad' the Dutch team was again beaten by the British, prompting Deterding to donate the money towards building facilities in the Netherlands similar to the Lensbury Club. In 1922 Te Werve was opened at Rijswijk, located on an estate near the head offices in The Hague.

Another feature of the emerging corporate spirit was the appearance of staff unions after the First World War. Indeed, the magazine *Minjak* was launched by a newly organized staff union in the Dutch East Indies, the 'Vereeniging van geëmployeerden bij de petroleumindustrie in Nederlandsch-Indië' (Association of oil industry staff in the Netherlands Indies). This association had its origins in calls for a union to represent staff interests with management, first aired in 1919. Initially the Bataafsche managers had reacted negatively and wanted only to organize regular talks with employees, but in May 1920 Bataafsche issued a circular welcoming the formation of associations to represent staff interests. By subsequent judicious manoeuvring, The Hague and Batavia succeeded in moulding the popular demand for staff unions to suit their purpose. There was great enthusiasm amongst the Dutch East Indies staff for the initiative. At Balik Papan, 96 per cent of the employees were said to have joined, and a federation of trades unions in the Dutch East Indies issued calls for a comprehensive union for all employees in the oil industry. This put managers in a quandary. Allowing outsiders to discuss employment conditions for Bataafsche employees was considered out of the question, but how could this threat be eliminated while meeting popular demand for staff representation?[83]

Writing to De Jonge in August 1920, Colijn set two basic conditions. In order to be accepted as a negotiating partner, a staff union would have to have only Bataafsche employees as members and only Bataafsche employees on the governing board. If those conditions were met and the union was admitted to talks with the management, employees would lose the incentive to join smaller trades unions, according to Colijn; after all, they simply wanted more power, and were less interested in the kind of specific interest group representation which trades unions could give them.

[16]

After presenting the De Kok Challenge Cup to Mark Abrahams as British representative in 1921, Deterding took the opportunity to replace De Kok Challenge Cup with another, the (naturally) much more ornate Deterding Cup.

Moreover, there was little chance that such an association would be admitted as a member of any wider federation of trades unions. Once the staff had appreciated the privileges of a dialogue with the management, they would think twice about endangering these gains.[84] The association accepted these conditions and had its first meeting with the Bataafsche general manager in August 1920, which subsequently developed into regular conferences about work and employment conditions.[85]

One characteristic difference between the British and the Dutch side of the business continued throughout the interwar period, and indeed well into the 1950s. Whereas British managers like Agnew and Godber tended to be trained on the job, Dutch managers usually had an academic background. This mirrored the very different recruiting patterns prevalent in the two countries. Group managers realized at an early stage that the business needed university graduates, and developed policies to attract them. Waley Cohen pioneered this policy on the British side, using his links with his own alma mater, Cambridge University, to spot talented students. His avid scouting led to him being appointed as early as 1910 to the Cambridge Appointments Board.[86] At times Waley Cohen's rational approach conflicted with existing traditions of network recruiting practised in the mercantile world to which, for instance, the Samuels belonged. As he put it in 1910, 'I am always treading on the toes of other Directors who offer me cousins and nephews, with the apology that that independent body works best'.[87] Similarly, Royal Dutch and more specifically Bataafsche developed close links with the Technical University in Delft. Already before 1914 Bataafsche's growing demand for geologists and mining engineers inspired attempts to convince the Dutch government and the board of Delft University to invest more in the training of specialists in these fields. IJzerman and Loudon were members of a state commission, set up in 1913, reporting on the need to streamline the curriculum and to focus on specialist subjects such as, for example, the geology of oil. In 1916, Erb wrote a memo complaining about the shortcomings of the curriculum which forced Bataafsche to give young engineers recruited fresh from Delft additional practical training for at least another year.[88] Bataafsche also considered creating a chair in oil technology in Delft, but the costs involved, estimated by the university to be the staggering sum of half a million guilders, was prohibitive.[89] In 1928 the ties between Delft and the Group were further strengthened when Deterding received an honorary doctorate.

The British part of the Group attracted a wide variety of candidates for jobs. Some of them, such as Waley Cohen, came from Cambridge and had a technical education, but the British companies were more interested in people with commercial and managerial skills and preferred to recruit candidates direct from public schools. Consequently, the London office always had a fairly large contingent of trainees on the staff, who would be groomed first in general administrative procedures, then sent overseas to acquire further skills. The number of trainees in individual departments ranged from almost a quarter to two thirds. Two departments handled almost 1,500 trainees together between 1919 and 1931.[90] From London these men would be sent to Asia, and only then to the US, which was regarded as too rough for unseasoned young staff members.[91] In 1937, Godber described the curriculum as follows: 'The procedure we usually adopt in connection with commercial staff is to engage promising youngsters from public schools, at or around the age of 18, and to a lesser degree, university men who would be around the age of 22. The latter, of course, can within six months immediately proceed to some destination,

[17]

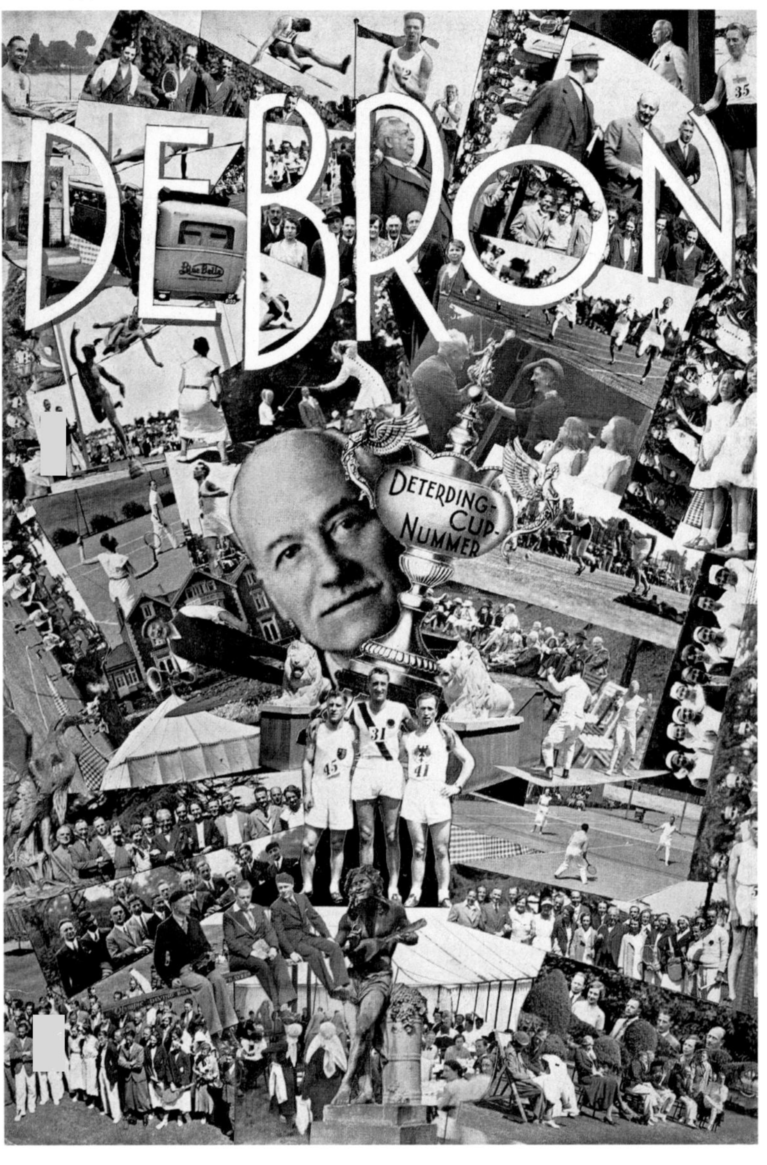

Sports galore on the cover of a special 1934 'Deterding Cup' issue of *De Bron*. The splendidly ostentatious Cup was given central position.

but the former must spend three or four years in the office in training of one kind or another, and particularly for the East, so that he does not leave for the East until he is a little more mature. This latter class are usually put into one or other of the departments, and during a short period given some sort of special courses, including a Fuel Oil Course, Lubricating Oil Course, and a three months course with the Shell-Mex & BP. This latter course includes a certain period on the roads, and in one of the courses, the man goes either to Shell Haven or one of the other plants in this country. In the course of time, as vacancies occur for juniors, these men are allotted to certain areas, but none of them proceed abroad until I myself and others of the managing directors, whose areas are affected, see them also.'[92]

By contrast, recruitment in the Netherlands focused on technical skills, and in recommendations by Delft professors the commercial abilities of candidates hardly played a role. Nor did the environment in which the engineers recruited came to work stimulate the development of such skills. Bataafsche's rather bureaucratic organization and the fact that the company dominated the market in its main area of operation, the Dutch East Indies, were hardly conducive to fostering initiative and marketing know-how. Reviewing Deterding's criticism about Bataafsche's overseas organization in 1927, Kessler wrote to Erb, 'the Administrators are rather inclined, because of their sense of discipline, to leave the initiative (...) too much to The Hague. It makes the organization work slowly'.[93]

Bataafsche's recruitment pattern and career development worried Kessler. He feared a growing managerial imbalance between the Dutch and the British parts of the Group because, in the race to the top, engineers without commercial skills would lose out to managers without technical know-how. Kessler at first sought a remedy in asking Delft professors to assess the commercial abilities of their students as well, but he failed to elicit a response, which he ascribed to the organization of the Dutch curriculum and the consequent lack of one-to-one contacts between professors and students preventing an all-round assessment.[94] During the 1930s, Kessler also began to worry about the fact that British trainees went all over the world, whereas the Dutch commercial staff remained tied to a fairly limited area:

'It has proved to be difficult to find the necessary training ground for Dutch commercial employees of the Group. The only opening which we have for these people is really the BIM and the Dutch East Indies. This is not a satisfactory position, as the result is that the choice of Dutchmen – who will, in due course, occupy important and responsible positions at the Group's headquarters – becomes very limited. The position of the British commercial employees is very much satisfactory (…) how can we improve this (…) we could make a point of appointing Dutchmen in places where there is no special reason why we should use one nationality and not another. I have in mind places like South Africa, Japan, North China and South America, etc. The normal procedure for youngsters might then very well be that they should first be, for

[18]

Deterding himself was a keen, lifelong sportsman, revelling not only in the country sports – hunting, shooting and fishing – but also running, swimming and riding. This picture of a jump at his Ascot estate was so characteristic that *De Bron* used it 'in memoriam' after his death in February 1939.

some time, with the BIM, after that they can be sent out to the Dutch East Indies, South Africa, Japan, North China, South America and other countries (...) we have some good Dutchmen (...) I understand that North China requires some overhauling and there may be openings for them at once.'[95]

Clearly Bataafsche's commercial staff were at a disadvantage which its technical staff did not have to quite the same degree. From the start, Royal Dutch and subsequently Bataafsche engineers were trained first in the Dutch East Indies and then sent to the other regions and countries in which the companies operated. After the First World War, the United States increasingly became the venue for both the last training stage of engineers in the early phase of their career, and for refresher courses for experienced engineers. The Group companies there were considered to be the technical advance guard, offering training in the newest cracking technologies at Wood River, or in the latest drilling and production techniques in California.[96]

Kessler's letter also demonstrates that, in the mid-1930s, Group employees were still not treated as a single pool. Staff policies and conditions became increasingly uniform for Group companies, but employees remained tied to the company they first joined. Bataafsche employees made their career within the Dutch parts of the business; British staff remained by and large on the British side. Even Group managers such as Kessler, Godber, or Erb, were primarily employed by one or other of the operating companies. There was another clear distinction between the British and Dutch sides. British employees of any operating company considered themselves to be working for Shell, whereas Bataafsche staff, notably in the Dutch East Indies, had a first allegiance to Bataafsche. To them, Shell was a brand.

An important ingredient of the emerging corporate culture was the conception of life time service, career moves to another company remaining a rare exception. This was the main point of building firm ties between employer and employees and amongst employees themselves, supported by common sports events, social clubs attached to major sites of operation, etc. For staff aspiring to the middle to upper reaches of the managerial ladder, a career with the Group was of course very different from a career with almost any other large company because of the constant shifting around between various parts of the world and the experience of very different working environments.

The first issue of *Minjak* in 1921 also offered an interesting discussion about what working abroad meant for the relationship between employer and employees. Under the title 'Are we "employees"? ' a certain G. argued that expats living in the Dutch East Indies had a dual role. They were both employees in the usual sense, and part of a European ruling class that could only maintain its position by being united and loyal to the employer who had put them there.[97] If they were to simply act as employees of the Group, they would seriously harm the position of their social group towards the Indonesian population. Even 'the semblance of antagonism could bring the Asian to dangerous conclusions'.[98] The article triggered a welter of reactions, some defending G.'s position, others attacking it. Paraphrasing an earlier article G. concluded the discussion by asking the fundamental question: 'And when the "class conscious" Asian employees will begin the "battle for freedom and for justice" against their white employees, on which side will we stand then?'[99]

Right, *Minjak* poked fun at its own staff ('Personnel are able to take control if need be'), but (far right) it also asked serious questions about the relationship between Europeans and Indonesians.

Fears about what would happen in the Dutch East Indies once the population started claiming ‘freedom and justice’ also inspired Bataafsche to take direct action to change the training of colonial administrators in the Netherlands. Within the Group there was growing concern about the ‘enlightened’ or so-called ‘ethical’ policies of the colonial administration in the Dutch East Indies, and about the discussions, still very tentative, about the future of the relations between the Netherlands and its most important colony. Owners and managers of colonial enterprises and politicians on the right of the political spectrum, however, regarded these ideas as a threat to the bonds between motherland and colony and firmly defended the indissolubility of the union. Prominent members of this lobby were two former Group employees: Colijn and Gerretson, one of the first management trainees selected by Colijn

[19]

„Personeel in staat desnoods zelf de leiding te nemen.”

[20]

Wij zijn er echter van overtuigd, dat bij de Europeesche werknemers langzamerhand het besef is levendig geworden, dat het zich inlaten met de op politieke aspiraties stoelende inlandsche vakactie, in den grond verraad beteekent aan de Nederlandsche zaak.

(Soer. Hbld.).

***

AFGEWEZEN SOLLICITANT.

Nu ik alles gezien heb, ga ik maar weer terug.

(van Lennep).

***

CONFERENTIES A³.

Ik kom tot de praters, de babbelaars bij uitnemendheid.

(Camera Obscura, Varen of rijden).

***

[21]

[22]

[23]

for Bataafsche. The official curriculum for colonial administrators at Leiden University was singled out for criticism as being too 'ethical' and thus responsible for undermining the bond between the Dutch East Indies and the Netherlands. To create a counterbalance against the Leiden training programme, Gerretson strove to set up an alternative institute for colonial administration at Utrecht University, to be financed by private enterprise. He convinced Bataafsche to contribute one-fifth of the total budget of 50,000 guilders, as a result of which the new institute became known as the 'oil faculty'.[100] He was himself appointed to one of its chairs, which enabled him to write his history of Royal Dutch as a professor of colonial history.[101]

The initiative demonstrates to good effect a blinkered and inflexible attitude towards the emerging signs of growing self-consciousness in the Indonesian population which became a distinct aspect of Bataafsche's corporate culture. It was shared by Deterding, who, for example, in a 1929 interview, expressed as his opinion that Dutch colonial policy was disastrous and that it was 'foolish to talk about an Indonesian people'.[102] This attitude, typical for the Dutch colonial elite, did not bode well for the immediate future.

In every empire, dress has been one element separating the colonists from the colonized, and the Netherlands East Indies empire was no exception. In 1919 daily formality in the dining room (main picture) was echoed even in the workplace, with elegant ladies visiting the construction site of an underwater pipeline (above), and was romantically reflected in the pages of a 1922-23 issue of *Minjak*.

SHELL
SHELL
SHELL
SHELL
SHELL
GIRLS BOWLING TEAM 1937

Conclusion

During the early 1920s, the Group's top management went through a prolonged power struggle over a much needed reform of administrative structures and procedures. Considering his position under threat, Deterding blocked the initiatives for reform, made the position of his opponents impossible, and clung to the original design of 1907 which happened to suit his own, very personal, erratic and instinctive style of leadership. Unfortunately this blueprint proved increasingly unable to meet the demands of the Group's sprawling empire. The evident solution, delegating authority to operating companies around the world, was applied to particularly good effect in the rapidly growing US operations. However, neither the repeated discussions about management frictions, nor the expensive mistakes in Russia and in Mexico, which the Royal Dutch board considered as prime evidence of organizational failures, resulted in decisive action to achieve the overhaul required. Time and again, in cases such as Deterding's handling of the Dutch East Indies export tax affair, or his wavering policies towards Russia, rational considerations were subordinated to strong, personal, emotions. After Deterding had tightened his grip on the Group no one, with the possible exception of Hugo Loudon, could restrain him. Gulbenkian, the outsider and one of the few people who knew how to handle him, also broke with Deterding, which meant that the Group could no longer call on this constant source of creative schemes and new opportunities. During the second half of the 1920s, however, a new team of managers emerged and the Group regained its balance and dynamism. The new team was strong enough to form a kind of protective cushion around Deterding and limit the harm his distinctive style of leadership could do to the company. In an era in which many a 'great entrepreneur' led his company to the brink of disaster as happened when William Lever almost killed Lever Brothers, and Henry Ford almost destroyed Ford, Deterding at his worst comes over as benign, because the organisation figured out a way to control the damage he could inflict, and to make many decisions without him.[103]

The rising importance of science and technology in the business restored The Hague to its place as a proper partner and some of the most glaring administrative inefficiencies were redressed. In the expansion of the provisions for staff members, the Group conformed to a trend which characterized the emerging modern business corporations, though its attitude to staff relations remained distinctly paternalist.

TONDO
SERVICE SHELL STATION
SHELL
DAY & NIGHT SERVICE

Chapter 6

The evolution of technology, research, and marketing

During the interwar period, the Group's commitment to science and technology deepened further at all levels of operations. The dissemination of knowledge and technical information became institutionalized in the Technical Department in The Hague, relieving the previous dependency on outstanding individuals and speeding up the exchange of ideas. The expanding US operations greatly contributed to the intensification of knowledge within the Group, but the pattern of innovation was a multilateral one, to which the American companies contributed and from which they benefitted in about equal degrees. Dedicated research helped the Group first to transform products and manufacturing processes, and then to diversify into chemicals. Meanwhile marketing and branding strove to give the business and its products a firm and instantly recognizable public image with every available means. Gradually the contours emerged of a set of maxims about the oil business which would guide Group policy for decades.

A Shell service station somewhere in the Philippines, 1930s

[2]

From skills-based to science-based in exploration and production

The advances made in E&P technology are a good example of the varying pattern of interaction between the Group's operations. In 1918-19, Dutch engineers at Shell California and at Roxana introduced a technique known as core drilling, which used a hollow bit to bring up samples of underground geological strata. Core drilling greatly simplified the process of getting accurate subsoil information at a time when there were few other means available of doing this.[1] At the same time paleontology, the study of fossils, developed a specialism focusing on the fossilized remains of micro-organisms in the earth's strata. In 1918 the Group commissioned American and Swiss micropaleontologists to work on fossils found in Mexico, Venezuela, and Trinidad. Within a few years the operating companies in Mexico and Venezuela had set up a paleontological service, and Bataafsche created a special subdepartment for the specialism in 1924, ending the Group's dependency on outside specialists. The technique spread more slowly to other areas, however. Astra Romana continued to rely on outside specialists as and when required until the early 1930s. Shell California experimented with micropaleontology in 1924-25 and then lost interest. The American companies finally adopted this method during the 1930s, establishing teams at Bakersfield, Long Beach, Tulsa, and Houston. In the Dutch East Indies an early interest in micropaleontology also failed to gain momentum; when general managers had become convinced of its usefulness, The Hague refused to grant the funds because of spending cuts. A proper service was only set up in 1932 by two specialists from Venezuela and from Mexico. Towards the end of the 1930s the Group pioneered the development of a new specialism, palynology, the study of fossilized pollen.[2]

In 1917-18 Roxana adopted the rotary drilling system for use in its Lucien Field, Oklahoma.

[3]

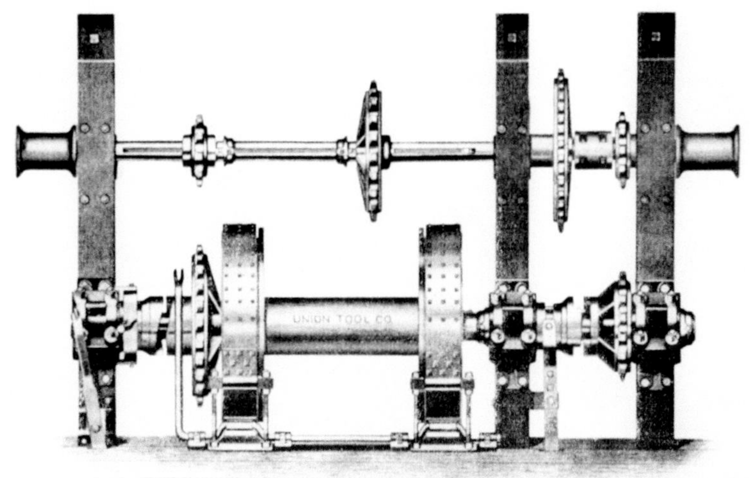

More or less parallel to paleontology, geophysics began to develop as a science. Gravimetric surveying with a torsion balance was introduced by Erb on a trip to Germany in 1921. A torsion balance can detect gravity patterns in the subsoil related to folded subsoil strata, thus helping to discover anticlines and salt domes. After the success of tests in Egypt, Seria (North-west Borneo) and Mexico, the Group trained survey teams in the use of torsion balances for all main areas of operation. Two such teams set to work in the US, one in California and one in Texas. The gravimetric method did not work for Shell California, who soon dropped it, but Roxana used it extensively. From the late 1920s the attempts to discover subsurface irregularities took a step forward with magnetometers and gravimeters, respectively to pick up variations in magnetism and in gravity.[3]

The benefits of rotary drilling over percussion drilling took years to develop, with early rotary drills frequently producing crooked holes. Matters slowly improved with the introduction of new designs. Clockwise from above, the 'Ideal Number 3' drilling rig was manufactured by the Union Tool Company at Torrance, California, and used by Shell Company of California from 1915 while other Shell companies were using different rigs. Next, the 'Holland core drill' (invented in 1908) is seen with its original fixed inner barrel, used experimentally by Shell Company of California in 1919. Later a loose inner barrel was developed. Next is the cutter head of a Holland core drill with interchangeable steel teeth, and lastly, the Elliott Alco core drill, which came on the market in 1925 after development by a former Shell employee, J. E. Elliott.

[4]

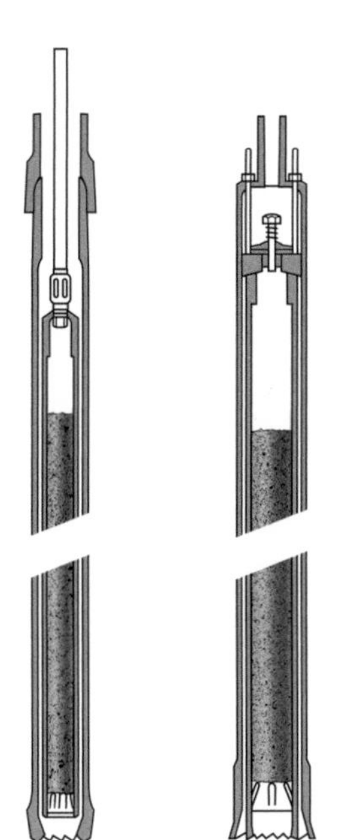

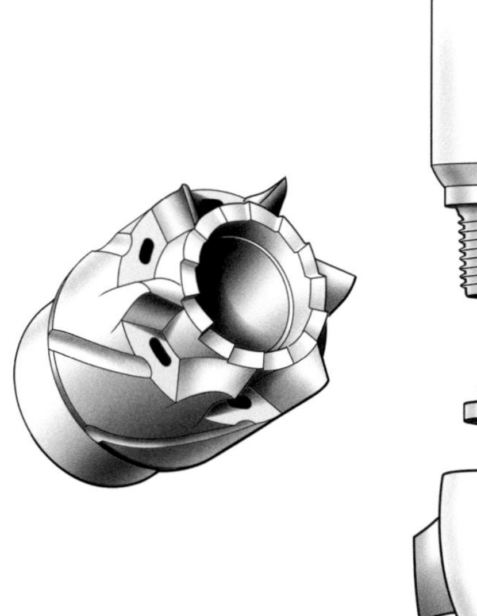

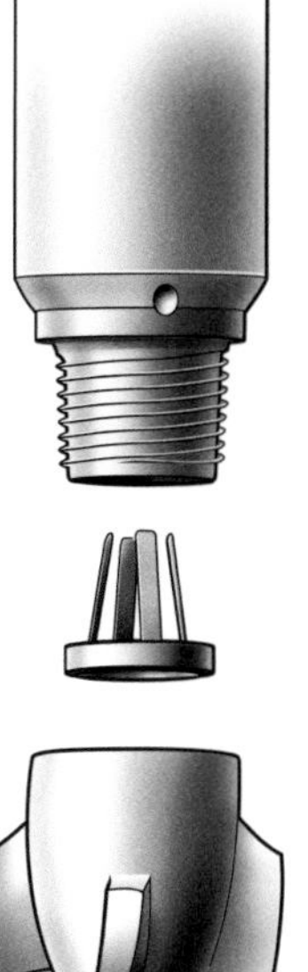

[5]

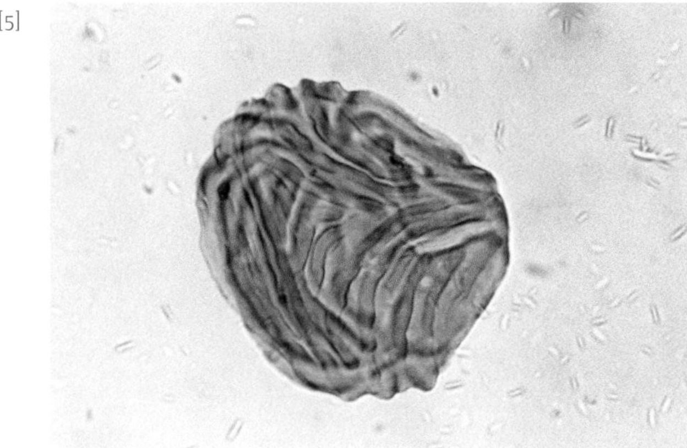

[6]

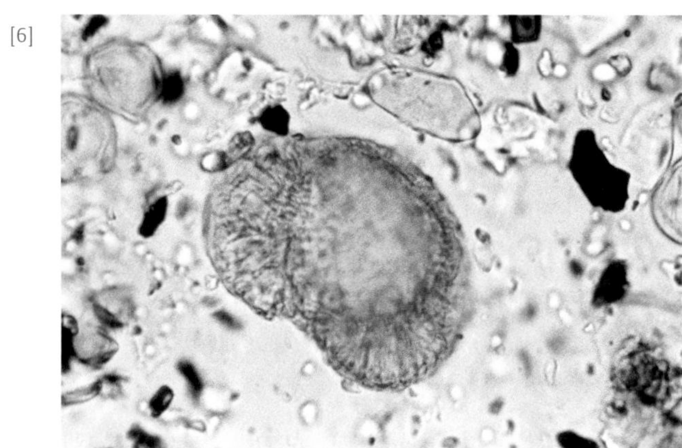

[7]

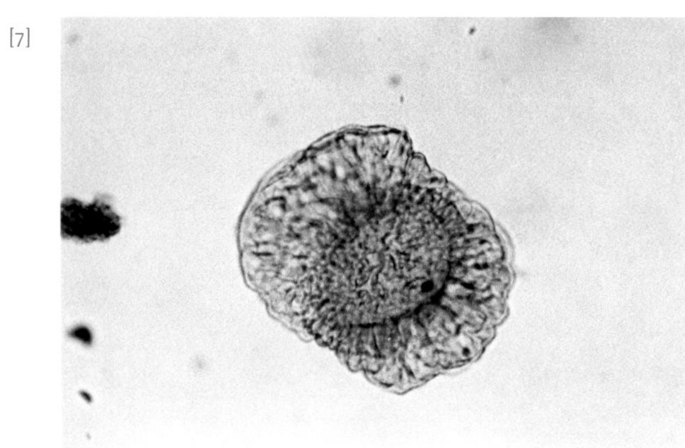

[8]

The bane of hay-fever victims could also be a useful clue in oil exploration. Above, magnified many times, are examples of pollen grains.

[9]

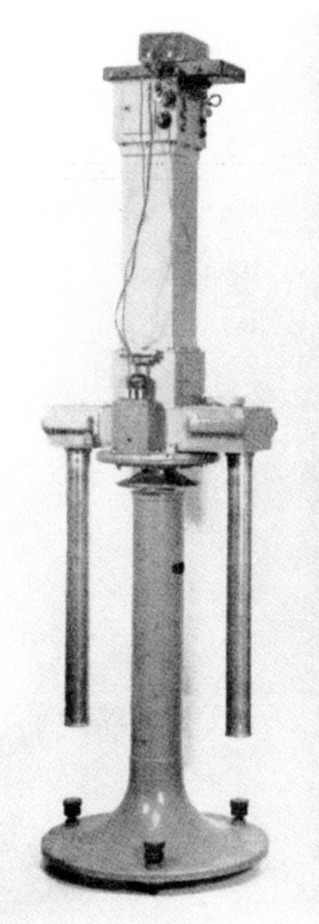

In 1888-96 the Hungarian Baron R. von Eötvös developed the principle of the torsion balance as a means of gravimetric surveying. By measuring changes in gravity, the instrument could reveal the nature of the subsoil. The basic principle is shown in the diagram (below). Left is one of the oldest types of torsion balance. The balance worked best when raised as high as possible above the ground, which in swampy terrain (centre) was not very high. Jungle terrain (far left) brought its own problems, especially the transport of heavy items such as iron scaffolding pipes, so this 'high station' was constructed entirely out of local materials.

[11]

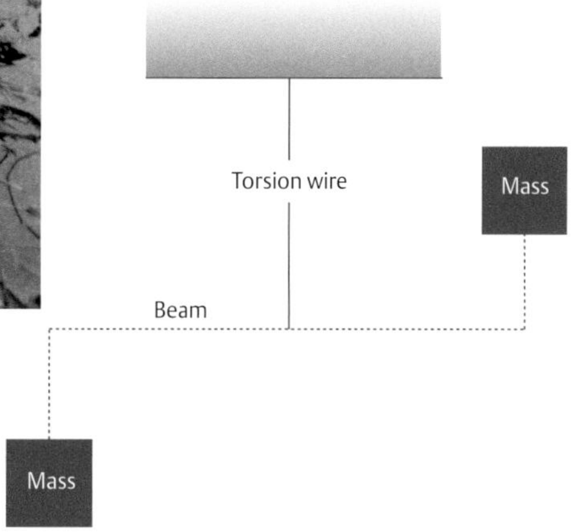

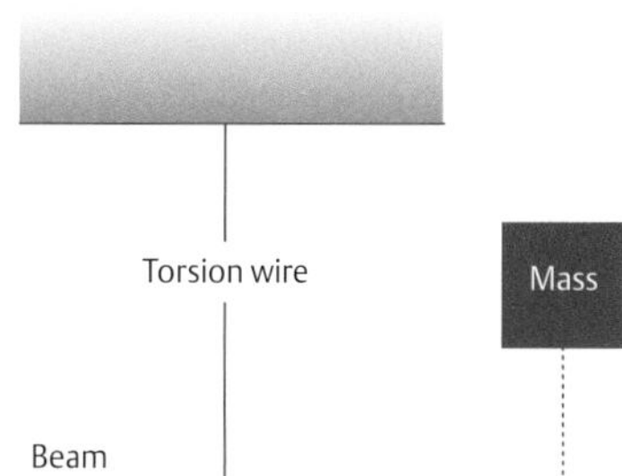

[13]

[12]

Spring

Mass

Transporting instruments through swampy country was a further difficulty. By 1940 torsion balances had been almost completely ousted by the modern gravimeter (left, in simplified diagram) which gave readings much more quickly and easily. Another tool for explorers, seismic surveying either by refraction (below) or by reflection (bottom), was widely adopted in the 1920s.

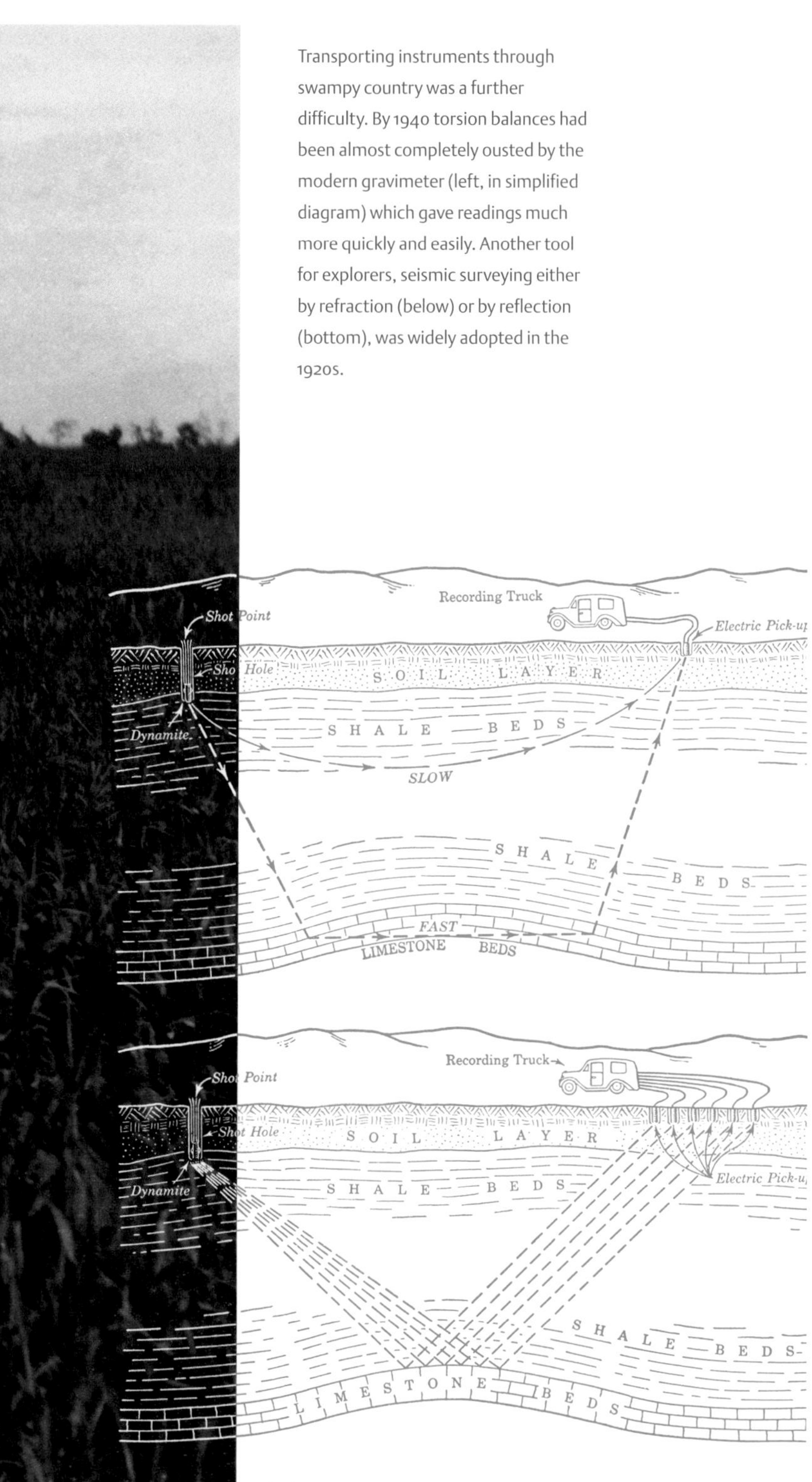

[14]

Seismic surveys were another innovation from Germany. Developed during the First World War as a way to locate enemy artillery positions from the pattern of air and ground shock waves sent out by firing guns, seismography measures the trajectory of vibrations generated by the detonation of an explosive charge through the earth's surface. The refraction method tracked how these waves were transmitted by various strata, hard formations returning them to the surface more quickly than soft layers such as shale. By contrast, the reflection method mapped the location and shape of hard formations by capturing how the shocks are reflected by subsoil strata. In 1922 the Group hired the pioneer of seismology, the German geophysicist Dr L. Mintrop, to do a survey in Mexico, where the method proved of sufficient value to stimulate further experiments. Two years later Marland Oil, the Group's E&P partner in the Midwest, successfully used refraction seismology to locate salt domes in East Texas and along the Gulf of Mexico coast. This triggered a rush to adopt the new technology by other companies, including Roxana, which in 1926 hired a German seismologist, Dr. Eugen Merten, for surveys. Merten's field work proved to be of great value, but his contribution to the development of more sophisticated seismic instruments at the engineering workshop which, in 1937, became the Houston Geophysical Research Laboratory, was possibly of even greater importance. One innovation was the Shell seismometer, a considerable improvement over the older seismographs, which the Group adopted in the early 1930s, when reflection seismography started to replace the refraction method.[4]

SEISMIC SURVEY IN THE JUNGLE

Seismic surveying involves sending sound waves through the earth and recording their echoes as they bounce back from underground rock formations. By using multiple seismographs placed at different distances from the source of sound, the data so gathered may be interpreted to give an approximate understanding of underground geological structures, possibly containing oil. The sound waves may be created by various means, an early common one being the use of dynamite buried in the earth, as here near Casigua in Venezuela in the 1930s.

[15]

[16]

[17]

[18]

[19]

[20]

After using machetes to clear a space in the jungle by hand (right), a local base camp is set up (main picture, far left). Centre, from top to bottom, the crew drill a shot-hole; a detector is placed in position; the crew leave the shot-hole site; the seismograph operator records results; and the long diagram of recorded sound-waves is spread out for interpretation.

[20]

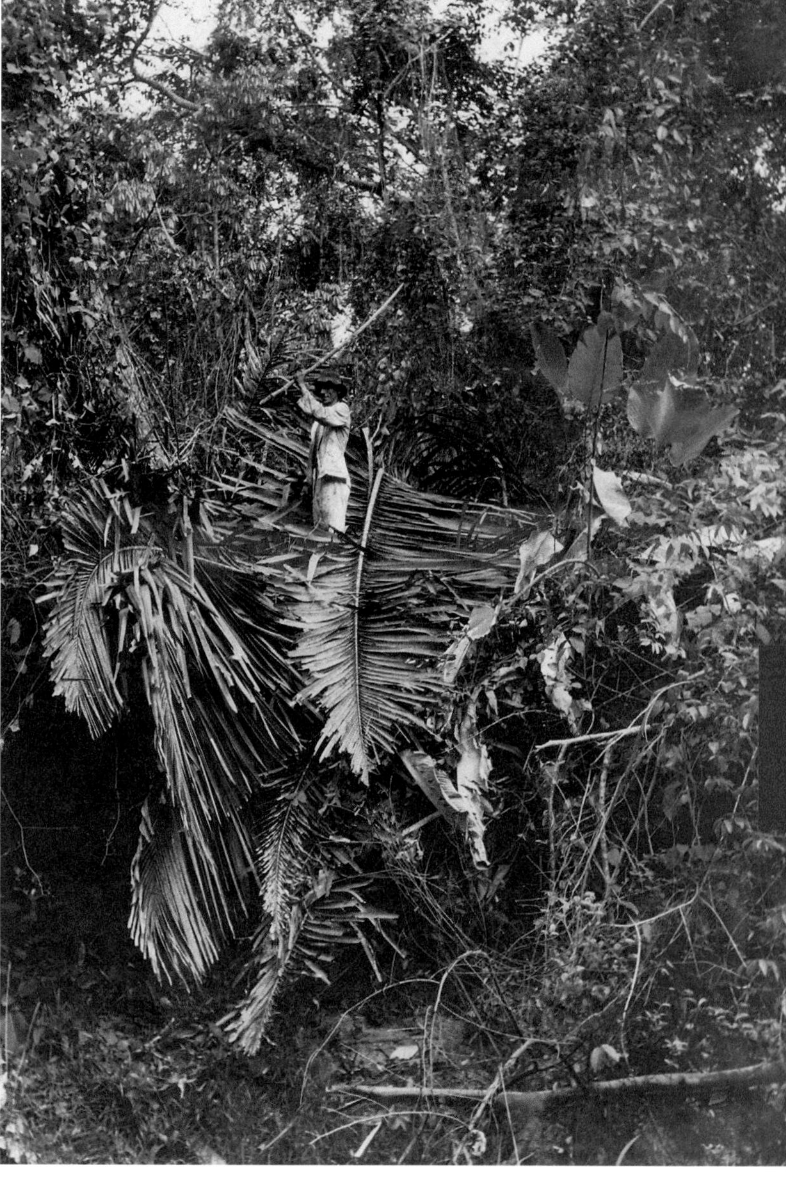

[22]

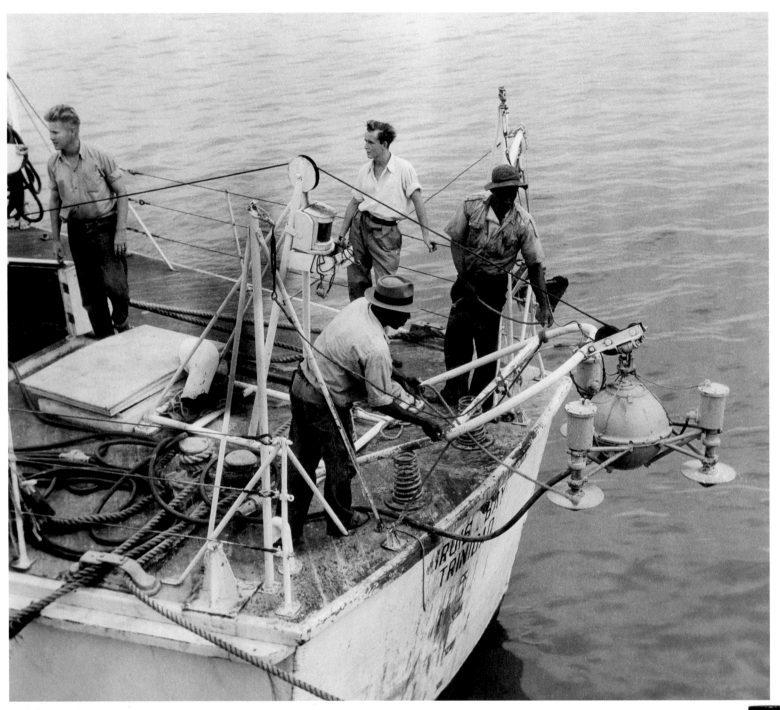

The remote control La Coste Romberg gravimeter (above) is being used to measure gravity differences on the seabed at Point Fortin, Trinidad, possibly during the late 1930s. Readings were transmitted by cable to dials in an instrument panel in the surface vessel.

A selection of crude oil samples from different parts of the world (right and overleaf) shows a wide variety of types and quality.

Seen here in its earliest form, with the camera attached to the outside of a British Royal Air Force aircraft, aerial surveying by photography was one of the many technical innovations of the First World War that later found benevolent peacetime use.

[23]

Surveying from the air also built on technology developed during the war, in this case automatic cameras and special lenses. Mapping an area with the aid of aerial photographs reduced the need for labour intensive terrestrial surveys to a minimum. Moreover, air surveys made it possible to map inaccessible regions such as tropical jungles at relatively low cost. In 1925 the Group started to experiment with air surveys, hiring an outside firm to photograph a large area of tropical swamp in Sarawak and Brunei. Two years later, Shell California also held trials. Once the method had proved its worth, air surveys were used in Sumatra to stake out new concessions during the 1930s, the Group collaborating closely with the Dutch colonial government and the NKPM to have the maps made and the geological information on them interpreted. With the Sumatra mapping project air surveying entered a new stage, for the photographs unexpectedly revealed geological details covered by the dense jungle. As a consequence, the priorities were reversed for the joint air survey of a huge uncharted area in New Guinea, which was primarily organized to obtain geological information, map-making coming second.[5]

Thus the innovations in exploration were very much the result of the Group's main areas of operation collaborating as a network of innovation, each one in turn contributing to initial development, field trials, or further improvements. By contrast, the Group drew heavily on the United States for innovations in production technology. Rotary drilling, for instance, was first developed there. The Group used the system on a modest scale in Romania from 1907 and tried it elsewhere but the indifferent success meant that percussion drilling remained the clear favourite.

[24]

Extensive trials in the Dutch East Indies during 1916-18 ended in failure and Roxana struggled with rotary drilling as well.[6] Meanwhile Shell California had adopted a new type of rotary rig which, especially when combined with a core drill, offered significant improvements over earlier models, including higher speed and much lower operating costs. In 1923 a drilling team from Shell California, led by its chief production engineer B. H. van der Linden, reintroduced rotary drilling in the Dutch East Indies, now with great success. When Van der Linden's successor at Shell California, W.C. McDuffie, was appointed as Group chief production engineer in 1924, the rotary system became official policy. Astra hired an American drilling crew with rotary rigs in the same year, followed by Venezuela and Trinidad. By 1930 the Group had ceased using percussion drilling almost entirely.[7]

The parade of oil samples continues. The paler light crudes, just as valuable as the heavier tarry crudes, can be turned into products such as gasoline, fuel oil, and chemicals.

From distilling to manufacturing Innovation in the midstream part of the business was also a network result, just as exploration was. The central importance of cracking, developed in the US, tends to overshadow the processing technologies which the Group drew from other areas of operation, as often as not exporting them to America. This happened with the Edeleanu process, with hydrogenation, and with manufacturing lube oil and asphalt.

Though the principle of using heat and pressure to break down heavy hydrocarbon molecules into lighter ones had been long known, industrial processes for cracking oil developed only during the 1910s, when the rapid spread of motor cars in the US prompted an urgent search for increasing the gasoline yield of crude oil. Early processes such as the Burton system cracked gas oil into gasoline, thus avoiding the problems of coke deposits involved with processing heavier oil fractions, but Deterding insisted that cracking only made sense if a cheap feedstock such as residue could be used.[8] Once the Group had bought the Trumble distillation patents in 1915, Pyzel immediately started to investigate the suitability of the converter unit for cracking heavy Mexican oil into gasoline. Despite early successes, a pilot plant built at Martinez did not come up to expectations in the end. The Trumble converter proved useful as a 'visbreaker', which upgrades heavy oils by reducing the vicosity and pour point, and for that purpose installations were built at Trinidad, Curaçao, and Suez.[9] From the welter of

competing thermal cracking systems which had meanwhile proliferated, Group engineers finally chose the process developed by the Californian refiner Jesse A. Dubbs, a man so devoted to oil that he named his son and later associate Carbon Petroleum Dubbs. This installation incorporated a feature which drastically reduced the formation of coke at an early stage, thus solving the crucial problem of deposits. A trial run fully convinced Pyzel that the Dubbs was superior to the Trumble converter and in December 1919 the Group took a Dubbs licence from United Oil Products (UOP), the company holding the rights.[10] The first unit came on stream at Roxana's Wood River refinery in March 1921 and proved so successful that the company ordered six more units in 1922. Roxana also built two Cross cracking installations to compare the rival system's merits relative to the Dubbs. That same year Balik Papan began to experiment with cracking. In 1923, some 20 per cent of the Group's gasoline supply was made by cracking; three years later it was 40 per cent. By 1927 the Group operated twenty-five Dubbs units around the world. Continuous experimenting led to further significant process innovations.[11]

By opening ways of rearranging hydrocarbon molecules and thus fundamentally changing the properties of a particular feedstock, cracking marked a watershed in oil processing. Until then the supply of products such as gasoline had largely depended on the availability of suitable crudes to make them, but now they could, in principle, be made from almost any feedstock. Major

Building on the work of Jesse A. Dubbs, in 1919 his son Carbon Petroleum Dubbs gained a patent for 'clean circulation', illustrated below.

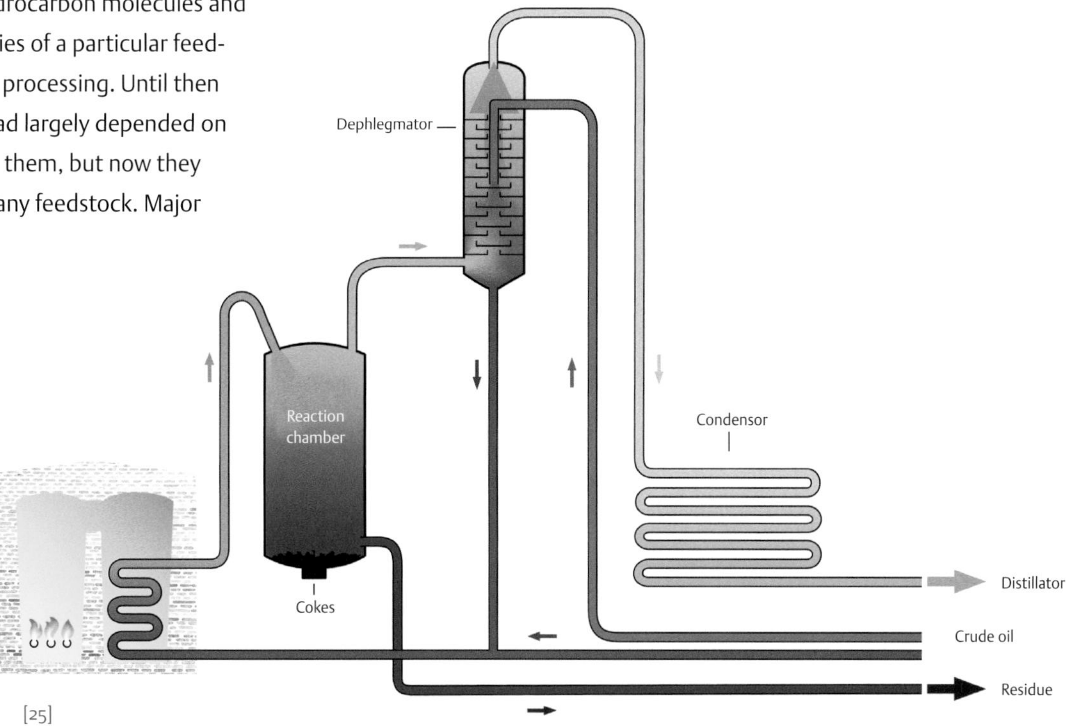

[25]

problems remained to be solved; cracked gasoline tended to form gum when stored and the complex structure of lube oils needed to be unravelled before the Group succeeded in developing a successful manufacturing process for it, but cracking remained the cornerstone. Processing became manufacturing, oriented at what was later to become one of the Group's business maxims, making quality products from cheap components. Moreover, cracking was particularly important for the Group, dependent as it was on a wide variety of crudes with very different properties, to serve markets with markedly different demand patterns. The business was always short on gasoline, long on asphalt, and bereft of Pennsylvanian crude for making lube oils. Such products could be purchased, of course, and they were, but cracking introduced both greater flexibility in supplies and the ability to balance output depending on market circumstances. This focus on crude flexibility became another one of the Group's business maxims. By generating large volumes of potentially valuable hydrocarbon gases, cracking also brought the industry to the threshold of the petrochemical era, for these gases provided the building blocks for a whole range of synthetic chemicals. Research played a crucial part in this transformation. We will now first take a look at the state of research at the beginning of the 1920s, and then look at the transformation of gasoline and lube oil, before discussing the diversification into chemicals from 1927.

[26]

With the growth of chemical engineering, pilot plants (such as this one in Amsterdam-Noord) became a vital link between laboratory research and industrial application.

The evolution of research

During the interwar period Bataafsche's Amsterdam laboratory was the hub of Group research. It was the largest single laboratory dedicated to a wide range of both applied and fundamental research. This scope set the Amsterdam laboratory apart from those attached to the refineries around the world, which were focused on monitoring products and processes, but as research interests widened and deepened, the Group's overall effort increasingly became a team effort, guided and coordinated by Amsterdam.

The refinery laboratories delivered substantial contributions in the form of test runs and detailed investigations of particular topics. The number of specialist research establishments attached to the Group rose as well, among them dedicated engine research departments at Thornton and Delft, a lube oil institute at Freital in Germany, and the Houston geophysical laboratory. Moreover, a growing number of outside experts was drafted in, to get the benefit of their knowledge on specific subjects or, as happened in asphalt research, to form a body of informed opinion about particular products and product standards so as to ease market acceptance of Group products.[12] Drawing on outside expertise was nothing new, of course. Some early fundamental work was outsourced in the UK. Researchers at Cambridge University spearheaded the investigations before and during the First World War on aromatic hydrocarbons in general and on their suitability for making synthetic dyes;[13] as we will see, research by the engineer Harry Ricardo transformed the understanding of gasoline composition and helped to open up new markets by improving diesel engines. However, these projects remained incidental, possibly commissioned to placate Waley Cohen's tireless attempts to have more research done in the UK without incurring material commitments. As a rule, such work was kept in-house.

LET OP
DRINKWATER
LET OP
DRINKWATER
EVERYWHERE YOU GO—
YOU CAN BE SURE OF SHELL

06|338

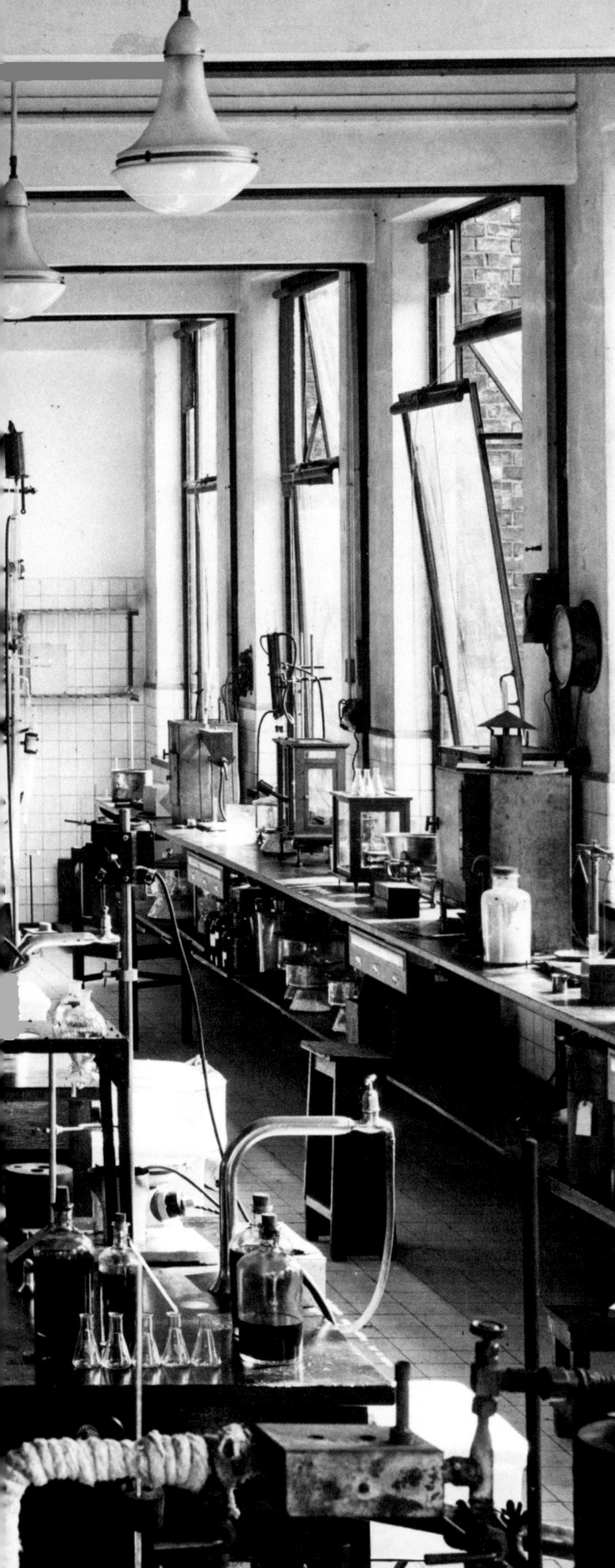

[27]

Bataafsche's oldest laboratory, Amsterdam-Noord.

In 1920 the Amsterdam laboratory was already quite a substantial organization employing 30 people, though with a puny budget of less than 0.05 per cent of Bataafsche revenues.[14] The product research occupied itself on the one hand with improving existing products and finding new ones, and on the other with making the most of the Group's supply position. Consequently, the laboratory had, in addition to the product analysis department, research teams for lube oil and crude composition, white oils, asphalt, and candles, the former two investigating alternatives to Pennsylvanian oil for the manufacturing of lube oils, the latter two endeavouring to turn the oversupplies of asphalt and paraffin wax into commercially viable products. During these years, research teams remained responsible for development of products and processes up to and including the pilot plant stage; the separation between research and development came only after the Second World War. Since the involvement with sulphonation and the Edeleanu system, process research had branched out into three teams: desulphurization, emulsions and extraction techniques, and cracking, which included catalytic hydrogenation. The laboratory also trained research staff for the various refinery laboratories, gave production and refinery managers a grounding in the basics of oil chemistry, and acted as the Group's central purchasing and testing department for laboratory equipment. In 1921 the Amsterdam laboratory issued its first booklet laying down standard methods for testing and analysis.[15] Central offices evaluated the research results at every step to make sure that commercial considerations retained priority over scientific ones. By 1930 the direct product-related work was done as service work under contract to operating companies.[16]

During the interwar period, the Amsterdam laboratory gradually evolved from methodical, empirical work to science-based research, and from applied product and process analysis for solving incidental problems to studying fundamental oil properties in order to develop new products and better processes. One stimulus in this direction came from the deepening of cracking and hydrogenation research following the involvement with the German chemical engineer Friedrich Bergius and his process to make gasoline from lignite and heavy oil fractions by hydrogenation, reacting the feedstock with hydrogen under very high pressures. Bergius had worked on this process since 1910 and by 1919 development had reached the pilot plant stage. Since hydrogenation appeared to be a viable alternative to cracking, Bataafsche decided to support Bergius by buying a half stake in the Internationale Bergin Compagnie (IBC), the company holding the rights with the exception of Germany, which Bergius kept, and the US, where all German patents had been expropriated during the First World War. In September 1920, the Amsterdam laboratory sent one of its experts to the pilot plant in Germany for observation and liaison. Four years later, however, the Bergius process was still no nearer to commercial application. Meanwhile the success of the Dubbs cracker caused Bataafsche to lose interest in Bergius. The company sold most of its share in the IBC to the German chemical company BASF, a constituent company of the IG Farben concern in 1925, but kept a minority stake at the insistence of Deterding, who thought the process would prove useful in countries with particular economic and trade policies, such as Italy. Hydrogenation research in Amsterdam was temporarily suspended and resumed only when unexpected developments, about which more below, suddenly propelled it back into the research programme.[17]

Though looking disparate, Group research really had a single purpose, to relate the properties of individual products to the hydrocarbon components of those products and of the crude oil from which they were made, and then to develop manufacturing processes for making them from the cheapest crude available. The coherence of this approach is highlighted by the fact that it yielded results for all main product groups within a few years of each other.[18] Making cheap components later became the third of the Group's business maxims. The first breakthrough happened in gasoline. Gasoline was the Group's most important product in terms of revenue, and the second after fuel in terms of volume. Throughout the inter war period Shell had difficulty in keeping pace with the fast expansion of demand for gasoline; that is, Asiatic consistently sold more than the business could produce. Despite growing crude supplies and a rapidly expanding Dubbs cracking capacity, the Group struggled to keep pace with the worldwide demand for gasoline. In 1922, its own gasoline supplies showed a shortfall of 200-250,000 tons; by 1929 this had risen to more than one million tons. Heavy investment in new cracking installations at almost all main manufacturing sites coupled with falling demand due to the Depression turned the shortage into a surplus during 1933, but two years later the Group again found itself some 300,000 tons of gasoline short.[19] Purchasing supplies served as a stopgap while the Group strove to become self-sufficient in gasoline. This ambition drove a continuous quest for ways to raise the volume of gasoline produced, improve its quality and, from the early 1930s, to increase its octane rating. Research enabled the Group to develop technological means for combating its shortfall of supplies by breaking up gasoline into separate components, mass-producing the parts in the most economic way from the particular type of feedstock available, then blending them to specification for individual markets. As a consequence, the entire process for making gasoline was transformed from simple distillation into a complex manufacturing operation.

Until the advent of cracking during the early 1920s, all gasoline was straight-run, i.e. distilled as one fraction from a sufficiently light crude such as Perlak produced in Indonesia. For marketing purposes, specific gravity served as the main quality criterion and gasolines were accordingly labelled as either 'light' or 'heavy'. This criterion had little relevance either for the product's suitability as fuel for internal combustion engines, or as a gauge for producing a standard quality gasoline, but the limited knowledge about gasoline and about what actually happened in engines prevented the development of better yardsticks. Motorists had to find out by trial

[29]

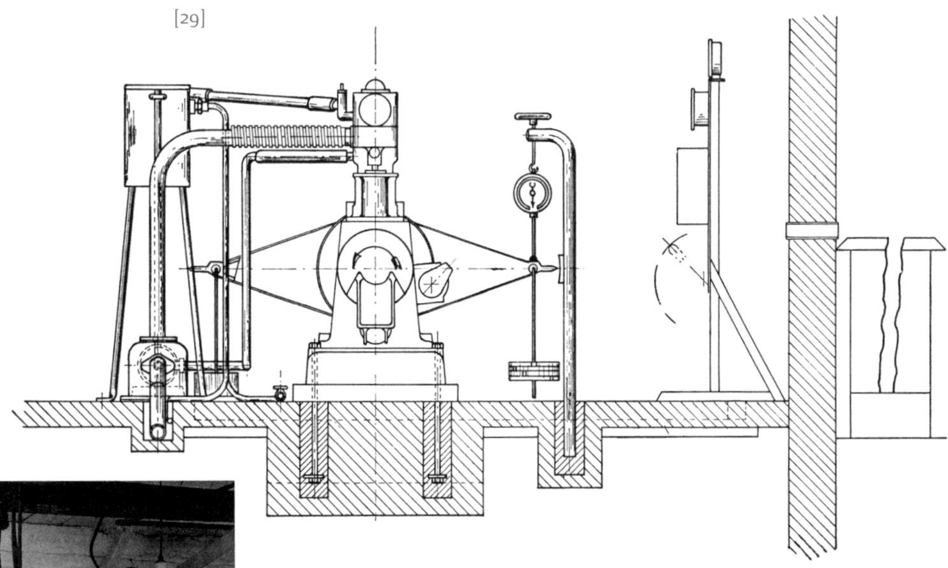

[28]

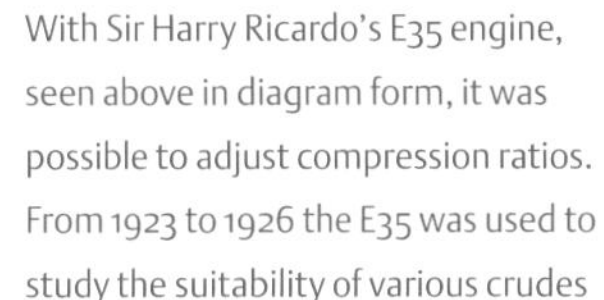

With Sir Harry Ricardo's E35 engine, seen above in diagram form, it was possible to adjust compression ratios. From 1923 to 1926 the E35 was used to study the suitability of various crudes for gasoline production. Ricardo's assistants (left) were apparently allowed very few sartorial concessions to their surroundings.

and error which gasoline suited their particular engine best, blithely unaware of the fact that manufacturers really had no way of ensuring that the product chosen would always and everywhere have the same desired qualities.[20]

The Group tackled this problem from two sides. In 1919 Anglo-Saxon commissioned Harry Ricardo to investigate the relation between engine knock and gasoline composition, following his earlier work on the role of aromatic hydrocarbons for the Group (see Chapter 3). At the same time managers set up a Standardization Committee which, in 1922, issued a set of instructions to Group refineries about gasoline production, so as to combat the diverging local distillation practices. To obtain up-to-date information about product guidelines and tests, the Group became a member of the American Society for Testing Materials (ASTM), considered keener in following industry trends than its British equivalent the Institute of Petroleum Technologists or IPT.[21]

Working with his assistants Henry Tizard and David Pye, who both became first-rank scientists in their own right, Ricardo began by focusing on the link between engine knock and compression ratio, i.e. the pressure created in the cylinder head by the piston at its zenith. Engines with higher compression ratios were more prone to knocking, but already during the war Ricardo had found that particular gasoline components such as benzol and aromatic hydrocarbons reduced this tendency. He now constructed an experimental engine, named E35, with a variable compression ratio, and used it to establish what he termed the Highest Useful Compression or HUC for any particular gasoline blend, testing hundreds of specimens supplied by the Group. Within eighteen months, Ricardo had found some ground rules for improving the HUC of gasoline. The presence of aromatic hydrocarbons and of volatile fractions, and cracking at high temperatures increased the HUC; cracking at low temperatures or treating gasoline with particular acids lowered it. Keeping in close touch with Group experts such as J. van Rijn van Alkemade, the head of the Singapore laboratory, Ricardo developed a standard figure for the distribution of different fractions of gasoline, first released in 1926, updated three years later and codified into a standard analytical method book by the Standardization Committee in 1931. E35 engines were purchased for Amsterdam, Emeryville, Wood River, and for the new Delft engine testing station set up in 1928 to continue Ricardo's research.[22]

Ricardo's pioneering HUC standard was destined to be overtaken by an alternative approach published in 1926 by two American scientists, that of octane numbers. This method used two pure hydrocarbons, heptane and iso-octane, as the opposite ends of a scale running from 1 to 100. Heptane makes engines knock, but iso-octane does not. The octane number of a particular gasoline blend could thus be determined by comparing its knocking tendency with that of the equivalent heptane to iso-octane mix, the number being the percentage of iso-octane in that mix. Thus gasoline with octane number 95, commonly found today, corresponds to 5 per cent heptane to 95 per cent iso-octane.[23] Adopted as standard by the US oil industry in 1930, the octane numbers method soon spread throughout the rest of the world, sweeping the HUC standard aside.

Though losing the standards battle may have disappointed Ricardo, it did not in any way diminish the crucial importance of his work for the Group. First of all, his findings laid the groundwork for the standardized production of gasoline based on a better understanding of both combustion engine performance and of gasoline as a composite hydrocarbon mix. Secondly, the HUC standard provided an analytical tool to maximize gasoline output and efficiency by organizing production as a supply chain of components, taking aromatic hydrocarbons as the basis. With its Edeleanu plants to extract aromatics with liquid sulphur dioxide, the Group already possessed the technology for this, since the plants produced extracts of concentrated aromatics, called kerex if made from kerosene, benzex from gasoline, gasex from gas oil, and resex from residue. These components were either sold directly to customers or used for upgrading other products. Its high concentration of aromatics made benzex into a very valuable blending component for making gasoline with a high anti-knock value. As a result, the production of extracts with the Edeleanu process gradually became more important than its original purpose, the upgrading of particular oil fractions.[24] Balik Papan's Edeleanu unit became the central installation for making extracts. By 1929 the Indonesian refineries operated a precise administration of their gasoline and kerosene fractions output in order to maximize the use of the available aromatic hydrocarbons. Three years later a central planning department located in London compiled monthly HUC balances covering all production centres so as to match worldwide capacity, supplies, and demand.[25] Edeleanu plants for making extracts were also built at Martinez and Wilmington.[26] After the agreement with AGEFCI, the patent holder, had expired in 1928, the Group negotiated a royalty-free use of the patents and started to build plants everywhere. At the outbreak of the Second World War the number totalled some fifty units.

Thirdly, Ricardo's discoveries gave the Group a good position in the rapidly rising demand for high-performance gasoline, which began with airlines requiring better performance for aviation gasoline in the late 1920s.[27] Engine manufacturers responded by raising compression ratios and oil companies rushed to increase gasoline octane numbers: from 60-63 for car engines around 1930, to 80 for cars and 87 for aircraft five years later, to 100 octane for aircraft in 1938.[28] The octane race stretched gasoline supplies and investment but, with production reorganized around the available HUC supplies, the Group was well equipped to compete, despite its gasoline shortage. Moreover, the reorganization enabled the Group to maximize the use of aromatic hydrocarbons to boost octane numbers, and to minimize the use of gasoline additives such as tetraethyl lead (TEL). This very effective anti-knock dope had been discovered in 1921; the patent was held by the Ethyl Gasoline Corporation, a joint venture of General Motors and Jersey Standard.[29] Group managers preferred hydrocarbon fractions to TEL, partly because the ingredient was toxic and dangerous to handle, partly because up to the 73 octane level blending high-octane hydrocarbon fractions such as aromatics into the desired grade of gasoline was cheaper. The desire not to pay royalties to a competitor may have played a part as well, plus the fact that an exclusive contract between Ethyl Corporation and Standard Indiana prevented Roxana from using TEL in most of its marketing area.[30] The professed liking for pure oil products appears to have mattered less, because for over a decade the Amsterdam laboratory conducted an expensive and fruitless campaign to find its own anti-knock dope.[31] By the early 1930s the octane race had passed the point at which TEL was more expensive than blending and, since Standard Indiana's exclusive rights had lapsed as well, Shell Union launched Super-Shell Ethyl in 1931.[32]

Ricardo also took a vital part in further developing the diesel engine. As mentioned in Chapter 2, the Group had an early interest in these compression-ignition engines, starting in 1912 with the pioneering motor vessel *Vulcanus*. In 1924 Anglo-Saxon commissioned Ricardo to design and build a high-speed diesel engine as a prototype for a range of stationary engines for industrial applications such as those used widely by the Group itself on drilling rigs. When this had proved very successful, the contract was extended in 1927 to develop an engine suitable for road use, where diesels had until then only found very limited application because of poor overall performance and excessive smoke in the exhaust fumes. Ricardo diagnosed these problems as originating from poor combustion and designed the so-called Comet combustion chamber to remedy them. His new engine was an instant success on London double-decker buses, leading to engine manufacturers around the world queueing up to take a licence for the Comet chamber. Diesel engines rapidly became the most preferred option for commercial vehicles, ranging from taxis to vans, lorries, and buses, opening up a huge new market for the Group.[33]

Lube oil research evolved along similar lines to gasoline, though here the Amsterdam laboratory took a leading role in the investigations from the start. During the First World War Bataafsche and Asiatic had succeeded in harmonizing and centralizing lube oil production in the Netherlands East Indies, but the Group remained dependent on purchasing large amounts of so-called bright stocks made from Pennsylvanian crude, the Penna oils, as the basis for its own production. As a consequence the lube oil research programme was designed to find proper substitutes and lessen, preferably end, this dependence. By 1922 the introduction of vacuum distillation and new refining methods had resulted in the development of a manufacturing process which yielded a fairly good base stock from several crudes, though not yet in the Penna class, notably because the oils discoloured with age. While the refinery laboratories at Shell Haven, Monheim, Martinez, and Balik Papan continued the search for the causes of discolouring, the theoretical work on lube oil progressed along two main lines. Ricardo was commissioned to do engine research; the Amsterdam laboratory would focus on the chemical composition of lube oils in collaboration with the Stern-Sonneborn lube oil works in Hamburg, also known as Ossag.[34]

This company had been taken over by the Group in 1924 and subsequently merged with the German subsidiary into Rhenania-Ossag Mineralölwerke. Stern-Sonneborn made a perfect fit both with Rhenania's Monheim lube works and with the Group's search for improving lube oil production. Ossag manufactured specialist oils from organic materials, had pioneered the making of emulsions for cylinder oils, and possessed a licence for the Voltol process for polymerizing rapeseed oils with an electric current to obtain a very flat viscosity curve, so the oils retained their particular resistance to flow across a wide range of temperatures. The flat viscosity made Voltol ideally suited for high-performance uses such as aircraft, Zeppelins, navy ships, and army trucks. Stern-Sonneborn had a Voltol factory in Freital near Dresden, which included a testing station fitted with a specially designed engine to test lube oil properties. Finally, the company had pioneered new extraction processes and held a patent for the extraction of aromatic hydrocarbons through the application of furfural, which promised to be better suited to lube oil production than the Edeleanu process. To secure the overall position in lube oil extraction, the Group made a patent exchange deal with Texaco to obtain the rights to the Duosol process as well. As the name implies, this system used two solvents rather than a single one.[35]

Analyzing the effects of the various manufacturing processes deepened the insight into lube oil properties and composition, and into making better lube oils, but by 1930 Group researchers were no nearer finding a substitute for bright stock, or even a recipe for making base stock from the various types of crudes available. A process which worked for the oil from one particular well in Venezuela did not necessarily suit the oil from another one. Hydrogenated lube oils were tested, but found not to be worth the cost of manufacturing them. Within a few years the combined efforts of the previous decade yielded the desired results. A test was discovered to assess the exact composition of lube oils, which revealed the links between properties and hydrocarbon components. This finally opened the way for organizing lube oil manufacturing as an assembly of components, by applying a combination of vacuum distillation, cracking, and extraction processes to the raw materials at hand. The result was called syntholub, or synthetic lube oil, and it could compete successfully with Penna oils, even in the United States. A pilot factory came on stream at Balik Papan in 1934; building on the experience gained there, new lube oil plants followed at Hamburg, Shell Haven, Wood River, Martinez, and, just before the war, Stanlow.[36]

Two other product segments, white oils and asphalt, followed a similar development trajectory to gasoline and lube oils: intensive experiments during the 1920s led to fundamental insights about the relationship between crude composition and product properties, resulting in a transformation of production from crude-based made via distillation to component-based, made by dedicated separation and cracking processes.[37] As a consequence, blending achieved a new prominence for all products. Originally used to upgrade poor quality products such as Borneo kerosene by mixing it with superior kerosene, blending now became the regular final stage of manufacturing, geared to finish products to the specifications required for individual markets.[38]

Pushing out on a new course The firm integration of research with the Group's business provided a good springboard when, during the mid 1920s, Group managers started to turn their attention to chemicals. The US companies had always obtained a large volume of natural gas with their crude and pioneered innovative ways of extracting the gasoline fractions from it.[39] Since most oil fields were located some distance away from conurbations where the remaining associated gas might be sold, it often went to waste. With the spread of Dubbs installations the companies started to produce even more gases by cracking. Both associated gas and offgases contained potentially valuable hydrocarbons, such as methane, ethane, propane, butane, pentane-heptane, ethylene, propylene, and butylene. During 1925-27, the Group's chief technical experts Kessler, De Kok, Pyzel, and Knoops, explored the feasibility of producing these components as pure chemicals on a commercial scale, concluding that this could be done provided that the Group invested heavily in the research required to pioneer the manufacturing processes. In April 1927, the four managers recommended creating a budget of $ 10 million to start research into the manufacturing of chemicals from hydrocarbons and to begin producing synthetic ammonia so as to gain experience with high-pressure catalytic hydrogenation processes. This was a very substantial sum. Net income for Royal Dutch and Shell Transport together totalled $ 55 million for 1927, Bataafsche netted 77 million guilders, or $ 31 million, the Group's entire American operations $ 11 million.[40]

At more or less the same time Jersey Standard entered into talks with IG Farben to get access to its hydrogenation techniques. The company first obtained a research licence and subsequently concluded, in August 1927, a comprehensive agreement for the worldwide use of the hydrogenation patents as applied to oil outside Germany. The chemical applications remained reserved for IG Farben.[41] Events now threatened to overtake the Group. BPM's IBC stake yielded dividends but no access to information about subsequent process improvements. Though publicly sceptical about the economic viability of gasoline produced by coal hydrogenation, the Group could not afford to remain on the sidelines if two leading industrial concerns joined forces to monopolize the process, if only because the high-pressure catalytic processes at the core of hydrogenation were likely to prove useful for treating oil.[42] An arrogant outburst by Carl Bosch, IG Farben's chief executive, raised another spectre. Kessler later remembered how, during tentative talks about the IBC between the Group, Jersey Standard, and IG Farben in August 1927, barely a week after the agreement between Jersey and Farben, Bosch suddenly switched to German for a tirade emphasizing his company's power to make products condemning other industries to extinction: first the traditional dye manufacturers, then the saltpetre fertilizer industry, and now the bell had tolled for the oil industry.[43]

This diatribe convinced Kessler that the Group urgently needed to return to hydrogenation, not in order to make synthetic gasoline, but as an entry into the petrochemical business. Jersey Standard already produced synthetic alcohol, insecticides, and a synthetic foam at one of its US refineries, though the difficulties encountered with obtaining alcohol of sufficient purity had rather dampened managerial enthusiasm for chemical manufacturing.[44] Chemical manufacturers in the US and in Germany were venturing into petrochemicals, buying oil fractions and natural gas to make solvents and a range of other products.[45] In Britain, Imperial Chemical Industries (ICI) had also begun probing the boundaries between chemicals and oil, buying control of the British Bergius Syndicate in 1927 and planning to build a coal hydrogenation installation at Billingham in County Durham.[46] IG Farben possessed both the financial and the scientific clout to become as dominant in petrochemicals as its constituent firms had been in carbochemicals, employing some of the best brains in the business active on a wide product spectrum.[47] Indeed, BASF had already started work on a process to upgrade gasoline by high-pressure steam reforming and, unknown to the Group, had markedly improved on Bergius by developing a two-stage process using tungsten sulphide as a catalyst.[48] Unless the Group acted immediately, it would find itself reduced to the position of a simple raw materials supplier. As an electrical engineer, Kessler sometimes protested that he really knew nothing about chemistry, yet he,

[30]

J.B.A. Kessler junior (left), 'founding father' of Shell's chemical enterprises, in conversation with Henri Deterding (centre). J.E.F. de Kok listens attentively.

more than anyone, was responsible for driving the Group's diversification into chemicals.[49] On 25 August 1927, he wrote to De Kok: 'I do not think it is right to say that we only want to go into this new chemical line if we cannot come to an agreement with the IG. It is not intended to build our industry into a complete chemical industry in order to compete with the IG, but it is because we feel that we must use to the fullest extent the very valuable gas that we have practically everywhere, and turn it into something that will give us the best results. That the resulting products will probably be those which are also made by the IG is possibly a little bit unfortunate for the IG, but it is certainly not intended to be a way to fight them. On the contrary, we want to co-operate with them if that proves to be possible, but before we can talk about co-operation we must first make the stuff ourselves. We have the enormous advantage of having our source of supply and energy everywhere in the world, and therefore we have an enormous advantage over the IG, because they have to send all their products from Germany, whereas we shall have the advantage of import duties, if there are any import duties on artificial manure.

As I told you over the telephone, I feel that we should look upon these new departures of ours in an entirely different light from the way in which we look upon building, say, another cracking unit or another lubricating oil bench. If the calculations show that, for the time being, a new chemical installation for making nitrogen [i.e. the binding of nitrogen from the air to hydrogen from natural gas or cracked gases] will not pay, I do not think that should be a reason not to build it. We should have the confidence, energy and courage enough to develop this new chemical part of our business, even if it does not give profits to start with. The Badische [i.e. BASF] have done the same thing and with very great success, and therefore, although we must have calculations, if we can, it seems logical that we should turn all this energy that is going to waste at present into something that we can put into packages and sell.

If we had a lot of gas somewhere in a very thickly populated area we might make electricity and sell it, but what can we do with the waste gas we have in the United States, the East Indies, Venezuela, Romania and so on? The only thing we can do is to make something which we can ship. (...) If we tackle the matter right from the beginning with very much enthusiasm, and fully convinced that it is absolutely necessary, in order to safeguard ourselves against a possible drop in prices of some of our oil products, to make whatever we can out of the products we produce, the larger the number of products we make, either in the form of oil or in some other form, say artificial manure, the safer we are against any drop in prices of some of these products.

As I told you over the telephone, I fully appreciate that this new departure is an extremely important one. It is not that we shall just make another product as we have made other products, for instance, by going into the white oil trade, but our business, I think, is going to be developed into a chemical business also, and, after ten or twenty years, I expect we shall be just as important in the chemical industry as we are in the oil industry (...).

Do not let us be frightened by some preliminary cost calculations which show that we may not make money in the beginning. I do not expect to make money; on the contrary, I expect that it will cost a good deal of money before we actually produce on a fairly large scale. The only thing we need is scientific knowledge and I do not see we should not be able to get that. We have got the money, the best raw materials and the best geographical position, and if we can work ourselves out of the routine to which we have become accustomed up to now in our rather simple oil business as we have been developing it hitherto, we shall certainly be in a position to develop our industry into a chemical industry on a very complete scale.'[50]

[31]

[32]

Chemicals from petroleum appeared to have boundless potential.
A marketing campaign in 1933 showed (far left) 'the grateful Max', happy to say goodbye to fleas, and (near left) a warning to all flying insects: 'Gasmasks on!' Below, around the same date, agricultural workers somewhere in England are seen spreading fertilizers by hand.

[33]

Six salient points stand out in this vision: a desire to turn waste into saleable products; the desirability of diversification into chemicals to counter price fluctuations in the oil business; the emphasis on substantial investments in chemicals requiring exemption from short-term profit considerations for reasons of business strategy; the urge for scientific knowledge; the expectation that, one day, chemicals would become as important as oil; and the conviction that oil is essentially a simple business.[51] These last two items, oil is a simple business and chemicals are the future, came to provide two more maxims in the Group's emerging business conception.

Kessler fully realized how different the chemical industry was, noting that, 'The chemical industry is, in contradiction to the oil industry, by nature revolutionary and the complete eclipse of products once profitable may occur at surprisingly short notice', but he thought that this specific property could be remedied by widening the product range.[52] Even so the contradiction of balancing a steadily evolving industry with one subject to sudden changes does not appear to have bothered Kessler or anyone else. Nobody seems to have fundamentally questioned Kessler's vision about the importance of chemicals for the Group's operations then or later, immediately after the Second World War, when he masterminded a further intensification of research efforts and a huge petrochemical expansion programme (see Volume 2).

Directors discussed budgets and raised objections to particular projects or items of expenditure proposed, but agreed on principles.

The reorientation towards chemicals manifested itself first of all in an intensification of Group research efforts. The Amsterdam laboratory drafted a new programme listing only one oil item amongst the top priorities, gasoline anti-knock dopes, against three chemical research fields: synthetic gasoline, the production of glycol from ethylene, and the study of hydrocarbons from natural and crack gases. In September 1927, a new Chemical Industry Department was set up under A. J. van Peski who, in close collaboration with the chemical engineering departments of Amsterdam and Delft universities, recruited a team of experts on high-pressure catalytic processes. As a consequence, the laboratory was considerably expanded and given new facilities, such as a pilot factory to test the production of chemicals on a semi-industrial scale, and a hydrogenation installation. From 1925 to 1929, the laboratory staff rose from 80 to about 300 people.[53]

Three considerations guided the initial chemical research aims, drawn up by Pyzel and implemented by Van Peski. Firstly, managers wanted to steer clear of hydrocarbons and processes protected by existing patents. Licences increased cost at a stage when the commercial prospects of a process remained unknown, so as a rule the Group avoided acquiring them until the benefits had become fully clear. In 1927, BPM set up a new Patent Department in The Hague to investigate the situation with regard to the components present in crack gases and natural gas, resulting in the decision to concentrate on propane, butane, propylene, and butylene, since ethane and ethylene, also present in the gases to be studied, were already ring-fenced with patents.[54] Secondly, the chemical composition of these hydrocarbons was known, but the technology to process them had not progressed beyond the laboratory stage. Starting to manufacture a new chemical, say acetone from isopropyl alcohol derived from propylene, on an industrial scale and to the high degree of purity required, meant inventing an entirely new range of factory installations. Lastly, there was the imperative for revenues within a reasonable span of time. For all his commitment to research as a driving force for the Group's business, Kessler never lost sight of the commercial dimension of such work, commenting after inspecting the Amsterdam laboratory's new buildings in 1928, that research proposals had to consider the likely impact on revenues and result in at least some earnings after about two years. Consequently, the priority given to propane, butane, propylene, and butylene derived partly from the intention to produce components such as glycol and other dopes for the Group's automotive products. Another early field of interest was pesticides. The first result, a flykiller named Shelltox, entered the market during the late 1920s.[55]

The Group also set up a new research organization in the US, for two reasons. It was first a matter of balance between fundamental and applied research. The Amsterdam laboratory operated in the European tradition of fundamental research, whereas Kessler, and presumably other managers too, fully realized the crucial importance of the innovations emanating from the applied research carried on in the US industry. That was why both the Group's chief technical experts, Pyzel for refining and McDuffie for drilling and production, were stationed in America.[56] The Group simply needed a fully-fledged research institute in the US to remain in close touch with developments there. The second reason was a practical one; the crack gases produced by the refineries and the natural gas from the fields had to be studied there, and not on the other side of the Atlantic. A site was chosen at Emeryville on the eastern coast of San Francisco Bay with sufficient space for the small pilot plants required. In 1927 the Simplex Refining Company,

As the Group started to learn how to work constructively with gases, new shapes appeared in its installations, such as spherical storage tanks, capable of resisting internal pressure equally in all directions. New shapes appeared in its formulae as well, such as the transformation of isobutylene to iso-octane.

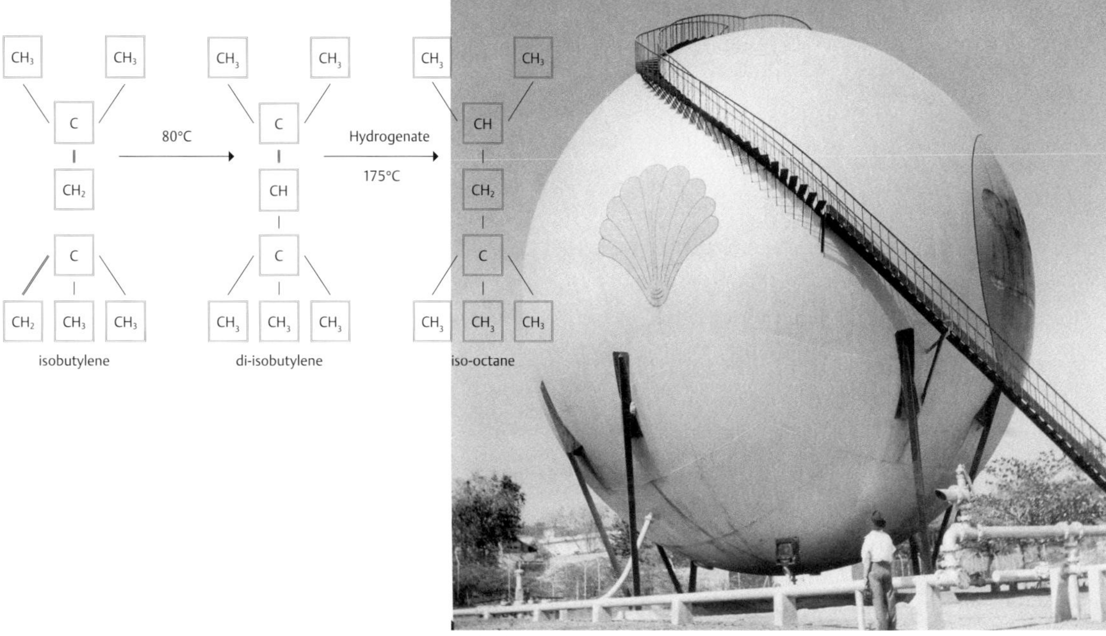

[34]

an engineering company owned by Asiatic Corporation and holding the Trumble patents, was given a new purpose as a corporation for research and patents management. Renamed Shell Development Company, Shell Union and Bataafsche became the joint owners to ensure close cooperation within the Group's research organization. A special contract, the so-called Simplex Agreement of 1929, laid down arrangements for the sharing of research costs, results, and patents.[57] Pyzel declined to shoulder the managerial burdens as president of the new company, preferring to devote his time to research as vice-president for technical matters. On the recommendation of Waley Cohen the Group appointed a British chemical engineer, Clifford Williams, to be director of research at Shell Development. Operations began in November 1928 with a staff of forty-two people. By 1938, the Group spent $ 5.9 million a year on research, rather more than Standard Jersey with 4.5 million. The Amsterdam laboratory and the Delft engine testing station spent 57 per cent of the total, Shell Development 43 per cent or $ 2.5 million, one per cent of Shell Union sales. Bataafsche's laboratory was still by far the biggest establishment, employing 1,355 people, against Emeryville's 490. For patent management, Shell Development worked in close collaboration with the The Hague patent department.[58]

Emeryville started investigating particular hydrocarbon fractions produced by the Dubbs crackers, the olefins butylene, ethylene, and propylene, at the time regarded as waste because little use was known for them. Manufacturing synthetic alcohol became the first research target. As a first step, the chemical engineers developed a two-stage process to separate isobutylene and the n-butylene adducts with sulphuric acid. The latter were then hydrolyzed into secondary butyl alcohol. A pilot plant for making this alcohol came on stream in 1930, enabling the Group to begin selling it in small quantities. In 1931 a pilot plant for dehydrogenating butyl alcohol into methyl ethyl ketone, a solvent, started operating. Since supplies from wood distillation had always remained small, this solvent found a keen market in a variety of industrial applications ranging from paints and lacquers via perfumes to printing ink and insecticides. Researchers then turned their attention to isobutylene and found a process for turning it into iso-octane, a compound with a momentous future, about which more below.[59]

Meanwhile Kessler's original wish to turn gases into 'something we can ship' had been fulfilled by setting up distribution networks for the sale of bottled butane. Marketing began during 1928 in the American Midwest using the product name Shellane; in 1929 the bottles were also introduced on the Pacific coast.[60] For reasons unknown, France was chosen for a trial in Europe and a dedicated sales company, the oddly named Société d'Utilization Rationelle des Gaz (URG), was formed. The installations were ready in October 1931 and the first shipment from Houston arrived five months later. Bottled gas became a spectacular success. The initial sales target was 6,000 tons a year; by 1939 the Group sold some 30,000 tons of butane a year in France alone. Supported by a dense sales network and imaginative marketing campaigns extolling the virtues of the new and easily portable source of heat and power, the Butagaz brand name had become generic for bottled gas, and the light blue bottles a familiar sight around the country.[61] Following the success in France, distribution networks for both products were set up everywhere.

After its introduction in France, Shell's portable Butagaz – 'Le gaz qui voyage' – was marketed heavily and successfully throughout Europe and (under the name 'Shellane') in the United States.

[36]

[35]

[37]

[38]

[39]

SHELLING THE LINE . . JANUARY 1, 1930

Shellane made hot "hot dogs" possible at this American Legion Auxiliary booth in Porterville, California

Rural districts like Shellane

From every section where Shellane has been introduced, we get enthusiastic and appreciative reports of this new Shell product. Originally it was intended for use chiefly in country dwellings—and hundreds of farm women are thanking Shell for this great convenience in their homes. As new uses continue to be discovered for the product, our market widens and energetic Shell employees are reaping valuable premiums on new business obtained.

One of the introductory stunts which has called Shellane forcibly to public attention was the installation of this new fuel for the operation of a "hot dog" stand by the American Legion Auxiliary in Porterville, California, during the Armistice Day celebration. A picture is shown here.

A new use for Shellane was demonstrated at the Fresno County Fair when this fuel was made to operate a dairy sterilizer. R. B. Schroeder of Fresno reports as follows:

"The installation cost of an electrically operated sterilizer was between $30.00 and $50.00. The installation cost of a Shellane operated sterilizer was between $2.80 and $4.20. The time required to raise a full load of steam for washing and sterilizing with electrically operated sterilizer was seventeen and one-half minutes minimum. The same test on Shellane in every respect required three and one-half minutes.

[12]

[40]

[41]

[42]

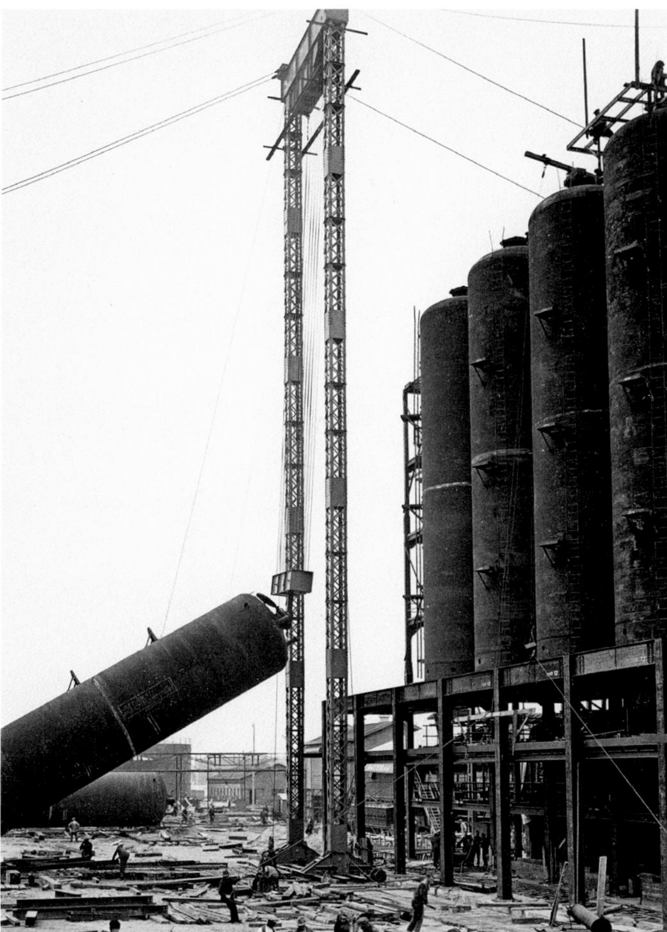

[43]

Preparing the site (main picture) for the Dubbs cracker at Pernis, the first in Europe. When the reaction chambers were delivered by sea (above left), their great weight caused the freighter to heel over considerably. Their enormous size is made clear (below left) by comparison with the human figures.

Exploring hydrogenation: fertilizers Both Emeryville and Amsterdam devoted a substantial part of their effort to supporting a new venture started by the Group, manufacturing synthetic ammonia by nitrogen fixation. The decision to take this direction originated in the need to acquire experience with high-pressure hydrogenation processes and technology. Nitrogen fixation, a famous milestone in chemical engineering pioneered by Fritz Haber and developed into an industrial process at BASF by Carl Bosch in 1913, used nitrogen from the air and reacted this with hydrogen into ammonia, which was then reacted with sulphuric acid to get ammonium sulphate. Pellets of ammonium sulphate found a ready market as fertilizer because of its nitrogen content.[62] Nitrogen fixation used very similar techniques to coal hydrogenation and thus offered the opportunity to build the experience required by making a commercial product, while at the same time carrying the competition with IG Farben to its own ground, nitrogen fertilizer being one of that company's core products. All that the Group needed was a hydrogen source, preferably close to the Amsterdam laboratory so as to facilitate the exchange of information between the installation and the research carried on there. The existence of large natural gas fields in the Netherlands remained as yet unknown. Nor did the Group then have offgases at its disposal, as the first Dubbs cracker at Pernis came on stream only in 1936. By talking to his brother Dolph, Kessler found the solution in the offgases from the coke ovens at the Hoogovens steel mill, situated on the Dutch coast at the entry of the canal linking Amsterdam with the North Sea.

As we have seen in Chapter 5, the two Kessler brothers had both worked for the Group. Both also possessed a dynamic drive very much reminiscent of their father though not, perhaps, to the same near manic degree which, considering his early death, was probably just as well for them. Even before the start of their careers, they shared a fervent ambition to fulfil their father's life's work, considering Royal Dutch to be almost a family firm, the right to lead it theirs by birth.[63] Deterding is said to have been ruffled by this attitude. Yet he did give the brothers their chance, and their regular promotion shows that he appreciated the Kesslers' superior abilities.[64] The brothers were also very strong willed but only Guus, the younger son, succeeded in controlling his emotions and avoiding coming into conflict with Deterding, in order to reach his ultimate goal.

After graduating as a technical engineer from Delft, Dolph Kessler became Deterding's personal assistant. After a tour around the world with Deterding, he was sent to work for Astra Romana in Moreni and Ploesti, becoming an assistant manager in 1910. He briefly conducted the company's commercial operations before his appointment to the board of Astra and a transfer to the Anglo-Saxon office in London, where he was entrusted with coordinating the Group's American, Mexican, and Russian businesses.[65] Though according to Dolph, Deterding had stated that he would not have two Kesslers in the Group, he took on Guus Kessler that same year, 1911, after his graduation from Delft. He worked first in Romania, then from 1912 at the St Petersburg office managing the Group's Russian companies, becoming the office manager two years later.[66]

[44]

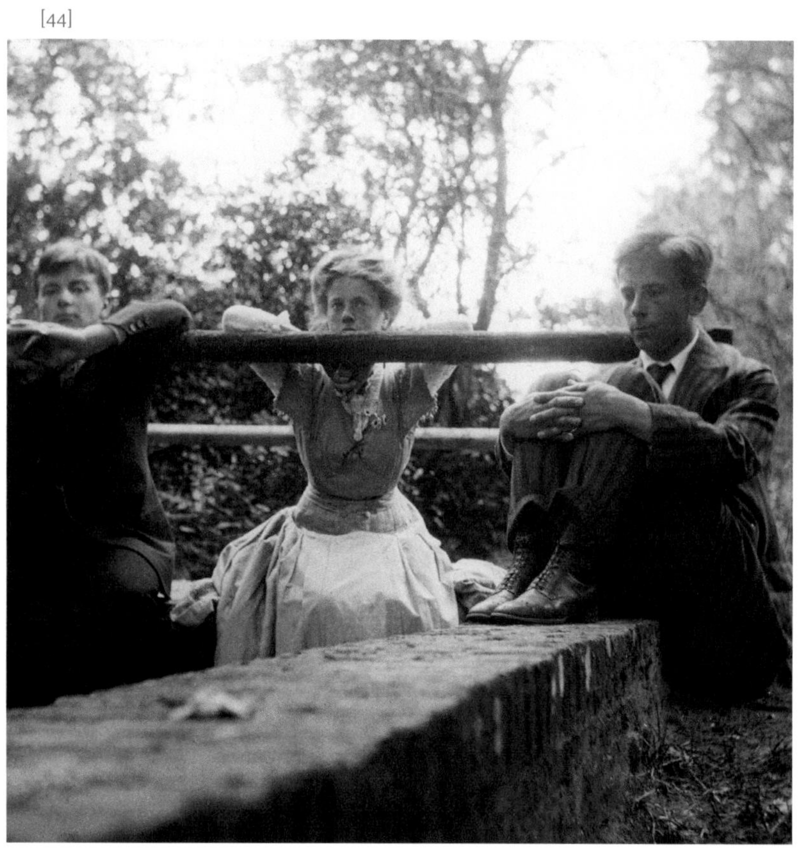

The Kessler brothers and their loves: Guus and Dolph Kessler wed girls who were cousins of one another. Left, photographed in 1910, a pensive Guus sits by his fiancée Anna Françoise (Mushe) Stoop, daughter of François Stoop. They married on 10 July 1911. Dolph and Elisabeth (Bep) Stoop are seen (right) on their wedding day, 14 April 1916, after he had chosen her in preference to a career in Royal Dutch.

Despite having a very difficult time in Russia, Guus built his career quietly, but Dolph chafed at the bit, constantly fretting about what he unjustly considered to be his slow progress up the managerial ladder. He lost heart when, in January 1914, he felt himself unfairly passed over in favour of Waley Cohen for an appointment to the board of Anglo-Saxon, even though he was seven years younger than Waley Cohen who, at thirty-seven, remained the youngest director by far. When in 1915 his fiancée Elisabeth Stoop, daughter of Dordtsche's founder Adriaan Stoop, put his love for her to the test by asking him to choose between her and the Group, Dolph Kessler resigned, returned to the Netherlands, and got married. He resisted Colijn's entreaties to reconsider and accept a managerial position elsewhere in the business, though he did take a directorship in a subsidiary operating company.[67]

[45]

That year Guus's career also reached a low ebb. He was recalled to the Netherlands and dismissed after a conflict between the Dutch and the Russian managers, but reinstated and appointed to a desk job in The Hague, at a sharply reduced salary, when he was proved innocent of the accusations levelled at him. After a short stint as acting manager of Astra Romana, Guus moved to London in 1920, succeeding Godber as head of the American Department and managed the remaining Russian affairs, overseeing the transfer of the Mazout installations to new Group companies in Poland and the Baltic Republics. With appointments to the central boards during 1922-4 he essentially became Cohen Stuart's successor as the second Dutchman in the London office, the position originally intended first for Colijn, then De Jonge. The contrast between Cohen Stuart and Guus Kessler could hardly have been greater: the former a suave, very able but finicky lawyer with an artistic bent, laid back to the point of indolence; the latter a brisk and forward-looking manager, totally committed to the business, modern in the sense of being science-oriented, unsettling fellow directors during meetings with quick calculations on his slide rule.[68]

His brother had meanwhile found, through Colijn, a job as secretary to the committee preparing to set up Hoogovens, which was formally constituted in 1918. Dolph Kessler became a managing director of the company in 1920 and its CEO four years later; his drive, entrepreneurship, imagination, and leadership secured the fledgling Hoogovens a firm foothold in a very competitive industry at a difficult time. Cokes ovens formed a central part of the works. These installations produced hydrocarbon offgases which Hoogovens found difficult to sell, and just the kind which the Group needed for the production of nitrogen fertilizer. Shortly after the April 1927 memo drafted by Kessler, De Kok, Pyzel, and Knoops, Bataafsche opened talks with Hoogovens to set up a jointventure to make synthetic ammonia.[69] To circumvent the Haber-Bosch patents owned by BASF, managers opted instead to use the rival Mont Cenis process of the German firm Gaveg. BPM acquired a preliminary licence in October 1927 and the world rights outside Germany six months later, intending to build several installations around the world: Romania, Indonesia, somewhere on the

Chapter 6

[46]

east coast of the US, Louisiana, California. Only the last location proved a viable proposition.[70]

The BPM board needed some persuading to accept the proposed joint venture with Hoogovens, but the arguments from Deterding, Kessler, and Waley Cohen won the day.[71] The company was constituted in December 1928 as the NV Maatschappij tot Exploitatie van Kooksovengassen or MEKOG. BPM supplied the capital of 1 million guilders plus a loan of 3 million, Hoogovens became the works manager. Underlining the experimental nature of the project, Mekog was to operate on the smallest scale considered commercially feasible, producing 10,000 tons of nitrogen a year.[72] The original plan envisaged two complete installations for producing ammonium sulphate side-by-side, one as a back-up to operate during breakdowns or maintenance but, in response to changing demand on the fertilizer market, Bataafsche decided during construction to add plants to process the ammonia into ammonium phosphate, and to make calcium nitrate, which in turn necessitated building a nitric acid plant. As a consequence building costs nearly doubled, from the initial estimate of 6 million guilders to 11.6 million. The MEKOG factory finally came on stream in September 1929.[73]

As MEKOG got underway, Bataafsche started preparations for erecting a similar plant, adapted to use crack gases and natural gas, in California, which appeared to offer the best commercial prospects for fertilizers. In November 1928 De Kok, Kessler, Pyzel, Knoops, and Herbert Gallagher, the president of Shell Develop-

The MEKOG factory in the 1930s.

ment, discussed the project in The Hague and decided on a factory with an annual production of 18,000 tons of nitrogen. Four months later Shell Development incorporated a subsidiary, Shell Chemical Company, to build and run the installations, and to sell chemicals, including those produced at the Emeryville pilot plants. Underlining the Group commitment to the new venture, Kessler took the chairmanship, De Kok became president, and as Vice-President Dan Pyzel provided a close link to Shell Development.[74]

A key feature of the Shell Point plant, the installation for producing hydrogen from natural gas, became a family affair, Dan Pyzel teaming up with his sons Fred, chief engineer of Shell Development, and Ewald, plant superintendent at Shell Point, as the site on San Francisco Bay came to be called.[75] No standard process existed. At the October 1928 meeting, managers had discussed a catalytic reforming process, which first removed the sulphur from the gas and then passed it, at high temperature but ambient pressure, with steam over a catalyst to form hydrogen and carbon monoxide. Initially rejected as too expensive for making hydrogen, the process was adopted when Shell Point had survived its teething troubles. As an alternative, the Pyzels adapted a process used by Californian utility companies for lowering the calorific value of natural gas to match the value of coal gas. With equipment and technicians borrowed from the Los Angeles-based Southern California Gas Company, the Pyzels successfully developed a pilot plant for cracking natural gas at high-temperatures into hydrogen and carbon. This was a very innovative solution, so much so that the German manufacturers Linde and Gaveg did not want to accept responsibility for the equipment supplied by them for the commercial scale installation. Shell Point started ammonia production in August 1931, just when the economic crisis caused a steep fall in fertilizer prices.[76]

Getting hold of hydrogenation

While Bataafsche wrestled with these new technologies, the Group re-entered talks with Jersey Standard and IG Farben about the hydrogenation patents. The 1927 agreement between Jersey and Farben had proved an insufficient basis for collaboration, since it left the partners uncertain about each other's intentions and will to share the fruits of further research. Jersey researchers had become very enthusiastic about the prospects of hydrogenating oil for a wide range of applications, increasing gasoline yields, improving the quality of lube oils, and for removing contaminants from oil products. After further negotiations in 1928-29, Jersey bought the world rights to IG Farben's hydrogenation patents for $ 35 million, Germany of course excepted, promising not to use them for manufacturing chemical products and to give Farben control of any discovery in chemicals not directly related to oil. The companies agreed to bring borderline cases between oil and chemicals into a joint company, Jasco, Farben to allow Jersey a minority share in new processes for making chemicals from oil or natural gas.[77]

Following its acquisition of the IG Farben processes, Jersey Standard started to recover some of its investment by setting up the Hydro Patents Company to license the patents in the US, selling shares to a total of seventeen other oil companies. As partners for holding the world rights, Jersey approached the Group and ICI, the latter now also deeply committed to coal hydrogenation. Jersey did this no doubt primarily to increase its revenues from the Farben deal, but perhaps also because, as Deterding put it, the company did not want to be left alone in the woods with such a big and devious wolf as IG Farben.[78] The Group paid $ 10.5 million for its share. This was a substantial sum when considering that in 1925 BPM had received only about $ 360,000 from BASF for a 60 per cent stake in the Bergin company, but not much given the huge prospects which catalytic hydrogenation appeared to offer in manufacturing oil products. The deal enabled Bataafsche to buy a complete pilot factory for coal hydrogenation from IG Farben.[79]

Assuming that ICI and the US companies paid similar sums in relation to their interest, we may take it that Jersey Standard ended up by having the patents free. The world rights outside Germany

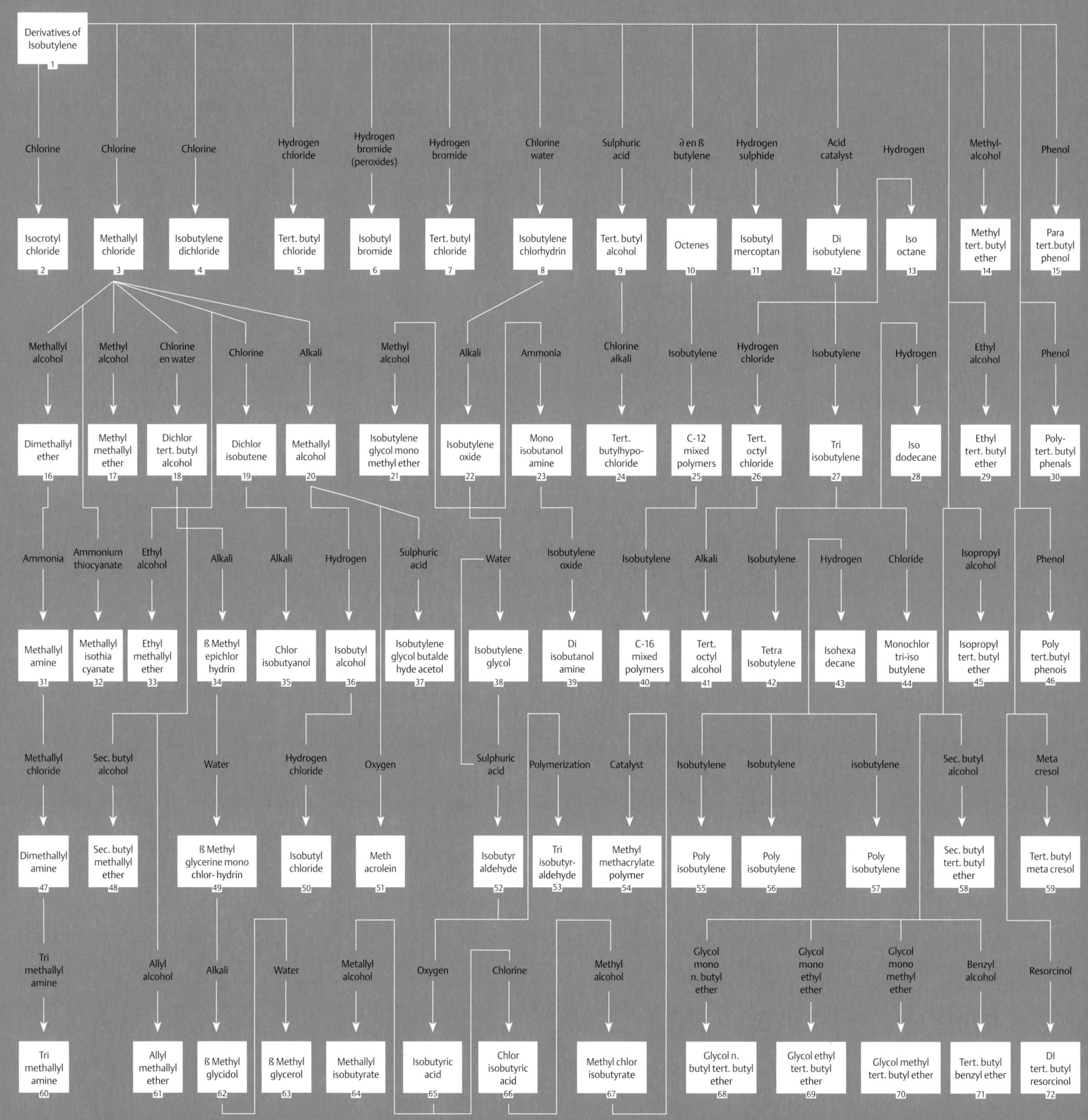

Derivatives of Isobutylene 1
Chlorine
Chlorine
Chlorine
Hydrogen chloride
Hydrogen bromide (peroxides)
Hydrogen bromide
Chlorine water
Sulphuric acid
∂ en ß butylene
Hydrogen sulphide
Acid catalyst
Hydrogen
Methyl-alcohol
Phenol
Isocrotyl chloride 2
Methallyl chloride 3
Isobutylene dichloride 4
Tert. butyl chloride 5
Isobutyl bromide 6
Tert. butyl chloride 7
Isobutylene chlorhydrin 8
Tert. butyl alcohol 9
Octenes 10
Isobutyl mercoptan 11
Di isobutylene 12
Iso octane 13
Methyl tert. butyl ether 14
Para tert.butyl phenol 15
Methallyl alcohol
Methyl alcohol
Chlorine en water
Chlorine
Alkali
Methyl alcohol
Alkali
Ammonia
Chlorine alkali
Isobutylene
Hydrogen chloride
Isobutylene
Hydrogen
Ethyl alcohol
Phenol
Dimethallyl ether 16
Methyl methallyl ether 17
Dichlor tert. butyl alcohol 18
Dichlor isobutene 19
Methallyl alcohol 20
Isobutylene glycol mono methyl ether 21
Isobutylene oxide 22
Mono isobutanol amine 23
Tert. butylhypo-chloride 24
C-12 mixed polymers 25
Tert. octyl chloride 26
Tri isobutylene 27
Iso dodecane 28
Ethyl tert. butyl ether 29
Poly-tert. butyl phenals 30
Ammonia
Ammonium thiocyanate
Ethyl alcohol
Alkali
Alkali
Hydrogen
Sulphuric acid
Water
Isobutylene oxide
Isobutylene
Alkali
Isobutylene
Hydrogen
Chloride
Isopropyl alcohol
Phenol
Methallyl amine 31
Methallyl isothia cyanate 32
Ethyl methallyl ether 33
ß Methyl epichlor hydrin 34
Chlor isobutyanol 35
Isobutyl alcohol 36
Isobutylene glycol butalde hyde acetol 37
Isobutylene glycol 38
Di isobutanol amine 39
C-16 mixed polymers 40
Tert. octyl alcohol 41
Tetra Isobutylene 42
Isohexa decane 43
Monochlor tri-iso butylene 44
Isopropyl tert. butyl ether 45
Poly tert.butyl phenois 46
Methallyl chloride
Sec. butyl alcohol
Water
Hydrogen chloride
Oxygen
Sulphuric acid
Polymerization
Catalyst
Isobutylene
Isobutylene
isobutylene
Sec. butyl alcohol
Meta cresol
Dimethallyl amine 47
Sec. butyl methallyl ether 48
ß Methyl glycerine mono chlor- hydrin 49
Isobutyl chloride 50
Meth acrolein 51
Isobutyr aldehyde 52
Tri isobutyr-aldehyde 53
Methyl methacrylate polymer 54
Poly isobutylene 55
Poly isobutylene 56
Poly isobutylene 57
Sec. butyl tert. butyl ether 58
Tert. butyl meta cresol 59
Tri methallyl amine
Allyl alcohol
Alkali
Water
Metallyl alcohol
Oxygen
Chlorine
Methyl alcohol
Glycol mono n. butyl ether
Glycol mono ethyl ether
Glycol mono methyl ether
Benzyl alcohol
Resorcinol
Tri methallyl amine 60
Allyl methallyl ether 61
ß Methyl glycidol 62
ß Methyl glycerol 63
Metallyl isobutyrate 64
Isobutyric acid 65
Chlor isobutyric acid 66
Methyl chlor isobutyrate 67
Glycol n. butyl tert. butyl ether 68
Glycol ethyl tert. butyl ether 69
Glycol methyl tert. butyl ether 70
Tert. butyl benzyl ether 71
DI tert. butyl resorcinol 72

and the US were vested in a Liechtenstein-based company, International Hydrogenation Patents (IHP), Jersey Standard and the Group becoming joint owners. As party to the agreement, ICI agreed to buy any oil it needed from Jersey or the Group, and sell any oil it produced to them. Subsequently the four participants also founded the International Hydrogenation Engineering and Chemical Company (IHECC) which, managed by BPM, provided design and engineering services for building hydrogenation installations, and organized the exchange of research results between the partners.[80]

This quadripartite arrangement between the two oil and two chemical companies did not aim to effect a containment of hydrogenation technology, as is sometimes asserted.[81] First of all, most of the major processing innovations during the first decades of the twentieth century were based on progress in the understanding of the chemical structure of oil and oil products. Such knowledge was impossible to patent and would always lead to a variety of technical options which companies could work out for themselves once they had got the basic idea. Consequently, in patent matters concerning such processes the oil companies tended to operate as a club with entry fees rather than as a closed shop. Licensing made sense. Sitting on rights bound sooner or later to become outdated by new technology could cost a fortune in defensive litigation, whereas licensing generated revenues with which to recoup some of the investment. A memo on Bataafsche's patent position in the extraction field neatly summed up Group licensing policy dilemmas: 'The Group also possesses several patents covering equipment improvements in the extraction field. Developments in this field are closely watched so as to negotiate these new inventions whenever this is deemed of interest. In view of the rapidly changing conditions in the industry and the recent intense activity in the extraction field, which may tend to saturate the market for new extraction processes, it seems advisable not to wait too long to make use of our new acquisitions by the exchange of rights or the sale of licences, so as not to miss the most favourable opportunity. In the licensing of our extraction processes conflicting interests had to be taken into account. Such licensing, on the one hand, gives competitors the opportunity of using our improvements, which we may wish to avoid. On the other hand, valuable rights and royalties may be derived therefrom. Our licensing policy in the extraction field has been a compromise of these contradictory considerations, resulting in our adopting a waiting attitude and granting licenses only when interested parties approached us.'[82]

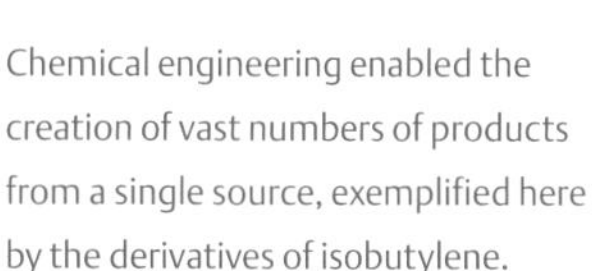
Chemical engineering enabled the creation of vast numbers of products from a single source, exemplified here by the derivatives of isobutylene.

Companies thus obtained access to the general process whilst striving to keep their competitive edge by reserving any technical improvements on it for themselves if possible, for as often as not the patent sharing agreements stipulated the sharing of subsequent improvements as well. With chemicals it was a different matter, since the steps required to turn laboratory experiments into industrial processes could often be successfully patented there.

The Group had managed Simplex and its Trumble patents as a club. Subsequently, in August 1930 Shell Union, together with six leading American oil companies, took the initiative for a patent pool covering cracking processes such as Dubbs and Burton, ending the tortuous litigation surrounding them. As a result, Shell Union ended up by holding 60 per cent of the shares in Universal Oil Products (UOP), the company originally set up to manage the Dubbs patents and now reorganized to hold all main cracking patents. The Group paid $ 10 million for its share, which yielded an annual saving of 3.4 million in royalty payments. Licensing fees were kept low by the fact that the Kellogg engineering company remained outside the UOP fold and offered its own cracking process at keen prices. The dangers of barring access to new technology by high fees were exemplified during the second half of the 1930s. When Jersey Standard found the rights to the new Houdry catalytic cracking process too expensive, the company started research into alternative solutions. Other interested oil companies, including the Group, soon joined this project, which succeeded in developing a superior fluid bed catalytic cracking process rendering the Houdry patents obsolete.[83] The IHP company considered its own patent position similarly exposed to potential technical alternatives.[84]

In keeping with the oil companies' general attitude towards processing patents, Jersey made no attempt at containment of hydrogenation in the US, granting licences to producers liberally in what was at that moment in time the world's biggest producer of oil and gasoline. Elsewhere, containment was neither feasible nor necessary. The German chemical firm Ruhrchemie AG possessed rival technology in the Fischer-Tropsch process, which converts the synthesis gas from coal gasification into pure liquid hydrocarbons in a high-pressure catalytic process. The oil companies would have had to acquire the Fischer-Tropsch patents as well if they really strove for containment. However, coal hydrogenation remained a white elephant living in a unique habitat beyond which it could not exist. For all the research and investment lavished on it by IG Farben, in peace time synthetic gasoline from coal was a doubtful commercial proposition inside Germany and really a non-starter outside it, even before world gasoline prices started falling at the end of the 1920s.[85] Synthetic gasoline owed its precarious existence to the determination of German captains of industry such as Bosch which could mobilize the research and the capital required, and to the support of the tax breaks which the German government created in a bid to reduce the dependency on imports. After six years of arguing, ICI obtained similar protection by import duty differentiation, the British government treating the project as a job creation scheme. The Billingham plant finally started production in 1935, but its volume remained small.[86]

The IHP once did refuse a licence. In December 1939 the company instructed IG Farben to tell the Japanese officials who had come to buy know-how and equipment that there could be no question of a licence, probably after American pressure to prevent the technology for processing gas oil to aviation gasoline from reaching Germany's Axis partner.[87] If the participants in IHP dragged their feet to construct hydrogenation plants in Germany, Italy, and France, they were driven less by a desire to restrict access to the technology than by their reluctance to support an uneconomical production process for gasoline. The true reason for the patent club was the shared access to the existing and future results of highly complicated and expensive research on the interface between oil and chemicals, from which they expected to obtain process improvements, and no more. In this approach the IHP participants were quite successful; during the 1930s successful applications of hydrogenating oil emerged, as we will see in due course.

The oil companies did contain sales of synthetic gasoline in Germany, however. In 1928 gasoline distribution there became a tight price cartel, led by Rhenania-Ossag and the Deutsch-Amerikanische Petroleum Gesellschaft (DAPG) as the two largest suppliers. On top of that each had a 25 per cent stake in another distributor, Deutsche Gasolin AG, IG Farben owning the rest. Deutsche Gasolin drew supplies from its owners according to strict quotas, giving Farben's synthetic gasoline a carefully controlled market. In 1931 the German government recognized the cartel in return for a quota agreement which gave the small local crude producers a chance to market their oil.[88]

Discovering the interface between oil and chemicals

The Group's sustained effort to master hydrogenation was based on the conviction that this technique held the future in manufacturing, which over time developed into another one of the core business maxims. However, hydrogenation initially found remarkably little direct application. Throughout the 1930s, Trumble and Dubbs plants remained at the core of manufacturing, supplemented from the early 1930s by True Vapour Phase (TVP) and Gyro crackers, each suited to processing particular types of oil. Jersey Standard developed a process to hydrogenate lube oils, resulting in a marketing offensive extolling the virtues of so-called hydrolubes. The Group started experimenting with them in 1930 and concluded four years later that the quality gain was not worth the extra cost compared with Syntholubes. Jersey had meanwhile come to the same conclusion.[89] A process for upgrading gasoline to a higher octane number by dehydrogenation was also considered, but rejected in favour of thermal reforming or gasoline cracking, developed at the East Chicago refinery in collaboration with UOP, a much more economical method using a modified Dubbs. The first installation came on stream at Curaçao in August 1933 and proved so successful that a second one was ordered the following month. It took some experimenting before reforming gasoline to produce 'reformate' had become a fully reliable process but, two years after Curaçao, reforming had also been introduced on Aruba, at the main Indonesian installations, and in Mexico, with Suez following in 1938 and Ploesti either in 1938 or in 1939. Reforming increased the volume of valuable offgases produced by these refineries and thus provided another stimulus to the chemical side of the business.[90]

The experience gained with hydrogenation acquired great importance, however, when the search for high-octane components took a new turn. Since 1929, Jersey Standard and the Ethyl Gasoline Corporation made synthetic iso-octane on a small scale, largely for lab-testing purposes. The developing octane race put a premium on producing iso-octane commercially for demanding customers such as the US Army Air Corps, which in 1930 raised its specification for aviation gasoline to 87 octane. Technically there

were two ways for making synthetic iso-octane. The first one, called polymerization, consisted of two steps. Heating a mixture of isobutylene (C_4H_8) and sulphuric acid turned the isobutylene molecules into di-isobutylene (C_8H_{16}), a compound which could be hydrogenated to obtain two more hydrogen molecules and become C_8H_{18} or iso-octane. Alternatively, butylene (C_4H_8) could theoretically be made to combine with butane (C_4H_{10}) molecules into iso-octane in a process called alkylation, again with sulphuric acid as a catalyst. The Amsterdam laboratory had already investigated alkylation processes during the 1920s. Emeryville conducted experiments as well and in 1932 Shell Oil and UOP began building a pilot plant at East Chicago. Alkylation offered better commercial prospects, since the process did not need hydrogenation, an important cost factor, and could be operated on a bigger scale than polymerization since butanes were in much greater supply than isobutylenes. However, making a stable saturated hydrocarbon like butane react in the desired way posed a formidable challenge in chemical engineering.

During the 1920s and 1930s the port at Curaçao developed enormously, as the great increase in shipping demanded larger docks and yards. The Group's Trumble installation (left) grew to include twelve units. Nicknamed Trumble Street, it was the largest such installation in the world at the time.

Consequently, polymerization yielded the first commercial results. Working closely with Van Peski's team at the Amsterdam laboratory, Shell Development constructed an installation at Martinez to produce di-isobutylenes, which were then hydrogenated over a nickel catalyst at Shell Point into iso-octane. In April 1934, Shell Chemical delivered the first truckload of iso-octane to the Army Air Corps, twelve months ahead of Jersey Standard. With commercial polymerization the price of iso-octane dropped from $ 20 to about 70 cents a gallon. Shell Development subsequently doubled the process efficiency by applying more heat to the acid-isobutylene mixture, resulting in a process termed copolymerization. Some minor problems remained to be solved, such as the yellow colour of the gasoline blended with polymerized iso-octane but, with the threat of war pushing up demand for aviation gasoline, Group managers immediately started planning polymerization plants at all four main refineries in the US as well as in Mexico, Curaçao, Romania, Stanlow, and on the new site at Pernis near Rotterdam. Within eighteen months of operations the Pernis plant had repaid the 4.4 million guilder investment, about £ 350,000, from the sale of octanes alone.[91] Negotiations with the French Government for a polymerization plant there, a planned joint venture with Jersey, had just been finished at the outbreak of war in September 1939.[92] Iso-octanes added another link to the gasoline supply chain. In May 1936, the Aviation Department wrote to Bataafsche that, 'pending further advice from America, our Shell Ethyl Aviation Gasoline 100 octane will have the following composition: 45% Iso-octane, 40% Shell Aviation Gasoline "C", 15% Californian Casinghead Gasoline, 3.6% ccs. T.E.L./Imperial gallon'.[93]

By the time the Group's first polymerization plants came on stream, the alkylation experiments had just begun to yield results. In the spring of 1936, the Amsterdam laboratory discovered a catalytic process using cold sulphuric acid to react isobutane and butylene into alkylate. The the Hague patent department immediately applied for a patent with the Dutch authorities, beating a very similar application from Anglo-Iranian submitted to the British patent office by thirteen days.[94] As so often, the main oil companies had all been working on very similar processes, with the usual result of tangled patent applications. It took another three years of careful negotiations to form a patent pool for alkylation, which came into effect in April 1939.

Without its earlier commitment to hydrogenation, the Group could not have developed gasoline polymerization and alkylation. Chemicals research had meanwhile moved on. In 1935, Shell Development discovered an efficient process for making butadiene, the key ingredient for synthetic rubber. Three years later Emeryville had a pilot plant on stream. The acid oils produced in refining were shown to contain phenols, a raw material for synthetic resins such as Bakelite. The so-called 'slack wax', a by-product of paraffin cracking, provided a good feedstock for lube oil manufacturing, but further investigations had revealed a range of alternative uses: alcohols, ketones, and ester salts, an ingredient for detergents. After obtaining a licence from IG Farben, which held a basic patent, the Group decided to start producing ester salts, under the trade name Teepol, from the alcohols and ketones produced at Stanlow and at Pernis. The Stanlow plant came on stream in 1942; construction of the Pernis installation was interrupted when the Germans invaded the Netherlands in May 1940.[95] Thus, by 1939, the Group was firmly set on a course of diversification into chemicals, producing pesticides and butadiene, building plants for detergents, and exploring the opportunities in resins. Compared to what would come later, it was still on a modest scale, but the intention remained clear enough.

[49]

The Teepol pilot plant in the laboratory at Amsterdam-Noord, 1939.

When a circus visited Malaysia, a member of Shell Staff saw the attendants washing the elephants in the river and gave them a sample of Teepol to try. It was so successful that the circus purchased a further trial quantity to use on all their animals.

[50]

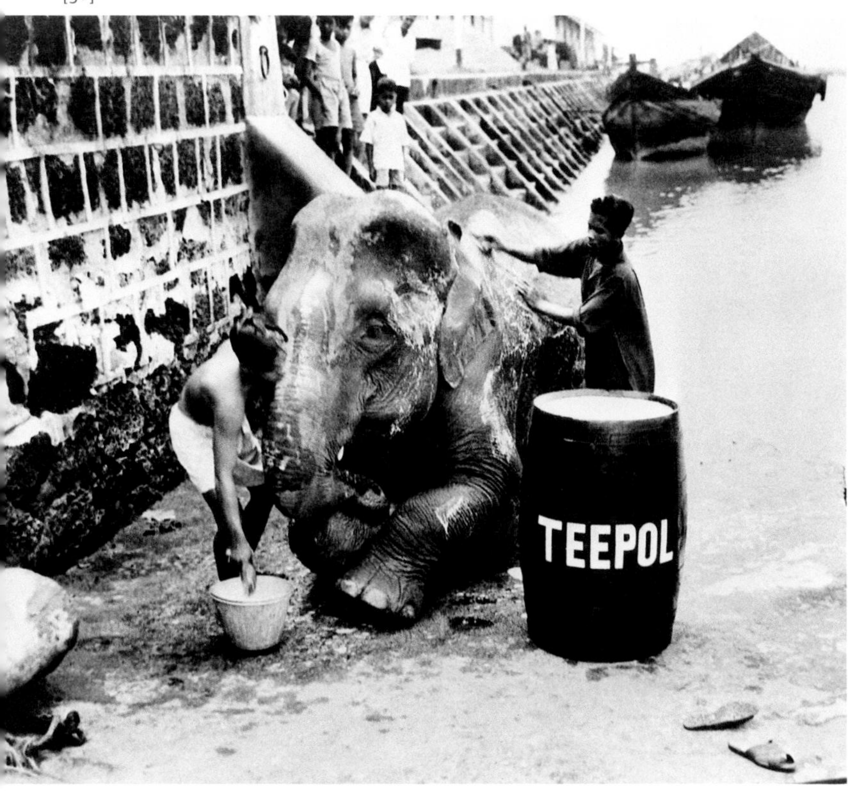

Branding and marketing: the reverse takeover The elder Kessler had realized, early on, the crucial importance of brand building and management. He carefully manoeuvred to position Crown kerosene as a premium brand amongst the other brands on the Asian market during the 1890s, and deliberately opted to sell the Russian kerosene sold during the Eastern Alliance years under a different label (see Chapter 1). Over time, product differentiation by distinctive branding grew into another one of the Group's business maxims, based on the conviction that commodities kill the trade. Shell Transport's red tins gave that company's kerosene an equally distinctive identity, but neither the name nor the pecten appears to have been used to promote products until 1904, when the board decided to start selling Shell Spirit gasoline on the British market via the newly set up General Petroleum Company. The red gasoline tins carried a Shell pecten so as to distinguish it clearly from the Pratt's Motor Spirit gasoline sold in blue tins by Jersey Standard.[96] Brand management was a key reason why Deterding, following the formation of Asiatic in 1902, immediately started replacing merchant firms as sales agents by an organization dedicated to selling oil products, thus building firmer links between the business and the consumer.

Branding took a further step forward in 1907 when the board of Asiatic decided to register the Shell name as a trade mark 'in all countries of Europe and the East', and use it as the main brand for gasoline.[97] During the next couple of years, both the name and the pecten were registered in all prospective markets, beginning with France and Italy.[98] This step could not be taken everywhere, however. The London-based international trading firm Ralli Brothers, a commodity trader mainly dealing in textiles, grain, and rice, also used a shell brand, though it is unknown for what kind of goods. To forestall confusion, Sir Marcus had come to an agreement with them which gave Shell Transport the use of the brand in the UK, leaving the rest of the world to Ralli; as a result, the Group was unable to register its trademark in Cuba, Argentina, Venezuela, and

[51]

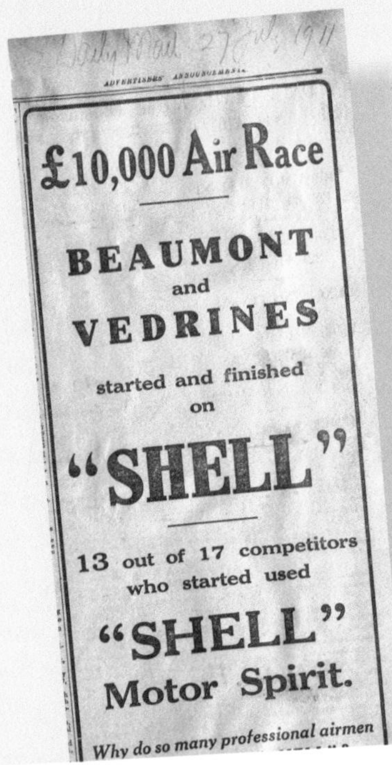

Daily Mail 27 July 1911
ADVERTISERS' ANNOUNCEMENTS
£10,000 Air Race
BEAUMONT
and
VEDRINES
started and finished
on
"SHELL"
13 out of 17 competitors
who started used
"SHELL"
Motor Spirit.
Why do so many professional airmen

[52]

BLOTTER
A SOUVENIR OF THE FIRST AERIAL MAIL IN AUSTRALIA
Between Melbourne & Sydney
CARRIED BY
M. GUILLAUX
on JULY 16-18, 1914.
Constituting A WORLD'S DISTANCE RECORD FOR ANY AERIAL POST.
"SHELL" BENZINE WAS USED IN THE TRANSPORT OF THIS MAIL.
THE FIRST AERIAL POST IN GREAT BRITAIN WAS ALSO CARRIED ON "SHELL".
PHOTO BY DARGE.
THE BRITISH IMPERIAL OIL CO. Ltd.

Shell quickly exploited the excitement of flight. Anticlockwise from left, Jean Conneau ('Beaumont') was the greatest racing flyer before the First World War, and won the *Daily Mail* round-Britain race of July 1911. Jules Vedrines, his main rival, came second. In 1914, on the eve of the outbreak of the First World War, this charming souvenir commemorated the first airmail in Australia. The war gave a huge impetus to flying technology, and Shell products were used in flights from London to Surabaya in 1919, to Cyprus in 1926, to Australia in 1928 and to Tehran in 1932. As Bill Lancaster's passenger in 1928, Jessie 'Chubby' Miller (1910-72) was the first woman to fly from Britain to Australia and later became a successful air racer herself. The Group's products also scored an aerial first in September 1930, when Errol Boyd (main picture) completed the first successful transatlantic flight outside the summer season.

[53]

[54]

SHELL
PETROL
& OIL
USED
EXCLUSIVELY
Paris-Téhéran
en 29 heures
avec l'huile Aéro-Shell

[55]

[56]

[57]

Chapter 6

At one end of the scale, Group advertisements of its aviation spirit ranged from the exuberant lorry poster of 1920 (above) to the Atlantic-class airliner design of 1932 (above right), with the unmistakable message that reliable lubricants were vital in keeping such a massive machine airborne. At the other end of the scale, grateful telegrams from famous users (right) provided valuable endorsements.

[59]

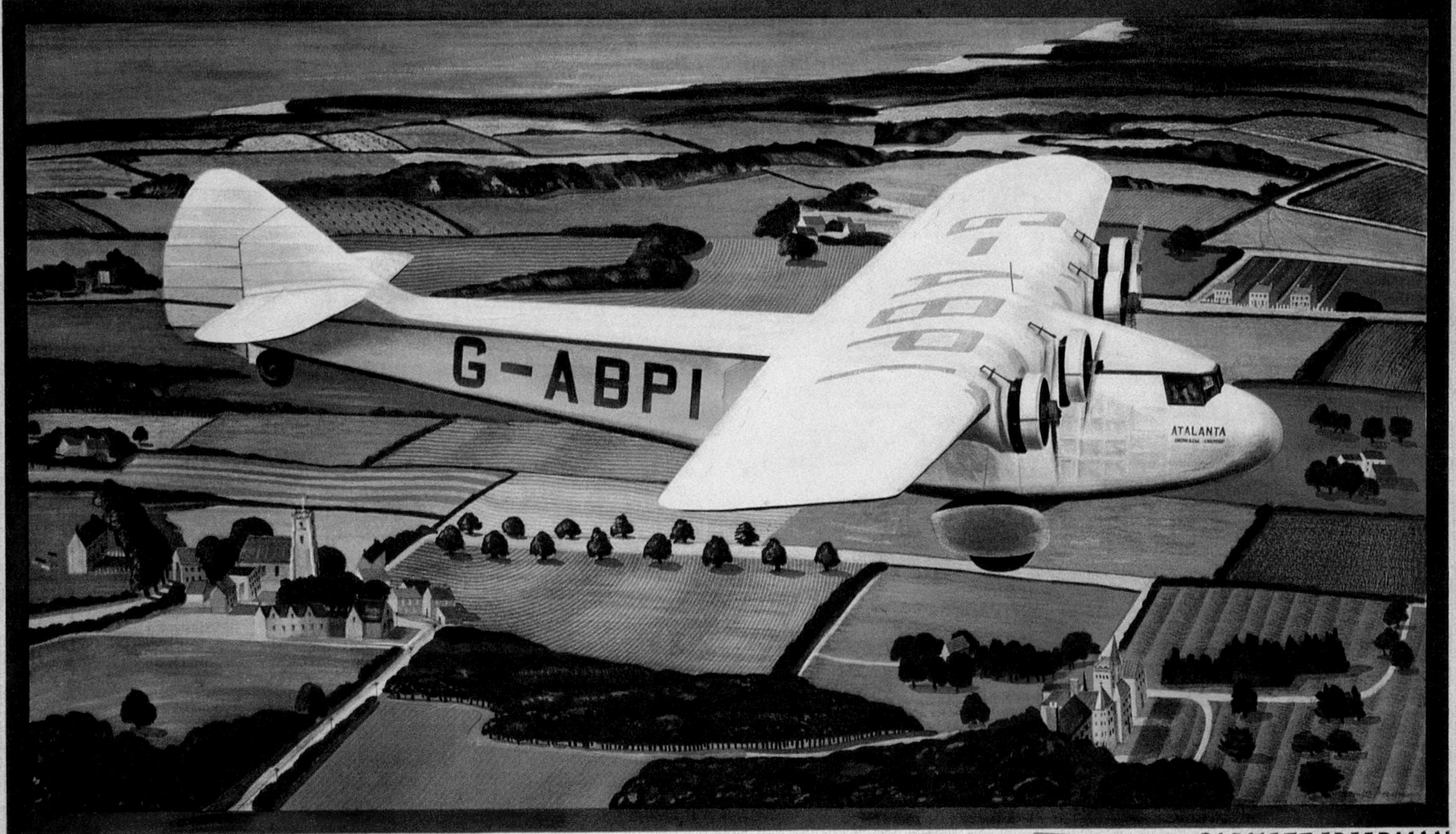

THE EASTERN TELEGRAPH COMPANY, LIMITED.

RECEIVED TELEGRAM

PLEASE SEND ANY REPLY TO THIS TELEGRAM *Via Eastern*

See back of form for list of Company's Stations in Great Britain.
Doubtful words should be OFFICIALLY repeated. See Rule Book.
No inquiry respecting this Telegram can be attended to without production of this Copy.

CIRCUIT	Clerk's Name.	Time Received.	
	B	12 5	11AUG1924

The first line of this Telegram contains the following particulars in the order named: Number of Message, Office of Origin, Number of Words, Date, Time handed in (in Letters: see Clock above), and Official Instructions.

SD 51 CALCUTTA 34 11 MY -

SHELL MEX SHELL CORNER KINGSWAY LN -

MY THANKS TO SHELL FOR FINE ORGANISATION OF FUEL AN
OIL SUPPLIES STOP QUALITY AND PURITY OF SHELL AVIAT
SPIRIT AND SUPER SHELL OIL UNIFORMLY EXCELLENT

MAJOR ZANNI .

[60]

SA

POST OFFICE TELEGRAPHS

Telephone: Central 3261

Date Stamp 29/9

IMPERIAL WIRELESS SERVICES

Via Empiradio

TO

AUSTRALIA, CANADA, INDIA, SOUTH AFRICA, ETC.

From VNB HALLGW

This form must accompany any enquiry respecting this telegram

The first line of this telegram contains the following particulars in the order named:— No. of Message: Office of Origin: No. of Words: Date and Time handed in: Service Instructions.

R1312 CAPETOWN 19 29 1526

KEROPAT LONDON

LONDON CAPE STAR MOTH FLIGHT SUCCESSFULLY
COMPLETED DUE TO EXCELLENT SHELL MOTOR SPIRIT
AND SHELL ORGANIZATION

BENTLEY +

23871

Please mark replies *"Via Empiradio"* and hand in at a Post Office. Ask for official repetition of doubtful words.
Books of forms, tariff lists etc., can be obtained from Post Office of delivery.

Brazil.[99] Similarly, the exclusivity of the brand had to be recovered from the Europäische Petroleum Union and its UK marketing company British Petroleum (BP), which used Shell for lube oil marketing. Any disputes over the trade mark were eliminated by the merger of General Petroleum and BP in 1907.[100]

Shell Transport's entry into the UK gasoline market in 1904 sparked an immediate price war with Standard Oil, which had pioneered European gasoline marketing.[101] Initially Shell imitated Standard's marketing methods such as the sponsoring of events – all drivers in the spectacular Gordon Bennett race of 1904 used Pratt's Motor Spirit – and the distribution of practical gifts to motorists, such as road maps.[102] General Petroleum set up a retail network, which at this stage largely consisted of garages and general stores, and placed advertisements for its products, mostly in magazines for motorists.[103] Brand reputation was further enhanced by sponsorships. The famous Paris–Peking rally of 1907 was won by Prince Scipione Borghese using Shell Spirit in his Itala motor car. Once firmly established as a trade mark, Shell became the brand for all new products issued by the Group, and the pecten the increasingly famous logo for distinguishing it. Consequently, when the Group entered the market for aviation fuel, in 1911-12, Shell Aviation Spirit became its top brand. To distinguish it from Shell Motor Spirit, the product was sold in gold cans, instead of red ones.[104] Sponsorships of races and special flights again helped to promote brand reputation around the world. Louis Blériot flew across the Channel in 1909 on Shell gasoline; the winner in the first race for the famous Schneider Trophy in 1913 was fuelled by Shell; the first flight carrying mail from Melbourne to Sydney in 1914 used it as well.[105] When the First World War broke out, the Shell cans had become so wellknown as a standard for reliable high-performance aircraft fuel that Royal Flying Corps pilots reportedly insisted on getting these rather than the standard issue khaki coloured ones, though the contents were the same, Shell Aviation Spirit.[106] As we will see, the Group's commitment to aviation would only increase during the interwar period.

[61]

Although the 'Shell' trademark was becoming famous and successful, within Royal Dutch there was some resistance to using it everywhere, and its 'Autoline' and 'Acetylena' brands continued for several years in the company's traditional outlets.

[62]

Chapter 6

[63]

Filling up with Shell oil did not have to be a problem, even for one of the white-gloved stylish 'bright young things' of the 1920s.

[64]

[65]

Whether in aviation or motoring, Shell products were marketed as 'the leading line' on lorry posters, with endorsement from drivers winning races at Reims (left) and Madrid (below).

[66]

THE LEAD

[67]

WINDSOR HOTEL & VOYAGEURS
SHELL
THOS. COOK & SON Ltd
WINDSOR HOTEL
DRAPERIE ELIAS HAKIM
DRAPERIE MODERNE
ELIAS HAKIM

[68]

Filling up with petrol in 1920s Cairo before starting for Alexandria. Choose Shell, of course.

The rise of the Shell brand pushed the Group's other trade marks increasingly to the background. Crown Oil, Royal Dutch's top kerosene brand in Asia, was demoted to the cheapest grade of gasoline as was an array of secondary products such as vaseline, candles, and turpene.[107] The UK marketing organization spearheaded this change, gradually followed by national marketing organizations elsewhere; in California, for example, the Shell brand was used from the start of operations there in 1912.[108] It thus comes as no surprise that the introduction of the Shell brand for gasoline occurred rather late in the Netherlands. Under C. A. G. Deterding, brother of Royal Dutch President Henri, the Dutch marketing organization, NV Acetylena, had begun selling gasoline with Autoline as the premium brand and Sumatrine for the cheaper product. It took until 1925, nearly twenty years, before these brands were replaced by Shell. The change was probably a sensitive issue for Dutch managers. After all, adopting the Shell trade mark for gasoline amounted to what one might call a reverse take-over, establishing the Group's public identity in the market with the widest consumer exposure as Shell, terminating customers' association of the established brands with Royal Dutch or Bataafsche. Autoline advertising sported the crown of Crown Oil plus an explicit reference to this being the gasoline of the Royal Dutch Petroleum Company.[109] The introduction of Shell was part of a complete overhaul of the Dutch sales organization, supported by heavy advertising, which included the building of a new type of service station with underground bulk storage, a retraining of sales managers, a switch to single-brand stations, and a makeover of forecourts and gasoline pumps in the then current red-and-yellow house style. Sales doubled within a few years. To emphasize the link between the new brand and the Dutch Group companies, Bataafsche's name was painted on the new Shell pumps and Acetylena was rechristened into NV Bataafsche Import Maatschappij or BIM, also in 1925.[110]

[69]

2,108 miles on two gallons of Shell motor oil – an impressive feat with 1920s automotive engineering. But there was no mention of the amount of petrol consumed.

The reverse takeover took a step further with the addition of Shell to the names of operating companies. During the early years of rapid expansion, Group managers preferred subsidiaries with names avoiding any suggestion of links to the larger concern, presumably because they wanted to keep a low profile at a time of intensive public debate about the economic power wielded by large corporations and trusts such as Standard Oil.[111] As a consequence, Group operating companies showed a colourful variety of names. Many of them reflected an original desire for companies to assume local colour as with the German subsidiary Rhenania, S.A. Italiana d'Importazione Olii, later to become Nafta Italiana, Lumina in Switzerland, Jupiter in France, or Rising Sun Petroleum Company in Japan. Other companies imitated the Jersey Standard habit of hinting at the affiliation by combining country names, as in for instance Svensk-Engelska Mineralolje Aktiebolaget and the other Scandinavian marketing operations. Still other names sought to link the company concerned with colonial power, for example Société Anonyme Française des Pétroles operating in Algiers, Compagnie Franco-Asiatique des Pétroles in Indo-China, and Anglo-Egyptian Oilfields Ltd. Some companies taken over by the Group simply continued under their original names, as was the case with the Caribbean Oil Company, Venezuelan Oil Concessions, Canadian Eagle, or Astra Romana. Finally, some companies carried names chosen for historical reasons of whim or expediency, such as Roxana or Burlington Investment Company. Out of more than a hundred affiliated companies in 1914, only two, Shell Company of California and Shell Company of Canada, had Shell in their names, while a third, La Corona (The Crown) in Mexico, referred indirectly to Royal Dutch. Managers had started to impose a degree of uniformity by giving new marketing operations the Asiatic name plus a term denoting the country or area of operations, but before the First World War there were only eight of them and there was no attempt to apply this policy to existing companies.[112]

The concern for the public expression of a common identity appears to have started during the early 1920s and received its initial impetus from the United States, where managers wanted their company to reflect the brand of products sold as closely as possible, objecting even to the suffix 'of California' which, following that company's expansion beyond the state borders into the Pacific North-west, was felt to be increasingly inappropriate.[113] With the 1922 consolidation the American subsidiaries all became part of the holding company Shell Union Oil Corporation. Three years later, Waley Cohen suggested renaming Roxana as Shell Petroleum Corporation. Deterding sounded out Kessler, but then the matter was allowed to rest.[114] Perhaps The Hague had objected, because when, in August 1928, De Kok got wind of imminent intentions to rename Roxana, he immediately wrote to Kessler: 'Since it is our express intention to fly the Dutch flag as much as possible in view of reciprocity, a rechristening into "Shell" seems to be unfortunate. I do not know in what stage the decision is, but if you agree with me, we can perhaps still take measures to stop it.'[115]

De Kok's plea came too late, for the wheels were already in motion. Roxana changed its name in October, setting a trend which led to a widespread renaming of operating companies. By 1940 only twenty-seven out of eighty-three operating companies did not have Shell in their name. Seven of those still used Asiatic. Amongst the exceptions were Rhenania, Jupiter, Bataafsche, and BIM. Rhenania and Jupiter changed their names shortly after the Second World War into Deutsche Shell and Shell Française respectively. Unsurprisingly, Bataafsche was only rechristened in 1967.[116] Products which, though hydrocarbon-based, were perceived as different from oil, such as Butagaz and asphalt emulsions such as Spramex did not always receive a Shell name, presumably so as to keep the brand pure.

The Shell trade mark, combined with the pecten and the dominant colours of red and yellow, had all the right characteristics of a good brand: it provided a close link between the company and its products, it had no political or religious connotations, it was short and easy to recognize around the world, and it could serve for the entire product range. None of the competing oil brands had a similar direct appeal or longevity. Jersey Standard introduced its Esso brand for premium gasoline only in 1926 and the Pratt brand lingered on until after the Second World War.[117] The changes in manufacturing and the rise of blending enhanced the importance of brand identity, for admitting that products could have varying ingredients depending on circumstances risked alienating customers. In February 1932, the London Aviation Department wrote to The Hague: 'I entirely share your opinion that in giving specifications of our Aviation Spirit to outsiders, we commit ourselves to a definite product. This we wish to avoid, as we wish at all times to be free as regard to the origin of the product we are supplying and as you know a more or less definite specification ties us to a definite origin. Apart from this I expect that Mr Moes (KLM) will enquire of other companies what they can supply by way of aviation spirit, and on paper the specification of our Benzine may not always be the most favourable-looking one and a lot of harm may be done in this way.'[118]

LAWN-BOY
AMF
YARD MAN
SMALL ENGINE SERVICE

For many years Shell's service stations in the USA featured an eclectic mix of designs, with one of the most extraordinary (main picture) being in North Carolina, around 1930. During the 1930s other less eccentric ones appeared (above, left to right) in the Midwest, with the Roxana name just visible, in Texas, and in California.

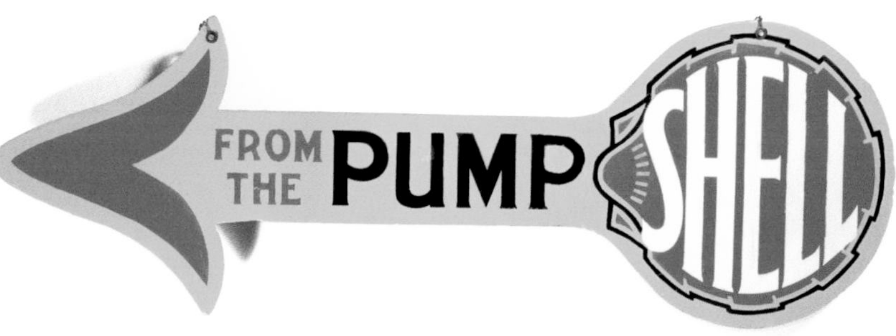

Nowhere did the strength of the Shell brand become more apparent than in the US. The first loads of Sumatra gasoline with which the Group entered the American market in 1912 were sold as Shell Motor Spirit and the brand was registered the following year.
The intense competition with other brands stimulated a rapid development of branding and marketing in the United States. As in Europe, the distribution of gasoline was still largely in the hands of shopkeepers and small garages. Consequently, when the Group arrived on the Californian scene in 1913, it had to start financing its own service stations in order to create a market. Service stations were then still a comparatively new phenomenon, with the first drive-in station dating from 1905, and they still catered to a small part of the market. Initially Shell California used a separate company, Omen Oil Company, to finance its service stations. From 1915 onwards Shell concentrated strongly on this market channel and assumed the financial burdens of setting up dozens of new stations in the north-west of California itself.[119] The advertising which accompanied the new strategy often showed the pecten, but used

[74] [75] [76]

From the 1930s, the widely varied styles of Shell's service stations in the USA gradually gave way to something more homogenous in appearance, strengthening brand recognition for interstate drivers.

a different colour scheme. Whereas in the UK red had become the typical colour of the Shell brand, Shell California used a combination of red and yellow, the latter dominating. One of the explanations suggested for this choice was that Meischke Smith, the then president of Shell California, selected 'these particular shades of yellow and red' because they were the colours of the Spanish flag, 'to which no Californian could object in view of the Spanish history of California'.[120]

So in California, far away from central offices in Western Europe, the Group experimented with new methods of gasoline marketing, and subsequently also of automotive lubricants, usually sold alongside and with the same trade marks. The number of service stations selling Shell products grew in a spectacular way, from a mere twenty-five in 1912 to 1,019 in 1919 on the Pacific Coast alone, then doubling in the next four years, to reach a total of 6,300 in 1928 and 13,512 in 1930.[121] Only a small minority of them were owned by the Group, 204 stations, or about 11 per cent, in 1923, but they accounted for 44 per cent of gasoline sales and functioned as 'the show-places, the pacesetters for dealers to follow'.[122] In addition to gasoline and lube oil, the stations also sold grease, sometimes tyres, patching materials, valves, and valve caps.[123] The name 'service station' was coined because part of the

SHELL SERVIC

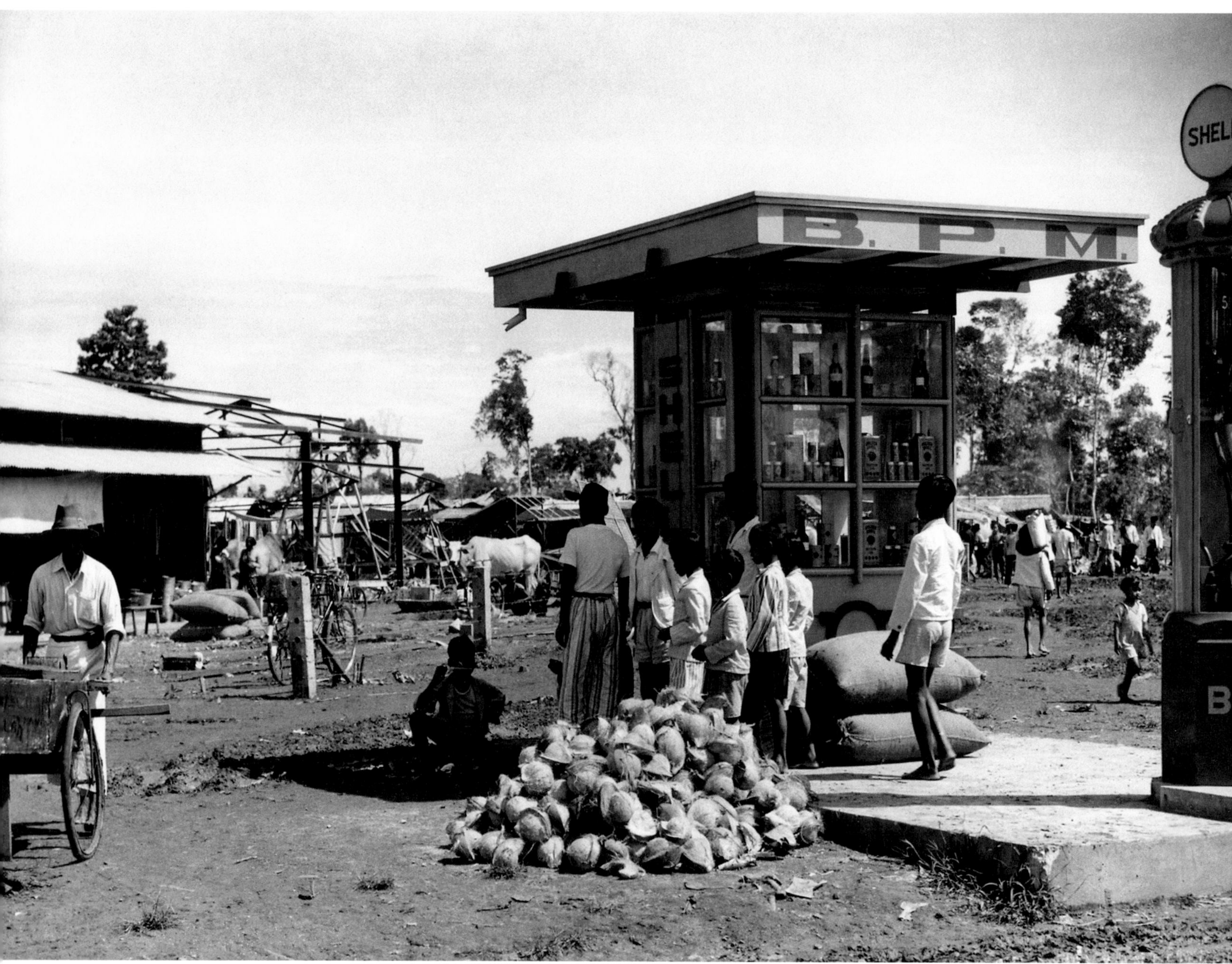

In the middle of a marketplace in 1930s Java, a Bataafsche/Shell petrol station seems like a precursor of the 'Shell Select' shops of later decades.

FARMERS PREFER SHELL
DENIS CONSTANDUROS.
DENIS CONSTANDUROS
YOU CAN BE SURE OF SHELL

package deal included a few 'free' services: checking oil and water, wiping windshields, inflating tyres. In return the motorist was supposed to tip the service station attendants.

This model emerged in the Los Angeles area during the 1910s and soon dominated gasoline marketing in the US, spreading on the back of the spectacular rise of the automobile, from less than 1.3 million in 1913 via 6.1 million in 1918 to 22 million in 1926.[124] Fully owned services stations were fairly capital-intensive, however, whilst competition kept the margins on sales to a minimum. A 1941 report on the US oil industry concluded that 'marketing is operated at a loss by the majors, but it does afford a necessary outlet for their products which they must control in order to ensure profits in other branches of industry'.[125] Standardization of station design and layout helped to keep costs low and to promote brand identification at the same time. Shell California decided to paint all service stations in the same bright yellow Shell colour. Special paint crews repainted entire stations from pumps to fences and even tree trunks, greatly enhancing brand visibility.[126]

Financing dedicated service stations was a feasible strategy only if they generated large and growing sales. This was the case in the USA but much less so in Europe, and consequently the European gasoline marketing organization lagged behind American developments. Kerbside facilities at motor garages or chemist's stores dominated the market well into the 1920s, and the practice of simultaneously selling several brands continued as well. By 1930 service stations were still a sufficient curiosity in Germany to provide the location for a hit operetta film, *Die Drei von der Tankstelle*. The US model of selling products through service stations was generally introduced as part of an entire network overhaul, as happened in the Netherlands in 1925, and in France three years later. In both cases, the Group began building service stations to meet competition from Jersey Standard, a Shell Transport memo arguing in the case of France that 'The S.O.C. companies in France are energetically advertising the "super" petrol service stations they have begun to erect, and are stressing in their publicity the service they offer to motorists.'[127] The Group set up a joint venture with the French company Société des Auto-Relais to build and operate a network of service stations dedicated to selling Shell products and painted in the house colour scheme. The Group paid for the pumps, the underground tanks, and the annual advertising budget. The initial seven service stations were soon increased to thirty.[128] Similarly, a campaign was set up to create a market for Shell motor oils. The budget of 2.2 million francs, the equivalent of 225,000 guilders or £18,600, was largely spent on 'triangular signs, show cases, and lubricating oil charts and booklets, and also on painting garages in the Shell colours'. Initial results were meagre, however, mainly because the Shell brand had only recently, in 1926, been introduced into France and still had a small gasoline market share of 4 per cent. In spite of the disappointing results of the first campaign, a new one was launched the following year.[129]

[80]

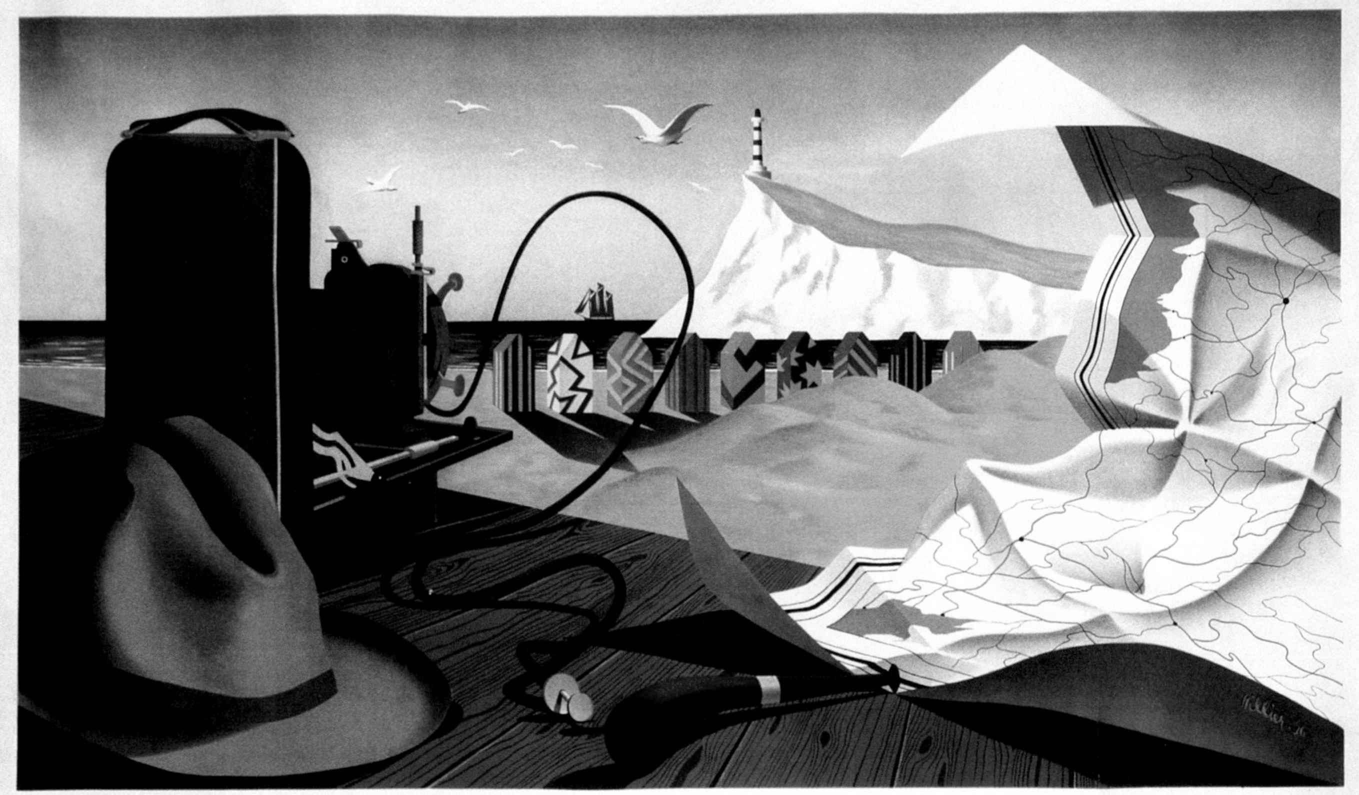
TOURISTS PREFER SHELL
YOU CAN BE SURE OF SHELL

The 1930s saw a flowering of Shell's advertising, with posters that have become sought-after works of art in their own right. These pages show a small selection.

THESE MEN USE SHELL
RACING MOTORISTS
R.GUYATT
YOU CAN BE SURE OF SHELL

DRAKE BROOKSHAW
JUST
OUT
SUMMER SHELL

Eckersley-Lombers
WINTER SHELL until next May

[79]

[85]

THESE MEN USE SHELL
SIGHTSEERS
C. MOZLEY
YOU CAN BE SURE OF SHELL

Thus the 1920s saw the emergence of a centralized branding and marketing policy revolving around the Shell brand in yellow and red. At the same time local operating companies retained considerable freedom to experiment with their own ideas and marketing models, in accordance with the Group axiom of balancing the global brand with strong regional and local differentiation. This freedom included, to a certain extent, the Shell pecten and colours, which were only gradually standardized. In the UK, red continued to be the dominant colour of the Shell brand, for example, whereas in the rest of the world the Californian combination of yellow and red was increasingly adopted. After the rebranding campaign in the Netherlands, the colour 'seen everywhere' was yellow.[130] In 1932 a Shell flag, red with a yellow pecten in a white circle, was introduced, operating companies receiving instructions to fly this flag from all installations, ships, and offices. Companies could still use their own pecten design for these flags.[131]

Though otherwise heavily influenced by trends in the US, European marketing did produce a highly innovative and characteristic advertising style of its own during the interwar years, notably in Britain and Ireland. Often designed by famous artists or young ones at the start of their careers, the campaigns inspired some of the best posters in the history of advertising. The emphasis of publicity shifted from wordy press advertising to posters on lorries and billboards, which required simple slogans but enabled a liberal use of colour to create a striking image underlining the message being presented.[132] Some campaigns emphasized the virtues of Shell products, their reliability and power, in images using a combination of wit and visual attractiveness, but the most innovative ones relied on the power of the images themselves which together formed a running theme chosen to inspire associations between the delights of motoring and sightseeing with the Shell brand. An excellent and well-known example was the 1925 series of posters commissioned from John Roland Barker, in which the simple message 'See Ireland first – on SHELL' appeared to celebrate both the natural beauty and the newly won independence of the country with grandiose pictures of the Irish landscape.[133]

[86]

Outside the USA, from Vyborg in Finland (above) to Homs in Syria (main picture), Shell's petrol stations in the 1930s remained blithely individualist...

[87]

شل
SHELL
06|397

[88] [89]

...and (above from left to right) in England, India, Germany and Malaya, the only shared characteristics of Shell petrol stations around the world were the pecten and the name.

Campaigns in Britain and in Germany used the same mild chauvinism in very appealing portrayals of natural beauty and tourist sights to seduce motorists to use their cars to tour the countryside and fill them up with Shell on the way.[134] The posters commissioned by Shell-Mex and subsequently Shell-Mex & BP during the late 1920s and 1930s were famous in their time as icons of their age, and became a prestige project of the Marketing Department of Shell-Mex & BP in the UK. The campaigns were so successful that in 1931 the New Burlington gallery organized an exhibition of Shell posters, which was repeated three years later.[135] Posters could also be bought by the public. Partly out of necessity, Shell California took up a related but slightly different advertising theme in 1924. Public opposition to roadside advertising inspired the company to make large shell-shaped billboards carrying, in addition to the company name, forest fire prevention warnings. The theme was used for one month every year into the 1950s. Shell-shaped billboards with city names, illustrated with a building or other landmark, were also used for marking city approaches and proved equally popular with motorists as windshield stickers.[136]

[90] [91]

The countryside theme broadcast by the posters was taken up by different media, notably touring guides, and movies. In Britain, Shell-Mex & BP started issuing county guides in 1934, enabling the poet John Betjeman, who edited them, to gather the knowledge which would much later bring him fame as the wistful keeper of 'Olde England's' charms.[137] For a long time the Group used films mainly for internal purposes, to record processes and events and to serve as a training medium, but the Film Unit set up in 1934 served another purpose.[138] During the 1930s, with cinemas beginning to reach mass audiences, the Group started producing films for publicity purposes as well. Some of the films told a technical story, such as the development of a new type of lube oil through engine research.[139] Others tied in with current advertising campaigns. In April 1937 the Dresden linguist and famous diarist Victor Klemperer, a keen movie-goer who had taken up driving after being dismissed from his professorship by the Nazis, watched a free showing of the Shell film 'Deutschland ist schön' extolling the beauties of the German countryside.[140]

[92]

In 1924, public dislike of roadside billboards caused Shell of California to change its shell-shaped billboards away from direct advertising and instead into the campaign against forest fires. These proved popular and were used annually for at least thirty years, and at the same time other shell-shaped billboards were introduced to announce the names of West Coast cities. Reproduced in miniature as windshield stickers (above), these too were popular with motorists, who collected them as souvenirs of their travels.

Producing films for public instruction and entertainment formed part of a new policy to enhance the Group's corporate image. In December 1938, Godber wrote to the general manager in Lagos about his contacts with officials there: 'It is our business and your business to prove to the people of each country that we are contributing in every way possible to the wealth and progress of that country, and that we are not just making money for our own purposes. So it seems we have two objects in view, the first negative and the second positive. We have to remove the unjustifiable reputation that we may have of being generally condemnable as a highly capitalistic concern, and to silence such commonly used statements as that "the country is dependent upon the international oil groups". We can do this best by putting different ideas there instead. Even people who ought to know better think that the oil trade exists purely to make profits and they do not realize that of all commercial undertakings it fulfils, and consciously endeavours to fulfil the biggest need of modern civilized existence. Obviously the first responsibility of our Representatives is to see that he himself and his own staff thoroughly appreciate the position. They also must feel themselves that they are in the country in which they are working for the good of that country, and so these ideas may permeate through the staff to the dealers and customers and the general public (...) I think I should here refer specifically to the O.P.Q. Dept. in this Office, which really was founded for this special purpose of removing the old incorrect impression of the oil companies and their dealings and substituting with the public the idea of the national value of the oil organizations in their midst. It is hoped that you will make all the use you can of the information and facts that the O.P.Q. Dept send you, and I would like particularly to ask you in turn to feed the O.P.Q. Dept with information which they can propagate and use. You probably know that this dept is now starting to produce films of general interest, not for the specific selling of our products but for purely educative purposes, and I am sure that the right films rightly used will prove very helpful to you.'[141]

The point was not immediately grasped by everyone in the organization. In April 1940, Asiatic wrote to BIM in The Hague, in response to a telex about the costs of a film on pesticides: 'I am sorry to say that what you write (...) only serves to show that there is still a considerable lack of appreciation at your end of the fundamental principles underlying film production and distribution (...) What you say in your telex is true if we look at films purely as a method of increasing our sales. To look at films in this way, however, is to fail to understand the wider possibilities of this medium, and moreover it seems to show the lack of appreciation of the important question of public relations. It is becoming more and more evident that a company such as ours, which in many respects is similar to a public utility company, cannot afford to think in terms of its direct consumers alone. We are dependent to an ever-growing extent on the goodwill of the general public or perhaps, more correctly, on the voting public which elects Governments and from which the Government itself is drawn. Unless we obtain the sympathy of the general public we shall always remain vulnerable to Government interference, etc. (...)

In films we have an exceptionally good medium for obtaining goodwill in the widest quarters and, moreover, we can obtain this goodwill at very low cost. Films distribution, such as being undertaken by the BIM, i.e. shows to invited audiences with lectures etc., is a matter of fact the most expensive way of distributing films. Whether or not this expenditure is justified must to a large extent be a commercial question as, by your own admission, this distribution is being carried out by the BIM in order to support sales. On the other hand, cultural distribution, i.e. distribution to schools, universities, clubs, associations, etc., only involves the cost of copies, which is relatively small, and the production of a small catalogue and postage. It is clear that petroleum matters must be of general interest in Holland owing to the important position of the Royal Dutch in the national economy. You will be showing a high sense of civic duty by placing the films dealing with these questions at the disposal of educational and cultural organizations

IMPERIAL AIRWAYS USE
THROUGHOUT EUROPE
G-E BOZ
DACRES ADAMS
SHELL PETROL EXCLUSIVELY

[93]

[94]

Little and large, Shell fuelled them both: (left) James H. Doolittle (1896-1993) and the famous *Gee Bee*, winners in 1931 of the Thompson Trophy race at a speed of 252 miles an hour, a record that stood for five years; and (above) the imposing front view of the Imperial Airways 'Argosy' aircraft *City of Arundel* (call-sign G-EBOZ) built by Armstrong Whitworth. After managing Shell Petroleum's aviation department in St Louis in the 1930s, Doolittle returned to the US Army Air Force and achieved international fame by leading a flight of B-25 bombers against Tokyo in 1942, decisively affecting Japan's naval strategy. From 1946 to 1959 General Doolittle was a vice president and director of Shell Oil Company.

in Holland – all the more so as our films are especially designed to conform to the idea of Public Relations and, with few exceptions, cannot be considered as true selling films (...) These opinions, I would add, are based on experience gathered all over the world and under the most varied conditions. Nothing has been more striking than the instantaneous appreciation of our work among Government, educational and other organizations which has reflected itself in the ever-growing demand for our films.'[142]

These two extensive quotations show that the Group adopted films precisely because managers recognized them to be the ideal medium for reaching out to a wider audience and building up public support, not just in times of crisis, or as a part of marketing policy, but as a matter of business principle. Moreover, the texts testify to a remarkable maturing of attitudes about the relationship between the Group and society as a whole, from the self-righteous tirades against Government interference and taxation which characterized the Royal Dutch annual reports during Deterding's tenure, for instance, to a wholehearted acceptance of the fact that the Group operated as a public, not as a private, company, with responsibilities towards society at large, not just to shareholders and customers. This shift foreshadowed the formulation of concepts such as the responsible corporate citizenship and general business principles which emerged towards the end of the 1950s.

Selling specialized business-to-business products was of course entirely different from marketing consumer products, more often than not requiring salesmen with technical training so as to ensure that product specifications continued to meet evolving technical requirements. Close relations with big customers were thus also important for the feedback which the Group obtained about product standards and performance under demanding circumstances. If such customers also provided good publicity opportunities, they could get preferential terms. For these reasons the Group had always been closely involved with aviation. With the development of civil aviation during the 1920s this involvement deepened further, in sponsorships of pioneering long-distance flights such as those undertaken by Charles Kingsford Smith in his Fokker F-VII *Southern Cross* from the late 1920s, but also in the development of close ties with airlines. To develop the aviation market, special departments were set up in relevant countries from the late 1920s. All three American companies set up such departments in 1930 and appointed well-known pilots to run them. The new recruits included James H. Doolittle, later to become famous as an air force general.[143]

[95]

[96]

Two of Europe's infant airports: at Budapest (left) in the summer of 1937, the KLM aircraft *Jan van Gent* is refuelled by Shell Aviation Services, while at Schiphol (below) handcarts of Shell lubricants are lined up outside the company's office.

KLM's brochure shows the extent of its summer services in 1937, and its front cover reminds readers that 'Shell-service' is available in every country at every airport..

[97]

The Group's relationship with Koninklijke Luchtvaart Maatschappij (KLM or Royal Dutch Airlines) shows to good effect the mix of motives binding supplier to customer. Bataafsche participated in KLM's original share issue in 1920 and later increased its holding on condition that the company would only buy Shell aviation spirit.[144] Despite chronic losses and consequent funding constraints, Royal Dutch Airlines built a substantial European network and from 1927 pioneered flights from Amsterdam to Asia, leading to the start of a scheduled service in 1928. This scoop had an interesting edge of Anglo-Dutch rivalry. Government officials and businessmen in the Netherlands were determined to beat Imperial Airways to Asia, fearing that the route might otherwise be closed to them. They achieved their aim; Imperial started its service from London to Karachi in March 1929, when KLM had already established itself along the line.[145]

There were thus good reasons for the Group to cherish the Dutch airline's custom. The company attracted a lot of publicity; its pioneering exploits required regular supplies of aircraft fuel and lubricants in uncharted territories, just the kind of combined logistic and commercial challenge which Group managers relished; and KLM was very keen on technology and performance to the point that, in 1938, the Group's only customers for 95-100 octane gasoline were the US air force and navy, the Royal Air Force, and KLM.[146] Its technical department proved a good partner in collaborating with Group engineers on investigating the effects of fuel and lubricants on engine wear and performance. But the airline was a difficult customer. Its manager Albert Plesman understood publicity and rarely let a photo opportunity pass without having Shell products prominently displayed, so contemporary pictures create the impression of KLM aircraft being refuelled all the time.[147] He was also a tough bargainer with his ear to the ground, always abreast of market circumstances and ready to exact more favourable terms from the Group by bringing up real or imaginary better offers from the competition. Indeed, up to a point KLM itself competed with the Group. Until the introduction of syntholubes in the early 1930s, the airline used Wakefield's Castrol oil and even appears to have had an agency for selling this at Schiphol and Rotterdam airports.[148] Under these circumstances, the Group tried to accommodate KLM in every way, providing refuelling equipment on the route to and at airports in the Dutch East Indies, giving extensive technical assistance, formulating special lube oils, supplying aircraft gasoline to the airline's specification, for which the company was charged a rock-bottom price, lower than that for the Royal Air Force. As a small concession in return the Group obtained a discount on air fares during the 1930s.[149]

Asphalt showed another interesting variation of the Group's general marketing policy. There existed a great urge to find outlets, for the Group produced a lot of asphalt from the heavy crudes of Mexico and Venezuela, but it was a difficult market to develop. As a leftover, asphalt had no product specifications or indeed anything in the way of international standards. The Amsterdam laboratory devoted considerable efforts to asphalt research and succeeded during the 1920s in getting some product uniformity and in developing better applications, publishing a reference booklet with data for the trade in 1927.[150] To stimulate the use of asphalt bitumen in road building, the Amsterdam laboratory collaborated closely with Asiatic in London. Road engineers and asphalt experts in various European countries were approached to promote a standardization of product norms and tests, which at the same time eased access to government departments responsible for road building. The Group initially made only hot asphalt bitumen, transported in heated vans to its destination, but switched to producing emulsions in 1924. Sold under the brandnames Spramex, Sproeiphalt, Mexphalt and later Shelmac, these emulsions were a 50:50 mix of asphalt bitumen in a water and soap solution. On coming into contact with the ground, the asphalt bitumen split from the mix to form a road surface.[151] In 1925, the Group bought the rights to a rival process and vested all patents in Colas Products Ltd., which established dedicated marketing companies for cold asphalt. Three years later the Group expanded its asphalt business by buying a controlling stake in the American company Flintkote, which owned patents for asphalt shingles, rolls, and paper used in roofing and other building applications. The Colas patents and companies were transferred to Flintkote with the intention of forming a dedicated asphalt marketeer, but the combination was not a happy one. Flintkote's own products were a commercial disappointment and the company's business collapsed with the onset of the 1930s depression. After a reorganization Flintkote was sold again, the Colas patents and companies returning to the Group.[152] Meanwhile the intensive asphalt research had generated a growing range of specialist products for new applications such as pipeline coating, rustproof bituminous paints, electric insulation, waterproofing, and hydraulic engineering.

[98]

The ANTI-WASTE Road Material

MEXPHALTE

MEXPHALTE

SPRAMEX
DE OUDE METHODE
DE NIEUWE METHODE
VERSCH GETEERDE WEG LANGZAAM RIJDEN WIELRIJDERS AFSTAPPEN
MET SPRAMEX BEHANDELD DOORRIJDEN
SPRAMEX
NEDERLANDSCH FABRIKAAT
waarborgt een goedkoop, waterdicht, stofvrij en duurzaam wegdek, dat langer stand houdt dan herhaalde toepassing van teer.–
Hecht zich niet aan banden en biedt een stroef wegdek voor paarden.
VRAAGT ONS INLICHTINGEN
DE BATAAFSCHE PETROLEUM MIJ.
CAREL VAN BYLANDTLAAN 30
'S-GRAVENHAGE

[99]

[100]

[101]

Stored in bulk at 'Spramexland' in Amsterdam-Noord (above left) and transported in bulk by rail (above), Shell's road coatings were produced in enormous quantities in the 1930s, and a charming advertisement for Spramex made clear the difference between 'the old method' and 'the new method'.

[102]

Conclusion

The Group's apparent consolidation during the inter-war period masks a profound transformation of the business, from a somewhat loose assembly of operating companies into a mature concern, firmly integrated along the axes of research, technology, marketing, and branding. This period also saw the emergence of a more or less coherent conception of the Group's position in the industry expressed in maxims such as oil is a simple business, the Group needs crude flexibility because the world and Shell are or will be short of crude, research focuses on making cheap components, manufacturing makes quality products from those components, hydrogenation is the best approach to manufacturing, the future lies in chemicals, products need differentiation because commodities kill the business, global branding must be balanced by strong local differentiation. Such terms of understanding helped to establish a corporate culture which enabled networks to operate smoothly, even though the individuals forming the nodes did not know each other personally. Many of the maxims would still be recognizable to employees who started their career in Shell during the 1970s.[153]

The Group's change of course at the end of the 1920s, recognizing hydrogenation and chemicals as the key to further development of the oil business, marks a clear break in its history. Managers perceived a need to keep pace with wider industrial developments and push out into uncharted waters almost regardless of cost, committing large resources to acquiring the necessary knowledge and technology as a leap of faith, without any prospect of immediate commercial applications. In exploring what we now call the interface between oil and chemicals, the Group followed industry trends, but it would appear to have done so with a deeper conviction about the future of the business being in chemicals rather than oil. With other oil companies, chemicals research remained subservient to the oil operations, but for the Group it served as the foundation for a diversification into agrochemicals, detergents, resins, and synthetic rubber. Though these activities would come into their own only after the Second World War, the start was already made during the 1930s. However, even for the Group the importance of chemicals at this stage was not in the products themselves, which generated very modest sales and profits, but specifically in the discoveries made at the interface between oil and chemicals, such as the various processes for making iso-octane.

The conversion to chemicals derived its initial impetus from the Dutch side of the business, managers like Kessler and De Kok, both engineers, Pyzel of course, Knoops, Van Peski, and a whole range of outstanding scientists committed to driving the business forward through pushing the boundaries of technology. Having such a strong group of dedicated experts made all the difference in persuading sceptical directors to allocate funding. The growing importance of research and technology enhanced the position of The Hague, as instigator and coordinator of innovation, and as the recruitment and training centre for engineers.

The pattern of innovation shows that the Group succeeded in maximizing the network effects of its global spread of operations. The Amsterdam laboratory acted as the hub of research, but successfully drew on the resources of other Group laboratories around the world. The balance of exchange would also appear to have been quite even. Some regions generated more innovation than others, but even the United States companies contributed and gained in about equal measure.

[1]

Chapter 7

Taxing times, 1929-1939

The world economic depression which set in after the 1929 Wall Street Crash hit the Group very hard, notably in the US and in the Dutch East Indies. Drastic reorganizations helped to restore the American business, but the Indonesian operations suffered from structural weaknesses which were more difficult to remedy. Though the Group and Jersey Standard adhered to fundamentally different business policies, the Depression reinforced the ties between them, providing the basis for regular talks and agreements on a variety of issues along the lines of the Achnacarry agreement. This collaboration proved successful in some cases, but not in others. When facing a common opponent, as with the resurgence of nationalist policies during the 1930s, the oil companies generally stuck together. In dealing with these threats to the business, the Group showed a far more balanced and mature attitude than during the 1920s. Deterding could not follow this evolution, however, which led to his resignation and the emergence of a new management team at the top.

Crowds gather outside the Paris Stock Exchange after the devaluation of the French franc in 1931.

N.V. De Bataafsche Petr. Mij.
's-Gravenhage.

KOPIE VAN TELEGRAM

AFGEZONDEN DOOR: DB Deterding La Haye 3637

GEADRESSEERD AAN: Agence Economique et Financière Paris 21-9-'31.

URGENT.

Je voudrais, que le public réalise, que Société et ses Associés ont des milliards de francs en or disponible et que même, si l' or cessait d'avoir de valeur, huile, source la plus idéale d'energie, restera toujours de plus grande valeur stop

Comme avec chaque crise, ce sera la même chose après la crise actuelle, notre Groupe en sortira plus fortifié qu'avant

-----DETERDING-----

Model No. 729 • 150.000 • Sept. '29.

[2]

In the international financial crisis of September 1931, Deterding's telegram to the press sought to assure French investors of the Group's continuing strength.

[3]

Na den oliestrijd hebben Socony en B. P. M. elkaar weer in roerende eensgezindheid gevonden op hetzelfde prijsniveau. Slechts de ontslagen employé's zijn minder opgetogen !

Navigating through the Depression As we have seen in Chapter 5, the 1920s ended on a high note for the Group, with bumper profits in 1929. For Deterding personally, the end of the decade enhanced his cherished position as the distinguished world statesman of oil, with the Achnacarry meeting in 1928 and, in 1929, an invitation to give the keynote speech to the annual meeting of the American Petroleum Institute. In his speech, he treated the audience to his favourite themes, the danger of overproduction in leading to dumping, the need to curtail production and conserve oil stocks, the desirability of investing in product quality rather than extensive distribution facilities, and the benefits of cooperation over competition, preventing the unnecessary duplication of expensive installations – in short, everything that Achnacarry stood for.[1]

The speech showed Deterding in the consolidationist guise which he adopted during the 1920s. Having achieved economies of scale in worldwide operations, he switched to a different strategy, emphasizing product quality over volume and stable profitability over growth, setting precepts which Group managers were to follow into the 1960s. It was no doubt this new conviction that made Deterding back the proposals to increase the research effort and go into chemicals. Even so he had not been entirely converted just yet. During 1928 and 1929, Shell Union conducted an aggressive marketing expansion in the US east of the Rocky Mountains, to the great annoyance of Jersey Standard and Standard New York (Socony).[2] Jersey and Socony responded by beginning a price war in the Dutch East Indies, the Straits, China, Japan, and Indo-China, hitting Bataafsche where it hurt most. Within months Bataafsche urged managers in Indonesia to start staff appraisals immediately with a view to cutting jobs and restoring competitiveness.[3]

These battles took place against the background of large oil surpluses and low prices which had dogged the US oil industry for the better part of three years. From 1927, Shell Union would have made a loss but for the profits generated by Shell Pipeline.[4]

[4]

DE VERZOENING.

SO CONY

SHE

— Conny dear, zullen we de scha inhalen? Shall we?
— O you SHELL! make it 26!!!

Roxana, the Group's Midcontinent company rechristened as Shell Petroleum in 1928, operated in the red for nine consecutive years from 1927 to 1935, and the new Atlantic seaboard organization never turned a profit. When demand began to slacken with the general economic downturn following Black Thursday in October 1929, crude prices fell precipitously. And then, in October 1930, the glut turned into a flood with the East Texas field discovery, which pushed Texas crude prices down from a dollar to fifteen cents a barrel, in some places even less. Jersey Standard had attempted, in the wake of Achnacarry, to restrict exports by forming a cartel of the main American companies, but this Export Association never succeeded in uniting even half of them. East Texas finished the attempt.

Since American exports supplied a third of the rest of the world's oil consumption, low prices there hit producers everywhere, as the frantic efforts to sell oil, any oil, began flooding markets first in the US, then around the world. Governments counteracted by regulation, putting up trade barriers and imposing inland sales quotas, but also by increasing taxes on oil products such as gasoline to meet budget deficits arising from crisis intervention.[5]

The demise of the gold standard and the return of currency fluctuations last seen during the First World War and its aftermath further handicapped trade. When in the Summer of 1931 a banking crisis triggered a flight from the Reichsmark, the German government imposed a moratorium on all international transfers, known

26 April 1930: After the oil price war in the Netherlands East Indies, the Batavian newspapers *De Volkscourant* (left) and *Nieuws van den dag V. N. I.* (far left) satirize the reconciliation between Socony and Shell.

[5]

[6]

as the *Stillhalte*. This action effectively took the mark out of circulation for payments abroad, thus freezing the assets and profits of foreign companies. The turmoil on the money markets then spread to Britain, which in September had to devalue the pound sterling. In the Summer of 1933 the United States followed suit; only the four countries of the so-called Gold Bloc, Belgium, France, Switzerland, and the Netherlands, held out until 1935-36.

As a consequence of the general overproduction during the 1920s the Group's profitability had declined, in line with the rest of the industry, despite sharply rising production (Figure 7.2). The absence of consolidated profit and loss figures means that we have to make do with the rough gauge of total revenues of Royal Dutch and Shell Transport combined. Measured by that standard, unit revenues showed a sharp drop, from the 1920 peak of over 47 guilders a ton to 8.28 in 1929, or from £ 4.36 to £ 0.68. The Royal Dutch annual report typically and disingenuously blamed rising salaries, distribution costs, and taxes for the fall in unit proceeds, omitting to link them to the overproduction and sliding prices denounced elsewhere in the text. Similarly, Deterding never tired of repeating that underconsumption and not overproduction was the true cause of the crisis, castigating politicians for imposing taxes and other restraints on demand.[6] With the onset of the crisis, Group revenues plummeted. From 1929 to 1931, the income of Royal Dutch and Shell Transport dropped by two-thirds. Bataafsche's operational earnings more than halved and continued falling before stabilizing in 1934 at 18 million guilders, less than 10 per cent of average earnings during the 1920s. The prices of Royal Dutch and Shell Transport shares went spinning to new record lows, even worse than during the companies' respective crises around the turn of the twentieth century (Figure 7.1). Deterding blamed a concerted bear attack on the Group, allegedly orchestrated by Gulbenkian from Paris, but the steep drop would seem to have been quite in keeping with the general collapse of share prices world wide.[7] The Group's fledgling chemical operations fully shared in the malaise. Overproduction and the breakup of the international nitrogen cartel sent fertilizer prices into free fall. Mekog's modest profits over 1929 and 1930 were wiped out by losses during the early 1930s, sparking discussions about converting the IJmuiden plant to hydrogenating oil. Before anything was done prices stabilized and revenues recovered somewhat during 1933-34, paving the way for very substantial profits during the late thirties.[8]

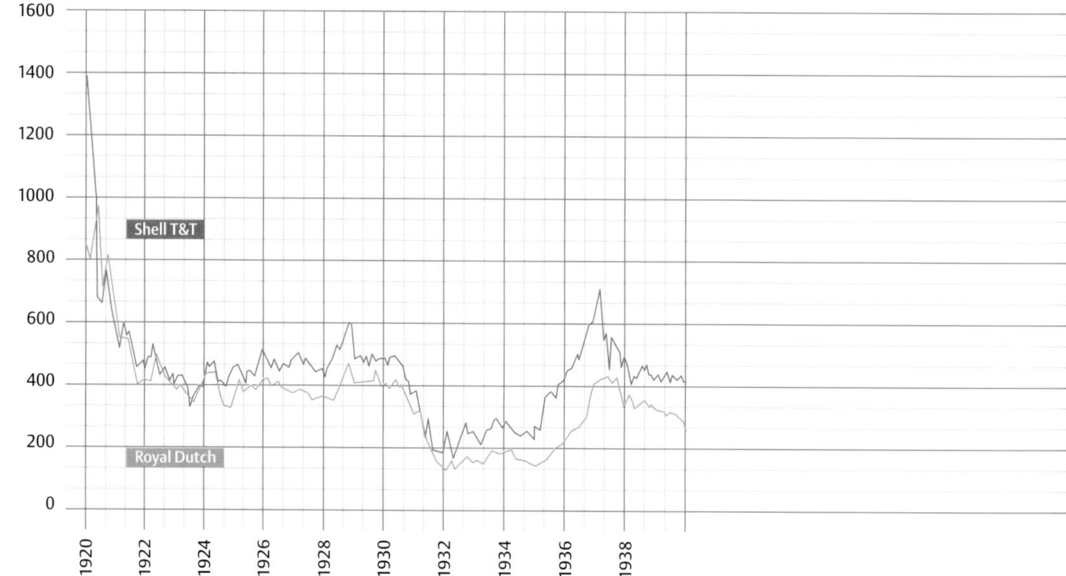

Figure 7.1
Share prices of Royal Dutch and Shell Transport, 1920-39 (per cent of par).

Pages 416 and 417 – Wall Street, 1929. As the New York Stock Market collapses, the crowds filling the roads are not busy shoppers but frightened investors.

Wall Street, 1929: Prolonged speculation sparks panic selling.

[7]

Fair, slightly cooler tonight. Friday fair. Fresh southwest to west winds.

BROOKLYN DAILY EAGLE

And Complete Long Island News

LATE NEWS

WALL STREET 1:15 PRICES ★★

89th YEAR—No. 295. ★ NEW YORK CITY, THURSDAY, OCTOBER 24, 1929. ★ REISSUE 32 PAGES THREE CENTS

WALL ST. IN PANIC AS STOCKS CRASH

Attempt Made to Kill Italy's Crown Prince

ASSASSIN CAUGHT IN BRUSSELS MOB; PRINCE UNHURT

Royal Suitor Was About to Lay Wreath on Unknown Soldiers' Tomb.

Princess, Deeply Moved, Falls Into Fiance's Arms and Kisses Him

Hollywood Fire Destroys Films Worth Millions

ATTEMPT MADE ON LIFE

Crown Prince Humbert.

Consolidated Studios Are Swept by Flames Fatal to One—Master Pictures Burned Include Many New Talkie Productions.

Please Turn to Page 15

FEAR 52 PERISHED IN LAKE MICHIGAN; FERRY IS MISSING

Wreckage Picked Up Indicates Craft Went Down With All Aboard.

PIECE OF PLANE LIKE DITEMAN'S IS FOUND AT SEA

Black and Orange Wreckage Indicates Daring Flier Went to Death.

High Duty Group Gave $700,000 to Coolidge Drive

Grundy Agrees Rates Went Up Due to His Activities as Propagandist. Favors More Lobbying to 'Carry Out Voters' Wish.'

FOR MORE LOBBYISTS

JOSEPH R. GRUNDY

By HENRY SUYDAM.

Please Turn to Page 2.

STOCKS CRASH IN RUSH TO SELL; BILLIONS LOST

Morgan, Mitchell Buying Stocks in Effort to Check Rush to Unload.

CARNEGIE CHARGE OF PAID ATHLETES ROUSES COLLEGES

Survey Finds Shopping for Stars Is General, Remedies Now in Force.

HOOVER'S TRAIN HALTED BY AUTO PLACED ON RAILS

2 Negroes Admit Scheme to Collect Insurance for Loss of Machine.

WARDER SOUGHT TO KEEP SEA TRIP SECRET, AID SAYS

Woman Testifies She Was Asked to Say Nothing About Plans.

SOMERS NAMED AS HEAD OF NEW EXCHANGE BANK

Damman Is Chosen for Chairman of Board, Miller Vice President.

10 ARABS SENTENCED

[8]

[9]

Pl. No. 182 TANKENPAR

Photographed in April 1930, the enormous oil-tank farm at East Plaju symbolized the problem facing the industry: was too much oil being produced, or were consumers buying too little?

[10]

As the ravaging effects of the Great Depression spread worldwide, the destitute were fed by charity in the USA (left), while in Holland (right) unemployed men queued for food stamps.

Shell Chemical was hit by the same storm. In 1931, the company's ammonium sulphate cost $30 a ton to produce, but it sold for only $16.50. Over 1931-33, the company lost a total of almost $1.6 million. Managers vigorously debated several options; keeping Shell Point open, mothballing it, or selling it to a chemical concern. Godber, Agnew, and Van Eck wanted to pull out of fertilizers, valuing the experience gained but unwilling to sustain further losses. De Kok appears to have wavered, before lending his support to Pyzel, Kessler, and Waley Cohen, who stuck by the original intentions of the diversification.[9]

The collapse of the gold standard added to the woes. The *Stillhalte* effectively took the Group's German operations out of the fold, necessitating increasingly complex manoeuvring to keep the business going without committing further money, and preferably getting some of it out. The devaluation of sterling dealt a direct and heavy blow to Group finance. As will be recalled from Chapter 3, Anglo-Saxon acted as the Group's treasury department and so managed Bataafsche's revenues, of which a large part had been invested in sterling assets such as tankers and installations. Consequently, Bataafsche faced a huge loss on its London-managed assets of 288 million guilders before devaluation. With some dubious financial engineering, managers limited the loss to 46.4 million guilders. By now financial relations between Group companies generated a substantial correspondence, suggesting that managers had some difficulty in keeping track of them.[10]

Managers reacted immediately to the crisis by imposing a rigorous efficiency drive, cutting exploration and production, reducing spending on inventories, drilling, maintenance, and the construction of new installations; by laying up ships and cutting fuel costs; by imposing wage reductions; and by massive redundancies. Here as elsewhere, the absence of consistent figures handicaps a detailed assessment of the impact on the Group as a whole, so we have to make do with impressions. In May 1930, the bonus payment to the Provident Fund was cut from 15 to 10 per cent.[11] The Central Office in The Hague, which had just been considerably enlarged, immediately laid off seventy people from a total of 930, with further redundancies taking the total to 135.[12] We have no comparative figures for the London offices, which

[12]

employed 2,100 people, but the plans for an extensive redevelopment of a site adjoining St Helen's Court to accommodate the steadily rising staff numbers were suspended.[13] In August 1931, managers introduced a salary reduction scheme for all Group staff, with pay cuts ranging from 5 per cent for the lowest scale to 20 per cent for the highest.[14] This would have appeared drastic enough to the people concerned, but the operating companies bore the brunt of the spending cuts and redundancies. Shell Union suffered losses of $27 million over 1930 and 1931, by far the largest of all American oil companies.[15] Godber took an active part in the reorganization efforts, travelling frequently to the US for inspections and progress reports.[16] As a consequence of the measures intro-

B12

OOST PLADJOE 7-4-30.

[13]

In Britain, mass unemployment gave strength to the trade unions movement.

duced, staff at the US companies dropped from 34,600 in 1929 to 19,500 two years later. Between 1930 and 1932, Shell Union also reduced its costs by 34 per cent.[17] Bataafsche took even more radical steps, cutting overall costs in 1932 by over 60 per cent compared to the 1929 level, from 174 million guilders to 67 million, or from £ 14 million to £ 7.7 million, the devaluation of sterling in September 1930 accounting for the slightly different percentage. The freezing of new investments yielded three-quarters of these savings, cuts in operational costs the rest. One simple expedient was ending the multiplication of inventory costs. The main operators in Indonesia all used to keep their own stocks of materials and spares. By making them share, and by selling equipment and hiring it back as and when needed, managers saved some 20 million guilders in 1931-32. In 1935, costs per ton were at their lowest since 1911.[18]

Bataafsche's dismissal of about 100 Europeans from its Indonesian operations in the winter of 1929-30 inspired hostile reactions from the local community, press, and government officials.[19] The board defended itself with the sanctimonious claim that the people concerned had been selected for their failure to perform, meanwhile instructing managers to be more discreet and spread redundancies over time to avoid undue attention.[20]

[14]

/GEM Telegram RECEIVED from SOURABAYA
to -
No. of Words 43 No. 11568 Despatched at 4.35 pm on 3.4.30.
Received on 3.4.30.

REPRESENTATIVE

852 Dismissal(s) Works Dept. European Staff on a large scale ?is bad(ly) antagonise(ing) public and some local newspapers are urging consumers to boycott B.P.M. Benzine fullstop

Local division(s) political party(ies) are contemplating combined action to further this boycott(ing) fullstop

Although we do not expect we shall lose much business in this way ?it may be well to refrain from rigorous(ly) dismissing or if this is impossible to explain clearly to the newspapers why this became necessary fullstop

In view of contradictory statement(s) that so far have appeared in the newspapers and with regard to fact that it is of the utmost importance ...

De Benzine vrede uitgebroken.

De oude prijs van de benzine 24½ cent per liter, is weer bereikt. Hoewel er nu gebleken is dat men desnoods tot 10 cent per liter kan gaan, met andere woorden, dat hier veel te veel wordt verdiend aan een gebruiksartikel van de allereerste sociale waarde, de beide groote concerns hebben wijselijk gemerkt, dat elkaar op de gevolgde manier dood te concurreeren een lange lijdensweg zou zijn, hebben een sigaar opgestoken en een glas gedronken en hebben de prijzen weer op den oude ... hersteld.

MASSA-ONTSLAG BIJ DE B. P. ...

De benzine-oorlog en zij[n] slachtoffers.

Represaille tegen de Bataafsch[e] van reactionnaire zijde.

De militairen beschermen de bronn[en]

The human cost of competition: newspapers in the Netherlands East Indies firstly fulminated against 'the petrol war and its victims' and the 'mass dismissals by BPM', and then, when 'the petrol peace' broke out, warned that the price would jump by as much as 10 cents a litre. Meanwhile, in telegrams to its representatives, Bataafsche sought to limit the damage to its reputation.

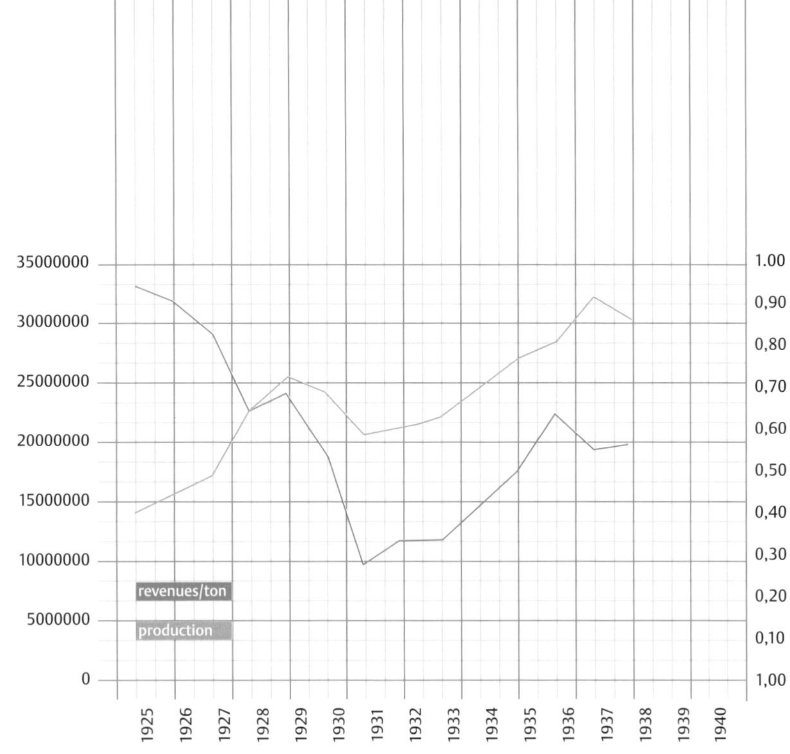

Figure 7.2
Group production (left scale) and net revenue per ton in £ (right scale).

Managers must have found it difficult to comply with this instruction. During 1929-33 Bataafsche cut its payroll in Indonesia from 65,600 to 27,350 people, in addition to terminating the contracts of another 18,000 casual labourers.[21] Between 1929 and 1934, the Curaçaosche Petroleum-Industrie Maatschappij and its shipping subsidiary CSM sacked two-thirds of their personnel, cutting employment from almost 11,000 to 3,412. Ten of the fourteen recently completed Dubbs crackers were shut down, and work on the new head office suspended. In 1930 and 1931, the officers and engineers of the CSM saw their wages cut by a total of 16 per cent, a far greater sacrifice than other shipping companies managed to impose on their staff.[22] The severity of the impact of redundancies and wage reductions on the Group's core staff may be gauged from the fact that the Provident Fund lost 6,463 members in three years, membership touching a low of 32,904 people in 1932.[23]

One dreads to think about the social devastation wrought by these massive redundancies. Communities like Balik Papan, Pangkalan Brandan, or Curaçao, largely built by and around the oil

[15]

Whether there to keep order or even hoping for some food himself, a policeman stands by a queue of hungry people at a soup kitchen, Los Angeles, 1920-38.

business, from the smallest shop to the inevitable police compound and such schools as there were, suddenly faced utter destitution and no viable alternative anywhere around. On Curaçao, for instance, about half of the total labour force was employed by the Group.[24] The staff cuts made a deep impact on public opinion. When recruiting picked up again during the late 1930s, the Group found it more difficult to hire university graduates in Britain because staff in the relevant departments advised students against a career in a company which treated its employees so badly.[25] If the boards concerned cared about the immediate consequences of the cuts, the records no longer show it. But then these were desperate times, forcing managers to extreme measures in order to keep going and reorganize the business. Moreover, because revenues fell almost as fast as costs, the draconian reorganization measures produced less effect than managers might have hoped. Revenues per ton recovered sufficiently for the Group to remain (marginally) profitable, but they failed to regain the levels attained during the middle 1920s (Figure 7.2).

A major cause lay in the Dutch East Indies operations, which never really recovered. Figure 7.3 gives unit returns for Bataafsche's operations there and for Shell Union, indexed at 1929 to avoid the noise from the exchange rate fluctuations of the 1930s. During the early 1920s the Indonesian part of the business was the most profitable by far, with unit revenues comfortably above those in the United States, the Group's second most important production area. The gap narrowed considerably, though, and by 1929 both Bataafsche and Shell Union earned about $ 20 per ton of crude supply. From then the two companies diverged sharply. Whereas Bataafsche's operational profits per ton plummeted from 1929 until 1934 to pick up only very hesitatingly, Shell Union's revenues per ton declined a year later and bottomed out earlier, to regain levels which would have looked very acceptable during the 1920s. In 1935, Bataafsche's crude yielded just over $ 6 a barrel, Shell Union more than two-and-a-half times that, $ 15.50.

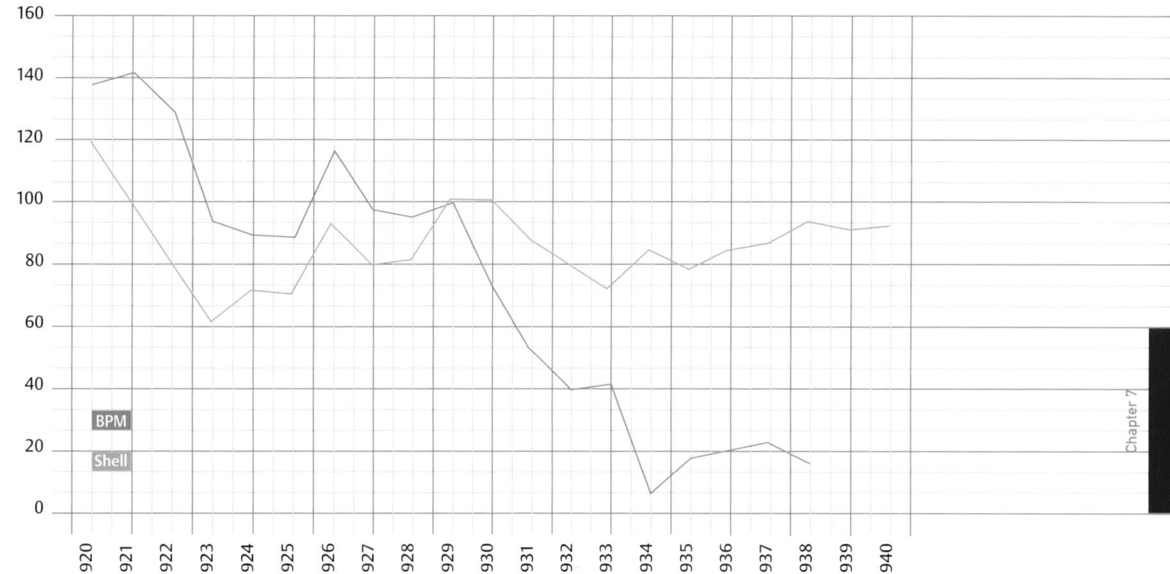

Figure 7.3
Unit returns per ton at Bataafsche in the Dutch East Indies and Shell Union, 1920-38/40 (1929 = 100).

[16]

[17]

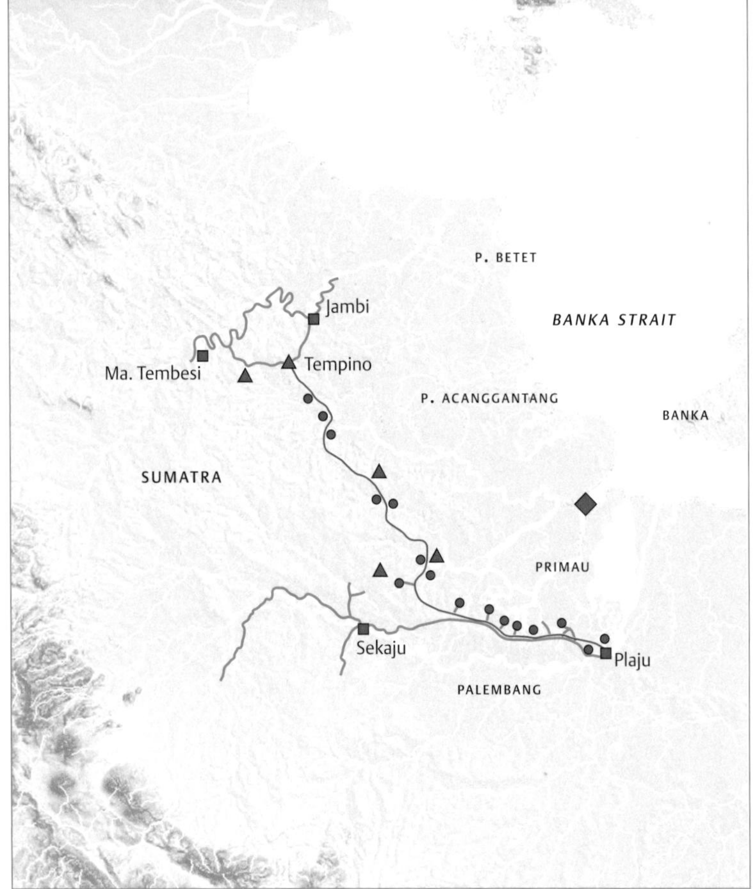

[20]

With the welcome shade of an incongruous parasol, European welders connect a section of the NIAM pipeline held in place by a caterpillar tractor (above left). Jungle clearance (above, right and far right) required great effort and was the inevitable price of gaining access to crude oil.

During the 1930s the NIAM pipeline in southern Sumatra was constructed from Plaju to Tempino, covering a distance of 265 kilometres.

● Pumping station
▲ Oil fields
◆ Harbour

[18]

[19]

Shell Union then managed to get on top of its problems in a way that Bataafsche did not. This divergence originated first of all in the changing pattern of exploitation in Indonesia. Production there continued to rise, but the Group derived an increasing proportion of its crude from new concessions worked with the colonial government in NIAM or under the so-called '5A' clause of the Mining Law. In 1931, 15 per cent of Group production in Indonesia came from NIAM or 5A fields; in 1936 it was 40 per cent and 45 per cent two years later.[26] The operations were barely profitable during the early thirties. When the situation improved markedly after 1935, the largest share of the dividends on nearly half of crude production went to the government, which goes a long way to explain why, despite the efficiency gains, Bataafsche's revenues per ton failed to recover during the second half of the 1930s. The NIAM arrangement gave the colonial government 60 to 80 per cent of profits, a very large share which spread to other countries only after 1973.[27]

There were more factors at work, however, another being the import levy on oil and oil products introduced by the US government in 1932. The Group was both helped and hurt by these measures. By stabilizing the American market, the import restrictions spurred Shell Union's recovery, but foreign oil previously sold in the US now started flooding markets elsewhere, further depressing prices and worsening the plight of companies such as Bataafsche. The Group also changed its supply pattern in response to the US restrictions, redirecting products from Venezuela to Europe, previously supplied from Romania and the Dutch East Indies. By 1938, 41 per cent of oil products sold in Western Europe came from Venezuela, against 27 per cent from the US, 19 per cent from the Middle East, and the rest from the Dutch East Indies.[28]

Then there was a fundamental difference in operations between Shell Union and Bataafsche. Shell Union's difficulties stemmed from the overambitious expansion at high prices during the 1920s, detailed in Chapter 4. These could be cured by the firm pruning that was exercised, initially along the familiar lines of reducing drilling programmes, surrendering leases, halting construction on refineries and new marketing installations, cutting back production, and a thorough overhaul of marketing. Refinery installations were partly shut down, since buying gasoline had become cheaper than refining it. By 1930 the east of the Rockies company Shell Petroleum, unique for a Group company, bought more than it produced.[29] Managers also explored the pros and cons of amalgamating the three operating companies, but dropped the idea as impractical at that moment.[30] A more fundamental quest to restore competitiveness began in 1935 with the first in a series of comparative cost studies to determine how the business actually performed in its various functions and many areas.[31]

This approach was surprisingly new for the Group.[32] Until then price had come first, volume second, and cost third. Managers were parsimonious from conviction but not cost conscious, that is, they held a firm grip on expenses as a matter of course and as a virtuous end in itself, but they did not use cost control as an instrument with which to analyze and improve operational performance across their sprawling business empire. For all Deterding's continuous carping about cost, nobody at Central Offices, not even he, knew how individual businesses compared with each other in terms of profitability and cost levels. It had not mattered so much when individual markets could be managed to sustain prices sufficient to generate operational profits. With the worldwide collapse of prices hitting all areas in the same spot, cost suddenly became of paramount importance, forcing a radical reordering of priorities. The vice-chairman of Shell Union, J.C. van Eck (1880-1965), who under the absentee chairmanship of Deterding was the CEO in all but name, appears to have been the first to order a comprehensive cost survey, in this case to establish ways of reorganizing Shell Petroleum so as to end its accumulating losses.

The result was the so-called Salmond Report which, starting from a five-year projection of expected demand, presented a functional analysis of how Shell Petroleum's supply chain might be rearranged so as to meet that demand profitably in all functions and in all areas of operation. The data presented enabled the board to identify marginal operations and switch resources to more remunerative ones. As Shell Petroleum president Alexander Fraser put it, the company marketed a mile wide but an inch deep, and could never hope to make money that way.[33] As a consequence, Shell Petroleum completely withdrew from areas in which it could not compete since its transport costs were too high, aiming to concentrate on the more populated areas.[34] Not all of the report's recommendations were accepted; nor were the measures adopted immediately successful, for Shell Petroleum again made losses in 1938 and 1939. Moreover, after a slight dip in the early 1930s the number of gasoline outlets continued to grow, so the withdrawal from the Midcontinental markets was more than compensated for by an expansion elsewhere. Even so the reorganizations which followed it profoundly reshaped the business, with the result that by 1943 the company matched Shell on the Pacific coast in profits, although not in profitability since its turnover was bigger.[35]

Within the Group, the Salmond report marked a breakthrough in exchanging the concept of overall stable profitability of operations for profitability in relation to actual cost of individual functions and areas. The concept caught on rapidly. In 1937 The Hague began compiling detailed cost comparisons for the upstream operations in Indonesia and repeated the exercise in the following year for all main production areas around the world and for manufacturing operations in Indonesia. Presumably London did the same for the downstream, but no instructions or data on it have survived. The tables drawn up in The Hague help to understand the underperformance of the operations in Indonesia.[36] Of the three main producing areas, Venezuela was the cheapest overall and it had by far the most productive wells. In 1938, Caribbean's 121 wells each produced over 13,000 tons a year for a total of 1.6 million tons. Shell Union worked 955 wells each yielding 3,296 tons per year, or 3.1 million tons. Bataafsche in Indonesia had no fewer than 1,717 wells giving only 2,500 tons a year each for a total of 4.5 million tons. In other words, Group production in Indonesia was about three times that in Venezuela, but from fourteen times the number of wells. These wells were rather deeper too, 1,000-2,000 metres in Indonesia against 700-1,000 metres for Venezuela.[37] The operations in Indonesia were far more labour intensive than those in the other two areas, but the use of cheap unskilled Asian labour kept its upstream costs reasonably competitive: at 7.5 guilders per ton the crude produced in Indonesia compared well with 7.4 per ton for Shell Union. Venezuela remained cost leader at only 5.7 guilders. Overheads made the main difference between the three areas. Bataafsche carried Central Office charges of 1.44 guilders per ton, almost 10 per cent of total cost and ten times those for Shell Union and Caribbean. Depreciation charges were also much higher, 3.20 guilders for Bataafsche against 1.49 for Caribbean and only 0.82 at Shell Union.

Consequently, the Indonesian operations found themselves between the devil and the deep blue sea, having the lowest revenues and the highest costs of all the Group's main producing areas. As for the cost side, the company's main problem would appear to have resided in its lopsided organization, with a heavily centralized administration running largely decentralized operations. The Hague continued to exercise close control over innumerable details which, if left to the discretion of the general manager in Batavia, would not have generated such a voluminous correspondence about labour relations, wages, individual employees down to clerical level, contracts with suppliers, office accommodation, discussions with the colonial authorities, and a huge range of other minutiae.[38] In his turn, the general manager kept a similar close rein on the area managers, but these managers were quite independent in operational matters such as barrel cut, crude yields, product balance and cost, or stores management, as long as they kept within their budgets. Regular meetings between them helped to coordinate operations, but evidently not to the degree required to create an integrated whole. The Bataafsche board scrutinized budgets, investment proposals, and annual results; details about costs and performance remained buried deep down in the organization and were unavailable for comparative analyses.[39]

NOTE ON THE TRAINING OF NATIVE OBSERVERS
FOR TORSION BALANCES.

In the Dutch East Indies and Sarawak attempts have been made to train natives for operating our gravity instruments.

The advantages of this system are obvious. Whenever native observers can be trained to shift the instruments, the cost of the survey may be considerably reduced. A larger number of instruments can be placed under the supervision of one highly paid European observer and this observer can make the necessary computations, study the results, determine the densities of the formations, etc. , devoting only a relatively small portion of his time to the supervision of the fieldwork and inspection of the instruments.

This system not only reduces the actual costs of a survey by decreasing the ratio of the number of European observers to the number of balances in operation, but also increases the efficiency of the work. When the observers have more time at their disposal to study the gravity method and are not constantly engaged on routine work, a better cooperation between the geophysical and field-geological staff can be reached.

In some instances the attempts made to train intelligent natives have been successful. We realize that field conditions and other special problems may make it impossible to introduce this system throughout the Malayan Archipelago, but we believed that it might be of interest to mention the attempts that have so far been made.

THE HAGUE, 27th January 1930.

[21]

Although training for local personnel began early, it was usually confined to basic tasks. However, torsion balances were delicate instruments, and as this Note of 27 January 1930 shows, local people could be entrusted with their erection, dismantling, and transport.

Moreover, Bataafsche's heavy reliance on cheap labour carried the penalty of low productivity. In 1929 Bataafsche produced nearly eighty tons of crude per head of staff, against 220 for Shell Union. Bataafsche managed to double its productivity to about 160 tons a head during the mid-1930s, by which time Shell Union staff produced 230 tons. The only way to increase productivity still further would have been to raise the skills levels of the Asian workforce. In 1930 Bataafsche did start with a programme, as successful as it was modest, to train Indonesians for geological field work, which was expanded into a more comprehensive course in production and exploration technology by 1941-42.[40] Otherwise the operational model remained very much the same, a small group of Europeans overseeing large numbers of Asians. In 1929, Bataafsche had about twenty-four Asians working for every European, not counting the contract labourers. The ratio then dropped because the redundancies effected fell disproportionately heavily on the Asian personnel, Bataafsche wanting to retain as many Europeans in managerial and skilled positions as possible. The European staff was halved, the Asian staff cut by 70 per cent. When in 1934 the trough had passed and recruiting resumed, more Asians than Europeans were hired, so that by 1938 the ratio had risen again to nearly thirty.[41] A fundamental shift in employment and training policy was probably unthinkable in the colonial situation and had to wait for the profoundly changed circumstances of the 1950s to become feasible.

Meanwhile Indonesia became less and less important for the Group as a whole. During the 1920s, Bataafsche's operational revenues, i.e. the income from Indonesia, had generated some 30 to 40 per cent of Group profits, peaking at 45 per cent in 1928. By the late 1930s, the amount had sunk to just over 20 per cent. At the same time Bataafsche changed from an operating company with some share holdings to a holding company with some operations.

In the 1920s dividends generated no more than 10 to 20 per cent of Bataafsche's earnings, operational revenues accounting for the rest; by 1936 it was half-and-half, and in 1940, 65 per cent to 35.

Thus the 1930s crisis exposed serious flaws in the Group's Indonesian operations. None of these could be remedied by the kind of pruning which Shell Oil applied; only decolonization would force the necessary changes. As political resistance grew against the modern imperialist model of colonial exploitation, so its economic foundation had started to crumble as well. We do not know what managers considered in the way of policy changes following the cost comparisons. By the time that these arrived it was too late for fundamental changes anyway, since the preparations for war imposed other priorities.

Joint efforts, diverging policies One obvious remedy for the drop in oil prices was to curtail production. As with other commodities suffering from overproduction and low prices, restricting oil production proved very hard to achieve for several reasons. It needed worldwide agreement between a huge number of producers plus the governments of exporting countries, none of them prepared to sacrifice tangible revenues for uncertain gains. Moreover, restriction required setting basic figures for production and exports, leading to endless debates about fair quotas and ways of monitoring them. Finally, if agreement could have been reached, it would have needed control over production backed by legal sanctions in each of the countries concerned, an impossible feat of coordination. In 1931-32 Kessler attracted considerable publicity with proposals to restrict production by a system of financial incentives managed by an industry body. However, for all Kessler's flexibility in meeting the inevitable criticism with changes to suit particular circumstances, his schemes were doomed to founder on the industry's sheer complexity as others had done.[42]

The majors were broadly in favour of restrictions but, as Kessler had written to Teagle immediately after Achnacarry, there were simply too many producers for the majors to be able to control production and markets in the way intended.[43] Godber was also sceptical about the scope for cooperation, because the Group and the American companies disagreed on policy. In November 1931, he wrote to Richard Airey, at the time a director of Shell Union and Asiatic Corporation in New York: 'There is a fundamental difference between our policy and that of most of the American companies. Particularly at this time we must at all costs maintain our position in the market. The Americans, on the other hand, are prepared to sacrifice to some extent their position in order to maintain high prices. We feel that this is not only wrong from the point of view that with so much production shut in and declining consumption generally we must keep the largest possible outlet for our supplies, but also from the point of view of holding our position strong against outside parties.

Undoubtedly the trouble in which we find ourselves in so many markets today of severe competition from bootleggers of one kind or another is due to having maintained in the past so wide a margin between c.i.f. prices and retail local prices. We have fought against it and have as a consequence become very unpopular with the Americans. They, on the other hand, have always tried to squeeze every possible penny from the pockets of the consumers, with the result that outsiders have been enticed into the market by the large profits and have built up permanent organizations, and now we are in the position of having to face bare costs in order to prevent them taking more of our trade (...). Reviewing our policy in the light of your letters and what Van Eck has told us, I cannot help feeling still that ours is the more logical and businesslike procedure. Why it should result in such hard feelings on the part of our friends I do not know. We have never hesitated to disclose this attitude. We have never reduced prices without giving them our reasons, and although, admittedly, they have often only followed us after a deal of persuasion, there is no doubt that in every case they have, even if reluctantly, finally agreed to follow our suggestions.'[44]

A few months later, Godber reiterated his reservations about cooperation in a letter to Shell Petroleum president R.G.A. van der Woude: 'There seems to be always a fundamental difference between the viewpoint of our Group and the Americans. We like to face a problem and take the consequences, whereas they nearly always prefer a compromise even if that means weakening their general position. I quite agree with you that cooperation is a good thing and a very necessary thing, and of latter years we have been and still are all in favour of doing everything we can to improve cooperation within the industry, and that being the case there is nothing much for us to do but to fall into the line on this particular problem. But I am not yet fully convinced that the policy itself is the right one.'[45]

Figure 7.4
Crude production, world total, Jersey Standard, and the Group, 1925-38 (1929=100).

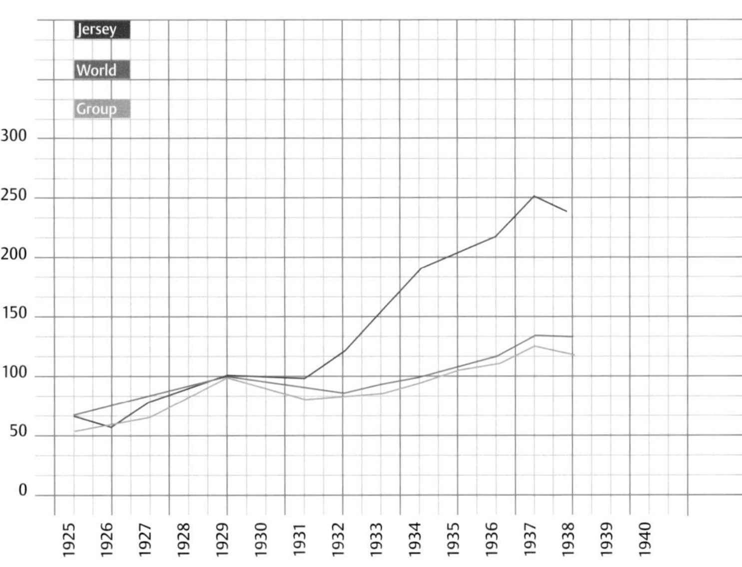

As we will see, the Group could be quite pragmatic when it came to applying its principles, but even so there was a marked difference with Jersey Standard in production policy. The Group tailored its production to follow demand, but Jersey did not (Figure 7.4). World production declined from 1929 until 1932 and then started rising again. The Group closely followed the overall trend, total production falling from the 1929 peak of almost 480,000 b/d or 25 million tons, to 390,000 b/d, 20.5 million tons, two years later, before slowly climbing back, achieving a new peak in 1935 and ending in 1938 at about 30 million tons, 570,000 b/d. Jersey, however, with Teagle determined to increase worldwide production and lessen dependence on purchases, held production at around 240,000 b/d or 12.5 million tons a year during 1929-31 and then began expanding at a fairly rapid rate. This was done to a large extent through acquisitions such as the Lago Petroleum Corporation in Venezuela, and thus not in conflict with the As-Is intentions. By 1935 Jersey's production had doubled; in 1938 the company drew level with the Group on a total of 560,000 b/d, nearly 30 million tons.[46] Since Jersey still bought around half of its crude, it now handled rather more crude than the Group. Shell Oil normally purchased almost two-fifths of its crude, Asiatic probably one to two million tons a year, so nowhere near the volume that Jersey did.[47] Most of Jersey's new production came from abroad; the percentage of US-sourced oil in total crude production declined from 55 to 60 per cent in the late 1920s to 30 per cent during the second half of the 1930s.

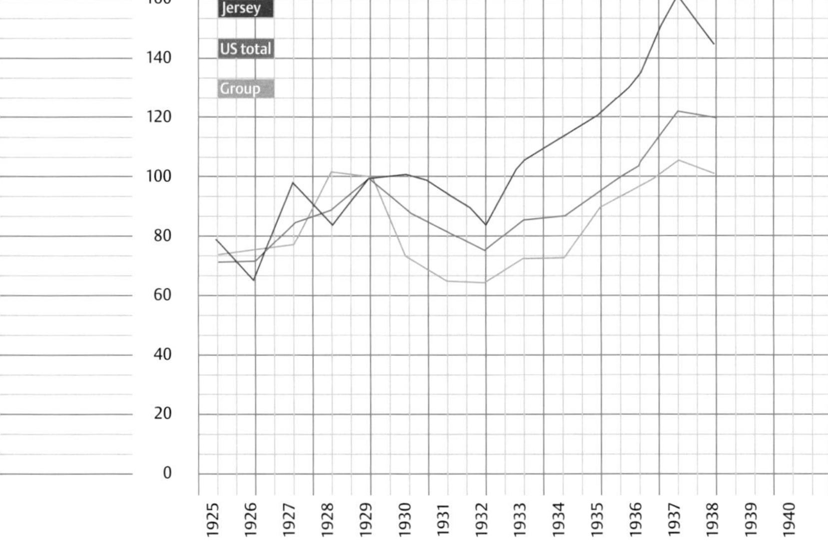

Figure 7.5
Crude production in the US total, Jersey Standard, and the Group (1929=100).

Chapter 7

[22]

[23]

[24]

[25]

Gradually introduced worldwide, rotary drilling required more men than cable or percussion drilling but could cope with more difficult conditions. This selection shows the massive size of the parts of a rotary rig, with drill pipe being run into the hole (top left), preparations to hook on the swivel (top right), the swivel and 'kelly' being raised from their storage in the 'rat-hole' (bottom left), and preparations to pull the kelly out. No one knows the origin of the name, but a kelly is a long metal bar that connects the swivel above to the drill string below. The worker with his foot on the equipment is contravening safety laws, since his foot could easily be trapped between the heavy metal components.

However, in the US Jersey also expanded more rapidly than total production, whereas the Group chose to follow developments at a distance, preferring to conserve cash and oil reserves (Figure 7.5). Low crude and gasoline prices inspired Shell Union to increase its purchases and shut wells. By contrast, Jersey expanded by buying up leases and properties. Let them spend, Godber commented, once they have used up their resources the Americans will come to see sense and curtail production.[48] Interestingly, from a practical viewpoint both strategies, though outwardly opposite, yielded roughly similar results. Over 1930-38 Jersey net income amounted to $ 506 million, net profits of Royal Dutch and Shell Transport to 494 million, a negligible difference of 2 per cent over a period of eight years.

For all their outward differences the Group and Jersey did agree on a policy for Venezuela. Production there had grown very rapidly during the 1920s, contributing to the developing glut, but contrary to the US, the big oil companies dominated the industry and there were hardly any independents to spoil their game. In December 1928 Teagle organized a meeting between representatives from the main producers Jersey Standard, Gulf, and Standard Indiana to discuss production restrictions in Venezuela. The Group did not attend, but was kept informed of the proceedings through Asiatic Corporation. Having long debated the legal implications of any agreement, the companies concluded that 'no agreement should be signed, but that all parties should meet round a table and verbally agree to observe the arrangement'.[49] The debate then moved to a consideration of production levels of the three leading companies, the Group, Gulf, and Lago/Standard Indiana, Gulf wanting to increase its share from 26.6 per cent to 30.8 per cent at the expense of the Group. Finally in November 1929, an agreement was signed aimed at stabilization of production levels during the first six months of 1930 at their 1929 level. Writing to De Kok about the agreement, Airey commented that 'the restriction is not drastic and depends largely on how the Agreement is carried out' but that 'we have temporarily at least prevented further expansion'.[50] De Kok clearly considered Venezuela as a test for the Achnacarry ideas, replying to Airey that 'I shall be very interested indeed to see how this Agreement works, as it has loopholes for adventurous experiments', and hoping that 'the scheme which has now become effective will open the way towards closer cooperation between the various companies operating in Venezuela so that they can add their quotas to the endeavour which is being made to get the petroleum industry as a whole on a more rational basis'.[51] Despite continuing friction between the participating companies, the restriction agreeement in Venezuela held. Output remained more or less stable for the next five years, as did the relative shares of the big producers. At one point the companies even started discussing bringing their interests under a single management in order to control oil production more effectively. The Group did not like the idea, because they considered their position superior to that of their American competitors, and because 'such a fusion would undoubtedly bring about some measure of preponderance of American management', which was considered to be 'objectionable'.[52]

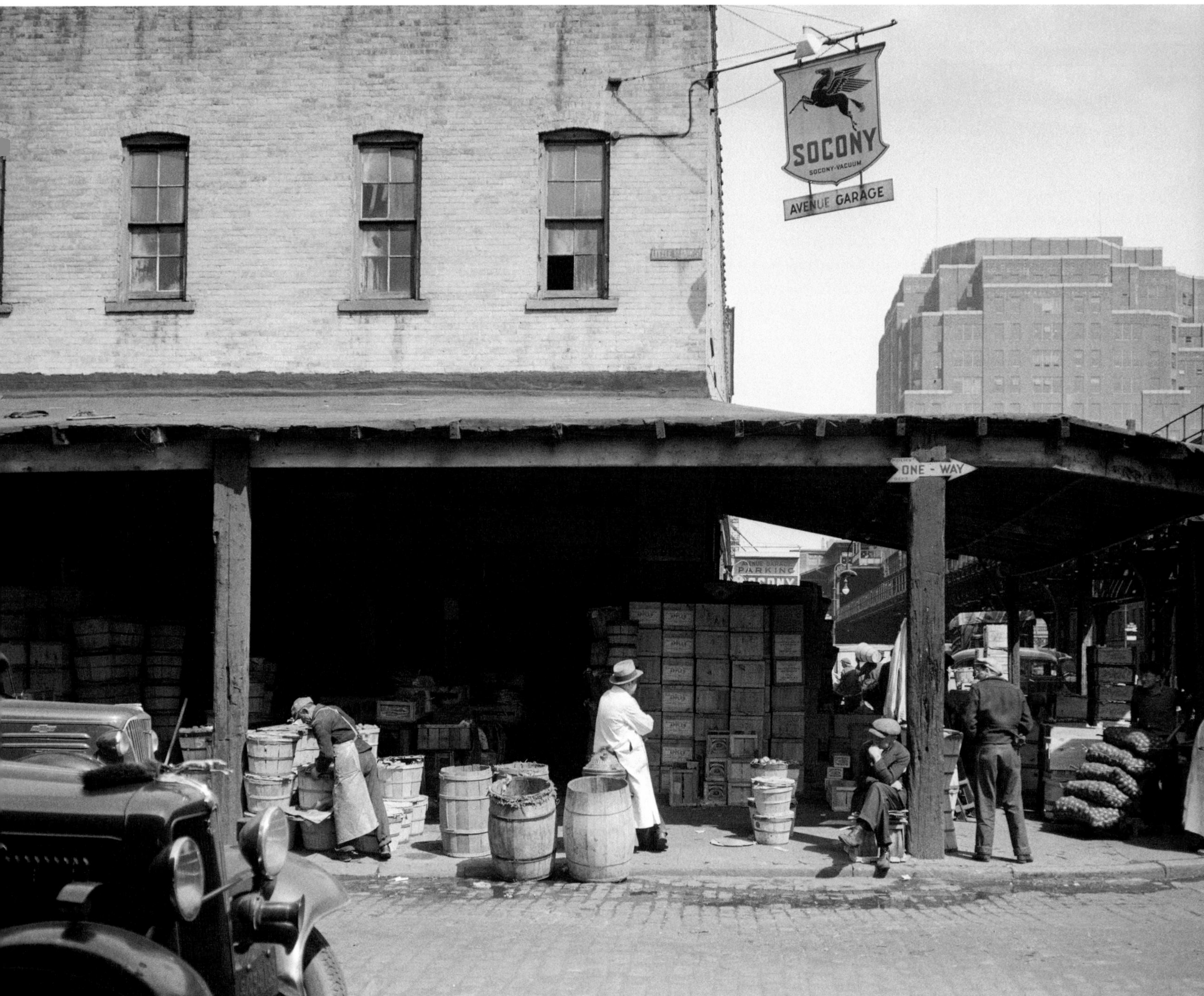

[26]

Taken somewhere in America in the 1930s, this atmospheric photograph shows a Socony outlet.

However, the success of production restraint in Venezuela remained incidental and largely due to the dominance of the majors there; world crude prices began to recover only when the US market gradually stabilized. Since the late 1920s, the major oil companies in the US had advocated prorationing, i.e. restricting production to market demand, and unit development, ending competitive drilling to exploit a given field by treating it as a production unit shared between the parties concerned. Some states had introduced legislation to curb excessive production and waste of resources, but such regulations had been swept away in the flood emanating from East Texas. In 1934 Roosevelt's incoming federal government introduced comprehensive measures to enforce state production quotas, combined with a tariff to keep imports from spoiling the efforts. Prorationing and unit development now began to make headway, resulting in prices rising to a fairly steady $ 1 a barrel. This stability created the necessary conditions for restriction initiatives elsewhere, for instance in Romania, finally to succeed, inaugurating a gradual worldwide recovery of crude prices.

For all the contrasting opinions and policies, the Depression failed to kill the spirit in which the major oil companies had met and agreed to divide markets, and notably strengthened the links between the Group and Jersey Standard. Since the days of Asiatic and Anglo-American having neighbouring offices in Billiter Street, the two companies had had personal ties between individual managers, fairly close ones in the case of Deterding and Teagle, building on the rapport struck up between them during the latter's stay in London as Anglo-American manager. Directors on business trips to London or New York would visit company offices as a matter of course; the disdain with which Archbold had treated Deterding on his first visit to 26 Broadway in April 1907 had gone and had been replaced by mutual respect.

In the wake of Achnacarry these links assumed the more definite shape of regular talks, often with Anglo-Persian as a third participant. In January 1929, the three companies signed an agreement with Russian Oil Products limiting the exports of oil from Soviet Russia. On the basis of As-Is, Jersey, Anglo-Persian, and the

[27]

Piping oil into a Socony tanker, 1939

07|442

[28]

In 1930, at Birkenhead in Australia, a new storage tank for 2.5 million gallons (11 million litres) of oil was big enough for a novelty cricket match; but like the enormous tank farm at East Plaju, it silently asked the industry's most troublesome question: was too much oil being produced, or were consumers buying too little?

Group established product quotas for European markets in 1930, revised twice in subsequent years to deal with various issues that had arisen. In June 1934, the companies accepted an overall As-Is agreement giving a framework for their local representatives to follow.[53] The arrangement included a special procedure, overseen by a Central As-Is committee, to deal with any differences of opinion. Jersey and the Group collaborated to achieve some control over the output of Romanian independents and concluded market agreements for various countries around the world, down to such small markets as Curaçao. An agreement covering fuel oil bunkers was also discussed, but the outcome is not known. Subsequently Socony Vacuum also came into the fold.[54] In October 1931, the Group and Anglo-Persian agreed to merge their British marketing organizations, forming Shell-Mex & BP Ltd (SMBP), a 60 to 40 joint venture with a combined gasoline market share of more than 45 per cent. Jersey was invited to join, but declined. By merging the installations, depots, and offices, of the two constituent companies, SMBP achieved substantial savings in UK marketing, but failed to stem the very gradual erosion of its market share.[55] On a different front, the Group came to an agreement with Standard of California and Texaco not to compete in the Asian markets on octane numbers, thus avoiding a strain on its stretched gasoline supplies.[56] In 1934, the Group took part in forming a Tanker Pool with three other oil companies and a group of independent tanker owners, aiming to stabilize charter rates.[57]

Moreover, the companies also began regularly discussing a range of subjects with each other, from relations with competitors to general policy on the building of refineries in consuming coun-

[29]

PERSONAL
27th January, 1933

Mr. J.C. van Eck,
NEW YORK

My dear van Eck,

For some long time now we have been thinking about your compensation. Much to our regret last year, as you know, we did not feel that we could recommend to the Shell Union a special bonus for you and your colleages. I do not know whether the results for 1932 will enable us to do so either, but I do feel that we cannot go on any loger on the present basis of your fixed salary, and that even if a bonus cannot be proposed we can at least make some adjustments in your fixed income to compensate to some extent.

I therefor propose, if it is agreeable to you, to suggest to the Executive Committee that as from 1st January 1933 your salary shall be at the rate of $ 60,000 a yeaar, and I suggest that you bring this formally before the Executive Committee, if it is the practice to do so, at their next meeting.

I enclose herewith copies of letters which I have addressed to Messrs. Legh-Jones and van der Woude, which will also need to be confirmed, and I enclose an original letter to Mr. Carter which, if you agree, I would ask you to hand to him.

I do not need to tell you how much we appreciate all that you and the other Representatives are doing under the mostdifficult circumstances, and how much we would like conditions to improve so as to give us the justification we [illegible]

tries, the use of flags of convenience, product swaps, aviation fuel quotas, and suchlike. Such conferences could last up to five days.[58] Product swap agreements, whereby the companies achieved economies of scale by sharing supplies of particular products between themselves, necessitated the exchange of specifications, grades, standards, and tests, leading to standardization. By 1937 Jersey and the Group had set up a joint Supply Committee in London to work out how they could best complement each other on markets across Asia. A typical swap would, for example, have Bataafsche supply gasoline from the Dutch East Indies to Jersey in Australia, to be compensated with similar grade gasoline in California.[59] With Anglo-Iranian the two companies formed a Coordination Committee with the aim of coming to a comprehensive agreement for sharing crude and products around the world, the US of course excepted since the anti-trust legislation made such arrangements illegal there. Jersey wanted a long-term deal, for five or even twenty years, but the continuous changes in crude supply and product specifications made this impracticable so in December 1937 the companies agreed on annual reviews of the situation carried out by the committee.

The As-Is collaboration had all the trappings of a cartel, but it was a far more sophisticated game than merely an attempt to corner markets. The global crude surplus set definite limits to the companies' market power and managers showed themselves acutely aware of those limits. Consequently, the effectiveness of joint action in supporting prices or in countering unfavourable legislation depended entirely upon the circumstances. The companies could cajole and persuade, but they could not push without opening the door to independent oil companies. Even the tight German gasoline cartel had to tack carefully in the face of constant pressure from outside suppliers.[60] Moreover, despite being in close touch and often broad agreement with each other, the Group and Jersey did not necessarily agree about the merits of a particular case, or about the best line of action. This became particularly evident in their pursuit of a common policy on refineries in consuming countries and on building hydrogenation plants, about which more below.

However, with their combined market power the companies probably did succeed in lessening considerably the impact of the price erosion during the Depression. Once again the absence of consistent figures renders an accurate assessment impossible, but the available fragmentary data are clear enough. During 1932 and

During the Depression, even Shell directors suffered cuts or stand-stills in their pay – but they remained very well paid. On 27 January 1933, the Depression's worst year, Godber wrote to senior directors, including (left) Van Eck in New York, regretting the lack of a bonus payment the previous year. The senior directors were offered an increased annual base salary of sixty thousand dollars. At that time, average annual per capita income in the US was well under one thousand dollars. By 1937 the worst was over and the launch (right) on 23 March that year of the Dutch-flag *Marisa*, one of the 'Triple Twelves', was greeted with cheers and traditional champagne. Overleaf, *Marisa* wears the house flag, introduced in 1932, and is seen at sea.

[30]

1933, crude oil prices dropped by almost 23 per cent, but revenues did not fall by anything like that percentage. Unit revenues at Shell Union, which operated on the most competitive market, fell by 17.5 per cent, but total income at Asiatic, the Group's main marketing arm, was down only about 7 per cent on a roughly similar volume of sales. A unit price comparison of kerosene in thirteen areas showed price reductions in four markets, but increases for the other nine. Similarly, gasoline unit returns were down in Asia, but up in the biggest market, Europe.[61]

The Dutch East Indies was definitely the least competitive market, dominated by the Group with a share of over 75 per cent. Consequently, price divergences were widest there. Between 1928 and 1934, the trough of the Depression, the cost of living for the area fell by over 60 per cent. However, the retail prices of Crown oil declined by only 35 per cent and the import prices of Jersey's Devoes brand by even less, 15 per cent.[62] From 1932 to 1933, Asiatic's kerosene unit prices halved in the eastern region, i.e. Asia-Pacific, but dropped by only 14 per cent in Indonesia.[63] In 1935, the Group and Jersey's subsidiary Koloniale started examining 'how far it might be possible to effect economies in the production and methods of manufacture of oil in the Dutch East Indies' by exchanging information on 'details of production, i.e. qualities of gasoline and kerosene in regard to all refineries in the Dutch East Indies', and by 'examining how far the requirements for Far Eastern countries can be supplied from the Dutch East Indies and how far a deficit in the requirements can be made good by supplies from elsewhere'.[64] As a consequence, unit returns remained high there. In 1939, Asiatic's gasoline revenues in fifteen areas averaged 0.35 guilders, but in the Dutch East Indies revenues were 0.52.[65]

We may thus conclude that the As-Is collaboration enabled the three oil companies to keep a hold over some of their most important markets. Crude oil prices may have determined their overall performance more than any gains from As-Is, but these gains were clearly very considerable indeed.[66] Moreover, the Group must have reaped additional efficiency benefits from the various marketing agreements, product swaps, and other coordinated actions, though we must keep in mind that there was also a hidden cost of the immense managerial and clerical effort required for drafting and monitoring the complicated contracts underpinning the deals. In 1938, Jersey Standard gave formal notice of withdrawal from all As-Is arrangements, signalling a clear change of policy following Teagle's resignation as chief executive.[67] Eighteen months later, war brought the various arrangements to an end, though a spirit of cooperation between the majors survived well into the 1950s.[68]

[31]

[32]

Turning the corner For the Group as a whole, the crisis bottomed out during 1932. By January 1933, the Anglo-Saxon managers had mustered sufficient trust in a recovery to give the London staff a modest pay rise, repeated the following year.[69] During 1933 business picked up to such a degree that ships which had been laid up were brought back into service; in November, the Group ordered twenty tankers of a new design, the so-called Triple Twelve class, which could carry a load of 12,000 tons, at twelve knots per hour, while consuming twelve tons of liquid fuel per day.[70] Three years later, both production and revenues surpassed the peak of 1929. The key to this quick recovery would appear to have been the rigorous efficiency drive. Though managers failed to restore the profitability of Bataafsche's Indonesian operations, they did succeed in reducing overall costs from 60 to 70 per cent of revenues to less than 20 by 1934, so the company could be profitable at almost any price level.

The renewed prosperity translated into rising dividends and the resumption of shelved expansion projects. From an unusually meagre 6 to 7 per cent during the early 1930s, Royal Dutch and Shell Transport dividends returned to double figures in 1935. As usual, Shell Transport paid slightly more, 12 per cent in 1934 rising to 24 per cent in 1937, whereas Royal Dutch remained cautious with its funds, awarding shareholders 10.5 per cent in 1935 and around 17 per cent in the following years. In May 1936, Anglo-Saxon decided to go ahead with the Central Office extensions suspended five years earlier. The new offices at Great St Helen's were to have a minimum of internal partitions, reducing cost and creating 'spacious light open offices (...) suitable for accommodating large departments, where staff may be more efficiently controlled'.[71] This was considered an interim solution; two years later St Helen's Estates Ltd. started to buy up more adjoining properties with a view to a comprehensive redevelopment.[72] Anticipating further substantial investments, the Anglo-Saxon capital was raised from £ 25 million (220 million guilders) to £ 40 million or 352 million guilders in 1937, two further increases taking it to £ 56 million (490 million guilders) in 1939. As a consequence, Anglo-Saxon drew nearly level with Bataafsche in terms of capital employed and in total assets. Bataafsche was still by far the most profitable of the three operating companies, generating almost 2.5 times as much as Anglo-Saxon and Asiatic each did, but Royal Dutch valued its investment in Anglo-Saxon twice as much as that in Bataafsche for reasons unknown.[73]

Once the trough had passed, the pace of manufacturing investment resumed as well. As related in Chapter 6, the 1930s were a very dynamic period in manufacturing, with the start of exploration of the interface between oil and chemicals. This boost-

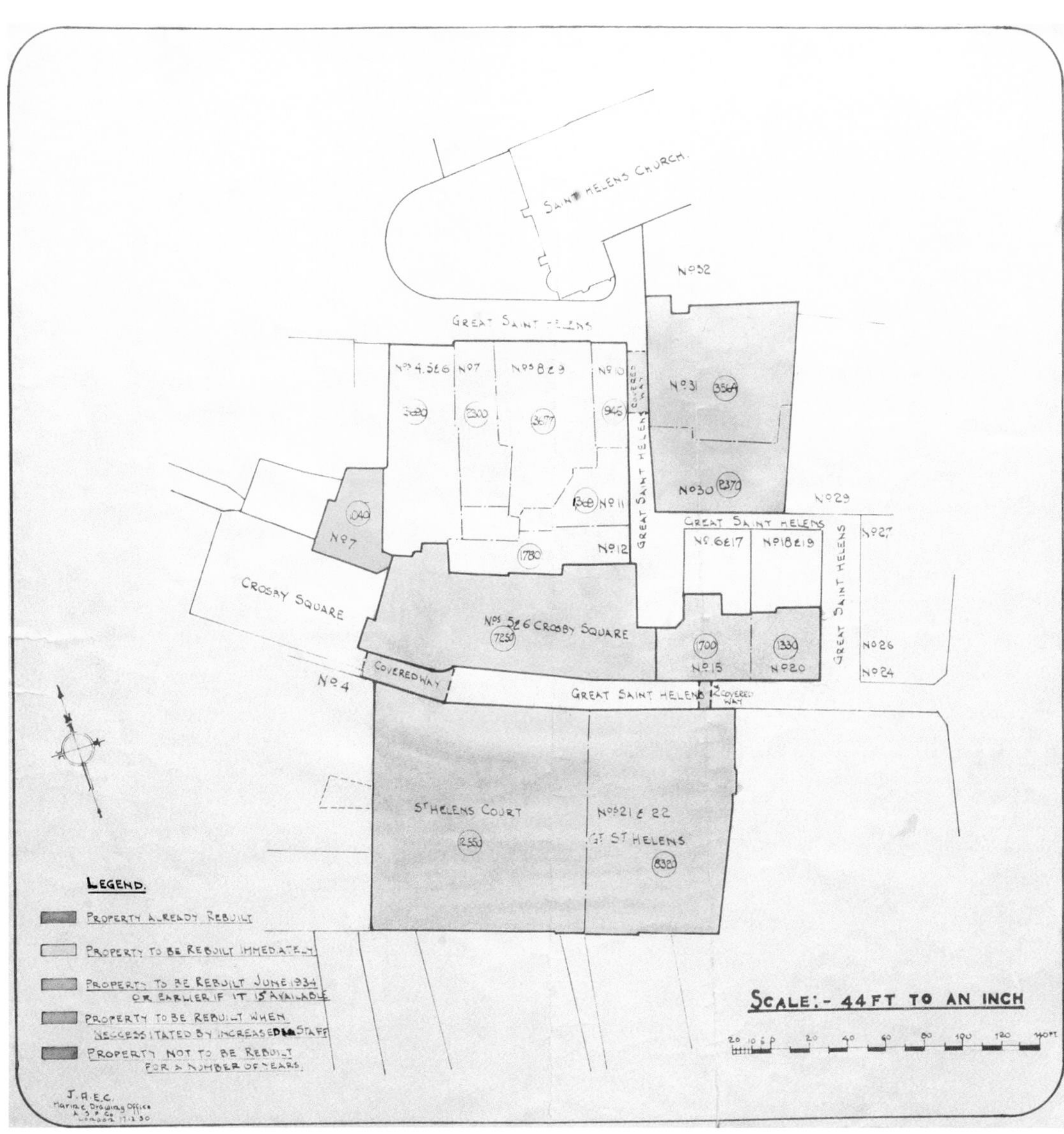

Ambitious plans were drawn up for the expansion of the St Helen's Court offices.

[33]

ed the pattern of investment at two major sites, Rotterdam and Curaçao. In 1927 the Group decided to close the Charlois installations and build a much larger complex at Pernis, further down the Nieuwe Waterweg and closer to the North Sea, part of a new harbour being developed by the Rotterdam city council to accommodate the oil industry. With a capacity of 1.5 million tons a year, Pernis was designed to serve as the Group's main manufacturing centre in Western Europe. In accordance with the desire for flexibility in supplies, the plant could process a variety of crudes into products ranging from aviation gasoline to blown asphalt. The complex incorporated the latest technology, including polymerization and hydrogenation units for manufacturing iso-octane.[74] Construction started in 1933, and the refinery came on stream three years later.[75] Reflecting the rising importance of Venezuelan crude in the business, the Group launched a comprehensive programme of modernisation and expansion on Curaçao in 1934, including the construction of three more Trumbles making a total of thirteen, two high-vacuum distillation plants for lube oil production, three Dubbs crackers, bringing the total to seventeen, a polymerization unit and an alkylation plant for iso-octane production, a butane-butylene plant, and an isopentane installation. With this expansion Curaçao became the Group's biggest refinery and the

[34]

During the 1920s and 1930s the port at Curaçao developed enormously, as the great increase in shipping demanded larger docks and yards. The Group's Trumble installation (above) grew to include twelve units. Nicknamed Trumble Street, it was the largest such installation in the world at the time.

Figure 7.6
The Group's sales by volume and by region, about 1938.

North America	36%
Europe	31%
Africa	5%
Middle East	1%
Asia / Pacific	13%
Central / South America	14%

Table 7.1
Group sales in 1926

	Tons sold	Proceeds £	Tons as percentage total	Proceeds as percentage total
Asiatic	6,070,500	69,631,500	26.3	42.3
Corona	985000	2,558,300	4.3	1.6
Astra Romana	543,700	2,155,300	2.4	1.3
Finska Shell	31,200	461,400	0.1	0.3
Baltic States	16,200	222,700	0.1	0.1
Jupiter	226,600	4,776,700	1.0	2.9
El Aguila	2,771,000	18,395,900	12.0	11.2
West Indies	124,800	455,800	0.5	0.3
Candles Ltd.	90,300	4,926,300	0.4	3.0
Venezuela	2,640,300	8,350,200	11.5	5.1
Central and South America	1,659,500	5,510,900	7.2	3.3
Shell Union	6,319,000	31,422,200	27.4	19.1
Bulk marketing	1,574,800	15,676,200	6.8	9.5
Total	23,052,900	164,543,400		

Table 7.2
Asiatic sales by product, 1939

Volume tons	Percentage	Total	Revenues	Percentage total	Revenues per ton, £
Gasoline	3,283,000	18.1	14,863,000	36.7	4.53
Aircraft gasoline	529,000	2.9	3,344,000	8.3	6.32
Kerosene	871,000	4.8	2,586,000	6.4	2.97
Dark oils	1,1865,000	65.5	13,327,000	32.9	1.12
Lube oil	516,000	2.8	3,734,000	9.2	7.24
Asphalt	597,000	3.3	788,000	1.9	1.32
Other	462,000	2.5	1,888,000	4.7	4.09
Total	18,123,000	100.0	40,530,000	100.0	2.24

third biggest in the world, after Jersey Standard's installations on the nearby island of Aruba and Anglo-Iranian's Abadan refinery in the Shatt El Arab river. In 1938, Curaçao processed 12.7 million tons of crude, over a third of the Group's total. The cracking capacity nearly equalled that of Shell Union and was six times greater than that installed in the Dutch East Indies, a consequence of the large volume of heavy oil coming out of Venezuela.[76]

Three documents at last enable us to get an impression of Group marketing during the interwar years (Tables 7.1 and 7.2, Figure 7.6).[77] Asiatic was both the biggest, and by far the most profitable of the Group's marketing companies, selling 26 per cent of the volume and earning 42 per cent of the revenues in 1926. By contrast, the overextension of the US operations stands out, Shell Union selling 27 per cent of the volume for only 19 per cent of the revenues. Two relatively modest activities did very well, candles and the bulk marketing of fuel oil.

Two other documents give details about the Group's sales pattern around 1940. We have not found data for total Group sales by product, but Table 7.2 shows the figures for Asiatic in 1939.[78] The company sold just over half of all products by volume.

Two-thirds of oil handled concerned dark oils, i.e. liquid fuel, gasoil and diesel fuel, which at the same time yielded the lowest revenues per ton and only a third of the total. Conversely, lube oil, aircraft fuel, gasoline, and 'other', i.e. speciality products such as candles and white oils, provided just over a quarter of the volume but almost 60 per cent of revenues. The Group's world market share for these products averaged 14 per cent, but for speciality products it was 24 per cent and for aircraft gasoline sold outside the USA even 50 per cent.[79]

Figure 7.6 shows sales by volume and by region.[80] Total sales amounted to some 37 million tons, i.e. an increase of about 10 million over 1926, quite good considering the intervening crisis years.

With 36 per cent of the total, North America took by far the largest amount, followed by Europe with 31 per cent. Central-South America and Asia-Pacific came joint third, the former figure flattered by the large sales from the Curaçao complex which were of course consumed outside the region. Africa and the Middle East together accounted for some 8 per cent of sales.

The figures for the individual countries show considerable differences, underlining the degree to which marketing remained fragmented, attuned to very different patterns of demand. With over 3 million tons annually, the UK was the Group's single biggest market after the United States, which topped the list. The singular importance of the British market comes out well when looking at France and Germany, in third and fourth place, both with just over 1 million tons. Even so sales had boomed in France; the Group had sold only 227,000 tons there in 1926. Marketing in Scandinavia had similar striking differences between countries; at 420,000 tons, sales in Sweden were nearly three times as high as those in Norway, Denmark, or Finland. As for Asia-Pacific, the Group now sold slightly more oil in Australia than it did in the Dutch East Indies though, given its dominance of this latter market, revenues were probably higher there.

Of course the volume of products sold gives no more than a tantalizing glimpse of deeper differences in market structure. That these could be considerable is clearly brought out by a comparison between sales during 1935 and 1936 in Germany and in France, which took roughly similar volumes from the Group. As a result of the French government's long-standing policy favouring indigenous refining, the Group refineries at Petit-Couronne near Rouen and Pauillac near Bordeaux nearly covered its market position, so sales to France consisted of almost 90 per cent by volume of crude, mostly from Iraq. By contrast, crude made up only half of the supplies to Germany by volume, intermediates such as lube oil distillates from Curaçao and finished products such as gasoline making up the other half. As related in Chapters 3 and 6, lube oil manufacturing was a speciality of Rhenania Ossag and this enabled the company to earn foreign currency by re-exporting part of its production, notably to the UK. Consequently the overall unit returns for Germany were rather higher than those for France, but this was compensated by the French operating company Jupiter generating higher profits than Rhenania-Ossag. In both cases profits were kept low by high depreciations, those in Germany notably high during the late 1930s as a consequence of the policy to build hidden financial reserves from profits which could not be repatriated under the foreign exchange moratorium. The individual products sold in each country differed to such a degree that Asiatic managers considered it impossible to use the c.i.f. cost of supplies made by the Group to determine which country was the better market.[81]

The volume of crude shipped into Germany and especially into France points to an interesting aspect of manufacturing policy. Supplies to France consisted of 89 per cent of crude oil; for Germany the figure was 50 per cent. Officially, the Group preferred to refine near source as much as possible and for that reason

managers strove to coordinate resistance to building refineries in consuming countries with other oil companies.[82] However, this policy ran counter to the increasing product differentiation. By the late 1930s it was fast becoming outdated for Europe, refining capacity covering nearly 60 per cent of Group sales there, largely a consequence of the 1927 decision to build Pernis as the central plant for Western Europe.[83]

Finally, one consequence of the overall marketing pattern must be pointed out, and that is the importance of marketing in the western hemisphere. In 1926, sales there amounted to 58 per cent by volume; by 1940, after the nationalization of El Aguila two years earlier, it was still well over half. As a consequence, the Group's dollar earnings gradually rose in relation to the sterling income. By the end of the 1930s dollar earnings amounted to some 80 million, or £ 20.1 million, against 34 million revenue in pounds sterling. The Group as a whole was probably a net dollar earner and Bataafsche already had a dollar surplus.[84] This aspect of the business would become very important during and after the Second World War (see Volume II, Chapter 1).

[35]

Miguel Primo de Rivera (1870-1930), Marquis of Estella, ruled Spain as a dictator under King Alfonso XIII (1886-1941) from 1923 until being dismissed by the king for breach of the constitution in 1930, shortly before his death.

New challenges from nationalist policies Since the First World War the oil companies had faced a rising tide of nationalism (see Chapter 4), which continued to swell during the later 1920s and 1930s. In consuming countries, the clearly vital importance of oil gave a powerful push to earlier inconclusive debates about government involvement with the industry, leading to the formation of national oil companies and in some cases even state monopolies. In 1924, the French government fused its wartime controls with its ambitions for an international oil policy into the *Compagnie Française des Pétroles* (CFP); two years later the Fascist regime in Italy, inspired by the French example, followed suit with the state oil company AGIP. Time and again industry insiders predicted the imminent demise of such companies, because of their supposed inefficiency and their lack of experience in the oil business. Sometimes they were right, but more often than not the state companies lasted much longer than expected, over time becoming serious competitors to the majors. The Group initially reacted to state companies and other nationalist oil policies with an attitude similar to that shown to Soviet Russia after the expropriation: a mixture of righteous indignation and contempt followed, wherever practical, by attempts to force a change of policy by refusing to supply.[85] However, the ultimately fruitless resistance against the imposition of one state monopoly, in Spain, led Group managers to develop a different approach which they implemented with varying success when, during the late 1930s, new challenges arose in Mexico and Germany.

Royal Dutch/Shell had only entered the Spanish market in 1920, investing heavily in a distribution network and acquiring a market share of about 40 per cent. The big competitor, as always, was Jersey Standard with a market share of more than 50 per cent. In 1927 Primo de Rivera, the Spanish dictator, announced the formation of a state oil monopoly with the aim of raising government revenues, setting a target of 30 to 80 million pesetas during the first year, climbing to 200 million pesetas after five years. De Rivera also wanted to set up an upstream business to increase exploration in Spain itself and acquire oil reserves elsewhere.[86]

Deterding wanted to counter this move constructively by offering to create a company, controlled by the Group and Jersey Standard, to realize the Spanish government's desired policy. Teagle refused to consider such a plan, however. He thought that it was 'highly inadvisable to make any offer to the Spanish Government', because he felt 'that the risk of losing the Spanish market is much less than the risk of establishing a bad precedent for monopolistic manoeuvres in other countries'.[87] Deterding then began to lobby against the proposed measures, sending a telegram to the Minister of Finance and a memo to the Prime Minister with warnings about, for example, the inability of the new monopoly to guarantee adequate and stable supplies. Such arguments failed to deter either government officials or Spanish business circles. It soon became clear that the government would accept the offer for running the monopoly scheme tendered by a consortium of Spanish banks. In January 1928 the state monopoly was established and conferred on the company established for the purpose, CAMPSA.

The imposition of a state monopoly meant that the Group's properties would be nationalized. The proposed compensation enraged Deterding. He put the cost price of the Group's properties at 37.5 million pesetas plus goodwill, estimated at 8 per cent of asset value. However, the Spanish government's assessment committee initially offered a meagre 30 million pesetas, which was raised, after heated debate, to 32 million, inclusive of goodwill.[88] The compensation terms themselves also generated sharp protests from both the Group and Jersey Standard, for they were not allowed to repatriate the pesetas received in compensation immediately after the transaction, the money had to remain in Spain. Deterding tried everything to turn the tide, visiting De Rivera, which was not a great success, and calling for a boycott of Spanish products in the UK, also to no avail.[89] In the end the oil companies enjoyed some form of revenge. When they were finally allowed to take their money out, the ensuing exchange rate pressure on the peseta triggered a currency crisis which helped to bring down Primo de Rivera in 1930.[90] By that time CAMPSA had succeeded in establishing itself. In their joint anger over the expropriation terms Jersey Standard and the Group had agreed not to supply the company, but it had no difficulty in buying oil from the Soviet Union and from the Texas Company, which latter concern became a particularly important source during the years of the monopoly. The return of democracy in 1930 did not really change things much. The Group tried to re-enter Spain by participating in a company bidding to run the monopoly, but to its chagrin CAMPSA, despite its association with the De Rivera regime, retained its position. When it became clear that the monopoly had come to stay, the Group resolved jointly with Jersey to refuse supplies to CAMPSA, only to find that, owing to the global glut, the Spanish company could buy oil at prices 10 to 15 per cent below market rates.[91]

The Group drew two seminal lessons from the Spanish nationalization. First, it could ultimately not hold out against governments determined to seize its local business; second, only the competition stood to gain if the Group stuck to principles, for there were always oil companies willing to step in.[92] Consequently, the only option was to offer dogged resistance, to keep talking and play for time, hoping for a change in official policy. This new approach was not notably more successful in averting government actions than the old moral straitjacket, but at least it had the merit of being more realistic, more flexible, and of holding out the possibility of getting better terms through continuing negotiations. The first serious test came in Mexico.

In the latter 1930s, Lázaro Cárdenas (1895-1970), president of Mexico from 1934 to 1940, meets some of 'his people'.

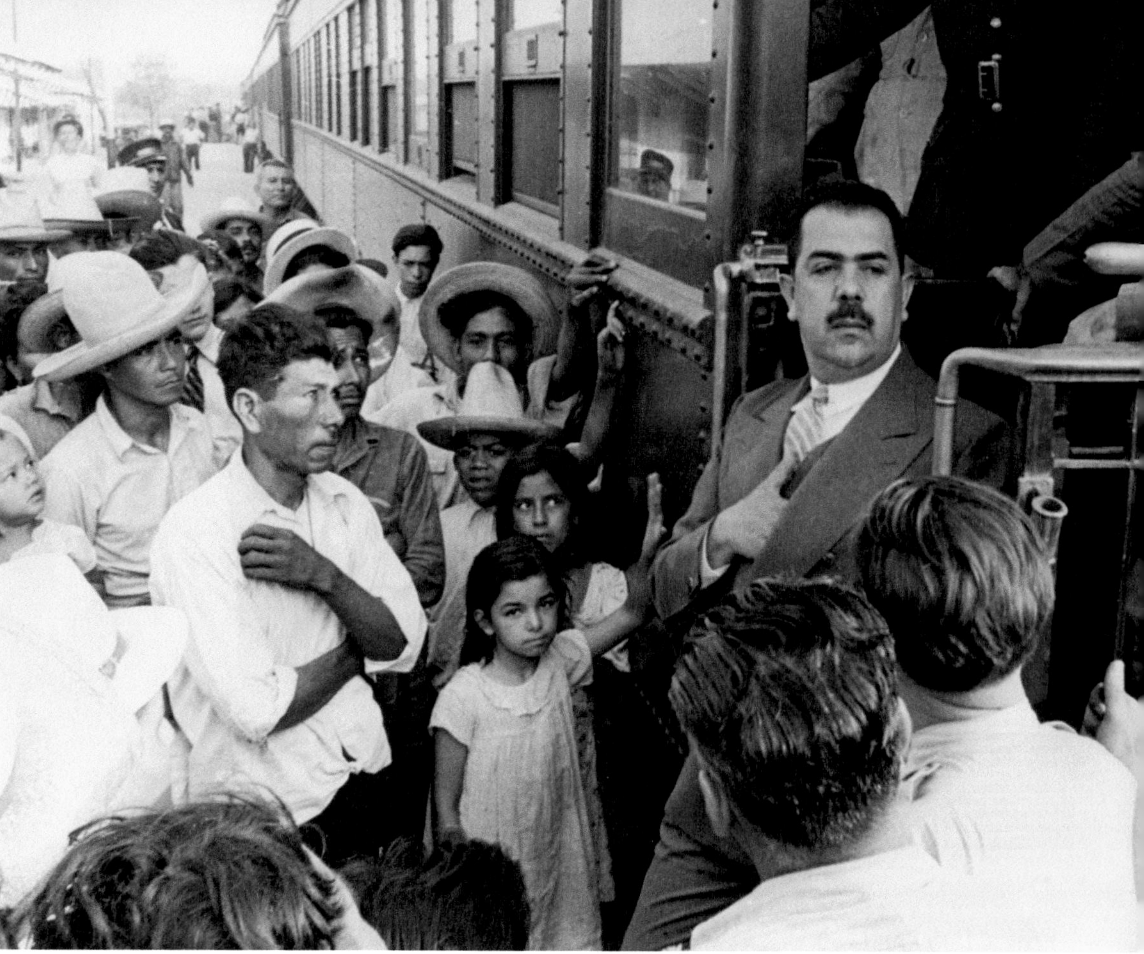

[36]

After the standoff between the oil companies and the Mexican government over Article 27 of the Constitution during the early 1920s (see Chapter 4), calmer conditions set in, allowing the industry to recover. In 1929, the two Group companies Mexican Eagle and La Corona were formally merged into El Aguila, like its predecessors a Mexican company with a capital denominated in pesos and staffed by a large number of Mexicans to comply with employment regulations.[93] The Group retained a minority share of 21.3 per cent in the company, with Bataafsche acting as manager.[94] Despite continuing difficulties, the Group remained optimistic about prospects in the country. The refineries at Minatitlán and Tampico were modernized and extended and a new one near Mexico City came on stream in 1932. That same year a second well came in at Poza Rica, then considered one of the most promising fields around. El Aguila also served as a very useful proving ground for the Group; new techniques such as micropaleontology and new technology like Dubbs crackers or thermal reformers were quickly applied in Mexico. From 1929 operations were reasonably profitable, El Aguila dividends averaging just over 4 per cent during 1929-36.[95] Crude production expanded from 1.5 million tons in the late 1920s to 3 million in 1935 and more than 4 million tons in 1937, nearly two-thirds of Mexico's output and 13 per cent of Group supplies.

However, the Mexican oil industry as a whole never recovered from the tumultuous 1920s. Global overproduction and the Depression reduced the appetite for investment in a country with a high cost level and an unstable political climate, the more so where neighbouring Venezuela offered both lower costs and stability. Consequently, Mexican oil lost its position on the world market and total crude output dropped precipitously from over 20 million tons a year during the early 1920s to around 5 million in the early 1930s. The decline of the oil industry helped to fuel a resurgence of nationalism and trade union agitation, both directed at the foreign oil companies as the supposed architects of Mexico's misfortune.

[37]

[38]

The nationalization of Mexico's oil on 25 March 1938 was one of the most momentous events in the nation's history, restoring the traditional Spanish custom by which all minerals in the sub-soil belonged not to individuals but to the State. Below he meets a delegation of oil workers; and above, standing second from left, he consults a map with officials and engineers taking him on a tour of the nationalized properties.

Riding on these forces, the government began to impose new handicaps on the oil companies, refusing applications for the confirmation of rights and concessions, turning down immigration permits for foreign staff, threatening to nationalize pipelines, and forming a national oil company, Petróleos Mexicanos (Pemex), which received as favours the concessions and permits which the other companies sought in vain.[96] High-level talks with President Cárdenas did not yield the desired results and diplomatic intervention was counterproductive.[97] By September 1934, the Group seriously considered pulling out, only to find that doing so would amount to suicide. It was expected that the Mexican government would frustrate any attempt to leave by imposing special taxes and other obstacles, taking further reprisals if the policy were still carried out. Patience and trying to create goodwill by cooperation was the only alternative under the circumstances which, it was hoped, would not last longer than five years. As a way of furthering El Aguila's standing in official circles, managers considered forming an alliance with Pemex to help that company get on its feet by giving much-needed technical assistance. Another way was improving labour relations, by training more Mexicans for staff and management positions, and by comprehensive housing and other welfare programmes.[98] Central Offices and the El Aguila management were in broad agreement on these points, resulting in a programme of what would later be termed staff regionalization.[99] To create a better atmosphere, the Group settled a long-running dispute about the rights to one concession at a substantial cost.[100]

It remains a moot point whether the foreign oil companies could have saved the situation by showing a greater flexibility towards the demands of the Mexican government.[101] As it was, Group managers in Mexico and in London felt that they were bending over backwards without getting anywhere nearer a solution. By 1934 it was already too late for long-term policies such as staff training programmes or building an alliance with Pemex to have sufficient impact on public opinion in Mexico. Moreover, by taking the handicaps imposed on the company as challenges to their competence, El Aguila managers may actually have increased the prevailing hostility. Overall, the mid-1930s were El Aguila's best ever; production attained a consistently high level and dividends over 1935 and 1936 amounted to a record 10.4 per cent a year.[102]

In 1936 the Mexican government showed its ultimate intentions by the introduction of a law allowing the expropriation of private property of a collective nature in cases of war or internal disturbances in the country. Upheaval was not slow to follow; six months after the law had come into force, the oil workers' union called a strike over pay and conditions in the industry. A government arbitration board found after a month's investigation of the dispute that the union's wage demands were reasonable given the companies' past financial performance, which according to the board had been better than reported by the companies themselves. This accusation of dubious accounting inspired further investigations by the Ministry of Finance and claims for back taxes.[103] On 22 December 1937, the Aguila managers Davidson and Van Hasselt cabled to London that it was better to shut down operations than to accept the award by the arbitration board. Though this might well lead to the government nationalizing the company, 'we really feel that any alternative would only result in our increasing our stake in the country without there being any real likelihood of the return of the financial reserves on which we would have to draw'.[104] The one option left was an appeal to the Supreme Court to overturn the board's decision. Meeting on the same day to consider the situation, the Anglo-Saxon board concluded that 'if we submit to the award we are definitely and certainly heading for bankruptcy and probably confiscation, whereas if we follow the line advocated by Van Hasselt and Davidson there is still a chance that we may in time find a change in the mentality of the Government and once more be given reasonable freedom to manage our own affairs.'[105]

These hopes for a turnaround at the eleventh hour were largely based on the special position which the Group considered itself to have in Mexico as the only foreign company which had consistently tried to cooperate with the government. El Aguila had made substantial investments and expanded production where other companies had reduced their commitments and run down production. Moreover, since 1935 El Aguila had been negotiating with officials to obtain better safeguards for its Poza Rica concessions in return for a loan to the state and increased royalties. During the spring of 1936 the two sides reached a provisional agreement which, despite being signed by a cabinet minister, was subsequently rejected by the government. In May 1937 El Aguila obtained another agreement, again signed by a minister, only to find that the promised signature from Cárdenas did not materialize.[106]

These special interests made the Group reluctant to antagonize the government in any way as long as a chance remained. Consequently, production continued while the oil companies lodged an appeal, but in March the Supreme Court rejected it. The companies then offered to accept the wage demands, but not the additional conditions laid down by the union, which included a far-reaching influence on the management. Two days after this offer, the Mexican government seized all properties of foreign oil companies and transferred them to Pemex. A long and tortuous process of negotiations about compensation for the nationalization followed which need not detain us now as it is covered in Volume 2. What matters for our story here is the patient diplomacy which Group managers displayed throughout the entire process, a conspicuous change from Deterding's disastrous handling of earlier confrontations with governments, and all the more remarkable for the fact that the main protagonists in Mexico were considered to be Communists. It demonstrates how the top management had matured by 1930, as a consequence of Deterding's colleagues coming to understand how to handle him (see Chapter 5).[107]

The Mexican nationalization once again showed the fragility of the As-Is collaboration. Jersey and the Group continued to have different perceptions of opportunities for and threats to their business, sometimes to the point of hampering joint action. This was also evident in the companies' policy towards Italy which, during the 1930s, extended its efforts to achieve greater self-sufficiency by building a national oil industry around Agip. In 1934 the Mussolini government passed legislation, introducing a system in which licences for oil imports were tied to concessions to operate refineries and retail trade access, with the object of expanding refinery capacity in the country.[108] The Group resisted the building of refineries in consuming countries as a matter of principle, arguing that small refineries were uneconomical to operate but, as we have seen, this policy was fast becoming strained in Europe. Location, hinterland potential, and the need for product differentiation mattered more than anything, and the prospects in Italy appeared not to warrant further substantial investment. The capacity of the Group's plant at La Spezia, on the north-western coast between Leghorn and Genoa, amply covered the volume of Shell products sold annually, and a doubling of the cracking capa-

Deterding vehemently opposed Cárdenas, declaring him to be an out-and-out Communist. Although that judgement has often been questioned, Cárdenas's sympathies were sufficiently left-wing for Leon and Nathalia Trotsky to be granted asylum in Mexico City, where they arrived on 9 January 1940.

[39]

city there scheduled to come on stream in 1936-37 would suffice for the sluggish growth of a heavily regulated trade.[109]
The two main As-Is partners dominated the Italian market, Jersey handling 170,000 tons a year, the Group 125,000 tons, leaving 120,000 tons to Agip.[110] Anglo-Persian had sold out to Jersey and Shell in 1930 as part of a collaboration agreement.[111] Yet their position was far from unassailable. During 1934 Socony Vacuum moved in by acquiring a refinery and cracking plant complete with refining rights and import quotas under the new system. Gulf applied for a licence to build a refinery at a location to be chosen by the Italian government. These companies stood to get marketing quotas to match their refining capacity, threatening to upset the cosy arrangements between Jersey and the Group. London immediately asked its representatives in New York to convince Jersey managers of the need for strong representations with the Italian officials, reminding them of the success of joint action against demands for local refineries in Ireland, Denmark, and Japan, and the crucial failure of cooperation over the same issue in Argentina. Jersey would have to convince Socony Vacuum and Gulf not to fall in with the Italian government's policies, since this would encourage other countries to follow the same course.[112]

Group managers clearly thought Jersey could lean on Socony Vacuum and Gulf to make them give up Italy just as Anglo-Persian had been made to do earlier. That was not a very realistic position to take, given the trend towards refineries in consuming countries and the apparent business opportunities in Italy. In contrast to Group managers, Socony Vacuum clearly considered the prospects there to be very good, having paid $ 1 million above the market price for its refinery. Jersey replied accordingly that there was no hope of coming to an agreement with Socony Vacuum given the price the company had paid; if the Group wanted to buy out Gulf, Jersey was prepared to go 50:50 on that deal.[113] This reply left London Central Office aghast. Throughout April, May, and June 1934, Group managers had negotiated with Jersey and Anglo-Persian representatives over the terms of the final As-Is agreement. Would Jersey *ever* understand the principles at stake? A sharply-worded cable went out to Asiatic New York, instructing them to tell Jersey managers that 'We are surprised and shocked friends [i.e. Jersey] callous and complete disregard interest of their friends in shortsighted attempt make capital out of entirely unsound law. They may now cause all of us waste many millions unnecessary plant and equipment. It is not first time friends show complete lack of knowledge of fundamental principles cooperation. We call to mind similar instance Argentine after which friends expressed regret and assured us cannot happen again. If this is to be attitude of major companies towards each other it is certain can expect nothing better from smaller concerns. It is in larger matters that we had expected much from friendly cooperation and if necessary joint action but if in such important matters friends take a decision without consulting and with complete disregard interests of partners we can only wonder whether such cooperation worthwhile.'[114]

Telegram SENT to NAFTA GENOA
IMH. from APCO.
No. of Words 82 No. 29606 Sent on 17.10.35.

288 We have read reports in press that Royal Dutch Shell Group Standard Oil Anglo Iranian and Soviet are considering mutually cutting off oil supplies to Italy in order help sanctions policy stop

We authorise you categorically deny that we have had any discussions or that we have had contact in any way with the above nor with any other parties to this effect stop

We are trying contact Stefani here in order deliver similar message to their representative

AUREOOL

In October 1935 Asiatic Petroleum swiftly issued a categorical denial that the oil majors were going to cooperate in support of sanctions against Italy.

The Jersey board now changed tack and started negotiating with Socony Vacuum about merging their Italian operations, so the Group opened talks with Gulf to buy its fuel oil and lube oil business.[115] While these were going on, Jersey suddenly lost patience with the situation in Italy, the slow progress over the new oil regulations and the increasing difficulties over obtaining foreign exchange payments for supplies. In March 1935 the company wanted to issue a joint ultimatum to the Italian government refusing further collaboration over the refinery expansions. The Group shared this concern, but urged caution, wiring to New York: 'Consider it would be most impolitic to present Italian government virtually with an ultimatum that programme of refineries which, however wrongly, they believe in as keystone of their national defence will be held up by oil importing companies, because it is quite evident that no country now going to give up so called defence proposals in which they believe, however uneconomic or shortsighted they may be. Consequently, if we give such ultimatum we should only be losers, because Italian government could in our opinion easily find, as proved by large number of refinery applications, either local or foreign groups, including other American companies, willing to spend requisite capital for refineries and ensure there is a regular supply of crude. Spanish monopoly at present able to buy their requirements far below market price, although friends and we refuse to supply them. Summarising then, feel that, while naturally very anxious over present situation, it would be very unwise, and probably result in losing valuable outlet, should we carry out suggested policy of New Jersey.'[116]

[41]

[42]

While inhabitants of a captured province of Abyssinia (Ethiopia) offered the Italian fascist salute to a gigantic portrait of Mussolini (left), others took up arms (below) in a vain attempt to resist the invaders.

Only now did the two companies start discussing possible joint representations to the Italian government. Socony Vacuum declined to participate, not wanting to endanger its recent investment; a suggestion from Jersey to bring the company into the fold by making its Italian refinery a tripartite joint venture came to nothing.[117] A memorandum outlining the companies' complaints and wishes was drafted which top executives would personally present to Mussolini.[118] However, another issue claimed priority when, during the summer, Italy's foreign exchange position worsened. In August, Jersey and the Group agreed on a common supply and payments policy towards Italy with restricted credit and payment in foreign currency. Socony Vacuum again remained aside.[119]

Events then took a new turn. Following Italy's invasion of Abyssinia (now Ethiopia) in October 1935, the League of Nations imposed economic sanctions on the country. The League achieved a decision on this embargo only by making an exception for oil, iron, and steel, rendering it largely useless as a means of coercion. The oil companies were thus free to supply, but still in a difficult position. Public opinion called loudly for the suspension of oil deliveries to Italy; curtailing shipments might have serious repercussions for the Italian operations and for the negotiations about refinery policy. On this issue the Group and Jersey quickly found common ground in continuing to supply at normal levels of demand for cash payment only, and to refuse any requests for additional cargoes. When it looked as if the US government might officially impose an oil embargo on Italy, London sent instructions that such a decision should be strictly observed.[120] In May 1936 the war came to an end when the Italians captured Addis Ababa and two months later the sanctions were lifted. The Group's business did not suffer any reprisals for the refusal to increase supplies during the war, quite the contrary, for by the end of the 1930s the volume of products sold had multiplied to over 900,000 tons.[121]

Only weeks after the defeat of the Abyssinians the supply questions raised by the Italian sanctions resurfaced when the Spanish civil war broke out. Jersey and the Group had maintained their refusal to supply CAMPSA, but now the country had two regimes both clamouring for oil. The Group remained steadfast in not supplying oil to any government in Spain, but some managers worried that the policy increased the kind of government interference which the industry tried to combat. A memo written in November 1936 by a senior marketing manager at Asiatic considered the dilemma facing oil companies, starting from the premise that the business had become so tied up with political considerations 'that nowadays we do not so much give service and sell to the public as give service and sell to governments'. In the ongoing confrontation with governments, the industry would lose out because it could never be sufficiently united. Supply restrictions imposed by oil companies could not work because of the competition and such measures only engendered undesirable reactions like the present oil fever, 'the excitable demands everywhere for the piling up of stocks, building of hydrogenation cracking plants and refineries, and the reckless and uneconomic support given to home-produced fuel'. There was therefore no good reason to continue refusing supplies to Spain. The monopoly had ceased to exist and not selling 'may lead to all sorts of unpleasant consequences, for it will be known within a very short time that the Shell, or say the Shell and the New Jersey, have taken on themselves to decide whether the war in Spain is a just or an unjust one, and against that accusation there is no answer'. The logical position was therefore one of neutrality, selling to both sides.[122]

[43]

[44]

[45]

[46]

[47]

[48]

This was the exact proposition which Asiatic had denied to Colijn at Bataafsche in August 1914 (see Chapter 3), but the delicious irony would have been lost on London managers. Though at first inclined to keep the informal embargo in place, the Group relented after talks with Jersey officials and the two companies agreed to joint sales, pooled in proportion to their respective positions on the Spanish market before the monopoly.[123] If the companies had hoped that this would facilitate their re-entry into the market, they were mistaken. By 1940, Group sales to Spain and Spanish Morocco were still negligible at a total of nearly 3,000 tons.[124] Nor did the memo alter the Group's position on supplying belligerents. When Japan started a full-scale war with China in 1938, Jersey and the Group adhered again to the policy of continuing supplies at peacetime level to Japan.[125]

In Germany the collaboration between the three principal As-Is partners appears to have been a little closer than in the other countries discussed here.[126] Again, these companies dominated the local market, in 1937 selling 55 per cent of the gasoline, 72 per cent of the diesel fuel, 76 per cent of the asphalt, and around 50 per cent of the lube oil. For the Group, these figures were respectively 22 per cent, 23 per cent, 39 per cent, and 25 per cent.[127] Two of the four main products, gasoline and asphalt, were tied up in tight cartels, the asphalt cartel being chaired by Rhenania-Ossag's general manager Dr. W. Kruspig. Germany was the Group's second biggest market in Europe and the fourth biggest overall, but it was still a comparatively underdeveloped market because of the low degree of motorization. During the early 1930s, the UK and France each consumed three times as much gasoline per head of population as Germany.[128] Rhenania-Ossag's manufacturing operations concentrated on the processing of imported intermediates, to produce gasoline, diesel fuel, lube oils, and asphalt. Imports from Rotterdam served the north and west of the country, while Romania supplied the southern part. As we have seen in Chapter 6, lube oils had become a speciality following the acquisition of the Hamburg Stern-Sonneborg works in 1925, and a considerable part of Rhenania-Ossag's production was exported to other countries, including the UK. Business boomed during the second half of the 1920s. Profits rose from RM 190,000 in 1926 to 2.4 million in 1929, and to accommodate the expansion the Group increased the company's capital from RM 40 million to 83.6 million. In 1930 Rhenania-Ossag moved its head office from Dusseldorf to Hamburg, into a splendid, modern building, just when the Depression had caused it to plunge back into losses.[129] The Group also had a small E&P operation in Germany, the Gewerkschaft Brigitta, a 50-50 joint venture with Jersey Standard acquired in 1931. After years of fruitless drilling the company was almost wound up when it finally struck oil in 1936.[130]

The Axis takes shape: In 1938 Japan's war against China is a year old (left and far left), and on 10 May Hitler and Mussolini drive joyfully through Florence (right).

[49]

The German foreign payments moratorium imposed in 1931 created the first of the difficulties which were to dominate the 1930s. Suddenly Rhenania-Ossag could no longer obtain pounds sterling to pay Asiatic for its supplies. As an interim solution Asiatic increased its credit to Rhenania to RM 18.1 million, the equivalent of 10.7 million guilders or almost £ 900,000.[131] Though a major nuisance, the freezing of all foreign exchange transactions did not materially affect the German operations in the long run, largely because the Group's global scope offered sufficient opportunities to offset the consequences. The bilateral clearing arrangements which Germany concluded with its main trading partners provided one way to obtain payment. Under this system the revenues from the trade between two countries were pooled in the exporting country for matching and clearing with the revenues from trade going in the opposite direction and pooled in the other country. Exporters received payments in their own currency from the respective pools, subject to the vicissitudes of international trade balances and any quotas imposed on particular categories of payments or goods for reasons of trade policy or equity. To minimize the dependency on individual clearing arrangements, the Group spread its exports to Germany over as many countries as possible. An attempt to use the dual nationality for obtaining a 60:40 split of this trade over the Dutch and British clearing systems appears to have failed, but the draw on the Netherlands was subsequently lowered by 'exporting to Germany via every available sidedoor and backdoor', as De Kok put it, i.e. by directing the trade via operating companies of different nationalities. The Group's belief in the need for flexible supplies paid dividend here. If necessary, supplies to Germany could be spread over at least eight different countries: the Netherlands, the UK, the US, Romania, Venezuela, Mexico, Palestine, and Iraq. Drawing gasoline supplies from Romania rather than, say, Rotterdam was one way of doing this.[132]

The Nazi regime which came to power in January 1933 progressively tightened foreign exchange regulations, but as late as 1937 Rhenania-Ossag experienced little difficulty in finding sufficient currency to pay for its imports.[133] The Group achieved this on the one hand by enabling Rhenania-Ossag to earn foreign currency through exports, putting the lube oil specialization to good use. By the mid-1930s Rhenania-Ossag produced 250,000 tons of lube oil a year, about 60 per cent of the German total. Around 30 per cent of this output went abroad, of which 50,000 tons went to the UK. The raising of import tariffs on lube oil in 1935 gave a further stimulus to production in Germany.[134] During 1937 and 1938 Rhenania-Ossag's lube oil production was further expanded and modernized. Since profits could not be sent out of the country, the company had enough cash for substantial investments over and above the ample depreciations exercised for the same reason.[135] On the other hand, the Group could create foreign currency credits

A Nazi election poster from 1932 (right) heralds Hitler's rise to power the following year, and warning posters (below) show anti-Semitism taking root in Germany: 'Ladies, girls – beware of the rapist! The Jew!' 'Watch out for Jews and pickpockets!'

[50] [51]

Adolf Hitler
hat euch in 13 Jahren zuſammengebracht!
Deutſche Kopf- und Handarbeiter!
Deutſche Bauern!
Laßt euch nicht mehr auseinanderreißen!
Wählt Nationalſozialiſten Liſte 1
FRAUEN, MÄDCHEN
HÜTET EUCH VOR
DEM SCHÄNDER!
DEM JUDEN!
vor
JUDEN
u TASCHENDIEBEN
WIRD GEWARNT!

for Rhenania-Ossag by having the company place orders for capital goods destined for use by other operating companies. During 1935–1939, Anglo-Saxon had a total of seven large tankers built at German shipyards, the largest number ordered outside the UK and the Netherlands during those years. At least three of those tankers were built as straight compensation for oil imported by Rhenania-Ossag.[136] By 1937 the Group ran a surplus on this type of internal clearing. Supplies to Germany amounted to £ 3.8 million pounds sterling or 34.4 million guilders, but the orders placed there during the year came to RM 52 million, the equivalent of £ 4.2 million or nearly 38 million guilders. Asiatic's credit to Rhenania-Ossag had fallen back to normal proportions and it had a regular and sufficient supply of foreign currency.[137]

The Nazi regime also confronted the Group with demands and restrictions of an entirely different kind. The first months of the Hitler government were characterized by great social turmoil. Brownshirt bands acting as auxiliary police roamed the streets, terrorizing their political opponents. The Reichstag fire in February 1933 was blamed on arson by Communists and taken as an opportunity to ban the Communist party and arrest its members and sympathisers. Outbreaks of violence against Jews and Jewish businesses also became widespread, culminating in an organized, national boycott of Jewish businesses on 1 April 1933. A week later the German government enacted a law barring Jews from holding civil service and university appointments, which was followed by similar regulations concerning the professions.[138] Rhenania-Ossag was also confronted with scattered actions. In February, a bomb was thrown at a service station in Königsberg (now Kaliningrad), but it failed to explode. On 31 March, Nazi mobs besieged three

den 3. April 1933.

D.E. -5. APR. 1933

Diverse D

1572

Herrn W. RUDELOFF,

Alsterufer 4/9,

HAMBURG 1.

DE

Sehr geehrter Herr Rudeloff,

Ich habe Ihren Brief über die Lügenmeldungen betreffs deutscher Judengreuel erhalten und sofort in den grossen holländischen Zeitungen bekanntmachen lassen, dass Sie mir mitgeteilt haben, dass in Deutschland die verschiedenen Vertreter der Rhenania-Ossag keine Ausschreitungen haben feststellen können.

Anbei sende ich Ihnen ein Exemplar des "Telegraaf", worin Sie die betreffende Mitteilung finden.

Was unsere Hauszeitschrift "De Bron" anbetrifft, so ist die März-Nummer derselben gerade erschienen, so dass die gewünschte Mitteilung erst im April aufgenommen werden könnte. Ich werde das natürlich gern veranlassen, falls es sich zeigen sollte, dass eine derartige Mitteilung gegen Ende April noch aktuell sein sollte.

Mit bestem Gruss

w.g. J. E. F. DE KOK

1 Ex. "De Telegraaf"

[52]

Early in April 1933, less than a week after Hitler ordered a boycott of Jewish shops, De Kok wrote to Rudeloff (above), saying that he had passed Rudeloff's comments about 'false reports of German atrocities against Jews' on to the main Dutch newspapers. At first the letter seems alarmingly naive, but it was extremely carefully worded and was more probably written in a mood of self-protection – Deterding was still in charge.

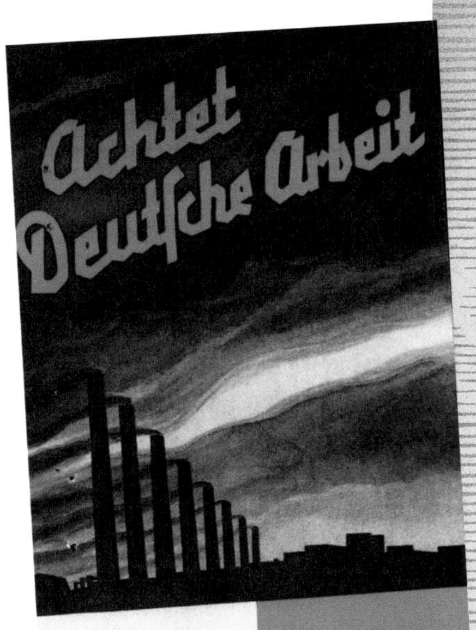

Achtet
Deutsche Arbeit

[53]

LEUNA
DEUTSCHES IG BENZIN
SENGER
BLN

[54]

Together with Jersey Standard, Shell. distributed the synthetic gas oil produced by IG Farben and was a partner in the extension of the project. The town of Leuna in eastern Germany was the hub of I. G. Farbenindustrie and gave its name to the company's synthetic petrol.

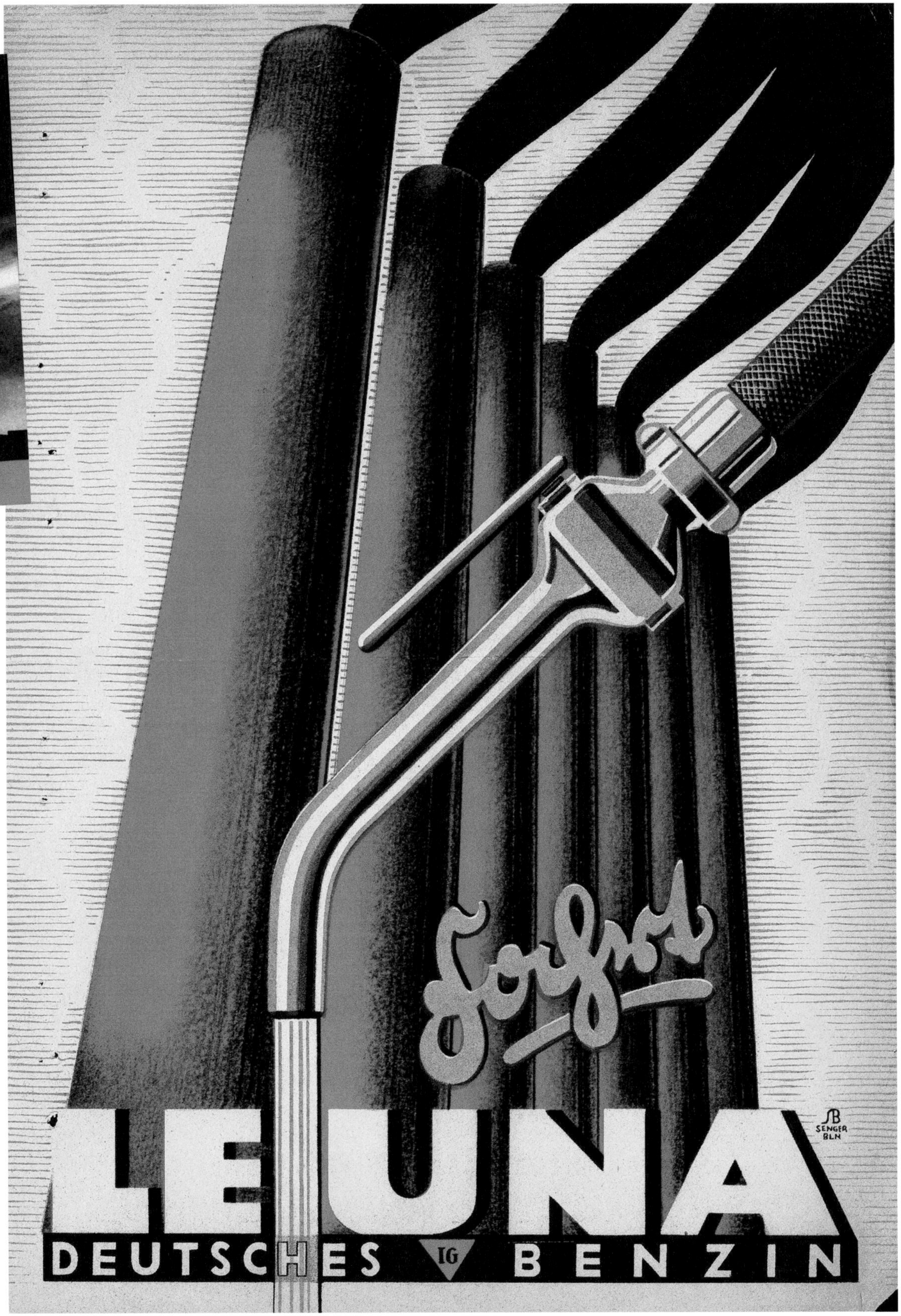

service stations in the Westphalia region, claimed the keys with the intention of closing them as belonging to a Jewish company. And on 3 April the Nazi representatives of Rhenania-Ossag's works council presented the management with demands for the immediate dismissal of all Jewish directors. Two Jewish directors of the Colas asphalt works near Dresden had already resigned after similar pressure there, and the Jewish lube oil sales manager in Berlin had taken immediate leave, shocked by what was happening.[139]

The events convinced the Rhenania-Ossag managers that the company needed to be aligned with the New Order in Germany, and quickly. Though they had earlier attempted to downplay the seriousness of the violence against Jews, asking Group directors to have denials placed in English and Dutch newspapers, such actions had clearly unsettled them, as evident from the relief with which chairman Rudeloff greeted the well-organized boycott as a sign that the government could control the Nazi mob's violent anti-Semitism.[140] In his view, the banning of Jews from the civil service and the professions would surely be followed by similar regulations for private business; the government had already barred Jews from membership of the semi-public branch organizations representing business interests at government level.[141] Accordingly, Rudeloff proposed to transfer a number of Jewish employees to other Group companies since he expected that this legislation would seriously limit their ability to work in the business.[142] During May and June, Rhenania-Ossag's *Aufsichtsrat* (board of directors) was thoroughly overhauled as well. The Jewish members resigned, among them the Stern-Sonneborn founders Jacques Sonneborn and Leo Stern. In at least one case, the resignation does not appear to have happened voluntarily. The appointments to replace them included a member of the Nazi party NSDAP, the second one on the board because Kruspig was already a member. Rudeloff postponed the appointment of a third because he wanted to wait and see how the situation would deve-lop. Rhenania-Ossag also put pressure on two Dutch members to make way for Germans, but this did not happen until 1935.[143] In retrospect the remodelling of the senior management to suit the New Order would appear to have happened with undue haste, long before the introduction of formal legislation, but Rhenania-Ossag did no more than follow the trend in German business. At about the same time, companies as diverse as Deutsche Bank and Unilever took identical action.[144]

The far-reaching changes to the Rhenania-Ossag board could not have taken place without the full consent of Central Offices. The record is incomplete and we have found no documents from which the attitude taken by London and The Hague might be construed.[145] Nor have we found any indications that Rhenania-Ossag took them under the influence of fanatical Nazis.[146] Moving managers and employees to other jobs was probably seen as a cosmetic exercise of the kind occasionally required to placate particular regimes, and no more. No questions of principle or moral judgements about the Hitler regime appear to have arisen and it bears pointing out that, whereas correspondence shows Group managers quick to identify and condemn Bolshevism, they appear not to have had the same sensitivity to Fascism or Nazism. This blind spot, quite a common affliction in the 1930s, may have impaired their vision when it came to perceiving the intentions of the unfolding Third Reich and the horrendous threat hanging over the Jewish employees in the Group's care. We do not know the Group's treatment of the staff members concerned, nor their fates.

The economic programme of the Hitler government had three priorities: reducing unemployment, achieving economic self-sufficiency, and rearmament. Though a serious drain on Germany's fragile finances, these policies boosted the economy and notably the automotive industries, because road building and stimulating car ownership and use were a key plank in the unemployment relief programme.[147] Rhenania-Ossag turnover shot up from RM

Late in 1933 Rhenania set up an information office for motorists and launched a large series of motorists' touring maps, emphasizing – despite some political doubts from The Hague – the importance of Shell as a contributor to the German economy.

180 million in 1932 to more than RM 500 million in 1938, the number of employees rising in tandem from almost 6,000 to 10,000.[148] During 1933 alone, asphalt sales increased by more than 35 per cent.[149] As for achieving self-sufficiency, Germany was almost entirely dependent on imported oil and had only a small primary processing capacity, because the fiscal regime favoured importing intermediates. Since indigenous production could never become sufficient to cover normal demand, let alone the requirements of rearmament, the German government launched a com-

Mit der Herausgabe unserer SHELL-Tourenprospekte, die sich bewußt von dem auf kartographischem Gebiet üblichen unterscheiden, wollen wir uns in den Dienst des Motortourismus stellen, indem wir mit diesen Karten Anregungen zu Wochenendfahrten geben. Die SHELL-Tourenkarten machen neben der jeweiligen Tourenroute durch Wort und Bild auf die Schönheiten der Landschaft und die Sehenswürdigkeiten, die an der Straße liegen, aufmerksam. Diese kleinen Führer sollen die oft gehörte Frage „Wohin wollen wir fahren?" beantworten und dem Lenker des Fahrzeuges selbst lebendige Wegweiser sein.
So, wie diese Karten Führer sein sollen, ist unser SHELL-Dienst überall Helfer unterwegs, dort steht man mit Rat und Tat dem Kraftfahrer zur Seite und hält die immer gleichbleibenden, hochwertigen SHELL-Produkte in verplombten, vorbildlichen Apparaten und Emballagen preisgünstigst zur Verfügung.
Frohe Fahrt mit
SHELL
BENZIN • AUTOOELE
DYNAMIN
das SHELL-Spezialgemisch für hochkomprimierende Motoren. Erhältlich an allen SHELL-Pumpen mit blauem Band.
RHENANIA-OSSAG MINERALÖLWERKE AKTIENGESELLSCHAFT

SHELL und DEUTSCHLAND

Soweit die deutsche Inlandsproduktion für eine wirtschaftliche Deckung des deutschen Bedarfs nicht ausreicht, versorgen wir Deutschland durch Einfuhr mit den für seine Wirtschaft und seinen Verkehr unentbehrlichen Erdölprodukten. Indem wir Deutschland mit der über die ganze Welt verteilten SHELL-Produktion der uns angeschlossenen Gesellschaften unmittelbar in Verbindung bringen, sichern wir Deutschland die wirtschaftlich günstigsten Belieferungsmöglichkeiten aus der Welt-Erdölerzeugung. Die SHELL-Organisation in Deutschland stellt sich gleichzeitig in den Dienst richtig verstandener Nationalwirtschaft. Sie bemüht sich um weitere Erschließung deutscher Erdölvorkommen unter Ausnutzung der in weltumfassender Tätigkeit gesammelten Erfahrungen. Insbesondere haben wir die Verarbeitung in größtem Umfange nach Deutschland verlegt. 6 Werke, darunter die größten und bedeutendsten ihrer Art in Deutschland, mit einer Verarbeitungsmöglichkeit von mehr als 600000 t jährlich, deren Erzeugnisse auch zu einem erheblichen Teil zur Wiederausfuhr gelangen, haben wir in technischer Vollkommenheit geschaffen. **Wir sind auf unserem Gebiet der größte Arbeitgeber Deutschlands.** Durch unsere umfassende Organisation von großen und kleinen Verteilungslägern und Tausenden von SHELL-Tankstellen sorgen wir für wirtschaftlichste und bequemste Befriedigung des Bedarfs. Durch die Ausgestaltung unserer Organisation führten wir der deutschen Wirtschaft seit 1924 Aufträge von mehr als 170 Millionen Reichsmark zu.

Die Ausfuhr von Erzeugnissen aus unseren deutschen Werken erbrachte seit 1924 einen Devisenerlös zugunsten der deutschen Zahlungsbilanz von 127 Millionen Reichsmark.

Die Verknüpfung mit der weltwirtschaftlichen Betätigung der SHELL-Gruppe sicherte Deutschland Lieferungen eines erheblichen Teiles des von der Gruppe in allen Erdteilen benötigten Materials — seit 1924 in einem Werte von über 144 Millionen Reichsmark.

Gegenüber diesen insgesamt 441 Millionen Reichsmark gab Deutschland seit 1924 durch uns für die Einfuhr von unentbehrlichen Erdölrohstoffen, Halb- und Fertigfabrikaten einen Betrag von 570 Millionen Reichsmark aus. Durch diese Verknüpfung und Wechselwirkung wirtschaftlicher Interessen sind demnach fast drei Viertel eines Betrages, der auf jeden Fall für den Import unentbehrlicher, lebenswichtiger Grundstoffe von Deutschland hätte aufgewandt werden müssen, dem Lande wieder zugeflossen.

RHENANIA-OSSAG
Mineralölwerke Aktiengesellschaft

ZRK 4140

prehensive programme to expand the production of synthetic gasoline by giving price guarantees to producers. In September 1933, representatives from the Group, Jersey Standard and APOC met in The Hague to discuss the threat of the new economic policies in Germany and once again agreed on the need to act in common. They would not expand their refining capacity and in August 1934 the companies also rejected a government proposal to build extensive new bulk installations and to maintain increased stocks of oil products there in return for an undertaking to limit the production of synthetic gasoline.[150] As so often, the German government could easily find alternative solutions. An American company, Crusader Petroleum Industries, built a new refinery near Hamburg and a government agency erected the bulk installations.[151]

With their refusal to fall in with the government's demands, the oil companies lost what little hold they had on developments. By September 1935, Kessler worried about Rhenania-Ossag losing its position on the German gasoline market. As things stood the

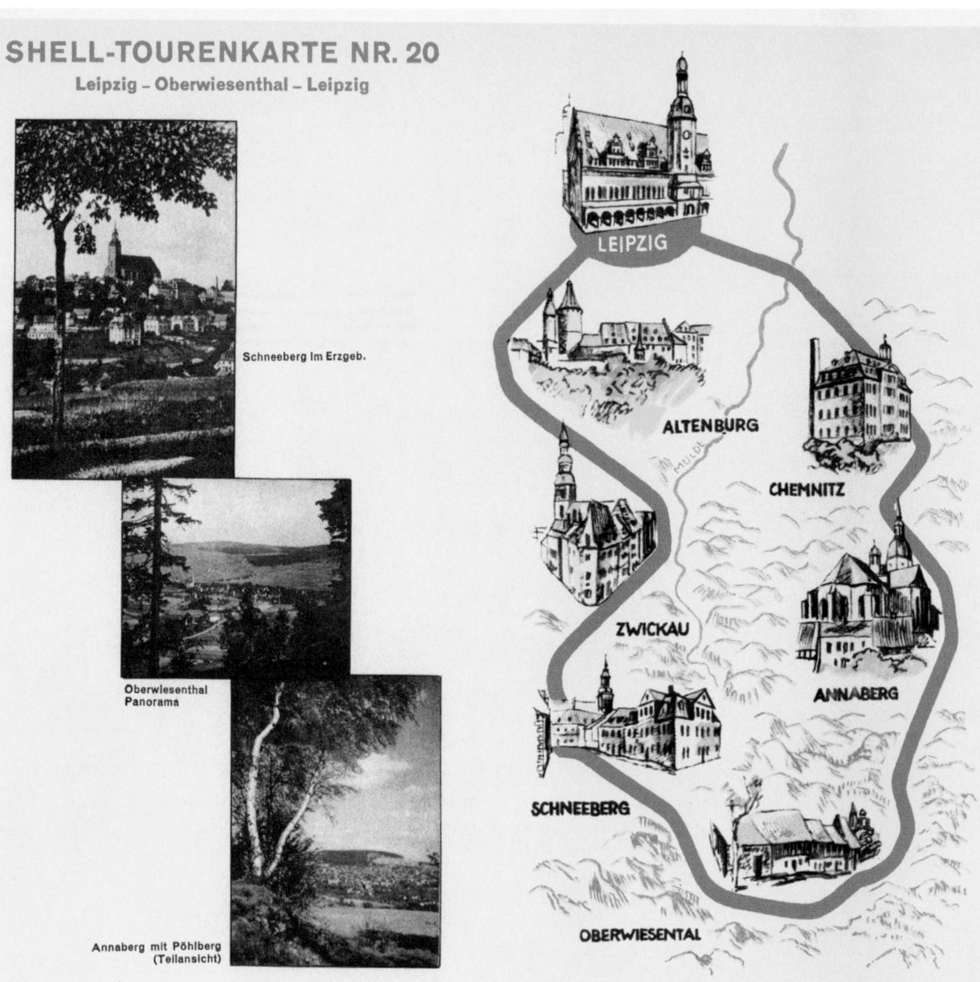

[56]

The shape of war to come: Germany's rearmament began in secret in 1934, and on 11 March 1935 Hermann Goering announced the creation of the Luftwaffe. Above, Heinkel HE-111 twin-engined bombers are seen in mass production in 1940, and above right, a disciplined formation of Junkers JU-87D Stuka dive-bombers fills the sky in 1943.

[57]

Group and Jersey Standard distributed IG Farben's output, but this was unlikely to be the case for the synthetic gasoline from other producers.[152] In August 1936 the Nazi government announced a Four-Year Plan to make the country entirely independent of foreign gasoline imports, putting Hermann Goering in overall charge. This direct threat forced the Group to rethink its strategy. In October, Kruspig had started negotiations with representatives of IG Farben and Jersey Standard about a joint synthetic gasoline plant to be built at Pölitz (now Police in north-western Poland, near Szczecin). The plant would produce 430,000 tons of gasoline from hydrogenated residue and later coal as well.[153] Kruspig warned that there was no way back. Given the aims of the Four-Year Plan, Rhenania-Ossag faced a loss of market share and trade in branded products. The company's public standing would suffer serious damage if it did not participate, for as a German business it had to support projects of such national importance.[154]

At first Group managers refused to let Rhenania-Ossag participate in the project, not wanting to let local considerations outweigh global policy considerations.[155] Governments around the world wanted to get hydrogenation plants, considering them to be the key for gaining independence from imported oil. Projects were under consideration in France, Italy, Japan, and the Netherlands, but the oil companies did not want to build hydrogenation plants because synthetic gasoline remained totally uncompetitive with the normal product (see Chapter 6). For that reason the Group had just declined to participate in building a hydrogenation plant in the UK.[156] By February 1937, however, the Group had reversed its decision, but we do not know what made managers change their minds: Rhenania-Ossag's market position, pressure from the German government, or the need to keep up with its hydrogenation patent partners IG Farben and Jersey Standard.[157] With Goering bragging about eliminating gasoline imports within eighteen months, the threat to gasoline sales was certainly taken seriously. The Group already handled a considerable volume of synthetic gasoline, 193,000 tons in 1938 against 551,000 of normal gasoline, or a quarter of the total.[158] The Pölitz project now rapidly acquired a definite shape. The plant would be built in two stages at a total cost of RM 150 million, the equivalent of 12.3 million sterling or 109.5 million guilders. The first stage cost RM 80 million, to be funded by the joint shareholders in the company, Hydrierwerke Pölitz AG, Jersey Standard and the Group supplying RM 27.5 million of capital each, IG Farben 20 million and the joint marketing company Deutsche Gasolin 5 million. This stage would come on stream in 1938. The second stage, scheduled for 1939-41, was to be funded with public bond issues. Because the Group wanted at all costs to avoid its commitment to a synthetic gasoline plant becoming public knowledge, two banks fronted as shareholders in Hydrierwerke Pölitz.[159]

During the summer of 1938, the Pölitz contract turned out to be a devil's pact. Costs soared, construction schedules proved too optimistic, and the projected bond issues now looked unlikely to find a market. On top of that the German government considerably raised the targets of synthetic gasoline production under the Karinhall Plan and now demanded additional installations for making aircraft gasoline which doubled the original cost estimates for the plant. The partners had to supply additional funding, but the Group flatly refused to contribute, arguing that it had agreed to take only the equity stake of RM 27.5 million and that Rhenania-Ossag could not afford more. A detailed memorandum was drafted to underline the company's precarious overall position, but managers privately admitted that an inspection of the accounts by an independent auditor would easily prove the opposite and thus had to be avoided.[160] With the shadows of war in Europe drawing closer, the Group wanted to reduce its financial and other commitments in Germany. Rhenania-Ossag's credit with Asiatic was already being liquidated, reducing the scope for oil imports.[161]

By now it was too late, however. From 1938 the German government took progressive control of the oil industry, laying down product specifications, restricting the use of certain products for private consumption, and introducing processing targets for oil companies. When Rhenania-Ossag's imports dropped following the curtailment of its credit with Asiatic, the government forced the company to process Mexican oil from the Group's newly nationalized wells, acquired under a barter scheme concluded between the two countries.[162] Officials at the *Reichswirtschaftsamt* recognized that the Group could justifiably claim not to have any further obligations to Hydrierwerke Pölitz but, with synthetic gasoline a top priority in the preparations for war, they were in no mood to let legal niceties obstruct their policy. In February 1939, the Group received an ultimatum: if Rhenania-Ossag would not take its full share in the costs, the government would interpret this as an infringement of the company's duty under law to act in the interests of the German people and the German State, and put it into administration.[163] This made Group managers capitulate and give Rhenania-Ossag permission to raise its financial commitment to Pölitz.[164] The plant finally came on stream in 1940 at a total cost of RM 270 million, the equivalent of roughly £ 28 million or 200 million guilders. Production peaked in 1943 with nearly 580,000 tons of gasoline, 80 per cent of which was aircraft fuel. During the war, Pölitz came to rely on large numbers of forced labourers housed in nine camps, which included a separate barracks from the Stutthof concentration camp. Thirteen thousand prisoners are said to have died there.[165] However, by that time the Group had lost all control over Rhenania-Ossag and the Pölitz works, as we shall see in Volume 2, Chapter 1.

[58]

Die Gerüchte um
Henry Deterding.

(Telephonat unseres Wirtschaftskorrespondenten.)

Auf die gestern besonders in Amsterdam, Paris und London
kursierenden Gerüchte, dass Sir Henry eDterding von seinem
führenden Posten bei der Royal Dutch und Shell Transport zurückgetreten sei, wird von der Gruppe offiziell erklärt, dass Sir
Henry nicht zurückgetreten ist. Ueber seine Zukunftspläne be-
sagt das Communiqué nichts. Von autoritativer Seite wird in-
dessen versichert, dass die Gerüchte wieder von der alten Seite
in der üblichen gehässigen Form mit Absicht verbreitet worden
seien und dass Sir Henry Deterding auch nicht zurückzutreten
beabsichtige. Sachlich lässt sich zu den Gerüchten sagen, dass
schon seit einiger Zeit Vermutungen über eine Verjüngung der
Verwaltung in der Royal Dutch-Shell-Gruppe aufgetaucht sind.
In diesem Zusammenhang ist auch wiederholt die Person des
Direktors Kessler in den Vordergrund geschoben worden, dem
man als einem der jüngeren Verwaltungsmitglieder der Gruppe
eine führende Rolle in der Zukunft zuschreibt, zumal da er auch
mit recht bemerkenswerter Initiative durch seine Regulierungs-
pläne für die Welt-Petroleumindustrie hervorgetreten ist. Man
sagt verschiedentlich, dass eine gewisse Umstellung in der Ver-
waltu[illegible] durchaus auch von der eigenen Initiative Sir Henry
Deter[illegible]gs ausgehen könnte, der heute selbst im 67. Lebensjahr
steht. Ganz abgesehen von den sicherlich sehr persönlich und
unnötig aggressiv gefärbten Gerüchten um den gegenwärtigen
Führer der Shell-Gruppe mag es also begreiflich sein, dass derartige Gerüchte, wie sie heute aufgetreten sind, überhaupt in
Umlauf gesett werden könen. Ausserdem hat auch der Abschluss
der Shell-Union Oil Corp. den heutigen Gerüchten einen gewissen

Newspaper about a possible stepping down by Deterding.

From distinction to disgrace Deterding reacted to the German events in a heavily compromising way. During the early 1930s his position as Group chief executive was still very secure, if anything strengthened by his gifts for financial manipulations and the kind of arbitrage needed to counter currency fluctuations.[166] His attendance at board meetings was regular and little different from that of other directors, even if his absences stemmed more from prolonged holidays than from business travel. Deterding's fellow directors also indulged his penchant for seeking their reassurance and testing their loyalty by announcements of retirement, documented for 1930, 1933, and 1935 but perhaps uttered more frequently than that.[167] It would seem, however, that he gradually lost touch with the Group's general direction. He was not involved with major new issues such as hydrogenation, chemicals, or iso-octane, rarely took part in As-Is discussions and, after one intemperate outburst over Mexico, remained on the sidelines in the development of the Group's attitude towards nationalist policies.[168]

Moreover, Deterding's opinions had started sliding from the off-beat to the downright cranky. His 1920s diatribes against taxation in the Dutch East Indies and the iniquities of Bolshevism had raised eyebrows; in the 1930s his views on the Depression and notably on currency policy invited scathing comments. His view that underconsumption kept down oil prices stood out from the general consensus which blamed overproduction. The debate on the merits or otherwise of the gold standard presented Deterding with a new hobby horse, mounted at every available opportunity. He ranted against what he termed the tyranny of gold, pleaded for

[59]

RATES OF EXCHANGE

SHANGHAI, Saturday 13th October, 1934

PARITY

BAR SILVER Spot......	$24\frac{7}{16}$	Import	1/8594
DO Forward .	$24\frac{6}{16}$	Export	1/7254

H. & S. B. C. Opening quotations 9.30 *a.m.*

Bank's Selling Rates

LONDON......T/T..............	1/6
NEW YORK .T/T..............	37
LYONS.........T/T..............	556
JAPAN.........T/T..............	128
STRAITS......T/T.............	$63\frac{7}{8}$
BATAVIA......T/T.............	$53\frac{1}{2}$
MANILA.......T/T.............	$73\frac{7}{8}$
INDIA..........T/T..............	$99\frac{1}{2}$
HAMBURG...T/T..............	$90\frac{1}{4}$

THE NORTH-CHINA DAILY NEWS, SATURDAY, DECEMB

MR. MORGENTHAU'S DISCLOSURE

America Buying Silver Every Day: Mexico's Total Output Last Year Taken Over

Washington, Dec. 19.

Mr. Henry Morgenthau, Secretary of Treasury, reiterated to-day that the United States Treasury was still buying silver, but he refused to say from where. He said he had bought silver every day this week. "There are many places—you would be surprised."

Mr. Morgenthau declined to say whether the purchases were from private individuals or governments.

He also saw no particular significance in New York's fixing a price in the absence of a London bid and declared that there was nothing else for N[illegible]rk to do inasmuch as there was [illegible]

In [illegible]

Treasury, urging that "every possible assistance be given, and all power and resources be pledged, to carry out the programme."

If foreign and domestic opposition to the silver-purchase policy is successful in defeating the programme, the Senator declared, the Treasury stands to lose vast sums on the metal already acquired.

"The Administration's silver policy," declared the Senator, "is being attacked by foreign and domestic reactionary influences through a concentrated drive to force down the price of silver, with the intent to discredit and terminate the programme."

China Ships Big Silver Cargo To American Banks

Something Over $10,000,000 Goes by Pres. Lincoln

EXPERIMENTAL CARGO MOVED SECRETLY

One of the biggest shipments of silver ever to be made out of Shanghai left last night. Late in the evening, shortly before the ss. President Lincoln was to sail, six trucks drove to the Customs jetty, laden with silver bars. To convey the shipment down the river required two lighters.

[illegible] total amount of the silver [illegible]ted is estimated at between $10,000,000 and $16,000,000. It is

Silver Shipments Explained

Shanghai, Dec. 20.

[illegible] reply to inquiries concerning silver shipments now going forward, a spokesman of the Central Bank of China informed the Kuo Min News Agency that this silver had been sold by the Government Banks to the Chase Bank and the National City Bank of New York, in order to provide additional reserves for exchange stabilization.—Kuo Min.

THE SHANGHAI TIMES, THURSDAY, JUNE 6, 1935

Important Statement On Silver Is Rumoured In New York As Imminent

Designed To [illegible]

Senator Thomas said that, having needed such propaganda in the past, the United States had witnessed not

U.S. Hits Britain In Silver Action

Purchase Suspension Is Seen As Demonstration Against Sterling

Newspaper reports (left and right) fuelled Deterding's anger over 'the silver question' in the middle 1930s. At the same time, frequent rumours that he was about to retire (far left) had to be officially denied.

bimetallism and a restoration of the silver price as a way out of the Depression, and advocated an early devaluation of the Dutch guilder. Though essentially right in blaming the gold standard for prolonging the Depression, Deterding, in this case as in others, isolated himself from potential supporters by defending his views with arguments as emotionally charged as they were economically nonsensical. He conducted a huge correspondence on the subject to newspaper editors, politicians, and business leaders all over the Western world, offering opinions and circulating articles by himself and others supporting his views.[169] In 1933 a letter expounding his views embarrassed the board of the Nederlandsche Maatschappij voor Nijverheid en Handel, an economic think tank. One of its members, the banker and expert on international monetary affairs J. W. Beyen, publicly berated Deterding for his ignorance of currency questions. Editorial commentaries in leading Dutch newspapers agreed with Beyen.[170]

Now in his late sixties, Deterding became increasingly resentful and suspicious of the outside world. For all his wealth and achievements, his ambitions remained unfulfilled and surfaced more and more in tirades against what he considered the lack of public recognition of the Group's great benefits to mankind and of his own merits in building it. His mind, always quick to reduce baffling complexities to a few basic truths, narrowed, reducing the world's colours, contrasts, and contradictions to a simple dichotomy of black and white, good and bad, us and them. In a typical transposition, he liked to portray the Group as enlightened, magnanimous, altruistic even, and any chance opponents as sinister money grabbers, part of a conspiracy to defraud honest business and to overturn the forces of order, progress, and prosperity. Deterding felt surrounded by enemies and he characteristically interpreted the social isolation caused by his temperament as a part of wider hostilities. A consummate schemer himself, he needed little

[60]

During a quiet moment at the World Economic Conference held in London in the summer of 1933, Deterding's critic Beyen (far left), director of the Rotterdamsche Bank, took the opportunity of a private conversation with Colijn and Trip, director of the Nederlandsche Bank.

persuasion to accept conspiracy theories which more sceptical minds might have found wanting. He typically saw the negative comments in the Dutch press about his currency opinions as a campaign against him.[171]

These attitudes were already noticeable during the 1920s, but from the early 1930s they began to dominate Deterding's outlook. At the same time he acquired a keen interest in Germany. Since under the *Stillhallte* he could not repatriate the returns on his private investments, he used the frozen Reichsmarks to buy a hunting estate in rural Mecklenburg, probably in 1931.[172] With the rise of Hitler, Deterding's feelings for Germany intensified. He does not appear to have had any great interest in Nazism per se. Deterding held few political convictions; indeed, he appears to have been totally unprincipled politically. His voluminous correspondence teems with extreme right wing opinions, but he rarely mentioned the words Fascism or Nazism in his letters. It was more a matter of personality. Nazism and the New Order in Germany were congenial to Deterding's authoritarian being and his fervent anti-communism; he does not appear to have understood its political or indeed social ramifications.[173]

Deterding now began to take a greater interest in the Group's German affairs, normally Kessler's remit. This brought him into contact with the new regime's leaders. He cautiously welcomed Hitler's appointment as Chancellor in January 1933, but two months later he sought an audience with Hitler, presumably to present a proposal for sponsoring a series of radio lectures on the evils of communism, to celebrate the coming to power of such a vociferous anti-communist. He was rebuffed, but in November Deterding claimed to have met Hitler the other day.[174] In May 1933 he discussed synthetic gasoline and monetary policy with Reichsbank President Hjalmar Schacht.[175] In April 1934, during the negotiations with the German government about increasing oil

[61]

[62]

Marinus van der Lubbe, a simple-minded young Dutch vagrant, was found guilty of starting the Reichstag fire on 27 February 1933 and was executed, but the blaze was widely seen as a conspiracy of right-wing agents provocateurs. Deterding, Hitler, and Johannes Bell, one of the German signatories to the Treaty of Versailles, were depicted as 'Van der Lubbe's accessories' (far left), while Deterding was rumoured to be Hitler's paymaster against Communism (left).

The Paris Rothschilds' restrained but indignant replies to Deterding.

product stocks mentioned above, he met Alfred Rosenberg, the chief Nazi ideologue and leader of the party department for foreign affairs. After the meeting Deterding wrote to Rosenberg offering his credentials as an anti-communist.[176] From September 1935, the German Foreign Office seconded one of its staff to Deterding as a personal assistant for political matters.[177] Deterding also met Goering, whose Karinhall estate was not far from his own, and went shooting with him. To facilitate proper entertaining, he exchanged his retreat for another and much grander one, obtaining tax advantages by threatening the authorities with the withdrawal of capital.[178]

Deterding clearly liked the New Order represented by the Nazis. He considered the notorious Night of the Long Knives in June 1934, when Hitler had a large number of his suspected party opponents brutally murdered, as a necessary step, confessing that it had increased his respect and veneration for the Nazi leader, if such were possible.[179] Nazi officials appear to have looked down on him with some disdain, disparaging his narrow interest in money making and his close association with Jews.[180] Since Deterding could still prove useful, the regime played up to him. Newspapers gave his economic views wide and favourable coverage, the *Reichskanzlei* sent him a signed photograph of Hitler.[181] Even before Hitler's appointment as Chancellor, Deterding's fervent anti-communism and opinions about general economic policy had received favourable treatment in the *Völkischer Beobachter*, the Nazi newspaper which also sported conspicuous advertising for Shell products. However, other producers of mass consumer goods did not discriminate in their advertising policy either. Indeed, the Shell ads even incited criticism from party stalwarts, prodding the paper's publisher to reply angrily that 'We accept Shell ads because not even we National Socialists can drive with water.'[182]

The propaganda to which he was exposed now began to exercise a subtle influence on Deterding's behaviour, inspiring actions which turned him into a managerial liability. A new row with the Paris Rothschilds, echoing the failed attempt to enlist their support over the Dutch East Indies export tax, provides a good example of his ill-considered moves and garbled motives. During the early 1930s France had sought a rapprochement with the Soviet Union, resulting in the signing of a non-aggression treaty in 1932. The following year rumours circulated that this was to be followed by a French loan secured on Caucasian oilfields and arranged by a group associated with Gulbenkian. In September De Kok, clearly acting on Deterding's instructions, wrote to the Rothschilds asking them to use their influence with the French government to abort the loan. The bank replied that it knew nothing about it, voicing surprise at Royal Dutch wanting them to intervene in a delicate political matter. De Kok responded with a tirade drafted by Deterding. As sellers of Russian properties to the Group, he argued, the Rothschilds owed it a moral duty of support in pursuing its rights. And the French state had also profited from that transaction so the Group could rightfully expect some loyalty there, too, so he reiterated his request for an intervention.[183]

DE ROTHSCHILD FRÈRES
R.C. SEINE 96.549
CHÈQUES POSTAUX 440.26

PARIS, LE 19 Septembre 1933

Cher Monsieur Deterding,

Nous regrettons d'avoir à vous importuner pour un incident un peu désagréable qui vient de surgir entre nous et la Royal Dutch, et dont la lecture de la correspondance que nous avons échangée avec votre Siège Social vous mettra au courant, car nous sommes convaincus qu'en l'occurence vous n'avez pas été consulté.

Nous tenons, aujourd'hui, à vous parler avec une entière franchise, ainsi que nous l'avons toujours fait. Nous n'hésitons donc pas à vous dire que la forme et le fond des lettres qui nous ont été adressées nous ont profondément choqués, et que tout particulièrement le ton et la teneur de la lettre du I4 courant sont aussi déplacés qu'inconvenants.

Vous partagerez, sans aucun doute, notre manière de voir à ce sujet et croyez bien que nous sommes persuadés que cette regrettable initiative n'émane pas de vous, cher Monsieur Deterding, avec qui nous avons toujours entretenu les relations non seulement les meilleures mais encore les plus cordiales.

Veuillez agréer, je vous prie, l'expression de nos très sincères sentiments.

[63]

DE ROTHSCHILD FRÈRES
R.C. SEINE 96.549
CHÈQUES POSTAUX 440.26

ontv. 13/9-33

N.V. KONIN
TOT EXPLOITATI

Care

Messieurs,

Nous avons bie
remercions d'attirer not
venus au sujet de divers
Gouvernement français et

Vous paraissez
car nous ne sommes aucune
question quelque indicati

En ce qui conce
lité d'une intervention de
çais, nous ne vous cacher
En effet, nous ne voyons p... quel titre nous pourrions aborder notre Gouvernement pour l'entretenir d'un aussi délicat problème de haute politique générale, qui n'est assurément pas de notre ressort, et dans l'examen duquel nous n'avons nullement le désir de nous immiscer.

Veuillez agréer, Messieurs, nos salutations les plus distinguées.

de Rothschild fr.

PERS-INFORMATIEN
JOHs. VAN HAARLEM Gzn., Directeur
Kantoor:
Spoorbaanstraat 8, bij de Grensstr. en Weesperzijde
AMSTERDAM-Oost
Opgericht 1 April 1895.
Opgericht 1 April 1895
Agentschappen: Parijs, Londen, Milaan, Brussel,
Lissabon, Weltevreden [Ned.-Indië]

Limburger Koerier (Maastricht)

29 DEC. 1932

SIR HENRI DETERDING

Geldschieter van de Duitsche Nationaal-Socialisten ?

Toen een paar weken geleden de heer Gregor Strasser, de voornaamste medewerker van den heer Adolf Hitler, zich uit de partij der nationaal-socialisten terugtrok, publiceerde dr. Otto Strasser, broeder van eerstgenoemde, in zijn blad „Die Schwarze Front" allerlei bijzonderheden nopens Hitler's legioenen, die voor den grooten leider nu niet juist vleiend konden worden genoemd; en hij deelde tevens mede, dat 'smans partij een schuld heeft van ongeveer 12 millioen mark.

Dit gaf velen aanleiding tot het opnieuw stellen van de vraag: wie verschaft den Nazi's toch de zeer aanzienlijke bedragen, die hunne organisatie noodig heeft?

Hitler houdt er een particulier leger op na. Het bestaat uit tienduizenden mannen, die allen uniform dragen en semi-militair uitgerust zijn. Vele duizendtallen worden gekazerneerd, gevoed en van soldij voorzien. De honderdtallen leiders van allerlei aard rijden en rossen in kostbare automobielen door het land en talrijk zijn de reizen per vliegtuig, die de heeren maken. In letterlijk alle steden ziet men de „bruine huizen", alsmede allerlei kleine en groot tuighuizen. Hitler en zijn staf logeeren te Berlijn en elders in de duurste hotels.

Wie betalen dit alles toch?

Dhr. H. Polak deelt in het (soc. dem.) „Volksblad" het volgende mede:

„Eenige weken geleden was ik in Duitschland en stelde toen allerlei menschen deze vraag. Men antwoordde, dat het ongetwijfeld de groote industrieelen en landjonkers zijn. Maar, zoo zeide men algemeen, ook buitenlanders, d.i. niet-Duitschers, dragen bij tot de instandhouding van Hitler's politieke en militaire org... ...n noem... de mij verschillende namen v... deze namen hoorde ik telke... belang inboezemde: Sir Hen...

[64]

This extract from the *Limburger Koerier* of 29 December 1932 was only one of many articles in the national and international press speculating about who might be funding the Nazi party, which was deeply in debt. Deterding's name came up more than once.

Did De Kok ever pause to consider the enormity of this letter? At that very moment Royal Dutch and Bataafsche were fighting tooth and nail against Dutch investors demanding that they honour the gold clause on their 1929 bonds issued in New York. With the gold standard suspended in the US, directors no longer considered the companies obliged to pay interest on the basis of the dollar's gold parity.[184] Would the boards have recognized moral obligations issuing from a transaction more than twenty years ago? Anyway the Rothschilds tactfully avoided a response in kind by writing directly to Deterding in London, professing their profound shock at the rudeness of recent letters received from Royal Dutch. In reply, Deterding went over the top. He listed the range of benefits which the Group had showered on France: the supplies during the war and after, the gold loan of 300 million francs, the access to Romania and to Iraq. Both the press and the government appeared to have forgotten them, he charged. The Group was surely warranted to protest against the unheard of loan to Russia, and was right to expect sympathy and support from the Rothschilds, and not complaints. Sending a copy of the letter to De Kok, Deterding vented yet another grievance, the fact that the Rothschilds had earned an average return of 50 per cent on their share in Asiatic. If such profiteers failed to cooperate, one ought not be afraid to confront them and, if necessary, sever relations.[185] Meanwhile Kessler had become involved as well. He discussed the matter with Baron Edmond and Baron Robert twice before suggesting to Deterding that it was better to drop the issue since they were unlikely to write something which he would find satisfactory. Deterding agreed, adding that it was better 'not to correspond with people whose only moral consists in money'.[186]

Deterding's handling of this sorry affair demonstrates just how much he could be buffeted by the emotions listed above, but in addition it betrayed a surprisingly deep rancour against the Rothschilds. It was as if, exposed to the rising barrage of anti-Semitic propaganda in Germany, he suddenly felt embarrassed, realizing how much of his achievements had depended on the support of this famous firm, and wanted to show strength. Deterding was not a patent anti-Semite; oblique phrases in letters from the mid-1920s criticizing Philips and Waley Cohen are as rare in his correspondence as they are cryptic.[187] His anti-Semitism was not of the racial type, seeking to discriminate against Jews on the basis of spurious shared ethnic characteristics. It was rather an extension of his anti-communism and rooted in his conviction that Jews were instrumental in building world communism. This counterfeit penny appears to have dropped in the spring of 1933, ostensibly as a result of hearing a radio lecture on the subject.[188] The particularity of his anti-Semitism explains why Deterding could remain on friendly personal terms with Jews in his acquaintance, including various Group directors, while venting strongly anti-Semitic remarks in his correspondence, and presumably also in conversation.[189] However, given his burgeoning sympathy for the New Order in Germany and his efforts to make himself agreeable to key party figures, he now had good reason for seeking to dissociate himself from the Paris bank. The refusal to grant him an audience with Hitler in March 1933 was motivated by two reasons for keeping him firmly at arm's length: his overriding interest in money and the fact that his business associates were almost exclusively Jews.[190]

It is more difficult to understand why De Kok and Kessler acted as they did. Neither shared Deterding's dubious political sympathies or sentiments. Perhaps De Kok simply misjudged the matter, ordering Deterding's letters to be typed out and sent as a matter of administrative routine.[191] Kessler probably knew little about the affair on his first meeting with the Rothschilds, since he went to see them when passing through Paris on other business. After that first meeting he requested copies of the preceding correspondence. Whatever the case, the fact that both went along with the affair highlighted the dangers of the managerial trade-off mentioned above, in which directors obtained their freedom of action by indulging Deterding's whims.

A similar mix of motives appears to have fed Deterding's growing resentment against NHM, the Royal Dutch house bank. He considered the bank in thrall to Fritz Mannheimer, a German Jewish banker who, as senior partner in the Amsterdam branch of the Berlin bank of Mendelssohn & Co., was a pivotal figure in the Dutch financial market and a close associate of NHM. Deterding felt betrayed by Mannheimer, whom he accused of taking part in a bear raid on Royal Dutch shares in 1932 and of devious manipulations to force Royal Dutch to pay coupons from the New York bond loan at the full gold exchange rate.[192] Unable to get at Mannheimer directly, he went for his supposed accomplice, NHM, moving heaven and earth to have Royal Dutch transfer its business to Hope & Co., an old Amsterdam firm with more standing than clout.[193] Typically, the Royal Dutch board minutes do not reveal anything of what must have been heated debates, spiced by the fact that the former president of NHM, Van Aalst, and the senior partner of Hope & Co., Luden, were both directors. Deterding's arguments failed to convince the board; Royal Dutch stayed put with NHM.[194]

Whilst the rows with the Paris Rothschilds and with NHM were played out behind closed doors, a great deal of public speculation went on about Deterding giving loans or donations, for amounts ranging from four million guilders to a fantastical £ 55 million, to the Nazi movement. In return, he was rumoured to have

ALS DUITSCHLAND ONZE EIEREN NIET
KOOPT, KUNNEN WIJ GEEN DUITSCHE
PRODUCTEN BLIJVEN KOOPEN!
De gezamelijke Landbouw organisaties

[65]

'Since Germany will not buy our eggs, we cannot buy German products!'

obtained promises of special advantages for the Group or even an oil monopoly under a Nazi regime. Such rumours circulated as early as 1931. They regularly resurface even today, but remain unsubstantiated.[195] The supposed quid pro quo does not appear a very plausible one. The Group's position in Germany, as one of two leading participants in officially sanctioned cartels, did not need support from any regime. Moreover, managers considered a monopoly there an impossibility since the government could never afford the sums needed for buying out existing businesses. Nor would the Group have wanted to spend large amounts on obtaining it, if only because it would have run counter to the close relationship with Jersey Standard.[196] It is, in fact, highly unlikely that either Deterding or the Group, used to getting very tangible returns for their money, would have spent substantial sums on promises. Official policy rejected political donations and Deterding also professed to be against them.[197] Press support was seen as another matter. From 1937 Deterding, now influenced by a personal secretary with outspoken Fascist sympathies, helped to finance the paper of a Fascist splinter group in the Netherlands, just as the Group sponsored a Russian anti-Soviet press service in London.[198] Finally, if Deterding or the Group indeed supported the Nazis during their wilderness years with any substantial sums, then one would have expected the *Reichskanzlei* to have acknowledged the fact in the minute considering whether or not to grant him an audience with Hitler in March 1933.[199] However, the document mentions nothing of the kind; Deterding was turned down without further ado.

The grand gesture Deterding made after his retirement illustrates how he liked to use his money. In December 1936 he donated, with much fanfare, 10 million guilders from his personal fortune to set up a fund for buying up surplus foodstuffs in the Netherlands and selling them in Germany, purportedly to ease the plight of Dutch farmers and German consumers, equally hit by the trade barriers and exchange restrictions between the two countries. The proceeds of this deal would be donated to the Nazi charity *Winterhilfswerk*. This was really an ingenious scheme to avoid taxes and currency complications in transferring part of his capital to

[66]

Sketched by Hitler in 1932 at a restaurant in Munich (left), a 'car for the people' – a *Volkswagen* – was a concept he valued highly. Admiring a model of the proposed vehicle (below), he was amused when Ferdinand Porsche pointed out that 'the engine was in the boot'.

[67]

Germany.[200] Deterding's financial support for the Nazis appears to have begun during his retirement in Mecklenburg, where he developed close ties with the local party organization and contributed to its charities. For all their praise of Deterding as a friend of the true Germany, none of the obituaries in the Nazi press mentions contributions other than those in his dotage, so the rumours about prior support were probably no more than that.[201]

However, as we have seen above the Group continued to do business with Germany, drawing lube oil supplies, ordering tankers, expanding the Rhenania-Ossag installations, and participating in the Pölitz works. Deterding presumably helped to arrange these transactions, though Kessler and Van Eck appear to have been primarily responsible for these matters. Both were fairly straight commercial transactions, given the currency situation and the threat to gasoline imports. In addition, Deterding is said to have promised Hitler a loan of RM 30–40 million to the *Volkswagenwerke* in January 1936, but this is a canard based on a wrong interpretation of the Goebbels diaries.[202]

Whatever the case, Deterding's actions in Germany created increasing friction with his fellow directors. He was ebullient about the country's future and about the Group's performance there, whereas his colleagues considered the likelihood of having to withdraw if the situation did not improve soon.[203] Moreover, Deterding was widely reported as making rash promises to German officials.

In June 1934, during the negotiations between the oil companies and the government over bulk installations, the British Ambassador to Germany, Sir Eric Phipps, heard information that Deterding had promised Hitler to have the Group supply one year's worth of products on credit to Germany should its economic situation become desperate. Foreign Office experts doubted the veracity of the report given the huge sum involved, estimated at four to £ 5 million.[204] The story also reached French and Romanian diplomats.[205] No records corroborating Sir Eric's information have been found in the company archives. However, Rosenberg noted in his diary that he had made a deal with Deterding in May 1934 that the Group would stock one million tons of oil products in underground tank farms built by the Group, only to see the transaction founder on red tape in German departments.[206]

The story remains puzzling indeed. During 1932-34, Group supplies to Rhenania averaged 650,000 tons, worth about £ 2.2 million.[207] Would managers, or Deterding for that matter, really have considered maintaining such a large stock on credit to a creditor which, under the *Stillhalte*, had effectively suspended payments? As things stood, Rhenania operated under a credit limit of £ 600,000, though this limit may have been set after the summer of 1934.[208] When later that summer the German government approached the oil companies with proposals to raise stocks, Deterding was reported to be firmly against them.[209] He may thus have made a rash offer, only to have had to retract it after protests from his colleagues against such a transaction; or he may have done no more than, in his usual way, dazzle his German counterparts with figures that showed Group credit would be cheaper than building the installations required, and then the officials had used the story to downplay Germany's very real currency difficulties. They may also have quoted it against other Group managers as a solid promise extracted from Deterding.

Whether true or not, the story reached Deterding's board colleagues, feeding a nagging suspicion about the soundness of his judgement. The passion glowing in his correspondence on the involvement of Jews in world communism, and on the virtues of Hitler in combating the same, would suggest that he must have

Deterding with his third wife, the German-born Charlotte Knaack, whose admiration for the Nazis coincided with and probably strengthened his own views.

[68]

harangued Group directors with his opinions on these subjects in the same way as he did with his currency convictions.[210] Now his infatuation with Nazi Germany made it appear that he was going native there. When in September 1935 Agnew again heard of Deterding promising Group supplies to Hitler, he asked the British government's Petroleum Department to have the Foreign Office establish whether it was true, so he could take appropriate action.[211] Deterding's temper helped to further undermine board-room trust. A few months previously, in April, he had created a terrific row when he felt that the decision to order three steamers for Anglo-Saxon from German yards had not been carried through with sufficient speed and diligence.[212] Exactly a year later Deterding blew his top again, now over an administrative matter, the volume of correspondence between London and Rhenania-Ossag, though there may have been a prior board confrontation over the continuous rumours about his various promises to German officials.[213] These conflicts brought to an end Deterding's career as Group CEO. By June 1936 he had effectively retired and gone to live in Germany.[214]

Did he choose to go, or did the other directors force him? There is no evidence either way, but it was probably a bit of both.[215] Giving up was not Deterding's style and, despite his de facto retirement, he continued to meddle in Group affairs throughout the summer. On the other hand, he had recently fallen in love with a young German woman, Charlotte Knaack, whom he wished to marry after divorcing his wife, life-changing events which may well ave induced him to step down. Moreover, given his record of announced retirements, Group directors will have had little difficulty in persuading him that it would now indeed be wise to do so. There were grand festivities planned for July to mark his seventieth birthday and his fortieth year in the service of Royal Dutch, so he would have a final, festive scene before bowing out. Deterding attended board meetings of Shell Transport and Anglo-Saxon for the last time in May, but it was only in September that chairman Loudon informed the Royal Dutch board of Deterding's wish to relinquish his position as from 31 December.

Forging a new team The news came as a surprise even for well-informed insiders such as Sandkuyl, the general manager in Batavia, otherwise kept abreast of developments by a frequent personal correspondence with his predecessor Oppenheim, now in The Hague as Bataafsche company secretary.[216] To the wider world, the appointment of De Kok as Royal Dutch Director General was something of a surprise as well. Kessler appeared the obvious candidate, having a more comprehensive grasp of the business than De Kok and a greater standing in the international oil business. Moreover, following Deterding's loyalty test in 1931, the Royal Dutch *commissarissen* made a clumsy attempt to prepare for his resignation from which Kessler emerged as the clear successor.[217] It must have been a huge disappointment to Kessler to see his life's ambition thwarted with fulfilment so near. Did Deterding perhaps play a final trick on him, out of spite against the shadow of his father? This cannot be ruled out, but the correspondence between them suggests a degree of confidentiality which renders this unlikely.[218] In the absence of firm documentation, we can only surmise that De Kok's appointment was a compromise solution to a formidable problem.

Formally the Group had never had a CEO, much less a procedure for succession. Under Deterding the Director-General of Royal Dutch, the Group's majority shareholder, had also chaired board meetings of the main operating companies. This situation could not now be prolonged without the risk of offending the British directors, some of whom must have felt that, after thirty years of Dutch rule and with so much of the business run from London, it was time for one of them to take the reins there. This made considerable sense. After Deterding's resignation, Agnew was the senior managing director and his wide experience made him well qualified for the job; in Deterding's absence, he usually chaired meetings. Since Erb's retirement in 1930, four managing directors had resided in London against only one in The Hague. On the other hand, having an Englishman as informal Group chief executive or indeed as Royal Dutch Director General was at that time, and for decades to come, out of the question. It would have caused a public outrage in the Netherlands, but it was also unthinkable. The internationalization of staff had largely bypassed Royal Dutch and Bataafsche. There were many Dutchmen at all levels in London and precious few Englishmen in The Hague. Only the Amsterdam laboratory had a contingent of foreign experts, mostly Germans and Swiss.[219] Both Shell Transport and Anglo-Saxon had Dutch directors; Bataafsche had English directors, but the executives were all Dutch, and since Lane's death in 1926 no foreigners had been appointed to the Royal Dutch board. The new general manager simply had to be a Dutchman and the Royal Dutch board may also have wanted to have him in The Hague, so as to reassert its formal power within the Group.

Choosing a successor to Deterding therefore required some judicious manoeuvring. The initial moves dated back to his 1931 resignation announcement, though the details remain unclear.[220] For all his qualifications, Kessler was junior to Agnew in London. Appointing him risked upsetting the British directors, while transferring him to The Hague and sending De Kok to London meant sacrificing a well-established position there. This may have been the reason why the Royal Dutch board chose De Kok, letting the London chairmanship fall to Agnew. It was a compromise designed to keep the two parts together, at the cost of continuing the split between them. Clearly thirty years of close collaboration had failed to bridge the differences between The Hague and London sufficiently for mutual respect to become trust when power came into play. In a further step taken to appease the London directors, Godber succeeded Deterding as chairman of Shell Union. When in December 1937 Agnew resigned as managing director of Anglo-Saxon, Godber became the senior English member of the management team. George Legh-Jones filled the vacancy at Anglo-Saxon and also took a seat on the Bataafsche board; as managing director of Shell Transport and director of Anglo-Saxon and Bataafsche, Agnew remained a powerful figure, however.

In October 1936 the rift between the Group's constituent parts was further accentuated by the procedure followed for two further appointments to strengthen the Dutch element in the managerial group. J. M. de Booy was promoted from Royal Dutch company secretary to managing director of Royal Dutch and

Zondag 5 Februari 1939 — DE TELEGRAAF — Derde Blad. 5

LEVEN VAN EEN GROOT NEDERLANDER

Sir Henry Deterding, de Napoleon van de Petroleum.

BOUWER VAN EEN WERELDCONCERN.

Zijn leven, zijn werk, zijn karakter.

ER is gisteren een groot Nederlander gestorven.

Met Sir Henry Deterding is een man heengegaan, die zijn volk onnoemelijk veel heeft gegeven. Hij heeft zijn vleugels over de geheele wereld uitgespreid en er is geen werelddeel, dat niet de sporen van zijn werkzaamheid vertoont, maar hij bleef Hollander en zoodoende kon hij uit ganscher harte Nederland laten profiteeren van hetgeen hij daarbuiten veroverde.

Veroverde, want in dezen man ziet men den vasten blik en de ijzeren hand van een veroveraar. Hij kon toegrijpen, hij kon beslissingen nemen. Mijn veld is de wereld, zei hij. Een verbeten strijd heeft hij gevoerd. Met zichzelf, met de omgeving. Langzaam opklimmende, en dan ineens, als het leven hem voldoende ervaring heeft gegeven, is het meesterschap bereikt.

De villa van Sir Henry Deterding in St. Moritz, waar de oliemagnaat gisteren plotseling is overleden.

Uitspraken van dr. Deterding.

Ik geloof, dat men in mij dikwijls een soort vechtersbaas ziet, die voor zijn plezier ten strijde trekt. Dit is geheel onjuist. Er is niemand, die minder graag vecht dan ik en die liever door vriendschappelijke besprekingen tot overeenstemming komt, en in dien geest heb ik ook steeds ten opzichte van de Standard Oil gehandeld.

*

Het succes van de Koninklijke is gekomen door de buitengewone samenwerking met mijn personeel.

*

Leert Indië kennen!

*

Breng jonge menschen, waar wat inzit, vroeg in de practijk. Tusschen 17 en 24 zijn de vermogens om op te nemen het grootst. Wat een jonge man weet, beteekent veel minder dan wat hij is.

Bataafsche, reducing De Kok's administrative responsibilities so he could devote his time to more general Group affairs, including frequent visits to London. Van Eck was transferred from New York to London as the second Dutch director there. The Royal Dutch board discussed and agreed the new appointments in private. The British Bataafsche directors were kept waiting in an adjoining room and informed whom their new colleagues were going to be once the decisions had been taken.[221] Though formally correct, this was surely not the way to create a team spirit, and quite at odds with the London courtesy of having visiting directors of Group companies attend boards on which they did not have a seat. The Anglo-Saxon directors understood the message that voting power now counted, whereas under Deterding's leadership it had not.[222] That same month they decided to exchange proxies for one another, ensuring that there would always be a proper balance between Royal Dutch and Shell Transport representatives.[223]

Subsequently De Kok's suave tact, an asset which Kessler may not have possessed to the same degree, did much to repair the damage done. He knew how to wield authority, but he preferred to be self-effacing, patiently bringing his fellow directors closer together into a single body which, by December 1938, referred to itself as the Group Managing Directors.[224] One practical way to forge closer links was the emerging practice of daily conference calls between directors in London and The Hague at noon. By 1938 this form of communication had become well established.[225]

Yet De Kok did not entirely succeed in closing the rift. The managerial power struggle between Royal Dutch and Shell Transport directors continued behind the new façade of unity, ready to break out at the next opportunity. Deterding was now relegated to the sidelines as non-executive director of Royal Dutch and Bataafsche, treated with respect but firmly kept at a distance. He fought one more round. On Boxing Day 1938, typically at the last possible moment, Kessler wrote to inform him about the terms for a 100 million guilders' bond loan to be floated by a syndicate led by NHM and Mendelssohn & Co. Amsterdam for Bataafsche,

La Depeche Dauphinoise.

PÉTROLE

Un de ceux qui connaissaient le mieux Sir Henry Deterding, magnat de la Royal Dutch, qui vient de mourir, et qui le redoutaient le plus, était le roi Feyçal. Un jour, Sir John Simon, secrétaire au Foreign Office, demandait au roi du Hedjaz ce qu'il pensait de Sir Henry. Feyçal répondit, avec ce sourire ambigu qui met tant de mystère dans ses moindres propos :

— J'aime bien Sir Henry. Il me rappelle mon pays. Mon pays est un désert où il n'y a qu'une ressource : le pétrole. L'intelligence de Sir Henry est comme mon pays.

L'Hebdomadaire de Carvin-Henin.

— Sir Henry Deterling, le magnat du pétrole, est mort à Saint-Mortiz.

— Une collision s'est produite entre le torpilleur « Bison » et le croiseur « Georges-Leygues ». Il y a trois morts et quinze disparus.

Cyrano.

Déterding, le financier d'Hitler

Sir Henri Deterding, le magnat de l'industrie pétrolière anglo-hollandaise, a eu de belles funérailles. Toute la presse a loué, comme il convenait, le grand capitaine d'industrie, le brasseur d'affaires et le financier tout puissant de la City. Mais il est un aspect de son existence qui a été passé sous silence :

Sir Henry Deterding a été l'un des financiers secrets d'Hiller.

A plusieurs reprises il a mis des sommes énormes à la disposition du Führer allemand. Tout récemment encore — en 1937 — il fit un don de 200 millions de francs au gouvernement hitlérien.

Sir Henry était un admirateur presque passionné d'Hitler, avec lequel il avait de fréquentes entrevues.

Son action occulte en faveur du Reich dans les milieux d'affaires britanniques, a été des plus importantes. Elle explique certaines complaisances anglaises à l'égard l'Hitler.

Le Führer n'a pas été ingrat.

A la mort de Sir Henry Deterding il a ordonné des manifestations de sympathie.

Son envoyé spécial, M. Hilgenfeld, déposa une magnifique couronne sur la tombe du potentat de la Royal Dutch, en disant :

« Au nom du Führer Adolf Hitler, je te salue Henrich Deterding, grand ami des Allemands ».

Un monument sera érigé dans le Mecklembourg, à Sir Henry et il portera cette simple inscription :

« A l'ami de l'Allemagne. »

Drapeau, pas drapeau...

Journal de Chateaubriant.

— **Sir Henry Deterding, célèbre financier, ancien directeur de la Royal Dutch, est décédé à Saint-Moritz, âgé de 72 ans.**

HITLER'S WREATH

GERMAN FUNERAL FOR SIR H. DETERDING

MEMORIAL SERVICE AT KELLING.

Herr Hitler and Field-Marshal Goring, and the Dutch Government, sent wreaths to the funeral on Friday of Sir Henri Deterding at his estate at Dobbin, Mecklenburg. A private memorial service in the house was followed by a public service and the interment. Simultaneously memorial services were held in all offices in Germany of the Royal Dutch-Shell group.

Herr Hilgenfeldt, head of the National Socialist Welfare Organisation, laid Herr Hitler's wreath at the open grave with the words: "In the name and by order of the Fuehrer, Adolf Hitler, I greet in you, Heinrich Deterding, a great friend of the Germans."

Mr. Frederick Godber, Mr. J. B. A. Kessler, Baron Van Eck, Mr. George Legh-Jones, and Mr. Andrew Agnew (directors of the Royal Dutch Shell Group), and several senior members of the staff attended the funeral.

It is announced that Sir Henri bequeathed the whole of his large estate at Dobbin—with the exception of the house and the immediately surrounding park—to the Friedrich-Heinrich Farming Institute, which he founded.

MEMORIAL SERVICES.

IN LONDON.

A memorial service was held on the same day at St. Michael's, Cornhill. Prebendary J H. J. Ellison officiated and gave an address.

Those present included:—

Mrs. Van de Weerd (daughter), Mrs. P. M Deterding, Mrs. Butterfield, the Netherland Charge d'Affaires, Baroness Van Eck, Lord Cadman, Lord Bicester, Marshal of the Royal Air Force Sir John Salmond, Sir Robert Waley Cohen, Sir William Foot Mitchell, Sir Peter Norton-Griffiths, Lieutenant-General Sir George Macdonogh, the Hon. Thomas Cochrane, Sir John Lloyd, Mr. H. W. Malcolm (secretary, Shell Transport and Trading Company), Mr. H. K. Stein (secretary, Royal Dutch Oil Company) Mr. S. Harmer (Midland Bank), Mr. L. Charles Wallach (representing Sterns, Limited, and Sternol,Limited), Mr. G. H. Bokker (Arbon, Langrish and Co., Limited), Mr. F. H. Coe (secretary, Association of British Creditors of Russia, and Oil Industries Club), Mr. O. J. Barnes (Hongkong and Shanghai Banking Corporation) Mr. J. Scholton (representing the Dutch Club), Mr. John Clark (secretary, Anglo-Iranian Oil Company), Mr. G. De Booer (Asiatic Petroleum Company), Mr. G. S. Engle (secretary, Asiatic Petroleum Company, Limited,also representing the Lensbury Club), Mr. A. B. Floyd (representing Messrs. Ricardo, Daniels and Co.), Mr. J. N. Tollenaar (Sedgwick, Collins and Co.), Mr. B. A. Roelvink (hon. treasurer, Dutch Club), Mr. Frank A. Horner (chairman, Airport Committee of the Corporation of London), Mr. E. W. James, and Mr. F. Newman (Davies and Newman, Limited), Mr. E. Sabline (president, Russian Refugees Relief), Mr. W. G. Hogarth (Eagle Oil Company), Mr. J. Heerbjes, Miss Webber, Mr. H. I. Caro (R. S. Laurie, Limited), Mr. S. H. Chapman (Iraq Petroleum Company), Mr. E. S. Shrapnell-Smith (chairman, Petroleum Industries Committee, British Standards Institution), Mr. B. W. Smith (Shell Magazine), Mr. J. A. Prosser (Lloyds Bank), Mr. R. A. McQueen (Trinidad Leaseholds, Limited), Mr. W. R. Harvey (Technical Products), and members of the staffs of the various companies with which Sir Henri was associated.

There was a large attendance of villagers, completely filling Kelling Parish Church, when a memorial service to the late Sir Henri Deterding was held there on Friday. Mrs. R. Deterding was present, and was accompanied by Mrs. M. E. Reid and Miss Cobbold. The service was conducted by the Rev. C. Swainson (Vicar), and Miss Forsdick was at the organ for the hymn "Fight the good fight."

Others present were Mrs. H. Digby, Mrs M Lee, Mrs. T. Kaye, junr., Mr. and Mrs. T. Kaye, Mrs. H. Read, Mr. F. Woodhouse, Mr. F. Holman, Mr. S. Adams, Mr. C. Duffield, Mr. J. Woodhouse, Mrs. G. Lubbock, Mr. E. Morse, Mr. and Mrs. A. Dewing, Mr. F. Starling, Mrs. G. Read, Mr. J. W. Otty, Mr. J. Gidney, Mr. R. Moy, Mr. L. Woodhouse, Mr. and Mrs. A. Lambert, Mrs. A. Lakey, Mr. F. Newall, Mr. and Mrs. F. W. Monement, Mr. P. Beckerson, Mr. G. Beckerson, Mr. J. Adams, Mr. G. Lee, Mr. H. Forsdick, Mrs. C. T. Crafer (representing Mrs. A. C. Barker), Mr. J. W. Beckerson, Mr. S. W. Moy, Mr. J. Gray, Mr. F. Fuller, Mr. I. Withers, Mr. H. Howell, Mr. J. Mann, Mr. J. Howell, Mr. W. Dawson, Mr. J. Mayes, Mr. A. Lakey, Mr. C. Allen, Mr. W. Otty, Mr. W. Dewing, Mr. E. Beales, the Rev. E. G. Peterson, Mr. W. J. Elsden (Holt), Mr. W. R. Woodhouse, Mr. J. Gallon, Mr. H Moy, Mr. C. Mayes, Mr. and Mrs. S. J. Bush, Mr. and Mrs. F. Smith, Mr. J. Hammond (Bale), Mr. A. Wordingham, Mr. and Mrs. W. Osborne, Mr. and Mrs. H. Read, Mr. R. Fuller, Mrs. Fuller, Mr. K. Wright, Mr. H. Vince, Mr. G. Pigott, Mr. W. Hill, Mr. W. J. Woodhouse, Mr. A. Woodhouse, Mr. T. D. Savory (Fakenham), Mr. W. P. Cubitt, Mr. H. Hancock, Mr. H. J. M. Baker (representing C. T. Baker, Ltd.).

Among those unable to attend were Mr. F. Andrews (Fakenham), Mrs. A. C. Barker, and Commander Champion de Crespigny.

olkszeitung, Wien.

Ein großer Freund Deutschlands

Trauerfeier für Sir Henry Deterding

Dobbin (Mecklenburg), 10. Februar

In der mit Tannengrün und rotem Stoff stimmungsvoll ausgeschlagenen Reitbahn des Gutes Dobbin fand Freitag mittag die offizielle Trauerfeier für Sir Henry Deterding, den Präsidenten der Royal Dutch Shell, statt. Auf schwarzem Podium stand der blumengeschmückte Sarg, zwei Förster hielten die Ehrenwache. Auf sechs roten Pylonen zu beiden Seiten des Podiums erstrahlte gedämpftes Licht. Den Hintergrund bildeten die Fahnen des Reiches und der Niederlande und Gebinde aus weißem Flieder.

Der Führer hat dem großen Freund Deutschlands, der diese Freundschaft auch durch die Tat bewies, einen prachtvollen Kranz gewidmet. Mit dem blauweißroten Farben sieht man einen Kranz der niederländischen Regierung, Kränze des Generalfeldmarschalls Göring, des Reichsstatthalters und Gauleiters von Mecklenburg Hildebrandt sowie verschiedener Persönlichkeiten des in- und ausländischen Wirtschaftslebens.

Landesbischof Schultz (Schwerin) würdigte in seiner Gedenkrede die Persönlichkeit Deterdings. Dieser habe mit der Kühnheit eines Napoleons und mit der Geisteskraft eines Cromwells gegen den Geist der Zersetzung und Entwürdigung allen Menschentums gekämpft, wie er im Weltbolschewismus wirke. Das Unrecht von Versailles habe ihn, den Mann des ausgesprochenen Rechtsgefühls, aufs tiefste verletzt. Die Wiedergutmachung sei ihm Voraussetzung für die Rettung von Alljuda und dem Bolschewismus gewesen.

Nach der Trauerfeier bewegte sich der Leichenzug durch den Gutspark zur Grabstätte, einem Eichenwäldchen im Dobbiner Gutspark. Hinter den vielen Familienangehörigen schritt Reichsamtsleiter Hilgenfeldt, der Reichsstatthalter und Gauleiter Hildebrandt und Freunde aus … Welt. Unter den … n Liedes und der … wurde der Sarg in …

… leiter Hilgen… rach: „Im Namen … hrers Adolf … einrich Deterding, … r Deutschen." … vollen Kranz des … r Luftwaffe über… des Generalfeld…

4843

FAMOUS OIL MAGNATE DEAD

Career of Sir Henri Deterding

MAN WHO HELPED ALLIES TO WIN THE WAR

THE HAGUE, Feb. 6.

Sir Henri Deterding has died at St. Moritz.—Reuter.

**Henri W. A. Deterding, the oil magnate, was born at Rotterdam in 1866, the fourth son of a sea captain. From 1882 to 1888 he was a clerk in the Tweentsche Bank. He then entered the service of the Nederlandsche Handelsmaatschappy, being stationed in Sumatra. Later he became manager of the firm's Penang branch. It was then that the oil industry began to develop in Sumatra. Its pioneer, Kessler, in 1896 induced Deterding to join the Koninklijke Nederlandsche Petroleum My. (the Royal Dutch). After Kessler's death Deterding became its chairman on Kessler's instructions. In 1901 he succeeded in bringing about a combine of all the oil concerns in the Far East, and in 1903 a common selling concern, the Asiatic Petroleum Co., of which he was chairman, was started.

STRUGGLE WITH ROCKEFELLER

In his struggle with Rockefeller and Standard Oil, Deterding enlisted the co-operation of Marcus Samuel (later Lord Bearsted), then an importer of shells. Thus the Shell Transport Co. became part of the great Royal Dutch-Shell combine. Rockefeller flooded Asia with oil sold far below cost and gave away 8,000,000 lamps, but in them the Chinese burned Deterding's oil. Later he defeated Standard Oil in Mesopotamia, held his ground in Europe, and secured a firm hold on the immense oil resources of Venezuela, as well as interests in the Caucasus, Rumania, Egypt, Central America and even the U.S. In 1903 he had got the financial backing of the Rothschilds in his fight with Standard Oil. Deterring became known to Admiral Lord Fisher, who described him as "Napoleonic in boldness and Cromwellian in depth." He secured the contract for the supply of the entire British navy with oil, with the consequence that his interests became linked up with those of Britain.

During the war he organised the supply of oil and petrol so excellently that the Allies "swam in oil," and Lord Curzon, after the armistice, declared that they had been swept to victory on a wave of oil. Deterding had moved to England during the conflict and become naturalised. His services were rewarded with a knighthood. He then acted as adviser to the British Government as regards oil at the conferences of San Remo, Genoa, Lausanne, London and Geneva, and played an important part in the Mosul question. The Iraq concessions were a feather in his cap.

From 1921 onwards he was engaged in a dispute with Russian oil interests, demanding compensation for pre-revolution oil properties. Early in 1926 he took strong measures to prevent the sale of Russian oil, which he succeeded in excluding from France, where he secured the contract for the supply of the army and navy. In 1927 an "oil war" broke out between the Royal Dutch-Shell group and the Standard Oil of New York, which, unlike Standard Oil of New Jersey, was dealing in Russian oil. But in 1928 he made an agreement with the rival combine which ended the price-cutting in India, the chief theatre of the struggle. Deterding managed to prevent Russia from selling cheap oil to his competitors and join with the Americans in reducing production with the result that in 1929 the cost of petrol in Britain was increased by 2½d. a gallon.

59 COMPANIES

Deterding was a director of over 59 companies, including the Anglo-Saxon Petroleum Co. and the Bataafsche Petroleum My., the chief one of course being the Royal Dutch with a capital of about £50,000,000. His combine controlled the bulk of the world's oil tanker tonnage and had 45 refineries in 25 countries. But despite its largely British Directorates the ultimate management of the huge concern remained in Dutch hands. In 1935 Deterding offered 2,000 of his employees and their wives a chance of travelling free of charge in Europe during their summer holidays, his object being to broaden their outlook.

In 1924 he married as his second wife a daughter of the Russian General, Paul Koudavaroff. They had two daughters. She secured a divorce in May, 1936, and in June he married his German secretary, Frl. Charlotte Knaack, aged 38. Deterding's fortune was said to be £65,000,000. He had a house in Park Lane, London, a country seat near Ascot, and estates in Holland and Germany, and he travelled incessantly.

Deterding's sudden death on 4 February 1939 was widely reported, amid gossip and speculation about its cause.

scheduled for board approval on 5 January. Deterding, stung into action by Mannheimer's involvement with the loan, immediately drafted an ingenious counterproposal and succeeded in convincing the Bataafsche board to postpone a decision on the transaction to the February meeting. This created great consternation on the Amsterdam stock exchange for, assuming that the board would nod the deal through, the syndicate had put the market on alert for a Bataafsche loan to be issued on the afternoon of the fifth. De Kok chose to ignore Deterding's obstruction and a week later the loan was successfully floated. On 4 February, Deterding died in St Moritz (Switzerland), knowing that he finally had lost his power over the business. The Bataafsche board meeting was cancelled so directors could attend the funeral on his Mecklenburg estate. This occasion was stage-managed as a dreadful Nazi show with swastika banners, wreaths from Hitler and Göring, tributes to the great friend of Germany, and Hitler salutes over the grave.[226]

Deterding's reputation has never recovered. If it seems unfair that Deterding's posthumous reputation remains deeply stained by this event outside his control, he must still bear the blame for keeping the evil company which organized it for him. In his Mecklenburg retirement he appears to have edged ever closer to Nazism, perhaps as a consequence of his wife's political leanings,

07|490

Dem Herrn Schriftführer des
Aufsichtsrates der Bataafsche Petr. Maatschappij.

Lassen Sie mich Ihnen meinen wärmsten Dank aussprechen u. dem Aufsichtsrat für die trauernde Anteilnahme um meinen lieben Mann u. unseren guten Vater, u. um Ihren großen Führer der Royal Dutch/Shell-Gruppe, Sir Henri Deterding.

In tiefer Trauer
Charlotte Deterding.

[71]

[72]

[73]

Deterding's funeral was taken over by the Nazis, whether the mourners approved or not. In the cortège (left), his favourite horse followed the coffin; the burial (top left and right) was marked by Nazi salutes and Hitler's personal wreath; and Deterding's widow wrote a letter of thanks (centre) to the BPM.

perhaps because of the Dutch Fascist private secretary whom he subsequently took on.[227] He entertained local party bosses and wrote articles for regional newspapers to support German rearmament.[228] Because of this close association with Nazism, the memory of Deterding became an embarrassment to the Group after the war and the many bronze busts, painted portraits and other images were removed from boardrooms and office entrance halls.[229] It would be wrong, however, to let the errors of his last years overshadow his earlier achievements. Deterding definitely belongs to the top echelon of international business leaders of the twentieth century, the master builder of a great corporation which, but for his uniquely long tenure at the top, might well have foundered on the strains of keeping the constituent parts together. He fully deserves to be remembered, for his gifts as much as for the character flaws which, in a terrifying balance, serve as a reminder that extraordinary qualities may come at the price of appalling failings.[230]

[74]

W. Rudeloff, Shell's top man in Germany, in conversation with Deterding, 1936.

Conclusion

The 1930s were years of extremes for the Group, of exciting new ventures and disastrous economic circumstances, great opportunities in products like iso-octane contrasting with mounting difficulties in countries with nationalist economic policies, the building of new installations and enlarged Central Offices overshadowed by the closing down of plants and devastating redundancies, the emergence of a tentative team of Group managing directors cast against the dismal aberrations of Deterding's last years. During these years, the Group showed both its strength, as a resilient and dynamic organization fully capable of dealing with formidable challenges, and its weakness, a brittle and inefficient top management structure. The inept handling of Deterding's succession showed that, almost thirty years after the merger, chauvinism was alive and well within the parent companies. The transformation which, during the interwar years, shaped the assembly of operating companies into a modern corporation with a single and strong brand, integrated policies for all functions, and a distinct corporate culture failed to reach the top. As a consequence, the Group somewhat resembled the heraldic eagle, a powerful body crowned by two heads looking in opposite directions, eminently practical for maximizing opportunities arising from the dual nationality, but a handicap for other purposes such as flying, or creating efficient management structures for that matter. One wonders what would have happened had Deterding persisted with his resignation in 1919-20, or had he retired following the bad riding accident which he suffered in 1923.

The Group undoubtedly owed part of its quick recovery to the As-Is collaboration, which in major markets acted as a brake on the steep price falls which the global glut created. As-Is was not effective everywhere nor to the same degree, however. It tended to work best in countries where the Group, Jersey Standard, and Anglo-Iranian dominated the market, but the common front was easily broken up by competitors offering to supply or to build the installations required. The companies could delay the construction of refineries or hydrogenation plants, but they could not prevent it. In the talks over As-Is issues, the Group tended to put principles above price and price above volume, though in practice managers were not above some pragmatic dealing, for instance in trying to obtain favours from the Mexican government while ostensibly maintaining a common policy with the other oil companies. The principled attitude resulted in a laudable restraint over supplies to Italy and Japan when those countries engaged in what were considered at the time to be aggressive wars, but it was not necessarily better than the more pragmatic approach favoured by Jersey Standard in meeting the challenges posed by the Depression and totalitarian dictatorships. Nor did the Group's principles, so evident in the cases of communism or state monopolies, enable managers to perceive the diabolical nature of the Nazi regime. In line with the general development of German business, Rhenania-Ossag quickly adapted to the New Order and grew luxuriantly until it was too late for a fundamental stand. When push came to shove, general considerations such as market position, coupled with a technocratic conception of the business as providing fungible services to governments, clearly won the day.

Conclusion

The Group's early history is a tale of spectacular success. General circumstances go a long way in explaining that success. The oil business boomed, prices and profits remained generally high, technology developed very rapidly, new markets for lucrative products such as fuel oil and gasoline opened as if by magic, and discoveries of exciting new fields were being made all the time and around the world. It would have been difficult indeed to fail; yet this does not explain why the Group did not remain a lucrative also-ran like Dordtsche, Burmah, or Anglo-Persian for most of their early history, but became a world leader instead. If it is trite simply to conclude that the Group mustered the right mix of entrepreneurship, know-how, technology, capital, and marketing skills to break away from the pack, we need to probe a little deeper and focus on the factors that explain the Group's success in particular terms, rather than general ones.

The first factor that stands out is the privileged access to the initially very cheap oil of the Netherlands East Indies in combination with a firm grip on markets across Asia. Until the mid-1920s, the Group derived most of its profits from the Indonesian operations, which to a large extent fuelled its early expansion. Group managers envied Jersey Standard's ability to fight price wars around the world thanks to its hold on the US market, but the Group had a very similar, though smaller, power base in Indonesia. For that reason the Group fought tooth and nail to retain its privileges and to keep out competitors, at the price of conceding 60 per cent of the profits of new fields to the Dutch colonial government, something no oil company was prepared even to consider until well after the Second World War. With the business partnership between the Group and the government in Batavia, the colonial exploitation entered its last phase. For all its outward allegiance to free trade, the Dutch government had been keen to foster and to protect, in true neo-imperialist fashion, the colonial oil industry from its inception. The resulting symbiosis was little disturbed by the purposefully delayed entry of Jersey Standard into Indonesia. The Jambi saga was about price, not about principle; neither government officials nor members of parliament seriously considered giving prime concessions to Jersey rather than to Royal Dutch. However, from a strictly business point of view the Group might have been wiser to embrace competition in Indonesia, as it did elsewhere. By accepting the partnership with the government, the Group perpetuated the colonial model of exploitation which was really unsustainable in the longer term, as the economic crisis of the 1930s demonstrated with devastating effect on primarily the indigenous employees. Moreover, this model sealed Bataafsche in its relative isolation as a Group operating company. Though the importance of the Indonesian operations declined, they continued to absorb a disproportionate share of management resources, to restrict Bataafsche's policy options, and to limit the career options for its staff. The attachment to Indonesia may thus be said to have retarded a closer Group integration. To name one telling example, the Anglo-Saxon and Asiatic staff around the world regarded themselves as Shell employees, whereas Bataafsche staff felt allegiance first and foremost to Bataafsche, that proud monument to the make-believe world of Dutch colonialism in which they lived and worked. For them, Shell was no more than a brand.

As for the second key factor, building an integrated company formed part and parcel of the vision that the elder Kessler laid down for Royal Dutch and then bestowed on Deterding. From its remote location in the Sumatra jungle Royal Dutch had to forge its own supply chain if it was to survive, but this beginning bred a firm determination to control the entire business from wellhead to consumer. It was the scope of operations which enabled Royal Dutch to survive its production crisis in 1899-1900 and which gave the company its decisive lead over Shell Transport as it struggled in vain to integrate backwards from transport and marketing. The rapid global expansion was equally motivated by the conviction about the competitive advantages of integration in beating Jersey Standard at its own game, for Jersey was, at that time, only a partially integrated company heavily reliant on buying American oil and shipping it overseas. Of all the integrated functions, however, marketing may be said to have played a leading role. Market presence ensured Royal Dutch's survival during its production crisis, and forced Shell Transport to start developing its own production. Royal Dutch's access to international markets finished the competition from other Indonesian producers such as Dordtsche with local sales only. It was Deterding's control over joint marketing in Asia which cornered Shell Transport, and his rapid expansion of gasoline sales in Europe and then in America which first brought Jersey Standard to heel. Having secured its worldwide marketing network, the Group moved to pioneer worldwide branding with the Shell name, emblem, and colours. The US operations, not exceptionally profitable in their own right during these years, paid handsome returns here by the generation of new concepts in marketing, branding, and advertising.

The drive for operational integration also created the third success factor, a strong commitment to leading-edge know-how and technology in E&P, manufacturing, and transport. Whereas marketing and logistics were very much the remit of the London companies, science and technology remained largely vested in the Dutch side of the business, Waley Cohen being a lone though powerful advocate for research in London. From the early days, Royal Dutch and then the Group quite deliberately nursed networks to acquire scientific knowledge and the latest technology through recruitment or commissioned research and to disseminate the results throughout the organization. Group directors appreciated early the virtues of rotating managers between operating companies, so as to maximize the exchange of knowledge and experience throughout the entire business. The global spread of the business materially helped to speed up this process. America may have been the most important source of innovations, in particular after 1914, but it was definitely not the only one. The company's international orientation also showed in its open attitude to hiring foreign specialists whenever needed. The geologist Erb and the E&P expert Macduffie were conspicuous cases, but lower down the ladder there were others, notably a sizeable contingent of German scientists and drilling engineers. Bataafsche also trained academics for the Group hierarchy, thus ensuring both a steady supply of Dutch managers for top positions, and a receptive audience for the interests of science and technology throughout the organization. The mid-1920s showed the vital importance of the commitment to technology for the Group's leading position. Having dropped coal hydrogenation research as a dead end after eight years and considerable expenditure, the Group suddenly found itself overtaken by Jersey Standard and IG Farben and nearly excluded from hydrogenation altogether. It testifies to the managerial flexibility and strength of the Group, however, that the danger was recognized and repaired in time, no expense being spared to recapture its position at the front of technological development.

Fourthly, the Group's managerial organization must have been very strong and efficient, surprisingly so given the unorthodox dual nationality structure with two head offices. The Royal Dutch/Shell Group was one of the first and definitely the most successful and enduring of all cross-border mergers in the corporate age, setting a standard which few have succeeded in matching. Such amalgamations face far greater challenges of integration than those between companies with the same nationality. All too often they fail even before they have been established, or the companies created run into major trouble afterwards. Moreover, the complications of dual nationality were not reluctantly accepted as a compromise between the constituent interests, but adopted as a deliberate policy to have the cake of the British Empire while eating that of the Dutch Empire.

As a case study of industrial organization, the Group defied categorization and the success of the dual structure remains baffling indeed. It did have two evident advantages. The ability to move subsidiaries to a holding company with a different nationality enabled the Group to deflect local chauvinism and, more importantly, to choose the most suitable fiscal regime for any particular subsidiary. With the tightening of company taxation during and after the First World War, tax became an overriding consideration in the administrative organization of the Group. However, the disadvantages were many. Royal Dutch, Shell Transport, and the three main operating companies, Anglo-Saxon Petroleum, Asiatic Petroleum, and Bataafsche Petroleum, all retained their separate corporate identities. Cross-shareholdings, overlapping directorships, and agency contracts provided the bonds between the Group's parts, generating vast amounts of administration simply to keep track of the arrangements between companies and to effect settlements. The dual head office structure and the lack of a single executive board for the whole business also created large overlaps, continuous conflicts of interest, reams of correspondence. Ultimately its success as a form of industrial organization was due to Deterding. Though himself the cause of serious managerial friction during the 1920s, it was Deterding who created the Group as a feasible corporation, who subsequently held the whole structure together, and whose dominating personality defused any Anglo-Dutch tensions. His resignation revealed how deep the chasm between the two camps still was after thirty years.

With Deterding we move to the fifth factor, entrepreneurship. The Group attracted an amazing number of outstanding managers, no doubt partly because of its success and the attendant rewards in the form of handsome salaries, bonuses, and fringe benefits such as the Provident Fund. This was true both for the top management, with the elder Kessler, Deterding, Marcus Samuel, and Waley Cohen as prime examples, and for the middle management and lower down the line. Pioneers like Mark Abrahams, equally able to start E&P operations from scratch in the thick jungle and in the American Midwest, people like Rudeloff or Meischke Smith, whose careers took them from selling Group products in China to Germany and the US respectively, or someone like Zulver, the creative Marine Superintendent in charge of the tanker fleet from 1910 to 1940. More importantly, Group managers succeeded early in creating recruitment and training procedures which ensured a steady flow of managerial talent into the business. Thus the transitions between successive phases were effected relatively smoothly: the extraordinary expansion following the 1907 merger, the change into a mature corporation during the 1920s, the diversification into chemicals from 1927, and the reorientation following the 1930s Depression.

That said, the years 1938-39 were clearly a watershed. The emergence of a collective body known as Group Managing Directors after Deterding's resignation showed the first tentative steps towards a new form of organization. It remained to be seen whether the Group would succeed in making this new transition.

Notes

Chapter 1

1 A map of refinery in Gerretson, *History* I, 141. According to him, the prison served to house forced labourers lent by the Government to the company in an effort to alleviate the chronic labour shortage: ibid. 143. However, one would assume it was also used to punish recaptured indentured labourers.

2 Gerretson, *History* I, 123 (400 labourers required), 130 (American technicians), 142 (housing); ibid. IV, 77 (shipping of women). SHA 190C/266, giving twelve Europeans and four Americans working at Pangkalan Brandan in 1891.

3 A vivid description of life at Pangkalan Brandan and the momentous event of February 1892 is given in SHA 190D/789, the memoirs of W. H. du Pon, who worked there at the time. Extracts from these memoirs were published in the staff magazine *Olie* 8 (1955) 82-5, 204-7, 296-9, and subsequently collected in a booklet entitled *Terugblik van een pionier*.

4 *The Times* reported the launch erroneously as having happened on 27 May, whereas because of the tide it had happened at 4 a.m. on the 28th. A facsimile of the launch report in the *Northern Daily Mail* in Howarth, *Sea Shell*, 14-5.

5 The official letter dated 18 April 1890 in Gerretson, *History* I, 99; J. R. van Zwet, 'Hoe de "Koninklijke" Koninklijk werd', De Nederlandsche Bank, Onderzoeksrapport WO&E No. 670, December 2001. Contrary to common opinion, the royal family did not buy or receive shares in the Royal Dutch at the IPO, but acquired them later: Gerretson, *History* I, 177.

6 For a modern overview of early prospecting for oil in Indonesia, see Poley, *Eroica*.

7 Gerretson, *History* I, 69.

8 Gerretson, *History* I, 64, 67, 103-9; *Jaarboek voor het Mijnwezen in Nederlandsch Oost-Indië* 25 (1896) 70; Royal Dutch, Annual Report 1895, 16; 1898, 24; 1904, 23.

9 Voorst Vader-Duyckinck Sander, *Stoop*, 56-65, 73-4, 81-95. The Dordtsche's initial nominal capital was 150,000 guilders, with no more than 35 per cent actually paid up: Gerretson, *History* II, 205, 227 with a typo giving the wrong figure of 15,000 guilders.

10 Gerretson, *History* I, 83-7; *Jaarboek voor het Mijnwezen in Nederlandsch Oost-Indië* 19 (1890) pt II, 10-91.

11 Gerretson, *History* I, 93; for a biography of Van den Berg Van Zwet, *President*.

12 Van Zwet, 'Koninklijke', 1-2; Gerretson, *History* I, 97, 102.

13 Gerretson, *History* I, 98-100.

14 Gerretson, *History* II, 243. To counter price fluctuations and keep shares in trusted hands, Royal Dutch board members actively traded in shares themselves. Such practice appears to have been regarded as perfectly normal. Cf. SHA 3/186, letters Kessler 19 February, 13 May, 1 October 1894, 9 September 1895, 7 November 1897; SHA 3/208, correspondence Lane, letters 9 June 1903, December 1904.

15 Cf. Royal Dutch, Annual Report 1890, 9-10.

16 Cf. Gerretson, *History* I, 189-90.

17 Gerretson, *History* II, 209-10.

18 Royal Dutch, Annual Report 1890, 10.

19 The plans in SHA 190C/230.

20 Gerretson, *History* I, 119-20.

21 *Research aan het IJ*, 10-11; the engineer hired, W. L. Sluyterman van Loo, initially concentrated on designing a lamp burner for optimum performance with Langkat oil: Homburg/Rip/Small, 'Chemici', 302-3.

22 SHA 190D/687, contract A. Gideonse, June 1890.

23 Cf. Gerretson, *History* I, 115 (plans sent out in October); Royal Dutch, Annual Report 1890, 11 (most stores in Langkat by December 1890).

24 These conditions were more or less similar on the Sumatran tobacco plantations. Gerretson, *History* IV, 75-6.

25 On the indentured labour system, see Breman, *Koelies*; Kampues, 'Na Rhemrev', Langeveld, 'Arbeidstoestanden'.

26 Cf. Gerretson, *History* I, 128-9, 255-7.

27 Gerretson, *History* I, 129-30; Royal Dutch board minutes, 14 March 1891.

28 Cf. for instance Gabriëls, *Koninklijke*, 28.

29 On Kessler, Gerretson, *History* I, 126-31, 179-80; De Vries, 'Kessler'; SHA 190D/733, with a brochure about his leaving the merchant firm.

30 Cf. SHA 190D/733, containing a letter from Kessler's father-in-law and Royal Dutch chairman De Lange with detailed calculations about the company's position.

31 Cf. Fennema's report in *Jaarboek Mijnwezen*, 70.

32 The text in Royal Dutch, Annual Report 1891, 25-30.

33 SHA 190D/766, draft contract between Royal Dutch and Martijn, April 1892; letter from a Yokohama firm to Martijn, 24 June 1892.

34 Gerretson, *History* I, 158, 168.

35 Gerretson, *History* I, 150-2, 196-7. Prices began to drop from May 1892.

36 SHA 190D/789, memoirs Du Pon, appreciates Waddell more than Gerretson's history.

37 Gerretson, *History* I, 147-9, 161-2, 260-1. On pp 147-9, and again on 161-2.

38 Kessler's calculations in Gerretson, *History* I, 162-4.

39 Gerretson, *History* I, 168-9, 173; *Jaarboek Mijnwezen*, 65.

40 Gerretson, *History* I, 169.

41 Royal Dutch, Annual Reports.

42 Gerretson, *History* I, 163, giving total cost as 1.45 guilders; however, Gerretson here omits to include the Sultan's royalties, which should be put at 0.10 guilders.

43 Royal Dutch, Annual Report 1891, 17-18; Gerretson, *History* I, 218.

44 Gerretson, *History* I, 206, 218-21.

45 SHA 190D/789, memoirs Du Pon.

46 Unfortunately only a fraction of the correspondence has survived. Cf. for instance SHA 3/186, Kessler to board, 2 September, 1 October, 28 October 1894; Kessler to Wakkie, 14 May, 11 July, 2 and 28 October 1894; Kessler to Capadose, 19 February 1894.

47 Gerretson, *History* I, 174; II, 41, 94-6; SHA 3/192, Kessler to Deterding, 8 December 1897, 20 December 1897. Standard Oil took secrecy so seriously that the company published its first full Annual Report only in 1918: Wall/Gibb, *Teagle*, 127-8.

48 SHA 3/186, Kessler to the board, 13 May 1894; SHA 3/191, Kessler to Loudon, 9 September 1895; Kessler to Deterding, 20 December 1897. A fifty-man police force suggested by Fennema in his report: *Jaarboek Mijnwezen* 19 (1890), 71. Gerretson, *History* I, 244, 282; II, 46-8; Voorst Vader-Duyckinck Sander, *Stoop* 127-32.

49 Cf. Henriques, *Samuel*, 149, 158, 179-80, 182, 184, 189-92, 195, 199, 209, 218-9, 221-2, 226, 229-30, 233, 234, 237, 239, 243-7, 293, 326, 389.

50 On Loudon, see Gerretson, *History* I, 236-40; Kessler at first thought Loudon would not have the required toughness and stamina: SHA 3/186, Kessler to Wakkie, 2 and 28 October 1894. He criticized him rather severely in letters to his wife, but by 1898 he clearly appreciated him: R. A. Alkmaar, De Lange papers, (no inventory numbers available at the time of writing) Kessler to Kessler-De Lange, 29 April, 2 July, 3 September, 13 November 1894, 3 January 1898. Typically, before his departure from The Hague to take up his job Loudon had had the tact to visit Kessler's wife so he could bring fresh news from home, Kessler to Kessler-De Lange, 29 April 1894. Loudon was appointed in February 1894, promoted to acting manager in January 1895 and to manager in February 1896.

51 Gerretson, *History* I, 277-8.

52 Royal Dutch, Annual Reports 1892-98; SHA 3/191, Kessler to Loudon, 9 September 1895.

53 Gerretson, *History* I, 227; cf. Van Driel, *Veembedrijf*, 79-80.

54 SHA 3/186, Kessler to The Hague, 28 October 1894; Henriques, *Samuel*, 176-9; Gerretson, *History* I, 280; II, 119-20.

55 Royal Dutch, Annual Report 1894, 10-1; 1898, 19; SHA 195/16, Deterding to Kessler, 22 February 1895 (first plans and expectations), Kessler to Deterding, 31 January 1896 (calming his dynamic approach to the business); SHA 102, Loudon papers, Kessler to Loudon, 1 September 1896 (Deterding's Asian market survey and plans for building installations); Gerretson, *History* I, 239. Kessler had wanted to raise finance and reduce the company's exposure by reorganizing it into a holding company controlling separate operating companies for the concessions and the refineries, but the board did not want to accept his scheme: cf. SHA 102, Loudon papers, Kessler to Loudon 23 August, 1895; board minutes Royal Dutch, 3 December 1895.

56 SHA 190D/759, Kessler to Loudon, 29 March 1900; 190A/116-2, report J. Pesch about a shooting incident involving Chinese labourers which left several dead.

57 Gerretson, *History* IV, 60.

58 Gerretson, *History* I, 264, and map on 257; cf. Royal Dutch board minutes, 3 December 1895.

59 Gerretson, *History* II, 46-7; cf. SHA 190D/759, Kessler to Loudon, 29 March 1900.

60 Gerretson, *History* I, 262-70; correspondence in SHA 3/186; Hidy and Hidy, *Pioneering*, 264.

61 Gerretson, *History* I, 280-86; Hidy/Hidy, *Pioneering*, 265-6; SHA 190B/211-1, memo Kessler; SHA 102, Loudon papers, Kessler to Loudon, 2 September 1897.

62 Gerretson, *History* II, 42-6, 79. Cf. Henriques, *Samuel*, 221-2; Hidy and Hidy, *Pioneering*, 265-6.

63 Gerretson, *History* II, 57-76; SHA 3/192, telegram Deterding to Kessler, February 1898 (no precise date given on the transcript), depicting the Dutch reaction on the Standard take-over threat as a national danger.

64 Gerretson, *History* II, 76-82, 94-7. Deterding was very much in favour of the prefs construction, but Kessler, away on a visit to Indonesia, did not like the idea at all: SHA 3/192, Kessler to Deterding, 4 January 1898; R. A. Alkmaar, De Lange papers (no inventory numbers available at the time of writing) Kessler to Kessler-De Lange, 15 and 24 January 1898.

65 According to Gerretson, *History* II, opposite 207 and 217, Dordtsche had a production of 80,000 tons, or 2 million cases, in 1896.

66 Henriques, *Samuel* 74-5, 367; 'Frederick Lane', in: Jeremy, *Dictionary*, III, 652-4; SLA GHS/2B/72 Working Papers Group Personnel, reminiscences T. E. S. Pate, T. Perry, H. S. Plante; photos of Lane in Howarth, *Century* 36, dated 1890; id. in Henriques, *Samuel*, dated 1920s, which would appear a bit late.

67 Gerretson, *History* I, 163, 169, 240; Chernow, *Titan*, 180.

68 'Frederick Lane', in: Jeremy, *Dictionary*, III, 652-4; Jones, *State* 32-46; Van Driel, *Veembedrijf* 79; Gerretson, *History* I, 227; II, 277-82. Hidy/Hidy, *Pioneering* 144-54.

69 Gerretson, *History* I, 214.

70 Henriques, *Samuel* 80-1.

71 Henriques, *Samuel*, 60-1, picture opposite p 21; SLA GHS/2B/72 Working Papers Group Personnel, reminiscences T. E. S. Pate.

72 Gerretson, *History* III, 213.

73 Henriques, *Samuel*, *passim* and 52-4.

74 Forbes/O'Beirne, *Technical development*, 529; Henriques, *Samuel*, 79, 91-5; Gerretson, *History* I, 233-4.

75 Henriques, *Samuel*, 32, 164.

76 SLA GHS/2B/72 Working Papers Group Personnel, reminiscences H. S. Plante.

77 Ibid., reminiscence T. E. S. Pate.

78 Gerretson, *History* I, 240.

79 Hidy and Hidy, *Pioneering*, 259.

80 Henriques, *Samuel*, 90-2, 100-2, 105-10.

81 Gerretson, *History* II, 222-4, 230; Henriques, *Samuel*, 104, 113.

82 Gerretson, *History* I, 112; Korthals Altes, *Prices*, 68-9.

83 Hidy/Hidy, *Pioneering*, 259; Gerretson, *History* II, 147.

84 Henriques, *Samuel*, 138-9; Gerretson, *History* II, 214-7. However, as late as 1897 Tank Syndicate sales in India were handicapped by difficulties in providing tins: SHA 3/192, Kessler to Deterding, undated, probably December 1897.

85 Suzuki, *Japanese issues*, 66-8, 70, 98, 151-2, 158, 200-2.

86 Henriques, *Samuel*, 159, 198; Chapman, *Merchant enterprise*, 210, 270, 313.

87 SLA MR/84 for a record of Group-owned

ships from 1892 to 1986 and includes details of building, dimensions, names and dates, with copies of the builders' certificates for *Trocas* (1892) and *Spondilus* and *Trocas* (1893); GHS3H/4-1, a folder with details about Shell Transport ships and the number of passages through the Suez canal.

88 One ton of carrying capacity equalled 333 Imperial gallons of kerosene: Samuel, 'Liquid fuel', 385; Gerretson, *History* II, 241 gives 35 cases per ton, or 350 gallons. At 4,000 tons the *Murex* could transport 1.3 million gallons, that is to say, the equivalent of 133,200 cases of kerosene, for there were two four-gallon tins to the case. There is confusion about the capacity of the six *Murex*-class tankers, which is variously put at either 4,000 tons (Henriques, *Samuel*, 119), 4,200 tons (Gerretson, *History* I, 217) or even 5,000 tons (Forbes and O'Beirne, *Technical Development*, 539; (Howarth, *Century*, 36; id., *Sea Shell*, 28). We have chosen to use the figures given in SLA GHS3H/4-1, which put the *Murex*-class tankers at 3,500 GRT, and assumed that the tankers would have carried 90% of their capacity. Cf. SLA 141/25/1, ledgers Shell Transport, book 2, 1899 fo. 17. If one takes a higher figure, the estimated volume of kerosene carried of course goes up (Figure 1.2), but the trend remains the same, and the sales performance of the Tank Syndicate/Shell Transport drops accordingly (see the discussion of Table 1.2).

89 Gerretson, *History* II, 147.

90 Cf. Henriques, *Samuel*, 140; Gerretson, *History* III, 135-6.

91 Gerretson, *History* III, 213; Sam Samuel's imports into Japan mainly consisted of Anchor brand tins, and the Bombay oil trade also appears to have continued taking a large volume of Anchor tins.

92 SLA 141/35/10-11, minutes committee meetings 31 December 1897; SLA 141/25/1, ledgers Shell Transport fo. 17.

93 Gerretson, *History* I, 233-4; idem, II, 307; cf. idem IV, 54. SLA 141/25/1, ledgers Shell Transport book 2, fo. 17.

94 Henriques, *Samuel*, 259.

95 In the summer of 1893, Marcus Samuel was diagnosed (wrongly) with cancer and given a year to live by his doctor, which must have given him an additional wish for consolidation: Henriques, *Samuel* 122-3.

96 SLA 141/35/10-11, minutes committee meetings, valuation committee 31 December 1897, 14 May, 25 June, 14 July, 17 August, 15 September, 5 November, 10 December 1898; Henriques, *Samuel*, 198; Howarth, *Century*, 46.

97 Gerretson, *History* III, 213.

98 Henriques, *Samuel*, 227-9, erroneously identifying the 5,000-ton tanker SS *Pectan* as a tug. According to SLA GHS/2B/72, working papers group personnel 1900-55, reminiscences T. E. S. Pate, Marcus Samuel wanted to claim heavy compensation, but his solicitors persuaded him to waive it and earn a title. The *Pectan* was indeed spelled like that; it was only in 1927 that the Group gave the same name to another ship, this time correctly named *Pecten*: Howarth, *Sea Shell*, 32.

99 Henriques, *Samuel*, 176-9, 223-6; Gerretson, *History* II, 119-20. The two men discussed a collaboration agreement in December 1896 and January 1897, but the outcome of these talks is unknown. The fact that Marcus Samuel initiated the talks by travelling to The Hague suggests that the initiative came from him: SHA 102, Loudon papers, Kessler to Loudon, 19 and 27 January, 23 April 1897.

100 SHA 3/192, Kessler to Deterding, 6 March 1898; Kessler expected a production of 10 million units a year and a potential of double that, plus 3,000 tons of liquid fuel.

101 Gerretson, *History* I, 276.

102 Royal Dutch, Annual Report 1898, 11.

103 SHA 102, Loudon papers, Kessler to Loudon 13 May 1897; Gerretson, *History* II, 46-8.

104 Voorst Vader-Duyckink Sander, *Stoop*, 149.

105 Gerretson, *History* II, 89-92; Wouters, *Shell tankers*, 16, dating the ship's arrival on 23 May; Henriques, *Samuel* 236, and no doubt following this Yergin, *Prize*, 118, wrongly situate the scene on New Year's Eve 1897. Cf. Royal Dutch, Annual Report 1897, 11. According to a speech by C. M. Pleyte to Loudon in 1913, the festivities and the discovery of water both took place in the night of 16 to 17 April: SHA 190D/757. Pleyte worked at Pangkalan Brandan at the time.

106 Royal Dutch, Annual Report 1898, 10-1, 15; Gerretson, *History* II, 95-6; the Kesslers had just inherited a considerable fortune from G. A. de Lange, the Royal Dutch *commissaris* who died in 1897. Kessler may also have hesitated because he had just invested a considerable sum in the syndicate which had vainly attempted to support the Royal Dutch share price: Geljon, *Algemene banken*, 372.

107 Gerretson, *History* I, 69-71; II, 122-4; Smits and Gales, 'Olie en gas', 68-72.

108 Gerretson, *History* II, 125.

109 Gerretson, *History* II, 93, Royal Dutch, Annual Report 1899, 9.

110 Gerretson, *History* II, 101-2.

111 Annual Report 1899, 14-7; Annual Report 1900, 16-19. Cf. SHA 8/1365, giving shipments of Russian bulk oil by Royal Dutch as 36,955 tons for 1901.

112 SHA 3/192, Kessler to Loudon, 13 August 1899, instructing him to keep the Russian contract secret so as not to let the Samuels know.

113 Gerretson, *History* II, 114-21, terming it a godsend on 118.

114 Henriques, *Samuel*, 296-7.

115 Gerretson, *History* II, 124.

116 R. A. Alkmaar, De Lange papers (no inventory numbers available at the time of writing), Kessler to Kessler-De Lange, 27 October 1899.

117 Gerretson, *History* II, 125-6; Forbes and O'Beirne, *Technical Development*, 65; Smits and Gales, 'Olie en gas', 68-72.

118 Royal Dutch Annual Report 1899, 12; Gerretson's special pleading in *History* II, 139-42; correspondence with Van Heutsz, SHA 3/209.

119 Gerretson, *History* II, 95, 126-38.

120 Royal Dutch, Annual Report 1901, 9; SHA 190C/266, memorandum history geological service. Expert advice was of course no guarantee for success, since even professors of geology would come to diametrically opposed conclusions, with one Royal Dutch manager complaining that even eminent geologists understood nothing of oil geology: cf. SHA 102, Loudon papers, Pleyte to Loudon, 3 June, 11 October 1900.

121 SHA 102, Loudon papers, Kessler to Loudon, 21 May 1896, asking Loudon to stop selling gasoline to India, since the fire risks were too great and the revenues negligible.

122 SHA 102, Loudon papers, Kessler to Loudon, 21 May 1896; Gerretson, *History* II, 254.

123 Contrary to Gerretson, *History* II, 73 (presumably on the strength of Deterding, *Oilman*, 62), Kessler did not leave instructions that he was to be succeeded by Deterding. Indeed, on his

deathbed in Naples he thought Loudon would succeed him, and worried that he would not be ruthless enough: R. A. Alkmaar, De Lange papers (no inventory numbers available at the time of writing) note dated Naples December 1900. Cf. SHA 190D/728 and 190D/733. The board nominated Deterding in its meeting of 15 December 1900.

124 Menten had offered them to Royal Dutch in 1892: Gerretson, *History* II, 161. Cf. SHA 190A/115, Kutei correspondence and deeds. A provisional agreement between Samuel & Co. and Menten covering his Borneo concession was signed on 12 September 1895, but in 1934 the document was reported as lost: SLA 141/11/15 Shell Transport board memos, folder 17-35, memo about the contents of the company's deed box.

125 Henriques, *Samuel*, 167-71.

126 SLA GHS/2B/72, working papers group personnel, reminiscences T. E. S. Pate.

127 Henriques, *Samuel* 155; SLA 141/11/1, Shell Transport Board memos 1901-02, letters 1 February 1897, 3 March 1897; Henriques, *Samuel*, 441, letter Lane in SLA 141/11/2, Board memos 1902, No. 1.

128 SHA 190A/115, contract between Menten and Samuel & Co., 12 September 1895.

129 Cf. Henriques, *Samuel*, quotes extensively throughout from the correspondence between the Samuels and Abrahams in SHA 190A/107.

130 SHA 190A/115, documents pertaining to Kutei.

131 Henriques, *Samuel*, 227, 239, 242-3.

132 Royal Dutch, Annual Report 1892; Shell Transport, Annual Report 1899; Henriques, *Samuel*, 259.

133 SHA 190A/116-2, an envelope with various documents about the early days of Group companies in the Dutch East Indies, one of them describing Sanga Sanga crude.

134 The 20-80 ratio given by Boverton Redwood, SLA 141/11/3 1905 Board memos vol. 3, 20-32. The practice of importing Devoes continued at least until 1914, when an enraged Deterding instructed Colijn to put an end to it: SHA 195/28, Deterding to Colijn, 30 November 1914.

135 Gerretson, *History* II, 320.

136 Gerretson, *History* II, 168.

137 SHA 190A/116-2, report Abrahams 1898-99, claiming KPM had bought a total of 36,000 tons by September 1899.

138 SLA GHS/B/72, memoirs James Kewley.

139 Henriques, *Samuel*, 306-13; SLA GHC/ID/BZ/1, Report Jago to NIIHM for 1905.

140 Samuel, *Liquid Fuel*, 393; Shell Transport, Annual Report 1899, 1; Henriques, *Samuel*, 315; Voorst Vader-Duyckink Sander, *Stoop*, 125.

141 Gerretson, *History* II, 171.

142 Gerretson, *History* II, 171, 315-6.

143 Published as Samuel, *Liquid Fuel*; about the campaign in detail, see Jones, *State*, 9-31, 38-43; id., 'Lost Cause', 139 (1902 publicity pamphlet for fuel oil); Henriques, *Samuel*, 387.

144 Gerretson, *History* II, 328. When Bataafsche took over Balik Papan after the 1907 merger, some 8m guilders were needed for the most urgent repairs and alterations in order to meet the company's safety criteria: R. A. Alkmaar, De Lange papers (no inventory numbers available at the time of writing) Dolph Kessler to Kessler-De Lange, 24 February 1908, published in *Tussen moeder en zoon*, 99.

145 Shell Transport, Annual Report 1905; SLA 141/11/3, Board memos 1905 vol. I, Production Figures Borneo 1904 (plus earlier years); SLA 141/11/3 Board memos 1905 vol. II, cost accounts Borneo 1904; SLA 141/11/3, Board memos 1905 vol. I, memo NIIHM 6 March 1905.

146 SLA 141/11/2, Board memos 1903, Report Sutherland, 10 September 1903; SLA 141/11/4, Board memos 1906, Jago to Benjamin, 26 July 1906; GHC/ID/BZ/1, Report Jago to the NIIHM, January 1906.

147 Henriques, *Samuel*, 227, 239, 242-3.

148 *Petroleum Review and Mining News*, 26 March, 2 April 1904, interview Gurgenian, in SLA GHS/2B/18, Group History Borneo; SLA GHC/ID/BZ/1, reports Kutei, report Jago January 1906. In 1908, working conditions at Balik Papan were still more relaxed than those at Pangkalan Brandan, work starting there at 8 o'clock rather than the 6 a.m. on Sumatra: R. A. Alkmaar, De Lange papers (no inventory numbers available at the time of writing) Dolph Kessler to Kessler-De Lange, 24 February 1908, published in *Tussen moeder en zoon*, 99.

149 SLA 141/11/1, Board memos, 1901-02, statement loans outstanding 9 July 1901; Henriques, *Samuel*, 348, giving an erroneous £132,000; id., 235.

150 SLA 141/25/1, ledger 1899 fol 226; 141/25/2, ledger 1901, fo. 367.

151 Gerretson, *History* III 260 puts the extra cost at 20 per cent.

152 Henriques, *Samuel*, 354-5, 366, 386-7 (gasoline attempt); ibid., 310, 387-8, 402, 438, 442 (Australia, with specific charges from Lane's December 1902 resignation letter); Murray, *Go Well*, 3, 13-4.

153 Annual reports; Gerretson, *History* II, 197 (sales 1902).

154 Annual Reports.

155 Sir Marcus still referred to his company as the Shell Line in the summer of 1898: Gerretson, *History* II, 120.

156 SLA 141/35/10-11, Minutes valuation committee 31 December 1897, calculation profits on steamers M. Samuel & Co., 1895-7. Samuel & Co.'s low profitability on shipping oil conforms to the general experience of British overseas trading firms engaged in trade and shipping: Jones, *Merchants*, 81-3.

157 Gerretson, *History* II, 101.

158 Freight rates on Russian oil were more than double that on Eastern oil: Archives Nationaux CAMT Roubaix, Banque Rothschild archive (hereafter AN/CAMT) 132 AQ 154, bundle March-May 1904, Deterding to Baer, 11 May 1904, taking freight on Russian oil shipped by Asiatic at 25 shillings per ton, against Eastern oil 11 shillings per ton. On the royalties see for instance Langeveld, *Colijn* I, 190-1.

159 Deterding to Pleyte, 18 August 1909, quoted in Gerretson, *History* IV, 167-8.

160 Shell Transport's tank terminals had 215,600 tons capacity, against 96,000 tons for Royal Dutch, so Royal Dutch pushed more kerosene through its installations, for at 1,680,000 cans Shell Transport's canning works had a similar capacity to those of Royal Dutch, i.e. 1,650,000. Gerretson, *History* II, 196. In 1903, Shell Transport sold or leased 44 ocean terminals and 190 subsidiary depots to Asiatic, with a further 48 under construction; Royal Dutch transferred only 22 terminals and 20 depots, plus 6 under construction: AN/CAMT 132 AQ 152, bundle Asiatic, definitive contracts and schedules of transfer, 1903.

161 Henriques, *Waley Cohen* 94; id., *Samuel*, 348.

162 Cf. SLA 141/11/1, Board memos Shell Transport 1901, committee reports 12 June 1901, 18 June 1901; SLA 141/3/2, board minutes, Shell Transport 26 February 1901. Henriques, *Samuel*, 437, 442-3; Lane's

December 1902 resignation letter in SLA 141/11/2, board memos, Shell Transport 1902, No. 1.

163 Henriques, *Samuel*, 444-5, for a description of the administrative arrangements in 1901.

164 SHA 190C/266; the exact years were 1902 for the head office and 1905 for the count in Asia.

165 Gerretson, *History* IV, 244-5.

166 Gerretson, *History* III, 296-8; Hidy and Hidy, *Pioneering*, 633. Standard began publishing regular accounts only in 1918.

167 Cf. the correspondence between Lane and Aron during December 1906-January 1907 in AN/CAMT 132 AQ 198.

168 AN/CAMT 132 AQ 199, bundle correspondence 1913, Lane to Baron Edouard de Rothschild, 13 October 1913.

169 Henriques, *Samuel*, 361; Gerretson, *History* II, 183-7; SHA, Board minutes Royal Dutch, 1 August 1901, termination contract association.

170 Gerretson, *History* II, 166-7.

171 Henriques, *Samuel*, 320; Gerretson, *History* II, 239.

172 Henriques, *Samuel*, 322-3; Gerretson, *History* II, 189-91; *The Statist*, 26 December 1903.

173 SHA 8/1365, letters of Lane and Emile Deen to Deterding, both 4 October 1901.

174 Gerretson, *History* II, 82-6; cf. Voorst Vader-Duyckink Sander, *Stoop*, 146-7.

175 Henriques, *Samuel*, 296-7.

176 Howarth, *Century*, 56, erroneously taking the AGM to have taken place on 19 June; Sir Marcus to the Shell Transport AGM on 18 June 1901, quoted in Henriques, *Samuel*, 353.

177 Henriques, *Samuel*, 355-6; Hidy and Hidy, *Pioneering*, 554; Seidenzahl, *Deutsche Bank*, 213-4.

178 Gerretson, *History* II, 197.

179 Henriques, *Samuel*, 355, 386-7.

180 SLA GHS/3F, a report by the Russian oil expert Gulishambaroff in *Petroleum Industrial and Technical Review*, 26 July 1902.

181 Henriques, *Samuel*, 355-6, 360-1, 369, 395-6, 407-8, 451; AN/CAMT 132 AQ 152, Lane to Aron, undated but written towards the end of January 1903.

182 Gerretson, *History* II, 193-4; SHA 8/1365, Loudon to Deterding, 4, 8, 9, 22, 23, 25, 28 October 1901.

183 SHA 8/1365, correspondence Lane, Deterding, and E. Deen.

184 SHA 8/1365, Emile Deen to Deterding, 4 November 1901, Lane to Deterding, 4 November 1901; Gerretson, *History* II, 197; Henriques, *Samuel*, 370.

185 Gerretson, *History* II, 118, 197-8; Henriques, *Samuel*, 372; the 7 shillings a ton a breaking point for Samuel, Deen to Deterding, 7 December 1901, SHA 8/1365.

186 Henriques, *Samuel*, 369.

187 Gerretson, *History* II, 171, 201, 230; Henriques, *Samuel*, 362.

188 SHA board minutes Royal Dutch, 3 December 1901.

189 SLA 141/11/1, Board memos 1901, letter Walton, Johnson, Bubb & Whatton 6 November 1901.

190 Henriques, *Samuel*, 373.

191 Henriques, *Samuel*, 373; Howarth, *Century*, 57; Shell Transport Annual Report 1901, the dollar taken as 4.87 to the pound.

192 Hidy and Hidy, *Pioneering*, 633.

193 Shell Transport Annual Report 1899; *The Statist*, 14 December 1901; Henriques, *Samuel*, 373; SHA 8/1365, Emile Deen to Deterding, 6 and 9 November, 11 December 1901.

194 Henriques, *Samuel*, 372-3, 380.

195 As early as 30 December 1901, he suggested bringing liquid fuel into the agreement: Henriques, *Samuel*, 377.

196 *The Financier and Bullionist*, 16 December 1903.

197 Henriques, *Samuel*, 392.

198 Gerretson, *History* II, 238, 241; SLA 141/3/3, board minutes Shell Transport, 29 April 1902; Dordtsche joined on special terms in September 1903, SHA 190C/64, minutes Syndicate 11 September 1903. The Syndicate met for the first time on 28 August 1903, so only after Asiatic had been formally constituted. Royal Dutch had 263 votes out of 391; SHA 190C/64 (minutes Syndicate).

199 AN/CAMT 132 AQ 154, bundle Asiatic 1902-03.

200 In September 1904, Deterding was glad to let a chartered ship take 3,000 tons of liquid fuel off his hands at cost price as a favour to the owners agreeing to a low charter rate: AN/CAMT 132 AQ 155, bundle July-December 1904, Deterding to Bnito, 19 September 1904.

201 *The Statist*, 28 June 1902, p. 1127-8; Deterding to Lane on 19 June quoted Gerretson, *History* II, 238.

202 Gerretson, *History* II, 238-9; SHA 8/1365, Lane to Deterding, 20 June 1901 bringing the news of Sir Marcus's agreement. Henriques, *Samuel*, 397-400 gives the erroneous impression of Sir Marcus signing because the navy fuel oil trial on the 27th had gone wrong, whereas he had already agreed a week earlier. Sam Samuel & Co. in Japan again remained out of the arrangements: Gerretson, *History* III, 213.

203 A map in Gerretson, *History* II, facing 173.

204 SHA 8/1365, Lane to to Deterding, 25 September 1902.

205 Sales prices of case oil from the Straits Times, published at *http://www.iisg.nl/hpw/singapore/prijs-east.xls*; Gerretson, *History* II, 344. Estimated sales from AN/CAMT 132 AQ 156, R2/154, Deterding to Bnito, 7 August 1903; Gerretson, *History* III, 50 for market shares in Asia; Gerretson, *Geschiedenis* IV, 76, 80, for a comparison of brands.

206 SLA GHS/2B/72, memoirs S. A. Ensor.

207 Cf. the correspondence in AN/CAMT 132 AQ 154, correspondence Asiatic 1902-3. Thus when in 1903 Deterding sidestepped Marcus Samuel's attempt to settle the dispute over tanker rates by arguing that Asiatic did not yet formally exist, so that this particular clause in its draft articles of association did not apply, he may have been right in law, but not in fact. He even went so far as to return letters from Samuel & Co. addressed to Asiatic, pretending that the company did not exist and that he traded as director of Royal Dutch: AN/CAMT 132 AQ 154, Samuel & Co. to Deterding, 9 January 1903.

208 Gerretson, *History* IV, 56-7; SLA 120/3/1, minutes board Asiatic, 4 August 1904; AN/CAMT 132 AQ 154 a heated correspondence between Deterding and Sir Marcus about the former's failure to produce accounts. SLA 141/11/4 board memos Shell Transport, 45-56 vol. I, 32/L, NIIHM to Asiatic, 28 February 1906 with angry queries about the accounts for 1902, 1903, and 1904. SHA board minutes Royal Dutch, 16 May 1903, 13 June 1904; 190C/64, minutes meeting Syndicate Eastern Producers, 14 January 1904. Shell Transport's 1902 accounts were only published in December 1903. The board blamed Asiatic's procrastinations in producing its accounts, which Henriques sees as another one of Deterding's ruses to undermine Shell Transport's position. The Royal Dutch 1902 accounts were published in June 1903, as usual, using

estimates of expected Asiatic proceeds, so it appears rather more likely that Sir Marcus's preoccupations as Lord Mayor delayed the production of Shell Transport's disastrous 1902 accounts.

209 A contemporary catalogue of Asiatic's failings in AN/CAMT 132 AQ 195, folder correspondence Royal Dutch, unsigned memo 25 January 1906. The rows can be followed at close range from the Asiatic board minutes in 132 AQ 153 ff and SLA 120/35/1 ff.

210 SLA 120/3/1, minutes board of directors Asiatic, 14 June 1905.

211 SHA 190D/766, correspondence with the Penang agent Martijn.

212 SLA 102/35/1 Asiatic Committee Minutes vol. 1, 14 July 1903, 15 July, 21 July 1903.

213 AN/CAMT 132 AQ 156, No. 155, Lane to Baer, 3 February 1906; SLA, 120/3/1, Minutes board Asiatic, 1 June 1906.

214 Wouters, *Tankers*, 20.

215 Gerretson, *History* II, 316-7.

216 SHA 190C/64, the first meeting of the producers' Syndicate on 28 August 1903 already considered a draft scheme for gasoline gravity standards plus incentives to producers for supplying the right quality.

217 Gerretson, *History* II, 305-6; Benjamin to Deterding, 29 May 1905, accepting the discount in SLA 141/11/3; Shell Transport's board approval in 141/3/4, minutes 30 May; AN/CAMT 132 AQ 156, No. 155, kerosene bulk proceeds 18.3 pence per unit in 1905. NIIHM sought to defend itself by lodging its own complaints; in an entertaining pot-and-kettle correspondence, the company accused others of supplying substandard kerosene, letter to Asiatic 12 September 1905, SLA 141/11/3, board memos 1905, vol. 3.

218 SHA 190C/64, Minutes Syndicate, 19 February 1904, 22 April 1904, 31 May 1905. The lab does not appear to have been an initiative of Royal Dutch. Submitting products to the lab became a standard precondition for Asiatic giving an advance on shipments. Gerretson, *History* II, 304.

219 Gerretson, *History* II, 236; cf. also III, 266. Cf. 190D/763, correspondence Macdonald-Deterding, 1900-06.

220 AN/CAMT 132 AQ 156 No. 154, memo Baron Edmond de Rothschild, 12 February 1906.

221 By 1922 it was still no more than a quarter: Fletcher, 'From coal to oil', 6-7. In 1898, Shell Transport had taken delivery of its first tankers capable of using either coal or oil, subsequently converting some, though not all, of its tankers in a similar way: Middlemiss, *Tankers*, 19.

222 Deterding wrote to Lane about bad news from Texas on 19 June: Gerretson, *History* II, 238.

223 Lane's letters of 11 and 29 December 1902, detailing a list of complaints, in SLA 141/11/2, No. 1.

224 AN/CAMT 132 AQ 156 No. 155, producers' account, division of kerosene bulk proceeds 1905; the same documents in SLA 141/11/4, board memos 1906, vol. I. AN/CAMT 132 AQ 156 No. 198, folder correspondence Royal Dutch 1906, Aron to Baer, 30 May 1906, 1905 a poor year. Asiatic sold 17.5 million cases in 1906: idem, statement sales and returns 1906.

225 AN/CAMT 132 AQ 155, folder comptes 1905; 132 AQ 156, folder correspondence Royal Dutch 1906, Aron to Baer, 30 May 1906; ibid. folder 1907, Baer to Lane 15 January 1907, estimate 40 dividend over 1906.

226 Gerretson, *History* 285-6, 351.

227 SHA 8/1365, E. Deen had applied for a permit to build the tanks with the Rotterdam city council, after having checked with Lane that he had no objection to Royal Dutch starting to sell gasoline: E. Deen to Deterding, 7 October 1901. The lab run by Sluyterman van Loo at the The Hague head office transferred to the new location of the tank farm at Rotterdam Charlois in 1902.

228 SHA 190B/138, memoirs Späth; *Shell Spiegel* July 1951, page 9 (obituary Rudeloff); SHA 102, Loudon papers, Deterding to Loudon, 2 February 1906, with copies of the proposed agreement, which originally was to embrace all Western Europe; Gerretson, *History* II, 343-4, with the footnoted Dutch edition referring to an August 1906 deal with Standard about the German market which lifted profits from gasoline sales by 4.7 m. guilders; Karlsch and Stokes, *Faktor Oel*, 88-9; Michielsen, *Van Ommeren*, I, 81-3, 85-93; Boele and Van de Laar, *Van Ommeren*, 22-3. Rudeloff had repatriated after a commercial career in China, where he had worked for Royal Dutch in expanding its sales network. SHA 190C/67A contains correspondence about the setting up of the Charlois gasoline installation; cf. De Goey, 'Deterding', 60-7 about Royal Dutch's entry into European gasoline sales, with particular attention to the Netherlands. AN/CAMT 132 AQ 155, folder correspondence 1906, Lane to Aron, 7 September 1906, a formal arrangement with Standard for Germany and an understanding for other countries not to compete on price.

229 AN/CAMT 132 AQ 156, minutes executive committtee Asiatic, 23 April, 15 October 1906; Asiatic to André Fils, 18 and 26 September 1907. Lane was against collaborating with the cartel since he thought it more profitable for Asiatic to operate on its own: ibid., Lane to Aron, 23 September 1907. Talks with an Italian group led by Agnelli appear to have stalled: ibid., minutes executive committee 25 April, 27 June 1906.

230 In 1899 there were already six Dutch companies working in Romania. Kessler had already visited the country in February 1900 to survey oil prospects there: R. A. Alkmaar De Lange papers (no inventory numbers available at the time of writing) letters Kessler 10, 12, 16 February 1900, Gerretson, *History*, II, 289, 290. Loudon was rather annoyed at Deterding's secretive and really clandestine Romanian operations: SHA 102, Loudon papers, Loudon to Deterding, 13 August 1906; reports by P. A. de Lange on Romania in SHA 3/209 and by E. Deen to Deterding 31 October and 1 November 1901 in 8/1665; Gerretson, *History* II, 293-7.

231 *Morning Post*, 10 October 1903, referring to the overdue payment of interest on the prefs: SLA, file news clippings. The 1903 profits were boosted by including an £80,000 interest charge on the NIIHM's debt for 1901 and 1902, with the company's debt to Shell Transport rising by the same amount, so there had been no payment at all, only a shift in balance items to create a false impression of revenues. The call for Sir Marcus's resignation in the *Financial Times*, 20 August 1903.The company ran a campaign to counter the negative publicity. During 1904 and 1905, the *Petroleum Review and Mining News* published illustrated articles about the Kalimantan enterprise and its liquid fuel. Other papers published stories about Shell Transport's heroic contest with Standard Oil in Europe. Clippings in the same file; Henriques, *Samuel*, 478-80 for quotes.

232 SLA 141/11/2, board memos Shell Transport

1903 Vol. 1, detailed costs and benefits Sanga Sanga; in December 1904, 141/11/3, board memos 1905, vol. 2, ibid. 1905; 141/11/3 vol. 2, and 141/11/4, vol. 1, estimates wax factory.

233 SLA 141/11/3, board memos 1905 vol. III, estimate of shipping revenues during January-September 1905.

234 Gerretson, *History* II, 328-30; Dutch control over Kalimantan figured prominently in the Admiralty's objection to Shell Transport as fuel supplier: Jones, *State,* 23. Middlemiss, *Tankers*, 20.

235 SLA 141/11/2, board memos Shell Transport, bundle 1, draft agreement between NIIHM and Royal Dutch, 3 November 1903; bundle 2, draft agreement ditto 1 December 1903. Henriques, *Samuel* 457, erroneously assumes the management of Balik Papan having passed to Royal Dutch.

236 Sir Marcus to the 1904 AGM in April 1905, a clipping in SLA 141/11/3, board memos Shell Transport 1905; Rudeloff to Shell Transport, 2 February 1906, in SLA 141/11/4, board memos Shell Transport 1906 vol. 1; Gerretson, *History* III, 92-4.

237 Seidenzahl, *Deutsche Bank*, 213-4; Gerretson, *History* III, 86-8, 92-4; Henriques, *Samuel*, 489-91.

238 SLA 141/11/4, board memos Shell Transport, estimate debt 12 March 1907 giving a total of £580,000; the same amount given by Lane to Aron, 17 May 1906, in AN/CAMT 132 AQ 155, bundle correspondence 1906.

239 SLA 141/3/5, board minutes Shell Transport 12 December 1905.

240 Minutes Royal Dutch board, 25 May 1905.

241 SHA 102, Loudon papers, Deterding to Loudon, 26 and 27 March 1906, writing in the latter that the Shell Transport board had accepted the proposed 60-40 that morning, though the day before Sir Marcus had favoured merging on principles of profit sharing, each company getting half of the first £300,000, the rest to be split 35-65 between Shell Transport and Royal Dutch. The 10 per cent represented 25 per cent of Shell Transport's 40 per cent share in the Group. The point was not lost on the Paris Rothschilds, though elsewhere few people appear to have noticed it: AN/CAMT 132 AQ 197, bundle 1903-09.

242 AN/CAMT 132 AQ 155, folder correspondence 1906, memo Baron Edmond de Rothschild, 12 February 1906.

243 AN/CAMT 132 AQ 155, folder correspondence 1906, Lane to Baer, 10, 30 January 1906. Deterding made a tentative offer of £ 1.5 m. claiming that on the basis of an understanding with the Dutch bank NHM he could raise the money within 24 hours. No further negotiations about price or conditions with Royal Dutch or with Deutsche Bank are mentioned in the correspondence between Rothschild and Lane, but the antagonism between Germany and the other European powers at the Conference of Algeciras in the spring of 1906 probably forestalled the possibility of selling Bnito to Deutsche Bank.

244 AN/CAMT 132 AQ 195, folder correspondence Royal Dutch, unsigned memo 25 January 1906.

245 AN/CAMT 132 AQ 282, comptes Bnito 1903-11.

246 SHA 102, Loudon papers, Deterding to Loudon, 27 March 1906.

247 AN/CAMT 132 AQ 282, comptes Bnito 1903-11. Dividends during the years 1887-98 averaged 5.5 %.

248 AN/CAMT 132 AQ 155, folder correspondence 1906, Lane to Baer, 30 January 1906.

249 AN/CAMT 132 AQ 156, folder correspondence 1907, Lane to Baer, 5 and 11 June 1907, Lane to his solicitor Colt, 5 June 1907.

250 AN/CAMT 132 AQ 155, folder correspondence 1906, undated memo written by either Aron or Baer. The memo calculated that, on the basis of an average 40% profit, the Rothschilds' share in Asiatic was worth four times its nominal amount of £300,000. The records do not show whether there were ever negotiations on this basis.

251 See for these negotiations AN/CAMT 132 AQ 198, folders correspondence 1906 and 1907.

252 AN/CAMT 132 AQ 155, folder correspondence 1906, Lane to Baer, 20 April 1906, with a scheme of the proposed amalgamation. The provisional merger agreement of 12 September 1906 still kept open whether Shell Transport and Royal Dutch would transfer their properties to one or more companies.

253 SLA GHS/3A/6, Royal Dutch Special Letter Book, 1906-07, gives a good impression of the huge amount of administrative details to be sorted out. This time, the firm of Sam Samuel in Japan was included in the merger: Gerretson, *History* III, 219.

254 Cf. e.g. Gerretson, *History* III, 239, quoting Embden, *Money powers* 385.

255 AN/CAMT 132 AQ 156, folder producers' accounts 1906; Gerretson, *History* IV, 56-7.

Chapter 2

1 Henriques, *Samuel*, 509-10.
2 He for example co-signed the Annual Report of Shell Transport from 1910 onwards.
3 Henriques, *Waley Cohen*, 79-143.
4 Ibid. 141.
5 Waley Cohen maintained in 1951 that votes were never taken, but that is an exaggeration; see Beaton, *Enterprise*, 54.
6 Henriques, *Waley Cohen*,151-3.
7 Ibid. 142.
8 The joint bonus from the two companies was according to the initial proposal of February 1907 limited to £40,000 or, in fact, £80,000 during a period of two years; SLA 141/11/4, board memos Shell Transport, February 1907. In 1913 this ceiling was lifted, SHA board minutes Bataafsche, 23 September 1913, a copy in 3/209, correspondence Lane.
9 The sums involved were initially not very big about 30.000 guilders between 1907 and 1911 but increased strongly after 1911 and would amount to almost 3 million guilders during the 1920s; see Ch. 5, for more details.
10 *Financial News*, 24 June 1909. This criticism led to some changes; in particular the bonuses of the non-executive members of Royal Dutch were limited in 1911.
11 Henriques, *Samuel*, 496, 506.
12 Henriques, *Samuel*, 499.
13 Ibid. 507.
14 Ibid. 506.
15 SHA board minutes Royal Dutch, 3 May 1913.
16 Gerretson, *History*, II, 348.
17 Deterding thought that Erb had worked so hard for the Group, visiting almost all continents for his geological surveys in a few years, that he was seriously overworked, and asked him to take four months' leave immediately after he received his 'definite arrangement' in January 1911; SHA 195/115, Deterding to Loudon 29 January 1911; the following year four recently graduated Swiss geologists were hired; SHA 195/119, Deterding to Loudon 29 February and 14 March 1912; see also Gerretson, *History*, IV, 3.
18 SHA 195/33-7 and 174-7, the Deterding-Gulbenkian correspondence is a fascinating source for the many and varied transactions arranged between the two men.
19 Deterding considered Gulbenkian a good negotiator but unsuitable as a manager: SHA 195/92, memo Deterding for Gerretson, 1932.
20 Staff numbers for The Hague in SHA 190C/266. Only a few of the clerical staff have left any traces in the records. A striking exception is Hendrika Troelstra, Deterding's secretary between 1903 and 1906, when she returned to the Netherlands to become the wife of a farmer in Drenthe; she was also the sister of Jelle Troelstra, the famous leader of the Socialist Party in the Netherlands, and became a well-known writer on socialist and feminist issues (see Meertens, *Biografisch woordenboek*).
21 SHA 195/112, Deterding to Loudon, 24 June, 2 July 1907.
22 SHA 190A/116; the Asian employees together earned 5.3 million guilders, the Europeans 3.6 million.
23 Gerretson, *History* IV, 60.
24 SLA GHS/3A/6, Royal Dutch special letter books, memo Benjamin.
25 Ibid.
26 SLA GHS/3A/6, Royal Dutch special letter books, memo Benjamin.
27 SHA 195/28, Deterding to Loudon, 9 August 1915.
28 SHA board minutes Bataafsche, 7 February, 20 December 1912, 10 August 1916.
29 SHA 195/113, Loudon to Deterding, 30 April 1909.
30 SHA 190G/64.
31 Gerretson, *History*, IV, 74-5; labour contracts with Chinese labourers remained important on Kalimantan, however.
32 Gerretson, *History*, IV, 60; it is unclear how the labour conditions of the labourers were affected by this change.
33 The net debts of NIIHM amounted to more than £108,000, almost 20 per cent of the negative balance of £557,000 buried in the accounts of Shell Transport; SLA 141/11/4, board memos Shell Transport, 12 March 1907.
34 SHA 195/112, Deterding to Loudon, 3 January 1907.
35 R. A. Alkmaar, De Lange papers (no inventory numbers available at the time of writing) Dolph Kessler to Kessler-De Lange, 24 February 1908, published in *Tussen moeder en zoon*, 99.
36 SHA 195/12, Deterding to Loudon, 26 March 1908.
37 SHA 195/113, Deterding to Loudon, 23 January 1909.
38 Royal Dutch Annual Report 1913, 15-6.
39 SHA board minutes Bataafsche 14 December 1911, 195/28, Deterding to Loudon, 23 November 1914, Deterding to Colijn, 15 February 1915.
40 SHA 195/12, Deterding to Loudon, 26 March 1908.
41 Henriques, *Waley Cohen*, 112.
42 SHA 195/113, Deterding to Loudon, undated memo about Penang and Java, (May) 1908.
43 Henriques, *Waley Cohen*, 112, 119.
44 Ibid. 110, 118.
45 Van der Putten, *Corporate Behaviour*, 72.

46 Ibid. 73.

47 SLA 120/35/1, Asiatic minutes executive committee, 14, 21 July 1903.

48 Van der Putten, *Corporate Behaviour*, 74.

49 Ibid. 75.

50 SLA 119/3/1, board minutes Anglo Saxon, 4 December 1907; Van der Putten, *Corporate Behaviour*, 76.

51 Van der Putten, *Corporate Behaviour*, 80.

52 SHA 190G/64, memos Dordtsche Petroleum Co.

53 SHA 195/113, Deterding to Loudon, undated memo about Penang and Java, (May) 1908.

54 One of the reasons for setting up the British Imperial was that the name of Asiatic 'did not apparently go down very well in Australia', Murray, *Go Well*, 20.

55 Henriques, *Samuel*, 594.

56 De Goey, 'Henri Deterding', 67.

57 SLA 141/11/4, board memos Shell Transport, Cohen Stuart to Benjamin, 7 June 1906.

58 Wouters, *Shell Tankers*, 32-3.

59 Ibid. 26.

60 Schenk, *Mergers*.

61 The strong increase in share prices of Shell and Royal Dutch was clearly linked to the merger in those years, and cannot be explained from the general prosperity of the oil business. Perhaps the best benchmark to compare with is the other oil company from the Dutch East Indies, the Dordtsche Petroleum Maatschappij; its share prices did not change much between 1905 and 1908 (May 1905: 125 per cent; December 1908: 127), and only began to move up in 1909.

62 This episode in Deterding, *Oilman*, 72-4.

63 Hidy and Hidy, *Pioneering*, 504-7.

64 Hidy and Hidy, *Pioneering*, 495; SHA 195/113, Deterding to Loudon, 31 May 1908.

65 SHA 195/113, Deterding to Loudon, 31 May 1908. Deterding added that '25 m. would be accepted', which meant that an even higher bid had been suggested. See also Gerretson, *History*, III, 297.

66 SHA 195/113, Deterding to Loudon, 31 May 1908; he also speculated on a weakening of Standard's power due to the anti-trust proceedings against them.

67 R. A. Alkmaar De Lange papers (no inventory numbers available at the time of writing) letters Kessler 10, 12, 16 February 1900, Gerretson, *History*, II, 289, 290, 293-7. Reports by P. A. de Lange on Romania in SHA 3/209 and by E. Deen to Deterding 31 October and 1 November 1901 in 8/1665.

68 Gerretson, *History*, II, 292-3.

69 Loudon did not like these really clandestine operations: SHA 102, Loudon papers, Loudon to Deterding, 13 August 1906.

70 Two Dutch banks, the NHM and the Twentsche Bank, also participated in the Consolidated: Gerretson, *History*, III, 107-8.

71 SHA 195/114, Deterding to Loudon, 24 February 1911.

72 SHA 195/113, Deterding to Loudon, 23 January, 14 April 1909.

73 AN CAMT 132 AQ 157, Lane to Aron, 12 and 22 August 1909, the quote is from the former.

74 Both quotes AN CAMT 132 AQ 157, Lane to Aron, 29 July 1909; only the UK was not covered by the alliance.

75 SHA 195/60, Deterding to Cohen Stuart, 18 June 1909 ('hooglopende ruzie met de Standard').

76 Hendrix, *Deterding*, 134.

77 Gerretson, *History*, IV, 130-1.

78 Hendrix, *Deterding*, 136-7.

79 SHA 195/115, Deterding to Loudon, 29 July 1910.

80 Gerretson's assessment (*History*, III, 304) that the price war did not lead to a decline in share prices is therefore not correct.

81 Shell's dividend: 22.5% in 1910, 20% in 1911; Royal Dutch: 1910: 28%, 1911: 19%; it had an immediate impact on share prices (see Figure 2.1) and Deterding was very unhappy with the decision of the Shell board, see SHA 195/117, Deterding to Loudon, 3 June 1912.

82 SHA 195/115, Deterding to Loudon, 10 November 1910.

83 Hidy/Hidy, *Pioneering*, 502.

84 SHA 195/113, Deterding to Loudon, 14 May 1908.

85 SHA 195/113, Deterding to Loudon, 2 July, 1 and 17 November 1909.

86 SHA 195/114, Deterding to Loudon, 27 and 28 July 1910.

87 Beaton, *Enterprise*, 64-5. Asiatic registered the Shell name and Pecten brand in the US for gasoline in June 1909, extending it in 1914 to cover a range of other products including kerosene, lubricants, greases, candles and wax: Shell Oil Houston, 'Legal Memorandum, Shell Oil Company/Scallop CorporationTrademark Matter; Use of the Pecten Symbol', by James J. Mullen, 3 September 1980, 1.

88 SHA board minutes Royal Dutch, 12 April 1912, 'en erop aan te dingen, dat noch Jambi noch een van de andere gereserveerde terreinen eventueel aan een andere maatschappij dan de Koninklijke worden uitgegeven en althans niet aan Maatschappijen die onder de invloed staan van de S.O.C.'

89 SHA 195/117, Rudeloff to Deterding, 28 March 1912. As early as 1911, Deterding stressed the strategic importance of Mesopotamia, SHA 195/117 Deterding to Loudon, 16 November 1911.

90 Gerretson, *History*, IV, 136-9.

91 The capitalized value of the Group had increased to the equivalent of about $250 m at this stage; $45m in 1912/13 would represent almost $900m in purchasing power in 2007 (using the cost of living to convert 1913 dollars to those of the present day; see http://eh.net/hmit/compare/)

92 SHA 195/37, Gulbenkian to Deterding, 3 and 4 May 1913.

93 SHA 195/37, Gulbenkian to Lane, 3 September 1913.

94 A good overview of the question in Jones, *State*, 191 (in 1913 Fisher called Deterding 'Napoleon and Cromwell in one. He is the greatest man I have ever met!').

95 Cf. DeNovo, 'Petroleum'.

96 SHA 195/117, Deterding to Loudon, 19 December 1911.

97 Jones, *State*, 143-4.

98 Jones, *State*, 151-5.

99 Henriques, *Samuel*, 572-89. Churchill was too good a politician not to realize that he might need Deterding at some point in the future, so he let Deterding know, via his brother who spoke to Gulbenkian, that what had been said in parliament was not directed at him, and that there was no reason to be insulted; SHA 195/37, Gulbenkian to Deterding, 23 June 1914.

100 Henriques, *Samuel*, 515-6. Lane was also an ardent promotor of the experiments to use the residue in diesel engines; see SHA 3/209, Lane to Loudon, 7 October 1912.

101 Net losses due to the price war in 1911 were generously estimated at £660,000; see Henriques, *Samuel*, 542.

102 In 1914 no more than 3 per cent of BPM's total income consisted of dividends from its subsidiaries.

103 SHA 195/60, Cohen Stuart to Deterding, 5 November 1909; by contrast, factoring the hidden items back into Anglo-Saxon's

profits raised them by only 25 per cent.

104 SHA 195/121, Deterding to Loudon, 8, 11, 15 July 1913; see also Gerretson, *History,* IV, 3.

105 Henriques, *Samuel,* 457; Gerretson, *History,* IV, 22-3; the most important publication about this research appeared in 1907: H. O. Jones and H. A. Wootton, 'The Chemical Composition of Petroleum from Borneo', *Transactions Journal of the Chemical Society*, XCI, 1907, 1146.

106 Gerretson, *History* IV, 25-7.

107 Cf. SHA 190Y/293, 190D/792, 49/160.

108 Beaton, *Enterprise,* 90. See SHA 195/117, Deterding to Loudon, 27 November 1911, Loudon to Deterding, 28 November 1911. On De Kok's involvement *De Bron*, December 1940, obituary De Kok.

109 SHA 195/209, copy of board minutes Bataafsche 12 February 1914; it was small consolation that of the total invested amount 'various outfits to a total amount of 425,000 guilders can be used for the Edeleanu plant'.

110 Schweppe, *Research aan het* IJ, 12, 19.

111 SHA 195/120, Deterding to Loudon, 28 and 29 April 1913, mentioning the resistance by board members and by the Shell board: 'also the Shell gentlemen (...) were not enthusiastic about the big expansion' ('ook de Shell heeren niet zeer enthousiast waren over de groote uitbreiding').

112 SHA 195/37, Gulbenkian to Loudon, 3 April 1913.

113 SHA 195/120, Deterding to Loudon, 1 May 1913.

114 SHA 195/37, Gulbenkian to Deterding, 9 October 1913, for the quote. During one of his visits to Paris, Sir Marcus Samuel visited Gulbenkian who 'again and again used [his] best endeavours to prevent him seeing anybody or from speaking to anybody'; SHA 195/37, Gulbenkian to Deterding, 9 April 1913.

115 The story of this share issue can be followed almost day by day in the correspondence between Gulbenkian and Deterding in SHA 195/37. Deterding's analysis of what went wrong in his letter to Gulbenkian, 15 November 1913.

116 SHA 195/37, Gulbenkian to Deterding, 28 November 1913.

117 SHA 195/119, Deterding to IJzerman, 10 October 1912 'In de eerste plaats spijt het mij meer dan ik u zeggen kan, dat, al zou *après tout* iedereen het met mijn betoog eens zijn, ik toch wel heftig geweest ben, en door mijn temperament heb ik toch steeds weinig vrienden, en kan dus slecht een der weinigen missen. Ik hoop dan ook, dat U mijn heftigheid vergeeft en vergeet'. That Deterding still wrote such a letter is a significant difference from the 1920s and 1930s when similar outbursts occurred quite often.

118 SHA board minutes BPM in 3/209, letter to Lane, 27 November 1911, and 195/121, Deterding to Loudon, 11 June 1913; also Gerretson, *History*, IV, 146-7.

119 SHA, Board minutes Royal Dutch, 3 May and 23 June 1913; CAMT Rothschild 132 AQ 199, bundle 1913, Lane to Baron Edmond de Rothschild 30 September 1913, emphasizing Sir Marcus's unease at what he regarded as an overly rapid expansion; Lane received English translations of Bataafsche board minutes since February 1913, which he passed on to Paris. The first batch included the minutes for September 1911 to October 1912: CAMT Rothschild 132 AQ 195, bundle Bataafsche, Lane to Aron, 13 February 1913.

120 SLA 119/3/2, board minutes Anglo-Saxon, 29 October, 20 November 1913.

121 SHA 195/119, Deterding to Loudon, 29 October 1912, and 195/120, Deterding to Loudon, 7 January 1913; 195/66, Cohen Stuart to Loudon, 19 June 1913.

122 SHA board minutes Bataafsche, 26 May 1913 in 195/209, stating only that 'Loudon and Cohen Stuart have expressed their desire to retire as delegate members of the Board of Bataafsche and respectively Managing Director of the Anglo Saxon on the 31th of December next and are therefore withdrawing from the daily management'. In August 1913 Cohen Stuart wrote to Loudon that his English doctor had told him to rest: SHA 195/66, Cohen Stuart to Loudon, 24 August 1913.

123 SHA 195/117, Deterding to Loudon, 24 February 1912 contains a letter apologizing for 'misunderstandings'; Hugo Loudon was not the kind of person who let himself be insulted by Deterding.

124 SLA 119/3/2, board minutes Anglo Saxon, 26 November 1913.

125 In June 1914 the Anglo-Saxon board suggested another joint meeting to review the Group's prospective financial commitments in Mexico and the US, but this did not materialize due to the outbreak of the First World War: SLA 119/3/2, board minutes Anglo-Saxon, 24 June 1914.

126 SLA 119/3/2, board minutes Anglo Saxon, 17 December 1913.

127 In 1911 Anglo-Saxon had acquired a controlling interest in Kotuku Oilfields which owned concessions in New Zealand, but the drilling there was unsuccessful, as a result of which exploration was discontinued in 1914; see SLA 119/3/1-2, board minutes Anglo-Saxon, 13 June, 18 July, 1 November 1911, 29 April 1913.

128 Pyzel's career as chief engineer is an interesting case in point. After his appointment in 1911, he immediately went on a world tour to visit all Group refineries to compare their design and suggest improvements. Arriving in California, he prolonged his stay on finding his know-how required for the design of the Martinez refinery. As a consequence, Martinez became the first continuous refinery in the US. Beaton, *Enterprise,* 90.

129 Deterding, *Oilman*, 77-80; the term merger appears on p. 79.

Chapter 3

1 SLA 119/3/2, board minutes Anglo-Saxon 29 July, 4, 5, and 26 August, 1 November 1914; ibid. 2 June 1915 (dismissal Asiatic employees). Cf. SLA 120/3/1, board minutes Asiatic, 4 August 1914, with an equal decision for employees regardless of their nationality. Archives Nationaux CAMT Roubaix, Archives Rothschild (hereafter AN CAMT) 132 AQ 162, correspondence 1914, Lane to Weill, 4 and 17 August 1914; Michielsen, *Van Ommeren*, i. 114; Karlsch/Stokes, *Faktor Öl*, 96. Asiatic also cancelled the agency of E.G. Rudeloff for Belgium and the Netherlands: SHA 190C/136, Asiatic to Colijn, 12 August 1914. In September 1914 there was a sudden panic on the discovery that Anglo-Saxon's shares in the German bulk installations were about to be sequestered in Germany, prompting Colijn to urge London to have them transferred in the name of Dordtsche: SHA 5/335, Colijn to Deterding, 8 September 1914.

2 SLA 119/3/2, board minutes Anglo-Saxon, 24 June 1914.

3 SHA 190B/137, Colijn to Rhenania, 17 August 1914. Asiatic had offered the load to Rhenania on 31 July; SHA 190D/637-1, Deterding to Colijn, 13 August 1914, Colijn to Asiatic, 13 August 1914, Waley Cohen to Colijn, 13 August 1914. Gerretson, *Geschiedenis* IV, 64-7.

4 The correspondence between The Hague and London in SHA 190D/637-1; AN CAMT 132 AQ 162, correspondence 1914, Lane to Weill, 26 August 1914; correspondence 1915, Lane to Weill, 19 July 1915; for the Dordtsche arrangment see 132 AQ 198, bundle 1912, Lane to Aron, 3 June 1912.

5 Gerretson, *Geschiedenis*, 92; Bérenger, *Le pétrole*, 177-8, mentioning gasoline supplies from October 1914; the Reisholz nitration installation was also shut down: SHA 190B/138, memoirs Späth, 12; SHA 49/567, internal memo Bataafsche, 17 August 1916.

6 SHA 190B/137, Colijn to Späth, 17 August 1914; ibid. 195/25, Colijn to Deterding, 17 August 1914, 'and I thought it better not to tell them the true reason [for sending no more gasoline]'. Cf. SHA 8/1866, memo Bataafsche Oil department, 17 August 1916, suggesting that Bataafsche's gasoline trade with Germany had come to a stop.

7 Michielsen, *Van Ommeren*, I. 126-128.

8 Import data in Friedensburg, *Erdöl*, 55; cf. ibid., 78, for vital lube oil supplies reaching Germany in barrels from neighbouring countries; SHA 195/167, Pyzel to Deterding, 28 January 1914, describing the system and the arrangement with the Stoomvaartmaatschappij Nederland through the shipping brokers Van Ommeren. The outcome of talks with two German shipping lines mentioned by Pyzel are unknown. Only heavy oils could be carried in this way, and not volatile products such as gasoline.

9 SHA 195/28, Deterding to Bataafsche, 10 December 1914, Bataafsche to Deterding, 14 December 1914, accepting the statement and assuring assent; 190A/146, Pleyte to Rudeloff, 23 February 1915 (Deterding resigns from Astra); Gerretson, *Geschiedenis*, IV, 198-9; AN CAMT 132 AQ 195, board minutes Bataafsche 11 March 1915; ibid. 132 AQ 162, correspondence 1916, Lane to Weill, 12 June 1916; ibid. 132 AQ 200, bundle 1918-19, Lane to Weill, 23 April 1919. Cf. SHA 190D/637-1, Colijn to Pleyte, 8 April 1915.

10 SLA 119/3/2, board minutes Anglo-Saxon 31 March, 23 and 30 June, 7, 14 and 21 July, 4 and 11 August 1915; SLA 119/11/3-4 bundle S-TU, memo 17 April 1917; the correspondence with the Foreign Office in National Archives Kew (NA Kew) FO 382/320; Jones, *State* 183-4.

11 Jones, 'British Government', 659-61, 667-8; ibid., *State* 182-6; Ferrier, *British Petroleum*, 244; rumours about the Group being German-controlled resurfaced as late as 1917 in Australia and the US, cf. for instance 141/3/7, board minutes Shell Transport, 7 March 1917.

12 SHA board minutes Royal Dutch, 23 April 1918; SHA 3/83A, Pleyte to Colijn, 29 March, 10 April 1915 (Deterding's involvement with decisions regarding the Ploesti lube oil factory); SHA 195/28, Deterding to Pleyte, 29 June and 22 July 1915 (discussion about merits of taking over the Roumanian Consolidated Oil Company); Gerretson, *Geschiedenis*, IV, 170-1, for Colijn consulting Deterding over Romanian affairs in April 1918. SLA 141/3/7, board minutes Shell Transport, 21 November, 19 December 1916 (instructions to Jacobson and board informed by Bataafsche about Astra stocks); SLA 141/11/9, board memos Shell Transport (Commons questions). Cf. Jones, *State*, 187-8.

13 SHA 3/209, Lane to Royal Dutch, 7 July 1916; SHA, board minutes Royal Dutch 25 July 1916

14 AN CAMT 132 AQ 195, board minutes Bataafsche 1 March 1915; cf. SHA 3/209, correspondence Lane with copies of Bataafsche board minutes in English circulated to Paris. On 12 March 1915 Colijn wrote a long letter to Deterding about the factory and the reasons for the large budget overrun, also omitting any direct reference to its location: SHA 190D/637-1. The same file has a letter from Colijn in London to Pleyte, 8 April 1915, instructing him to suspend sending the Bataafsche minutes. On 28 April, Colijn wrote to

Deterding that he had informed 'the gentlemen of the lub oil factory' of Deterding's decision: SHA 5/335.

15 Cf. for instance SHA board minutes Bataafsche 23 January, 3 March 1913, decision to begin the lube oil factory and Loudon replying to a question by Sir Marcus about the Monheim works and the Group's lube oil policy; SHA 8/1154, reference to the contract between Anglo-Saxon and Bataafsche, 24 March 1916. The German Government refused to accept this transfer, however: SHA Country Files Germany vol. III. s.v. Fusion. In July 1915 Pleyte had a meeting with Späth and Rudeloff about the German operations, at which A. de Jongh, an Asiatic employee entrusted with its German business, was present to liaise with Deterding: SHA 190D/637-1, Pleyte to Colijn, 9 July 1915.

16 NA Kew FO 382/320, minute 15 June 1915; FO 382/792, minute 29 September 1916.

17 Gerretson, *Geschiedenis*, IV, 64-9; SHA 195/25, Colijn to Deterding, 13, 15, 17 August 1914; ibid., 190A/146, Colijn to Rhenania, 12 August 1914; cf. Jones, *State*, 182-9.

18 Sparse details on Pleyte in SHA 190D/787; Van Soest, *Olie*, 181.

19 On 26 February 1916, Colijn reiterated his pledge and wrote to Deterding that he could count on him until 1925: SHA 5/335.

20 Gerretson's protestations that hiring Colijn had nothing to do with his political background simply fail to convince: Gerretson, *Geschiedenis*, IV, 53, echoed by Gabriëls, *Koninklijke Olie*, 64-5; cf. Langeveld, *Colijn*, 180-2.

21 Langeveld, *Colijn*, I. 199-203; Vrije Universiteit HDNP Colijn papers no. 352 box 1, Colijn to the British envoy Johnstone, 25 February 1916.

22 Gerretson, *Geschiedenis*, IV, 64-6, 68-9.

23 NA Kew FO 382/792, No. 163006, No. 165258.

24 SLA 119/3/2, 3 and 4, board minutes Anglo-Saxon with weekly statements of the investments held for Bataafsche; SHA, board minutes Royal Dutch, 12 August 1915 (surplus cash sent to London), 29 June 1916 (decision to invest in Treasury bills), 8 August 1916 (lombard loan on collateral of £1 million of Treasury Bills, and £500,000 worth of bills received from British Legation); AN CAMT 132 AQ 162, correspondence 1916, Lane to Weill, 28 April 1916 (£7 million of bills sold in the Netherlands); *The Times*, 21 June 1918, for Sir Marcus parading at the Shell Transport AGM with the Group holding £20 million of British Government paper. Some correspondence between The Hague and London went by the Dutch diplomatic bag, cf. for instance SHA 190D/637-2, Colijn to Deterding, 22 November 1916.

25 SHA board minutes Royal Dutch, between the pages with the minutes for 29 June and 13 October 1915 a summary of the Bataafsche minutes for 12 August 1915 with the formal decision to transfer all Group funds in the Netherlands to London for Anglo-Saxon to manage.

26 Homburg, Small and Vincken, 'Carbochemie', 335-6; SHA 190B/138, memoirs Späth has 1911 as the year in which the factory came on steam, but this must been earlier since building started in the summer of 1908; SHA 8/1585, board minutes Bataafsche, 28 November 1907, 12 May, 14 July, 5 August, 8 September 1908.

27 According to Bérenger, *Le pétrole*, 177-8, supplies started in October, but Gerretson, *Geschiedenis* IV 88-91, puts the first talks with the refiners in December.

28 Lévy, *Commerce*, 37. French coal output amounted to 41 million tons in 1913, of which the provinces occupied by Germany produced 30 million tons: Jensen, 'Importance of energy', 539; Friedensburg, *Erdöl*, 95, for import data. De Kok was first entrusted with overseeing the toluol operations in France; he subsequently transferred to Britain: SHA 3/83A, Deterding to Pleyte, 22 July 1915; Waley Cohen to Pleyte, 9 July 1915. For Italy SLA 119/11/3-4, board memos Anglo-Saxon, bundle N-P, memo 23 May 1917; Friedensburg, *Erdöl*, 100.

29 SLA 119/3/2, board minutes Anglo-Saxon 27 January, 17 March, 7 July 1915; AN CAMT 132 AQ 195, board minutes Bataafsche, 11 February 1915; Henriques, *Waley Cohen*, 92-3, 200-7 and id., *Samuel*, 597-601, drawing rather heavily on the propaganda brochure by Smith, *'Shell'*, but see Henriques, *Waley Cohen*, 190-9 for the tussles between the Group and the Admiralty over chartering conditions; Homburg, Rip and Small, 'Chemici', 302-303; cf. especially note 13; Gerretson, *Geschiedenis*, IV, 22-6, 86-93; Homburg, Small and Vincken, 'Carbochemie', 336-7.

30 Henriques, *Waley Cohen*, 202; Gerretson, *Geschiedenis*, IV, 92; SHA 190D/637-1, Colijn to Van Karnebeek, 3 February 1914.

31 Homburg, Small and Vincken, 'Carbochemie', 338; the stills were quite solid, two of them being used until the 1950s: ; SLA Sc46/1, typescript J. W. Vincent, 'The History of Shell's United Kingdom Oil Refineries 1914-1959' (June 1959), 1.

32 Smith, *'Shell'*, 36; cf SLA 119/3/2, board minutes Anglo-Saxon, 3 March 1915.

33 Henriques' claims in *Samuel*, 601-2, echoed by Yergin, *Prize* 175, that the factories produced 80 per cent is based on a misreading of Smith, *'Shell'*, who claims this percentage only for 1915, and gives the correct half share on 18-19. The Portishead plant was subsequently transferred to Rouen: GHS/2B/83. *St Helen's Court Bulletin*, 6 April 1918.

34 AN CAMT 132 AQ 195, board minutes Bataafsche, 11 March 1915; Bosboom, *Moeilijke omstandigheden*, 138-40, echoed by De Leeuw, *Nederland*, 190-2; cf. the weak defence of Gerretson, *Geschiedenis*, IV, 18; Homburg, Small and Vincken, 'Carbochemie', 336-7; SHA board minutes Bataafsche, 10 February 1916 for the contract with the Dutch Government.

35 Henriques, *Waley Cohen*, 206-207. German steel supplied to shipyards came with similar restrictions, preventing the Group from having ships built in the Netherlands: Royal Dutch, Annual Report 1917, 25.

36 Gibb and Knowlton, *Resurgent Years*, 122-6.

37 SHA 3/83A, Pleyte to Loudon, 5 July 1915; Waley Cohen to Pleyte, 9 and 15 July 1915. The suggestion that this was the first step to a complete transfer of the research activities to the UK would appear to be too strong: cf. Homburg/Rip/Small, 'Chemici', 303-4; Homburg, Small and Vincken, 'Carbochemie', 337-48. SLA 119/3/3, board minutes Anglo-Saxon, 26 March, 14 May 1919, for the £50,000 donation to set up a chemistry school at Cambridge in conjunction with Anglo-Persian, Burmah, and Mexican Eagle; cf. for that particular initiative Henriques, *Waley Cohen*, 247-8. In 1915, Royal Dutch had considered giving 500,000 guilders (£40,000) as part of its silver jubilee celebrations to Delft for setting up a chemistry department, only to reject the idea since the board thought the Group's staff would not agree with it: SHA board minutes Royal Dutch, 16 June 1915. In preparation for the dyestuff business, Anglo-Saxon started looking around for

suitable factories in 1917: SLA 119/3/3, board minutes Anglo-Saxon, 4 April 1917. Colijn was against rushing ahead, however, considering the capital requirements too big and the German competition too strong: SHA 5/335, Colijn to Deterding, 9 January 1917.

38 Forbes and O'Beirne, *Technical Development*, 294; Reynolds, *Ricardo*, 106-8, 129, 142-3, 157, 159, 163, 168, 220. Ricardo was a direct descendant of the economist David Ricardo, and thus of distant Anglo-Dutch descent. Cf. SLA 119/11/17, Anglo-Saxon board memos, 15 November 1927 (£19,573 spent by the Ricardo High Speed Diesel project, request for another £18,000 granted); 119/11/23, Anglo-Saxon board memos, 18 December 1933, £86,685 spent since 1925, rising royalty earnings and successes with the introduction of new engines. According to Forbes, Waley Cohen had argued to employ Ricardo by tactlessly remarking that 'we suffered greatly in the past relying too much on our experts for suggesting ideas'. This opinion enraged De Kok, who had just moved from Amsterdam lab to manage the Technical Department in The Hague, and he made Waley Cohen retract his ill-considered statement. SHA 190C/34A, Forbes 'Benzine', 7. Forbes' reference to Asiatic commissioning Ricardo must be an uncharacteristic slip.

39 A description of the festivities in *The Times*, 22 November 1918, *The Petroleum Review*, 23 November 1918 pp. 329 ff, and Bérenger, *La pétrole*, 170-5.

40 Friedensburg, *Erdöl*, 14; Sumida, 'Naval logistics', 465.

41 Friedensburg, *Erdöl*, 14.

42 Ibid., 39-41.

43 Ibid., 59; AN CAMT 132 AQ 161, correspondence 1914, Lane to Weill, 17 and 19 August 1914.

44 Friedensburg, *Erdöl*, 15.

45 Sumida, 'Naval Logistics', 479-80.

46 Van Creveld, *Supplying War*, 141, quoting Henniker, *Transportation*.

47 Singleton, 'Military Use', 194; SLA 119/3/2, Anglo-Saxon board minutes 4 February, 2 April, 16 December 1914.

48 DiNardo and Bay, 'Horse-drawn Transport', 130; Deterding knew this, cf. SHA 195/28, Deterding to Loudon, 23 November 1914. See also Schouteete, *Chevaux*.

49 Becker, *La première guerre mondiale*, 209.

50 Laux, 'Trucks', 68-9 ; see also Heuzé, *Camions*.

51 Friedensburg, *Erdöl*, 9.

52 Singleton, 'Military Use', 194, 195; Jones, 'British Government', 655.

53 Fenton, 'Ambulance drivers', 328-30; Singleton, 'Military Use', 194.

54 Morrow, *Great War*, 297.

55 Becker, *Première guerre mondiale*, 32-3.

56 Friedensburg, *Erdöl*, 15.

57 Bérenger, *La pétrole*, 42, 77-80.

58 Singleton, 'Military Use', 182-3

59 Friedensburg, *Erdöl*, 14, 61.

60 Ibid., 14, 51-2, 59-61.

61 Bérenger, *La pétrole*, 22-3.

62 Jones, 'British Government', 655-7. The liquid fuel shortage was unnecessarily exacerbated by the tight viscosity specification of the Admiralty. Royal Navy ships did not have heating coils in the bunkers, unlike for instance those of the US Navy: Foley, 'Petroleum problems', 1829-30.

63 Bérenger, *La pétrole*, 69-71.

64 Yergin, *Prize*, 176-7; a detailed discussion from the French perspective of the wrangles over transport in Clémentel, *La France*, 95-247.

65 Jones, 'British Government', 655-6, 665.

66 Jones, 'British Government'. 659-61; ibid., *State* 182-4.

67 Jones, *State*, 184-185; Henriques, *Waley Cohen*, 228-9.

68 AN CAMT 132 AQ 162, correspondence 1914, Lane to Weill, 3 September 1914; 132 AQ 163, correspondence 1916, Lane to Weill, 28 April 1916; Henriques, *Waley Cohen*, 239-41.

69 Smith, *'Shell'*, 36-8; Foley, 'Petroleum Problems', 1811 outlines the bulk gasoline delivery network in France.

70 Foley, 'Petroleum Problems', 1814-15, recognizing that in fact the tin system was more suited for war requirements.

71 Deterding received an honorary KBE and allowed himself to be addressed as Sir Henry Deterding, though, as a foreigner, he did not have the right to use the title in that way. Corley, *Burmah Oil*, I, 258, erroneously states that Deterding was naturalized, which the author later corrected in his entry on Deterding for the 2004 *Oxford Dictionary of National Biography*.

72 NA CAMT 132 AQ 162, correspondence 1915 for a list of shareholders in the holding company which held the Asiatic shares. Two Deutsch brothers owned 2,000 out of the 10,000 shares, Lane and Macandrew 268, two private investors 202, and the Rothschild bank and family the rest.

73 SHA 195/33-2, correspondence Deterding-Gulbenkian, 1916; ibid. 195/175, Gulbenkian to Deterding, 12 May 1918.

74 SHA 190B/138, Memoirs Späth, 4, quoting a range of uses for gasoline before finishing with the motor car. Späth, the director of a gasoline works in Southern Germany, set up the Rhenania for Royal Dutch and was the firm's first director. The number of cars in Mitchell, *European Historical Statistics* 668, 670; Karlsch and Stokes, *Faktor Öl*, 93, also mentions 22,000 motorcycles.

75 Friedensburg, *Erdöl*, 13-14, and Table 15 on p. 70 showing no imports of liquid fuel at all in 1913. Cf. DeNovo, 'Petroleum', for the US Navy's similar conservative stance, more surprisingly given the country's share in world oil production.

76 AN CAMT 132 AQ 161, Lane to Weill, 30 March 1914; the agreement was the result of negotiations by Deterding and Teagle on behalf of the two companies.

77 Friedensburg, *Erdöl*, 67-8.

78 Ibid., 14.

79 Pearton, *Oil*, 93; quote from Friedensburg, *Erdöl* 126-7; ibid. 77 for the rationing measures.

80 Friedensburg, *Erdöl*, 70, 73.

81 Karlsch and Stokes, *Faktor Öl*, 94; Forbes and O'Beirne, *Technical Development*, 473; Homburg, Small and Vincken, 'Carbochemie', 344-5.

82 Friedensburg, *Erdöl*, 71-2.

83 Homburg, Small and Vincken, 'Carbochemie', 341.

84 Friedensburg, *Erdöl*, 71, 75-6.

85 Bérenger, *La pétrole*, 95-136.

86 British experts, with the notable exception of Ricardo, considered benzol unacceptable as engine fuel due to its high specific gravity: Reynolds, *Ricardo*, 104-5.

87 Friedensburg, *Erdöl*, 72; Karlsch and Stokes, *Faktor Öl*, 100.

88 Friedensburg, *Erdöl*, 9-10, 74-5.

89 Laux, 'Trucks', 68-9.

90 Friedensburg, *Erdöl*, 15, see however 74; Corum, *Luftwaffe* 37; Morrow, *German Air Power*, 101, 102, 119.

91 Morrow, *German Air Power*, 112, 124, 130, 133; Morrow, *Great War*, 300-1.

92 Jensen, 'Importance Energy', 540-2 ;

Mitchell, *European historical statistics*, 383; Morrow, *German Air Power*, 75, 102.
93 Singleton, 'Military Use', 189.
94 SHA 195/28, Deterding to Loudon, 23 November 1914.
95 AN CAMT 132 AQ 161 and 162, accounts 1913 and 1914.
96 SHA 190A/146, Pleyte to Rudeloff, 23 February 1915, telling him to approach Jacobson for oil supplies. Astra sent its reports to the Group's bank in Berlin, Delbrück Schickler & Co., which passed them on to The Hague: SHA 8/2210, three files on Astra during the First World War.
97 Pearton, *Oil*, 70-6; Gerretson, *Geschiedenis*, IV, 198-201.
98 Friedensburg, *Erdöl*, 70.
99 SHA 8/2138, profit and loss statements Astra. The sterling figures calculated at the official exchange rate of 44.25 lei to the pound in 1919 given in this file. Translated into guilders, Astra's 1913 profit would have been 3.6 million guilders in 1913, rising to 6.4 million in 1914 and 8.1 million in 1915.
100 Gerretson, *Geschiedenis*, IV, 203-5; Pearton, *Oil*, 78-9.
101 SLA 141/3/7, board minutes Shell Transport, 21 November 1916. On 4 December, Colijn sent Deterding Astra's latest inventory of stocks, so London would know what had been destroyed: ibid., 19 December 1916.
102 Pearton, *Oil*, 82. In its usual style, the Royal Dutch Annual Reports emphasize the damage sustained and the outstanding claim for compensation, but omit to mention either the amount of the claim or its settlement.
103 SHA 8/2210-3, statement of damages 31 December 1918; see also 8/2173.
104 Royal Dutch Annual Report 1917, 17-8; Pearton, *Oil*, 83-4.
105 Pearton, *Oil*, 106-7.
106 Gerretson, *Geschiedenis*, IV, 124-31.
107 AN CAMT 132 AQ 161, correspondence 1914, Lane to Weill, 21 July 1914m
108 SHA monthly report Bataafsche, January 1918.
109 Royal Dutch Annual Report 1916, 16-8.
110 Gerretson, *Geschiedenis*, IV, 144.
111 SHA Country Volumes Russia, vol. 6, statement of Russian claims, 18 July 1922. Gerretson, *Geschiedenis*, IV, 11, only gives Royal Dutch's investment and not those of the London companies as well. The £32 million figure did not include Bataafsche's investment of 20 million guilders (£1.6 million) in Benzonaft and Argoun during 1920, since it was thought that, coming as it had long after the establishment of the Soviet regime, the inclusion might harm the prospects of the larger claim: SHA 190A/200-2, Van Wijk to Price, 9 July 1930.
112 NA CAMT 132 AQ 162, correspondence 1915, Lane to Weill, 18 November 1915.
113 Beaton, *Enterprise*, 149; this office became the Asiatic Petroleum Corporation in 1920.
114 Friedensburg, *Erdöl*, 90, 96.
115 AN CAMT 132 AQ 162, accounts 1915; SHA 195/51, reorganization Bataafsche, for the 1918 Asiatic sales data.
116 Gibb and Knowlton, *Resurgent Years*, 676-9, 681; Ferrier, *British Petroleum*, 271; Corley, *Burmah*, 320-1.
117 SHA 190B/138, memoirs Späth.
118 Cf. SHA 190B/137, urgent telegrams Rhenania to Colijn, 13 August 1914, to send more of this type of oil from Amsterdam or Rotterdam now that import duties had been suspended.
119 Karlsch and Stokes, *Faktor Öl*, 100-1; Gibb and Knowlton, *Resurgent years*, 229-233.
120 SHA 8/1585, board minutes Bataafsche, 9 March, 13 July 1916; SHA Country Files Germany vol. 3. s.v. Shareholders, and Fusion. Fifteen million Reichsmarks would have been about 400,000 guilders or 35,000 pounds at 1917 exchange rates. The original capital of 840,000 RM was the equivalent of 14,000 guilders or 1,200 sterling at prewar exchange rates.
121 Forbes and O'Beirne, *Technical Development*, 368-370; SHA 190C/251, Forbes 'Lube oil', 10 and SHA 190C/250, Forbes 'Extraction', 8 (viscosity improvement); Karlsch and Stokes, *Faktor Öl*, 102; SHA 190B/138, memoirs Späth; after consultation with Deterding in London, Bataafsche instructed Astra to increase its lube oil production capacity: SHA 190D/637-1 Colijn to Pleyte, 23 March 1915.
122 SHA 190G/245, Annual Reports Rhenania Ossag, 1917-20.
123 Gerretson, *Geschiedenis*, IV, 266-388; AN CAMT 132 AQ 163, bundle correspondence 1918, Lane to Weill, 26 February 1918.
124 SHA 190C/250, Forbes 'Extraction', 4 (rise of kerosene output Balik Papan).
125 Beaton, *Enterprise*, 97; SHA 195/51, reorganization Bataafsche, for the 1918 Asiatic sales data; SHA 195/25, Deterding to Loudon, 23 November 1914, Deterding to Colijn, 15 February 1915, Colijn to Deterding, February 1915. Forbes and O'Beirne, *Technical Development*, 368-70; a similar installation in Romania had only taken eight months to build; SHA 195/100, report R. A. Wischin February 1913 on the importance of lube oil for Bataafsche; SHA 190D/637-1, Van Tienen to Colijn, 21 July 1915, cost-price calculations for lube oil; SHA 190D/767, Colijn to Agnew, 1 February 1916, appointing the liaison manager; SHA 8/1872, memo by Dubourq 16 March 1917 about the charts and his visit to London, plus Colijn to Dubourq, 18 June 1917; SHA 190C/251, Forbes 'Lube oil', 3-4; SLA GHS/3A/4 binder 2 memo no. 62, estimated profit and loss Asiatic 1919.
126 Beaton, *Enterprise*, 69-71.
127 Ibid., 125, 130-139.
128 Gibb and Knowlton, *Resurgent Years*, 106-7, 666, giving sales figures only from 1919; Beaton, *Enterprise*, 782.
129 AN CAMT 132 AQ 162, correspondence 1916, Lane to Weill, March 1916.
130 AN CAMT 132 AQ 162, correspondence 1916, Lane to Weill, 14 February 1916.
131 Beaton, *Enterprise*, 88-93; Forbes and O'Beirne, *Technical Development*, 306-9; SLA 119/11/1-2, Board memos Anglo-Saxon, memo Pyzel, 19 February 1915, describing the original system.
132 Forbes and O'Beirne, *Technical Development*, 309-10.
133 SLA 119/3/2, board minutes Anglo-Saxon, 19 and 26 August, 9 and 23 September, 28 October 1914; SLA 119/11/1-2, board memos Anglo-Saxon, bundle 2, agreements 28 July and 22 October 1914 plus undated memo Kessler; SLA Sc46/1, Vincent, 'History of Shell Refineries UK', 3-4, erroneously dating the decision to build a Trumble on 22 July 1914, when the licence had not yet been taken; the Anglo-Saxon board specifically postponed a decision to September in order to obtain the guarantee; SHA 195/28, Loudon to Deterding, 16 November 1914, Deterding to Loudon, 23 November 1914. Trumbles would have been very useful as well in Russia, and it was considered to build one to treat the paraffinic Grozny crude, but the plans came to nothing owing to the war: SHA 190A/203-3, report technical development Russian companies, 11-3.
134 SHA 195/167, Pyzel to Deterding, 28 January 1914.
135 Beaton, *Enterprise*, 91-2.
136 Colijn considered Meischke Smith's arguments for buying the Trumble patents

more important than Pyzel's report: SHA 5/335, Colijn to Deterding, 13 December 1916. The Trumble company was reorganized as a patent holding company and renamed Simplex Refining.

137 SLA 119/3/2, board minutes Anglo-Saxon, 17 February, 24 March, 28 April, 26 May 1915; SLA 119/11/1-2, board memos Anglo-Saxon, bundle 2, Meischke-Smith to Deterding, 16 February, 15 June 1915, memo Pyzel 19 February 1915; SLA 141/11/9, Board memos Shell Transport, Meischke Smith to Deterding, 2 January 1915, Deterding to Meischke Smith, 6 February 1915; Beaton, *Enterprise*, 91.

138 Forbes and O'Beirne, *Technical Development*, 309-10; SHA, board minutes Bataafsche, 14 March 1918.

139 Royal Dutch, Annual Report 1918, 28 (total capacity); Henriques, *Waley Cohen*, 208-10 (75,000 tons in official use).

140 Bosboom, *Moeilijke omstandigheden*, 138-40.

141 Henriques, *Waley Cohen*, 211-2, 214, 222-3; cf. Royal Dutch Annual Report 1917, 8-9.

142 AN CAMT 132 AQ 162, correspondence 1915, Lane to Weill, 18 November 1915, memo Weill November 1915.

143 SLA 119/11/5-6, board memos Anglo-Saxon, undated memo of the marine superintendent Zulver, probably July 1918.

144 Wouters, *Shell Tankers*, 47; SHA, minutes Royal Dutch, 17 December 1914 (changed shipping schedules); SLA 119/11/1-2, board memos Anglo-Saxon, bundle A-B, Memo 26 April 1916.

145 SLA 119/3/2, board minutes Anglo-Saxon 28 April 1915; ibid. 119/3/3, 19 January, 5 April, 10 May, 5 July 1916; SLA 119/11/1-2, board memos Anglo-Saxon, bundle A+B, report Shell California, 25 January 1916; ibid. 119/11/3-4, bundle A-C, report Shell California 31 December 1916; Beaton, *Enterprise*, 97-9.

146 Foley, 'Petroleum Problems', 1821-22, 1824-5.

147 Henriques, *Waley Cohen*, 207-11; Friedensburg, *Erdöl*, 90, for British import data.

148 SHA 195/167, Pyzel to Deterding, 28 January 1914.

149 Foley, 'Petroleum Problems', 1804, 1819.

150 SHA 195/33/3, memorandum Zulver for the French Government, 30 October 1917, claiming the system had been used since the beginning of the war; SLA 119/3/3, board minutes Anglo-Saxon, 5 December 1917; SHA monthly report Bataafsche 18 May 1918; *St Helen's Court Bulletin*, 14 December 1918; Henriques, *Waley Cohen*, 214-7, quoting a January 1916 letter from Waley Cohen to the Admiralty that the Group had used it 'for years' in Dutch ships and giving a total of 761 ships converted; Howarth, *Century*, 106 and Smith, *'Shell'*, 28, give a total of 1,280 ships for the UK alone. Sir Marcus Samuel had already suggested using ballast tanks in his 1899 Society of Arts paper on liquid fuel: Henriques, *Samuel*, 606-7. SLA GHS3H/4, first folder, gives a synopsis of the correspondence between Anglo-Saxon and the Admiralty. At the Transport Department of the Admiralty, George Legh-Jones oversaw the implementation of the double bottom scheme. He then transferred to the British War Mission in the US, and joined Asiatic in 1919: *Petroleum Times*, 8 January 1938.

151 SLA 119/3/3, board minutes Anglo-Saxon, 16 August 1916 (contract with the Admiralty for freighter conversions); 27 September 1916 (six bought, five under offer); 11 April 1917 (Admiralty buys ship to be managed by Anglo-Saxon) 11 July 1917 (19 freighters under construction placed under Anglo-Saxon's management); ibid. 24 May 1916, administration Dutch ships returned to The Hague.

152 The term 'Rothschild interests' is strictly speaking not correct. The Asiatic's 'C' shares were held by the Commercial and Mining Company. The Paris Rothschild bank owned just over half of that company's 10,000 shares, with family members possessing another 2,345 for a total of 7,363. Emile and Henry Deutsch each had 1,000 shares, Lane and Macandrew 268, and various other individuals owned the rest. A list in AN CAMT 132 AQ 162, correspondence 1915; cf. SHA 8/1366, files Asiatic 'C' shares.

153 SLA 131/3/3, board minutes Anglo-Saxon, 5 July 1916.

154 AN CAMT 132 AQ 161, correspondence 1914, Lane to Weill, 17 September, 5 October 1914.

155 Ibid., Lane to Weill, 12 August 1915.

156 Ibid., Lane to Weill, 19 July 1915 (extra funding estimated by Deterding at £800,000); ibid. correspondence 1916, Lane to Weill, 15 January 1916 (£4.5 million).

157 SLA 119/3/2, board minutes Anglo-Saxon, 14 April 1915; ibid. 120/3/1, minutes Asiatic 17 May 1915; AN CAMT 132 AQ 162, correspondence 1915, Lane to Weill, 31 March, 8 April 1915.

158 AN CAMT 132 AQ 162, correspondence 1915, Lane to Weill, 15 October, 18 November, 2 and 9 December 1915.

159 SLA 119/3/3, board minutes Anglo-Saxon, 10 November 1915; AN CAMT 132 AQ 162, correspondence 1915, Lane to Weill, 18 November 1915.

160 AN CAMT 132 AQ 162, correspondence 1915, Weill to Lane, 24 November 1915; Lane to Weill, 9 December 1915.

161 AN CAMT 132 AQ 163, correspondence 1918, memorandum probably from Weill, 30 April 1918.

162 SLA 120/3/1, board minutes Asiatic, 19 August 1914, adopting the principle; SLA 119/3/3 board minutes Anglo-Saxon, 19 January 1916.

163 AN CAMT 132 AQ 162, correspondence 1915, Lane to Weill, 15 and two letters of 18 January, 16 February 1916.

164 AN CAMT 132 AQ 162, correspondence 1916, Lane to Weill, 16 February, 8 April 1916.

165 Ibid., Weill to Lane, 11 March 1916; ibid. 132 AQ 163, correspondence 1918, memorandum probably by Weill, 30 April 1918.

166 SLA 119/3/2, board minutes Anglo-Saxon, 28 April 1915 (decision to set up Shell Marketing), 119/3/3, board minutes Anglo-Saxon 2 February 1916 (Shell Marketing to start operating 1 November); 141/11/9, board memos Shell Transport (history BP); AN CAMT 132 AQ 162, correspondence 1916, Lane to Weill, 1 April 1916 (Deterding no longer wants BP). Ferrier, *BP*, 218.

167 SHA 8/1366, Deterding and Cohen Stuart to Loudon, 23 January 1918; SHA board minutes Bataafsche, 14 March 1918.

168 AN CAMT 132 AQ 163, correspondence 1918, Lane to Weill, 4 and 8 January, 16 April 1918. Anglo-Saxon chartered Dutch vessels at 35 shillings a ton, however: SLA 119/3/4, Board minutes Anglo-Saxon 15 January 1919.

169 AN CAMT 132 AQ 163, correspondence 1918, Lane to Weill, 16 April 1918. Anglo-Saxon and Asiatic only modified the freight rates clause of the 1908 contract from a fixed rate to actual cost at market rates in 1921, SLA 119/11/9, board memos Anglo-Saxon, draft letter to Asiatic agreed by the board on 8 December 1921, and calculation of freight rates to be charged in 1922, 2 March 1922. At the same time the rental conditions for Anglo-Saxon's

installations was modified to take into account the substantial expansions to them: SLA 119/11/9, draft letter Anglo-Saxon to Asiatic, agreed by the board 8 December 1921.

170 SHA board minutes Royal Dutch, 16, 29 June 1915.

171 SHA 195/28, Deterding to Bataafsche, 4 August 1915, using US conditions to argue that staff in the Netherlands East Indies had nothing to complain of and should take to heart that they were no longer supplying the bulk of the Group's profits.

172 SLA 119/11/2, board memos Anglo-Saxon, Bundle A-B, Meischke-Smith to Deterding, 16 February 1915, with memorandum Pyzel 19 February 1915; SHA 195/28, Deterding to Pleyte, 11 November 1914, Loudon to Deterding, 16 November 1914; Beaton, *Enterprise*, 130, 158, 200, mentioning the training of an American geologist in The Hague after he had first worked in Venezuela, Pyzel's Yarhola casinghead gasoline plant, and Van der Gracht's introduction of core-drilling. On Van der Gracht see Van der Veen, *Van der Gracht*.

173 For the strains on Deterding, see RA Alkmaar De Lange papers (no inventory available at time of writing), diary Dolph Kessler, entry for 11 June 1913 commenting on stress wearing down Deterding's capacity for work.

174 AN CAMT 132 AQ 162, accounts 1915; SHA 195/51, reorganization Bataafsche, for the 1918 Asiatic sales data.

175 SHA 195/28, Loudon to Deterding, 16 November 1914, complaining about his inability to follow developments in California; Deterding to Loudon, 23 November 1914. Colijn had aired the same complaint a few weeks earlier: SHA 5/335, Colijn to Deterding, 6 November 1914.

176 SHA 8/1863-1, undated manuscript memo Colijn and typescript memo 24 April 1917, laying down the tasks of the statistical department; 195/110, Colijn to De Jonge, 28 May 1920.

177 SHA, board minutes Royal Dutch, 17 December 1914.

178 Ibid.

179 Ibid.; SLA 119/3/2, board minutes Anglo-Saxon 30 December 1914, amounts to be transferred; the Bataafsche board had earlier opposed the transfer: AN CAMT 132 AQ 195, board minutes Bataafsche, 10 and 22 December 1914.

180 SHA 190C/266, a file with various staff statistics.

181 VU, Historisch Documentatiecentrum van het Nederlands Protestantisme (hereafter VU Colijn Papers) Colijn no. 352 book 1 with regular complaints to private correspondents about his having to interview applicants; SHA 8/1863-1, reorganization Bataafsche office, memo 7 April 1917.

182 Gerretson, *Geschiedenis*, V, 34-5; SHA 8/1863-1, reorganization Bataafsche, an organization chart dated 19 December 1916; NA The Hague 2.21.095, De Jonge papers No. 58, a memo written by Gerretson for De Jonge to prepare him for his directorship of Bataafsche and thus probably dating from the winter of 1921. For the reorganization of the information flows and record keeping at Bataafsche see the manuscript study by Vos, 'Archiefwezen'.

183 SHA 8/1863-1, reorganization Bataafsche, organization chart 1916 and undated memo Colijn about the AG department; SHA 8/1420, Asiatic, report on a visit to St Helen's Court, October 1915. The preparations for the reorganization clearly predated the move to the new head office, and did not follow it, as Gerretson, Geschiedenis, V, 34-5 has it. Cf. Gabriëls, *Honderd jaar*, 71, for a description of the new building and its amenities.

184 The Bataafsche football club had been set up in September 1916; similar clubs for tennis, fencing, gymnastics (notably tug-of-war) and athletics soon followed. The fields were located more or less at the site of the present building C23. Cf. *Clubhuis Te Werve 1922-1982*, unnumbered pages; *Zestig jaar Te Werve*, unnumbered pages. Other operating companies soon followed suit and founded their own sports clubs, see for instance Pradier, *Shell France*, 24-5, for the development of the French sports club since 1924.

185 SHA 8/1863-1, organization Bataafsche, Colijn to Deterding, 3 November 1915.

186 RA Alkmaar De Lange papers (no inventory numbers at time of writing), memo Guépin, 8 April 1936.

187 SLA 119/11/2, board memos Anglo-Saxon, Russian department presenting a new style of report, August 1915; 141/11/9, board memos Shell Transport, a letter from the Russian Department, April 1915, with an improvised letterhead; cf. SLA 119/3/2, board minutes Anglo-Saxon, 26 May 1915, sanctioning Godber to spend $140,000 on four Trumble plants.

188 Godber had been appointed head of the American department in 1912, overseeing two clerks; see Rady, 'Godber', *British Dictionary of Business Biography*, and Barran, 'Godber', *Oxford Dictionary of National Biography*, online at www.oxforddnb.com.

189 SLA Sc46/1, Vincent, 'History of Shell Refineries UK', 3.

190 SHA 8/1420, Report on the organization of St Helen's Court, October 1915; SLA GHS/3A/4, binder 1, memorandum no. 53, floor area and staff of Asiatic; SLA 119/11/7, board memos Anglo-Saxon, undated memo Staff and floor area Anglo-Saxon, 1914 and 1920, April 1920; SLA 141/3/7, board minutes Shell Transport, 18 January 1916.

191 Annual contests between London and The Hague sports clubs appear to have started shortly after the war at Teddington; when in 1921 London beat The Hague in all competitions save the tug-of-war, Deterding resolved to let The Hague have facilities similar to the Lensbury Club, which became the Te Werve complex in Rijswijk. From 1923, The Hague team for these events included Group employees from Sweden, Denmark, Italy, and France. Cf. *Clubhuis Te Werve 1922-1982*, unnumbered pages; *Zestig jaar Te Werve*, unnumbered pages.

192 AN CAMT 132 AQ 162, correspondence 1916, Lane to Weill, 18 January, 16 February 1916; ibid. 132 AQ 163, correspondence 1918, Lane to Weill, 26 February 1918.

193 SHA 195/28, Deterding to Loudon, 9 August 1915. One staff member, Van Rees, died when the ferry on which he travelled was sunk. SHA 8/1872, Dubourq to Bataafsche, 16 March 1917 (visit to Asiatic).

194 SHA 195/33-1, correspondence Gulbenkian to Deterding, 1915-1916. The general correspondence on the New York issue in SHA 8/1657.

195 SHA, board minutes Royal Dutch, 8 April 1916. Earlier, Stuart had written privately to Loudon about the matter: SHA 190D/637-2, Stuart to Loudon, 1 March 1916.

196 SHA, board minutes Royal Dutch, 4 December 1916. Agreement with Kuhn Loeb was reached in November, but the shares were to be handed over and paid in January 1917. Deterding's other objective for a New York issue, stabilizing the

market in Royal Dutch shares by creating another market, was foiled by speculators arbitraging between New York and Amsterdam, leading to a large but unspecified number of the American certificates being exchanged for Royal Dutch shares: SHA 195/22-4, Deterding to Gordon Leith, 10 March 1934.

197 SHA 8/1657-1, Deterding to Loudon, 23 November 1916.

198 SHA, board minutes Royal Dutch, February 1916–January 1917.

199 Waley Cohen to Philips, 27 December 1923, quoteed in Jones, *State*, 223.

200 Ferrier, *British Petroleum*, 243; NA Kew ADM I 8537/240, memorandum Board of Trade, 'The future control of oil supplies', 12 August 1916.

201 Ferrier, *British Petroleum*, 243-4; Corley, *Burmah Oil*, 244-7; SHA 190D/637-3, Deterding to Colijn, 4 January 1917, reporting discussions with Sir Marcus, Waley Cohen, and Stuart about the question of whether Asiatic could ever become a British Government controlled company, and dismissing the idea with some contempt.

202 SLA 141/3/7, board minutes Shell Transport, 15 January 1918; Jones, *State*, 201-2; Ferrier, *British Petroleum*, 250; NA Kew POWE 33/13 p39.

203 Ferrier, *British Petroleum*, 252-253; Jones, *State*, 201-3; the committee's report and proceedings in NA Kew POWE 33/13.

204 Cf. for the Mosul question Fitzgerald, 'France's Middle Eastern ambitions'.

205 SHA board minutes Royal Dutch, 4 February, 8 March 1919; Ferrier, *British Petroleum*, 197, 258-9; Jones, *State*, 212-4 links the episode also with the developments in Russia and Romania, where the Group needed British support to recover lost assets. However, Deterding's sudden switch of position and the continuing emphasis in the Royal Dutch board discussion suggests that access to oil concessions was uppermost; the Harcurt Committee even appears to have threatened to put a stop the Group's expansion in Egypt, cf. SHA board minutes Royal Dutch, 8 March 1919.

206 Thus Deterding held on to a joint venture with Burmah in Trinidad, since selling would help Burmah's expansion: AN CAMT 132 AQ 163, bundle correspondence 1918, Lane to Weill, 26 February 1918.

207 SHA board minutes Royal Dutch, 8 March 1919.

208 SHA board minutes Royal Dutch, 4 February 1919; the Harcourt report states unequivocally that Harcourt and Deterding initialled a provisional agreement on 31 January, followed by an amended agreement which the committee approved on February 7: POWE 33/13, 13-4.

209 AN CAMT 132 AQ 162 correspondence 1915, Lane to Weill, 8 April 1915. Colijn became a Royal Dutch director in 1921.

210 SHA board minutes Royal Dutch, 4 February 1919; SHA 102, Loudon papers, memo Deterding explaining the agreement, 1 February 1919. The board's decision was immediately transmitted to London by diplomatic telegram from The Hague Chargé d'Affaires to the Foreign Office, a copy of which in SHA 102, Loudon papers, telegram 4 February 1919; quoted in Henriques, *Samuel*, 625.

211 SHA board minutes Royal Dutch, 8 and 11 March 1919; the conditions summarized in Henriques, *Samuel*, 626-7.

212 SLA 141/11/10, board memos Shell Transport 1917-20, list of the biggest shareholders, 27 July 1917. Cf. NA Kew POWE 33/13, report PIPCO, 4.

213 The text of the two agreements of January and February 1919 in NA Kew POWE 33/13; see also Kent, *Oil and Empire*, 178-82. The Anglo-French memorandum was confirmed at the 1920 San Remo Conference.

214 SHA board minutes Royal Dutch, 25 July 1919; SHA Country Files France, vol. 1, s.v. Régime National; Nouschi, *La France*, 25-6; Jones, *State*, 215-6; Kent, *Oil and Empire*, 155, erroneously puts the Société's foundation in 1924; Jones, *State*, 213, 216, gives the correct sequence. Interestingly, Bérenger defended the agreement by pointing out that French investors owned almost 40 per cent of Royal Dutch's capital, so that the Group was almost as French as it was Dutch or British: Bérenger, *Pétrole*, 290.

215 Jones, *State*, 215-6; cf. Friedensburg, *Erdöl*, 91, Mexico having been the second biggest supplier of oil to Britain during the war, after the US. SHA 102, Loudon papers has some correspondence about the further complications surfacing after the initial agreement.

216 AN CAMT 132 AQ 161, correspondence 1914, Lane to Weill, 4, 26 August 1914, 12 August 1915.

217 SHA 195/28, Colijn to Deterding, 13 November 1914.

218 AN CAMT 132 AQ 162, correspondence 1915, Lane to Weill, 6 January 1915.

219 SLA 141/11/9, board memos Shell Transport, Waley Cohen to Sir Marcus, 31 May 1916.

220 AN CAMT 132 AQ 162, correspondence 1916, Lane to Weill, March 1916.

221 Calculated as 18,881,800 shares issued at par times 5.3 is 100,073,540 guilders plus 29,688,336 guilders ordinary dividends divided by 120,000,000 guilders of ordinary capital=108 per cent. AN CAMT 132 AQ 162 correspondence 1916, Lane to Weill, March 1916; SHA board minutes Royal Dutch, 3, 16 and 28 March, 8 April, 2, 11, 17 May 1916. The Rothschilds held shares for a total of 10.1 million guilders, which entitled them to 3,373,700 guilders worth of bonus shares: AN CAMT 132 AQ 200, bundle 1918-19, Lane to Weill, 18 December 1918, plus copy of Rothschilds to Royal Dutch, undated; Lane to Royal Dutch, 4 February 1919.

222 Shell Transport had issued shares to shareholders at a discount in 1912 and 1913, but the 1917 issue had been at about market price: cf. *The Economist*, 1 July 1922, 7.

223 Cf. AN CAMT 132 AQ 198, bundle 1910, Deterding to Aron, 31 May 1910, explaining the Group's depreciation policy.

224 Waley Cohen had already pleaded for writing off the Russian investments in 1916 at the prevailing low exchange rates, which nobody would notice from the balance sheet: SLA 141/11/9, board memos Shell Transport, Waley Cohen to Sir Marcus, 31 May 1916. AN CAMT 132 AQ 200, bundle 1918-19, Lane to Weill, 7 November 1918 (Bataafsche writing off its Russian investments). The Royal Dutch Annual Report for 1918 still printed a concise balance sheet for Bataafsche showing a reservation of 60 million guilders (about 5 million pounds) for anticipated Russian losses, but the 1919 report no longer gave details on Bataafsche, so the rest was probably written off during that year.

225 Cf. a long report dated 15 October 1917 by G. S. Engle about costs of living of the London office staff in SLA 119/11/3-4, Board memos Anglo-Saxon.

226 Cf. e.g. Henriques, *Samuel*, 581-592.

Chapter 4

1 Source: Annual Report Royal Dutch 1927 p. 13, quoted from 'South American Oil Reports', March 1928.We were able to check the crude production data of this table for Royal Dutch Shell, Jersey Standard and Anglo-Persian.
2 Sources Figure 4.1: See appendix.
3 Sources Figure 4.2: See appendix.
4 Sources Figure 4.3: See appendix.
5 SHA 102, Correspondence Loudon, memo 'Finantieel voorstel van den heer Deterding aan Lord Cowdray'. Unfortunately the official conditions for buying these shares were not published; the current share price was 530 per cent, Deterding's proposal was to pay 550 per cent, but probably he paid up to 600 per cent or £6 per share. According to Gerretson, *Geschiedenis*, IV, 317, the Group paid 98 shillings per share.
6 Howarth, *Century*, 120-1.
7 SHA 190C/386 gives estimates of market shares of the Group on for example Java between 1912 and 1929 which varied between 78.5 and 91.9 per cent.
8 Source: Annual reports Shell Transport, Royal Dutch, and Bataafsche.
9 Before the Harcourt Committee, for example, Deterding stated that 'he was chiefly concerned in avoidance of waste and was therefore anxious for co-operation (...) He favoured amalgamation because of the economies which would result both in production and distribution'. NA Kew CAB 27/180, meeting 30 March 1922.
10 CAMT, Archives Rothschild 132 AQ 166, memo 21 June 1927.
11 Gerretson, *History*, IV, 241: 'Deterding simply *had* to get to America'.
12 SHA 195/115, Deterding to Loudon, 18 February 1911.
13 Gerretson, *Geschiedenis*, IV, 504.
14 AN CAMT 132 AQ 162, correspondence 1916, Lane to Weill 19 April 1916.
15 See Van Veen, *Van Waterschoot van der Gracht*. After leaving the Group, he became the most prominent proponent of the theory of Continental Drift. Originally formulated by Alfred Wegener in 1915, this theory was further elaborated and given wider berth by Van Waterschoot van der Gracht during the 1920s and 1930s. The idea was then largely forgotten, however, until it was rediscovered in the 1960s.
16 Beaton, *Enterprise*, 159-66.
17 SHA 8/204-3, Deterding to Van der Gracht, 15 January 1919.
18 SHA 8/204-3, Deterding to Colijn, 16 January 1919.
19 SHA 8/204-3, Deterding to Van der Gracht, 15 January 1919.
20 SHA 8/307, memo 24 February 1919.
21 SHA 8/307; memo 5 March 1919.
22 SHA 8/204-3, Van der Gracht to Deterding, 25 December 1918.
23 SHA 8/204-3, memo 24 February 1919.
24 SHA 8/204-3, Van der Gracht to Erb, 9 August 1921.
25 Beaton, *Enterprise*, 172-90 gives all the details about this episode.
26 Ibid., 206.
27 Gerretson, *Geschiedenis*, IV, 510-12.
28 Beaton, *Enterprise*, 197.
29 Ibid., 299-300.
30 At one point it was suggested to let Marland become president of the new company: Beaton, *Enterprise* 215.
31 Beaton, *Enterprise*, 228.
32 Ibid., 206.
33 SHA 8/204, Erb to Godber, 13 February 1922; Beaton, *Enterprise*, 219.
34 Beaton, *Enterprise*, 218-19.
35 SLA 119/3/5, board minutes Anglo-Saxon, 26 January 1922.
36 SHA 195/22-1, memo Deterding, 24 December 1923.
37 SHA, board minutes Bataafsche, 10 January 1924.
38 SHA, board minutes Bataafsche, 10 January 1924; the following November, Philips resigned from the board of Bataafsche.
39 SLA 141/11/11, 22/l Board memos Shell Transport, Deterding to Shell Transport, 19 November 1921; note that he interpreted these structural problems very much in personal terms: HE (and HE alone) 'had to fight for the interest of the parent companies' etc.
40 SHA, board minutes Bataafsche, 25 September 1925.
41 SHA, board minutes Bataafsche, 30 October 1925.
42 Beaton, *Enterprise* 360.
43 Ibid., 315-6.
44 Larson, Knowlton and Popple, *New Horizons*, 134.
45 SHA 8/790, Erb to Deterding 26 August 1925.
46 McBeth, *Gómez and the Oil Companies*, 38.
47 Ibid., 37.
48 SHA 8/772, 8/774, Colon Development Company, correspondence related to the possible sale of CDC.
49 SLA 102/3/4, board minutes VOC, 14 November 1924.
50 Ibid., 27 March 1925.
51 Ibid., 17 September 1925.
52 McBeth, *Gómez and the Oil Companies*, 64; Van Soest, *Olie als water*, 234, mentions that in 1922 there were only six oil companies in Venezuela, all related to the Group; in 1926 there were 75.
53 Van Soest, *Olie als water*, 162-77, 233-42.
54 Ibid.
55 Ibid., 240.
56 Larson, Knowlton and Popple, *New*

Horizons, 134-7.

57 AN CAMT AQ 132 161, Copy of Agreement Turkish Petroleum Concessions.

58 The text of the two agreements of January and February 1919 in NA Kew POWE 33/13; the latter published in Kent, *Oil and Empire*, 178-82. There is also a copy of the memorandum in SHA Loudon Papers. See also Jones, *State*, 193.

59 SHA 102, Loudon Papers, memo Cohen Stuart 15 August 1919; in a related letter dated 24 August 1920, Cohen Stuart even argued that he felt 'more and more that our association with the Shell (...) does not seem to be a happy one. The whole world is getting more and more afraid of British Government influence in oil matters (...) I have, until of late, never been very much in favour of the idea which has sometimes been promoted by Sir Marcus Samuel, namely that the Shell Company should merge in the Royal Dutch in such a way that the existing Shell shareholders would receive a certain proportion of Royal Dutch shares in exchange for their Shell shares, after which the Shell company would be wound up. But I have gradually been led to think that this would not be a bad solution'.

60 SHA 102, Loudon Papers, J. B. Body to Deterding, 25 July 1919.

61 NA Kew CAB 27/180, interview with Walter Samuel, 5 April 1922.

62 Yergin, *Prize*, 199.

63 Jones, *State*, 220; Yergin, *Prize*, 194-5.

64 Jones, *State*, 237.

65 SLA RA 262/1, Gulbenkian to Deterding, 25 July 1924.

66 Ibid., 31 July 1924; he considered the working agreement 'wholly unacceptable' and 'disastrous as regards my interest'.

67 Gulbenkian confidentially told his solicitor that he 'was going to make great use in Court of the VOC agreements', meaning that he would also use his inside knowledge of the problems within the VOC between the majority shareholders and the minority to put pressure on the Group (see chapter 5 for details); SLA RA 262/1, Cochrane to Agnew, 13 September 1924; Cochrane added that 'it would be very difficult to face them in Court', meaning that the threat was quite credible.

68 Howarth, *Century*, 149.

69 As early as 1913, Erb warned about the risks of investing in Mexico because of the geological circumstances: Gerretson, *Geschiedenis*, IV, 543-4.

70 Gerretson, *Geschiedenis*, IV, 321.

71 Brown, *Oil*, 241.

72 Van Vuurde, *Los Países Bajos*, 28-33; Brown, *Oil and Revolution*, 164.

73 The debate about the causes of the decline of the Mexican oil industry in Haber, Mauser and Razo, 'When the law does not matter.'

74 Pearton, *Oil*, 105, 110-1.

75 Ibid., 110.

76 Ibid., 129.

77 Gibb and Knowlton, *Resurgent Years*, 91-3, mentioning a sum of $6 million spent between 1912 and 1918 for getting a production of 135 barrels a day.

78 Gerretson, *Geschiedenis*, V, 169-73.

79 SHA, 190C/76, Colijn to Idenburg.

80 SHA, 190A/120-1, Colijn to De Jonge, 1 March 1921.

81 SHA 141/11/11, containing the translated minutes of the Bataafsche board of 6 May 1920; Sir Marcus's remark was certainly an exaggeration but, as noted above, oil from the Dutch East Indies was indeed the main source of profits in these years.

82 AN CAMT 132 AQ 164 Lane to Weill, 26 January 1920.

83 Ibid., 15 June 1920.

84 Ibid., Lane to Weill 28 January 1921.

85 Ibid., Deterding to Baron Edmond de Rothschild, 2 September 1921.

86 Taselaar, *De Nederlandse koloniale lobby*, 226-7.

87 *Minjak*, 3, November 1921.

88 AN CAMT 132 AQ 164, Deterding to Baron Edmond de Rothschild, 16 September 1921: 'In strict confidence, I can tell you that I have been asked to meet Her Majesty the Queen on the 21th, and I may have the opportunity at this private interview to say a word in that regard'.

89 AN CAMT 132 AQ 164, Deterding to Lane, 5 October 1921.

90 SHA 195/176, Gulbenkian to Deterding, 30 August 1921 and Gulbenkian to Colijn, 21 September 1921.

91 SHA 195/168, Philips to De Graaff, 25 January 1922.

92 SHA 195/168, Philips to Deterding, 22 November 1922

93 Source for yield of export taxes: *www.iisg.nl/indonesianeconomy/index.html* (database of reconstructed financial accounts of the Dutch East Indies); in 1922 the yield of the export tax increased to almost 12 million guilders.

94 SHA, 190A/120-1, undated and unsigned letter, probably from November 1921. Deterding was criticized by colleagues for his linking the export tax to the Jambi agreement; another undated document from the same files, probably written by Colijn , argued that at that stage it was impossible not to continue with the signing of the agreement with the government, because of the damage to the reputation of both the Group and the Minister of Colonial Affairs.

95 SHA, 190A/120-1, minutes meeting with De Graaff, 29 November 1921.

96 The story of Jambi has received a lot of attention in the Dutch colonial literature, most recently by De Graaff, *Kalm*, 453-493.

97 Wilkins, *History* II, 102.

98 Gerretson, *Geschiedenis* V, 231.

99 De Graaff, *'Kalm'*, 475, 477.

100 SHA 195/34.

101 SHA 195/25, Gulbenkian to Deterding, 22 September 1920; the connection with Wrangel was supposed to be secret, cf. SHA 195/34, Gulbenkian to Deterding, 2 October 1920, 'I regret that I mentioned the name of General Wrangel in connection with the 100,000,000 pouds transaction. I really gave you more information than I was authorized to give, for that part dealing with General Wrangel was given me under great secrecy'.

102 SHA Country volumes Russia, vol. 4, Gulbenkian to Philips 28 October 1920.

103 SHA Country volumes Russia, vol. 4, letters by Philips to Colijn and Gulbenkian, 2 November 1920.

104 SHA Country volumes Russia, vol. 4, BPM Board minutes 11 November 1920; see also Gerretson, *Geschiedenis*, IV, 150-3.

105 The first letter, from Gerretson to Krassin, is from 20 October 1920, and refers to 'our recent conversation'; SHA Country volumes Russia, vol. 4, entitled Negotiations between Mr Krassin and Dr Gerretson.

106 SHA Country volumes Russia, vol. 4, Colijn to Sir Robert Horne, 22 December 1920; the text of the draft agreement between Krassin and Anglo-Persian, for which again Gulbenkian was the source of information, in SHA 195/32, Gulbenkian to Deterding, 20 December 1920.

107 SHA Country volumes Russia, vol. 4, the most detailed proposals, notably a memo 'Main conditions of Concessions for Exploiting oil fields in the R.S.F.S.R.' date

from March 1921. During that month Krassin was also negotiating with Anglo-Persian and Jersey Standard about identical plans, see Gibb and Knowlton, *Resurgent Years*, 337.

108 Cited in Yergin, *Prize*, 238.

109 Yergin, *Prize*, 238; Gibb and Knowlton, *Resurgent Years*, 334.

110 Gerretson, *Geschiedenis*, IV, 159.

111 SLA 141/11/13, memo 8 June 1927; £27 milllion was a formidable amount, but only slightly more than 10 per cent of the total capitalized value of the Group in these years. In 1922, the figure was put at £33 million: SHA Country Volumes Russia, vol. 6, statement of Russian claims, 18 July 1922.

112 Gibb and Knowlton, *Resurgent Years,* 340.

113 SHA 195/39, N. Gulbenkian to Deterding, 27 March 1923.

114 SHA 195/39, Deterding to Gulbenkian, 21 March 1923; the volumes involved were 70,000 tons with an option on 200,000 tons; on 30 January 1923, Deterding wrote to Gulbenkian that he was not in favor of the boycott of Russian oil.

115 Gibb and Knowlton, *Resurgent Years*, 344.

116 Ibid., 346-7 for these negotiations; they also mention his marriage.

117 AN CAMT 132 AQ 166, memorandum Weill 28 July 1927; already in 1925 Deterding had warned Cadman of Anglo-Persian not to buy Russian oil (SHA 195/39, Deterding to Cadman, 3 December 25).

118 AN CAMT 132 AQ 166, memorandum Weill 28 July 1927; according to Deterding they replied that Deterding had an agreement with the president of Socony, not with the Managing Director, and that the former did not have the power to bind the company ('n'avait pas qualité pour engager le Standard Oil de New-York').

119 AN CAMT 132 AQ 166, Weill to Baron Edmond, 19 May 1927; as usual, this scheme had been developed by Gulbenkian, idem memorandum Weill, 21 June 1927. The Rothschilds were concerned that a conflict between Poincaré and Deterding might harm their interests.

120 SLA, 119/11/72, board memos Anglo-Saxon, memo Russian Department, 14 June 1932; it contributed to the publication of Anglo-Russian News by Dr. Edouard Luboff, who was 'a dangerous critic of their [= Soviet] regime'.

121 AN CAMT 132 AQ 166, Weill to Baron Edmond, 19 May 1927, 'Il estime que c'est pour lui une question de prestige de maintenir le point de vue qu'il a pris et qu'il ne peut pas accepter qu'on manque de parole à une Société de l'importance de la Royal Dutch'. Weill also mentions that his colleagues were surprised by the violent nature of his outbursts ('étonnés de la raideur de ses notes'); on 15 February 1926, Deterding wrote for example to Nobel: 'The Bolsheviks *must be defeated*, but the defeat will never come if everybody is frightened into buying Russian oil'.

122 Yergin, *Prize,* 243.

123 AN CAMT 132 AQ 166, memorandum Weill, 9 March 1928.

124 Bamberg, *History BP*, ii., 107.

125 Howarth, *Century*, 164.

126 AN CAMT 132 AQ166, memo Weill 15 July 1928.

127 SHA 195/30-2, Deterding to Agnew, 13 August 1928.

128 See also Bamberg, *History BP* II, 107-9, who claims that Fraser had prepared a first draft. In his regular conversation with M. Weill from Rothschilds, Deterding also claimed to have drafted the agreement: 'J'ai rédigé, d'accord avec eux, un projet de contrat de 8 á 9 pages qui fixe les bases d'une entente entre les gros producteurs'. To stress his predominant role, Deterding added that 'ces conversations ont duré 11 jours pendant lesquels M. Deterding aurait parlé 8 heures par jour'. AN CAMT 132 AQ 166, memo Weill 1 October 1928.

129 Bamberg, *History BP*, II, 110.

130 AN CAMT 132 AQ 166, memo Weill, 1 October 1928.

131 Yergin, *Prize*, 265.

132 SLA SC7/A22/22/1, letters concerning negotiations with Teagle and Mellon.

133 SHA 195/23, Deterding to Teagle, 17 May, 19 August 1932.

134 SLA 119/11/26, board memos Anglo-Saxon, 29 August and 16 November 1938; cf. Corley, *Burmah*, II, 56-7.

135 SLA SC7/92/1/1, Godber to Airey 23 August 1929, about the boycott of Spain 'because if by any chance and of course it is an exceedingly small chance the Monopoly eventually becomes successful (...) it will encourage other countries like France and Italy, who are all the time on the verge of creating monopolies, to do so'.

136 SLA 119/11/15, board memos Anglo-Saxon, memo 12 August 1926.

137 Larson, Knowlton and Popple, *New Horizons*, 39-41, 51.

138 Ibid., 52.

139 SLA 119/11/23, board memos Anglo-Saxon, memo 20 April 1933.

140 SLA 119/3/9, board minutes Anglo-Saxon, 26 April 1933.

141 SLA 119/11/23, board memos Anglo-Saxon, memo 20 April 1933; 119/11/24, memo 10 July 1935.

142 Cf SLA GHC/COL/A1-A19, an extensive file on the Colombian operations.

143 SHA 15/133-1, statement of Group sales 1926; Van Wijk to Engle, 27 June 1927; Engle to Van Wijk, 4 July 1927. Bamberg, *History BP*, II, 120-2; the volume rose from 340,000 long tons in 1928 to just over 400,000 tons by the late 1930s, all from the Abadan refinery. Warwick BP Archive 109194. We are indebted to Jim Bamberg for this reference.

Chapter 5

1 NA CAMT 132 AQ 159, Lane to Aron 21 September 1912. This solution created a new complication of its own, because some of the profits previously realized by Asiatic were now made by Bataafsche as formal owner of Dordtsche. Consequently a special compensation account had to be created to ensure that the third partner in Asiatic, the Rothschilds, received their due.

2 SHA, board minutes Royal Dutch, 30 September 1919.

3 SHA 195/175, Gulbenkian to Deterding, 17 November 1919.

4 SHA 8/1863-1, Colijn to Deterding, 3 November 1915.

5 NA CAMT 132 AQ 164, Lane to Weill, 26 January 1920.

6 The Rothschild archives contain a draft letter (NA CAMT 132 AQ 164) 'que suggeré mr Gulbenkian', dated 26 April 1921 containing his views; the crucial passage runs as follows: 'après tant d'années d'admirable et de si fructueux labeur consacrées par vous à ces entreprises auxquelles vous avez donné votre vie, il est indispensable, dans l'intérêt même de ceux que vous représentez, que vous vous ménagiez des périodes de détente et de repos: ceci est une nécessité absolue. Si vous ne le faisiez pas vous-même, nous serions les premiers à vous presser de le faire.' Lane and Gulbenkian must have been in contact about this, since they held almost identical views; the letter cited here suggests that they have tried to convince the Rothschilds to put pressure on Deterding.

7 In the event, Cohen Stuart died in March 1921 before he could take up the appointment.

8 SHA, board minutes Royal Dutch 10 February 1921; De Jonge, *Herinneringen*, 69-70.

9 SHA, 195/51, Philips to Capadose. In Dutch, Philips' complaints were 'gebrek aan goede organisatie der werkzaamheden, onvoldoende afbakening van ieders taak (...) vermindering van verantwoordelijkheidgevoel bij sommige leiders, usurpatie van macht en invloed bij sommige ambtenaren (...) Vaak wordt hetzelfde onderwerp door een bureau in Londen behandeld, waarvan ook iemand in den Haag overtuigd is dat hij het behoort te behandelen (...). De verwarde verdeeling van terrein tusschen de verschillende boards (...) het ontbreken van een systematische en doeltreffende werkwijze in de alleropperste leiding'. The complaint about 'the management in The Hague has also become rather autocratic' probably alludes to Colijn's rise to power within the BPM.

10 In 1908 Sir Marcus had even suggested to liquidate the two companies and 'form one Company to hold the entire business', NA CAMT 132 AQ 157, Lane to Aron 30 September 1908. However, the proposed compensation for the management was so generous they would receive '1 million sterling 5 per cent preference shares' that the plan was rejected as too expensive.

11 SHA 195/51, Capadose to Colijn 3 October 1921 ('commissarissen (...) waren het, meen ik, allen eens dat het critische gedeelte van Philips epistel voortreffelijk was').

12 SHA, 195/51, anonymous author to Capadose, 4 October 1921; Colijn, Deterding, and Philips are mentioned in the letter, so it was probably written by Loudon.

13 SHA, 195/51 notes in margin letter Philips to Capadose ('Hoofdzaak is dat den laatsten tijd geen personen maar departementen aansprakelijk worden gemaakt wat m.i. geheel fout is')

14 Ibidem, 'Kwaal van de Hollanders die te veel persoonlijk crediet willen in plaats van zich zelf weg te cijferen + de zaak alleen alle crediet te geven'.

15 SHA 195/176, Gulbenkian to Deterding, 6 January 1922; he added 'that perhaps you should not exaggerate and smash too hard in order to gain your ends'.

16 SHA 195/51, memo 6 September 1921.

17 Langeveld, *Colijn* I, 243.

18 SHA 195/168, Colijn to Deterding 30 March 1922: 'Steunend op wat men meende te zijn uw anti-Hollandsche gezindheid, voelde ik overal, natuurlijk beleefden en uiterst correcten, maar toch ook standvastigen tegendruk (...) Tegen de "stille kracht" om dat te beletten heb ik het feitelijk afgelegd'. Colijn's assessment of Deterding's attitude was shared by others. Sir Philip Lloyd Graeme, member of the Harcourt committee, said at a meeting of the committee in March 1922 that Deterding's 'sympathies were very pro-English (...) as long as he remained Managing Director it was probable that the British share in management would tend to grow', NA Kew CAB 27/180, meeting 10 March 1922.

19 See Klein, 'Colijn', 105. Deterding did reply to Colijn's letter, clearly showing his anti-Dutch sentiment in this period: SHA 195/168, Deterding to Colijn, 3 April 1922: 'als die exportrechten gehandhaafd blijven, het Hollandsch element geheel en al moet plaats maken voor het Britsche' 'Indien de tegenwoordige positie bestendigd wordt, zullen zeker nog wel Hollanders gebruikt kunnen worden, doch deze moeten dan beginnen met zich te verengelschen of als u wilt te ver-internationaliseren'.

20 De Jonge, *Herinneringen* 70-1. De Jonge became a Royal Dutch non-executive director in 1923 and a Bataafsche director on his return to the Netherlands in 1930. His frequent attendance at London board meetings belies his self-denigrating portrait of an ineffectual stranger in the oil business.

21 SHA 195/177, Gulbenkian to Deterding, 28 December 1922.

22 SHA 195/92, memo Deterding for Gerretson, 1932.

23 SHA 195/177, Gulbenkian to Deterding, 4 January 1923.

24 SLA SC7/A22/11, De Kok to Agnew, 6 July 1923.

25 SHA 195/177, Gulbenkian to Deterding, 17 July 1923.

26 SHA 195/177, A. Cull of Cull & Co. to Deterding, 18 December 1923. The letter's postscript reads: 'I have just heard from Mr. Gulbenkian that he has declined to arbitrate with you, but I hope you will persuade him to alter his mind.'

27 SHL 102/3/4, minutes VOC board, 25 April 1924.

28 SHL 102/3/4, minutes VOC board, 10 November 1924.

29 SHL, 102/3/4 minutes VOC board, Meetings 21/12/1925 and 11/1/1926.

30 SHA 195/33/4, Gulbenkian to Deterding, 30 September 1925.

31 NA CAMT 132 AQ 164, Lane to Weill, 21 November 1921.

32 Health reasons may have contributed to Philips's decision. He suffered from recurrent depressions, which in 1918 had forced him to resign as Dutch envoy to Washington before he had even presented his credentials: obituary Philips in *Olie* 8 (1955) 66; NA The Hague 2.21.095 De Jonge papers No. 25, De Kok to De Jonge, 23 December 1931.

33 Henriques, *Waley Cohen*, 296.

34 SHA board minutes Royal Dutch, secret annex to the meeting of 9 October 1924, 'Indien in het college van Directeuren der Koninklijke de meerderheid een besluit zou willen nemen waartegen de heer Deterding bezwaar heeft, zal dit niet worden uitgevoerd, maar worden onderworpen aan het oordeel van de Raad van Commissarissen'. The first version of the proposal reads as follows: 'De heer Deterding wenscht dat indien' (Mr Deterding wishes that if the board of directors), which clearly indicates that Deterding was behind this proposal.

35 Colijn's profit share had been paid out of Deterding's share, to the amount of one half percent of the dividend of Royal Dutch; SHA 15/13, Philips to Deterding, 28 May 1923.

36 SHA 15/13, Philips to Deterding, 28 May 1923.

37 SHA Royal Dutch Board, 9 October 1924 and 12 February 1925.

38 SHA 15/13, Van Wijk to BPM, 1 July 1929; Deterding's income from the British companies was about 600,000 to 700,000 guilders.

39 SHA 102, Loudon papers, and 8/1648, contract of sale between Mrs. Kessler-De Lange, Deterding, Loudon, and Capadose, the last clause reading 'De ondergeteekenden ter andere nemen de zedelijke verplichting op zich, elk voor zich, tegenover de ondergeteekende ter eenre om, wanneer een of meer harer zonen mocht wenschen om als Directeur of Directeuren der K.N.P.M. op te treden en deze daartoe de noodige bekwaamheid mochten beschikken, hem of hen in de vervulling van dat verlangen met raad en daad bij te staan, echter alleen voor zoverre de vervulling van dien wensch door hen geacht wordt in het belang der K.N.P.M. en hare Aandeelhouders te zijn'.

40 There were four sons; the other two showed no inclination for joining the Group, one becoming a lawyer, the other an opthalmologist.

41 SHA board minutes Royal Dutch, 27 September 1923; significantly it was added that Kessler's remuneration would not be subtracted from Deterding's. The first mission of the newly appointed Kessler was to discuss the preference shares with his mother: board minutes Royal Dutch 10 January 1923 .

42 SHA, board minutes Royal Dutch 14 February 1924. Additionally, all owners of preference shares had signed a blank agreement to transfer their holding upon retirement, board minutes Royal Dutch, 19 September 1929.

43 SHA 15/134 (correspondence between Kessler and Waley Cohen about financial relations within the Group); SLA 119/3/6, board minutes Anglo-Saxon, 2 December 1925 (Kessler to approve plans for Belgium garage and workshops), 22 September 1926 (oversees building tank installations Switzerland), 3 November 1926 (supervises purchase motor cars and lorries Germany), 119/3/7, board minutes Anglo-Saxon, 4 July 1929 (must decide on Brussels office building), 2 October 1929 (idem Hamburg building); SHA 10/541-1, A. de Jongh to Rudeloff, 23 May 1930 (Kessler's instructions for a commercial audit in the Netherlands, Belgium, Germany, and Switzerland); CAMT Rothschild 132 AQ 168, notes from a conversation between Baron Edmond and Kessler, 19 December 1933 (French organization performs so much better since Kessler has taken over its management). SHA 190C/251, memo Forbes 'Smeeroliebedrijf ', 7; 190C/34A, memo Forbes 'Benzine', 22-23; SHA 11/22-1, Kessler to Fenwick and others, 19 January 1939 (gasoline cost prices and selling policy).

44 *The Pipeline*, September 1922 announced Kessler's appointment to the board of Anglo-Saxon, so this was not 1923 or 1924, as stated in his obituaries and other texts.

45 SLA GHS/3E/1, group directors; NA The Hague 2.21.095 De Jonge papers No. 18, Deterding to De Jonge, 13 July 1932 and 2 January 1933, No. 25, De Kok to De Jonge, 16 March 1932, all three letters professing shock and grief over Debenham's sudden death.

46 SLA GHS/3E/1, group directors; *ODNB*, 'Agnew'.

47 *ODNB* 'Godber'; Rady, 'Godber'; one of Agnew's sons married one of Godber's daughters.

48 SHA Royal Dutch board minutes, 13 March 1930; Erb regretted having to give up his London directorships, since these had enabled him to keep abreast of the production in Iraq, Sarawak, Egypt, and Trinidad, figures which he would not otherwise receive. This underlines the continuing lack of integration at board level: SHA 195/172, Erb to Deterding, 5 March 1930.

49 Obituaries in *De Bron*, December 1940.

50 Lane's assessment of Deterding in 1913 in NA CAMT 132 AQ 199, Lane to Baron Edouard, 21 October 1913.

51 RA Alkmaar De Lange papers (no inventory numbers at time of writing), Deterding to Kessler 7 December 1930.

52 Interview with J. B. A. Kessler III, June 2004.

53 For a remarkable testimony of Deterding accepting this state of affairs, see SHA 195/22-6, Deterding to Kessler, 20 July

1932, emphasizing the need for co-operation and the joint responsibility of directors.

54 When Agnew wrote to De Kok in May 1930 asking his approval for a circular letter about Provident Fund contributions over 1929 to be sent out to representatives overseas, he stated that 'Godber will deal with America; Kessler with the various European countries; Debenham with the Asiatic representatives; whilst I will deal with Venezuela, Sarawak, and Egypt', so this division of tasks by area was not an established matter. Moreover, in June Engle wrote to Van Wijk asking him to write 'as usual' to the companies in Romania, Germany, the Netherlands, and others under control of Bataafsche, at which Van Wijk noted that this was supposed to have been done by Kessler, but that the companies were to be notified by Bataafsche anyway. Thus there was even no clear demarcation between London and The Hague. SHA 15/227, Agnew to De Kok, 19 May 1930. Deterding wrote separately to the chief executives in the States, informing them of their bonuses: SHA 195/22-4, Deterding to Airey, Daly, Van Eck, and Legh-Jones, 6 May 1930.

55 De Jonge, *Herinneringen*, 70.

56 SHA 190C/449, survey of the most important Group companies, listing Royal Dutch, Shell Transport, Anglo-Saxon and Bataafsche on p. 2, and Asiatic amongst the large body of operating companies on p. 8.

57 SHA 8/1869, organization patent department; Homburg, Small and Vincken, 'Van carbo- naar petrochemie', 351.

58 SLA 119/3/7, Anglo-Saxon board minutes, 12 May 1927, 2 May 1928; 141/3/9, Shell Transport board minutes, 12 May 1932, 11 May 1933, 17 May 1934, 15 May 1935.

59 Cf. SHA 15/2, De Kok writing to the Governor-General in 1935 on the announced gasoline excise increases; Taselaar, *Koloniale lobby*, 416-7.

60 SHA Royal Dutch board minutes, 28 March 1930 (undue haste with the bond issue). The same happened in 1939 (see Ch. 7).

61 The tone of the articles in which the leading Dutch financial magazine *De Kroniek van Sternheim* discussed the Annual Report of Royal Dutch became increasingly critical during the 1920s and 1930s, pointing to the deficiencies of its financial reporting and the repetitive nature of Deterding's complaints about the Soviet Union.

62 *Kroniek van Sternheim*, 1 April 1930, 308, 1 June 1931, 45-6, 1 July 1934, 60-1; *Naamlooze vennootschap*, 15 August 1939, 157; *De Maasbode*, 14 January 1939, criticizing the lack of information in the prospectus for Bataafsche's 100 million guilder bond loan.

63 SHA 15/112, correspondence with Bianchi. The auditor had asked for better data on the holding companies so as to be able to give an accurate picture of Royal Dutch's investment in them, but managers refused to give precise figures.

64 SLA 119/11/16, board memos Anglo-Saxon, January 1927; cf SHA 15/217, for a run of estimates 1932-8; SHA 15/133-2, Colquhoun to Van Wijk, 8 November 1933, for the compilation and use of the material.

65 SHA 15/133-1, Statement Group Sales 1926, dated 20 April 1927; Van Wijk to Engle, 29 June 1927, Engle to Van Wijk, 4 July 1927.

66 The existence of the Group Finance Committee can be deduced from SHA 15/131, Van Wijk to Engle 13 February 1931, about appointing Colquhoun to the committee as successor to Anglo-Saxon's company secretary Price. In 1930, Group production data were still not shared between the boards of Bataafsche and Anglo-Saxon: SHA 195/172, Erb to Deterding, 5 March 1930.

67 SHA 49/23-2, Godber to Van Wijk, 18 May 1937.

68 SHA 49/23-2, Van Eck to Van Wijk, 21 September 1937.

69 SHA 15/114, Van Leeuwen to BPM, 29 February 1940.

70 Deterding, *An International Oilman*, 15; according to Deterding the Pope answered: 'The material rewards from such a vast business don't impress me in the least (...) What impresses me is the intense happiness you must draw from the knowledge that your life-work provides the wherewithal for so many thousands of families'.

71 SHA 190A/116; the figure for 1929 shows a considerable increase from 1913, when Bataafsche employed 23,167 Asians and 825 Europeans.

72 Beaton, *Enterprise*, 352.

73 Royal Dutch Annual Report 1935.

74 SLA 119/3/2, Minutes Board Anglo-Saxon, 8 April 1914.

75 The variety of Engle's tasks explains his description as having 'no particular title in the company and no precise designated function' in Howarth, *Century*, 134-5.

76 For instance SLA 119/3/5, board minutes Anglo-Saxon, 16 December 1922; 141/3/9, board minutes Shell Transport, 15 July 1930, 15 March 1932; the words 'at the board's discretion' were probably added for tax reasons, gratuities not counting as taxable income.

77 SLA 141/11/15, board memos Shell Transport, memo 12 December 1932.

78 SHA 15/238, De Booy to Kessler, 4 March 1937; SLA 141/3/9, board minutes Shell Transport , 1 February 1938; 119/3/10, board minutes Anglo-Saxon, 19 January 1938.

79 SHA 1585, board minutes Bataafsche, 18 November 1919, proposal by Colijn of the new rules for staff salaries: 'a. Op 21-jarigen leeftijd behoort de ongetrouwde employé zelfstandig te kunnen leven, waarvoor een salaris van f. 1200 wordt noodig geacht. b. Op 25-jarigen leeftijd behoort het salaris tot F 2000 gestegen te zijn, opdat de employe in staat zij een gezin te vormen. c. Naast het vaste salaris wordt een gezinstoeslag toegekend, zijnde 10 per cent voor de echtgenoote en 2 per cent voor elk ten laste van de employé zijnd kind. d. de duurtetoeslag en extra-duurte toeslag, zooals tot heden genoten, vervallen'; Bataafsche accepted the new rules, but also advised to consult the London office in order to harmonize employment conditions. The London boards already operated a family and child allowance. See SLA 119/3/3, board memos Anglo-Saxon, memo G. S. Engle 15 October 1917, for a detailed description of salaries and employees' budgets to argue for raising salaries across the board.

80 SLA 119/11/13-3 and 4, board memos Anglo-Saxon, report of the St. Helen's Housing Co. 1923; 119/11/24, board memos Anglo-Saxon, report for 1933.

81 SLA 119/11/22, board memos Anglo-Saxon, report St Helen's Court Stores for 1930.

82 See e.g. 'The De Kok challenge competition' in *The Pipeline*, 2 (1922), 170-7; 'Deterding challenge cup competition', *The Pipeline*, 5 (1925), 174-81. In 1922 membership of the different sports clubs was 4,000 in Britain and only 400 in the Netherlands.

83 SHA 190C/276-1, Wurfbain to Pladju, 17 July 1920, and a loose memo with notes

about the early history of the association; *Minjak*, 1921, 2-8.

84 SHA 190C/276-1, Colijn to De Jonge, 7 August 1920. In earlier talks in Indonesia, managers had put a third condition, that a union would have to represent 75 per cent of the European employees.

85 Trade unions were in general not influential in the oil industry, perhaps because of relatively high earnings. Mexico was the exception, of course; unions played a large role there, see Brown, *Oil*, 307 ff.

86 Henriques, *Waley Cohen*, 166-8.

87 Ibid., 166. The Group later developed a special procedure for recruiting relatives to prevent strings being pulled, see SHA 11/22-1, Godber to Engle, 6 July 1939.

88 Henssen, *Geschiedenis*, 81.

89 Ibid.

90 SHA 15/126, report four accountants on cost cuttings in the London office, 3 March 1931.

91 SHA 11/22, memo Gray, 13 February 1939, Gray to Legh-Jones, 22 February 1939.

92 SLA SC7/42/9/2-2, Godber to De Booy, 26 January 1937.

93 SHA 8/962, Kessler to Erb, 10 January 1927.

94 SHA 15/242, Kessler to De Kok, 27 September 1934.

95 SHA 15/223 memo Kessler to Deterding, 30 January 1935; on reception of a copy, De Kok replied next day that he entirely agreed with it; the B.I.M. was the marketing organization in the Netherlands.

96 SHA 8/1865 Bataafsche to Pladjoe, 15 October 1928.

97 G., 'Zijn wij 'werknemers'?' *Minjak*, 1921, 13-14. The identity of 'G.' is unknown. Gerretson always signed his memos as G., but it appears unlikely that a staff member of Bataafsche in The Hague would publish an article in the first issue of the staff magazine for the Netherlands East Indies. On the other hand, the views expressed in the article could well have been those of Gerretson.

98 G., 'Zijn wij 'werknemers'?' *Minjak*, 1921, 13-14, 'want ook de schijn van antagonisme zou den Asiaat tot gevaarlijke conclusies kunnen leiden'.

99 *Minjak*, February 1922, 161.

100 SHA, board minutes Bataafsche, 9 October 1924, which leave no room for doubt about the motivation of this decision: 'zulks met de bedoeling om, tezamen met andere belanghebbende ondernemers in Nederlandsch-Indie, tegenover de overwegend-LINKSCHE en weinig praktische opleiding der Oost-Indische ambtenaren te Leiden een tegenwicht te hebben'.

101 Henssen, *Gerretson*, 43-63.

102 Interview Deterding in 'De Mijlpaal', printed in *De Telegraaf*, 28 February 1929, 'Nederlands politiek in Indië funest' and 'dwaasheid te spreken van een Indonesisch volk'.

103 We thank Geoffrey Jones for pointing this out to us.

Chapter 6

1 Forbes and O'Beirne, *Technical Development*, 174-5; Beaton, *Enterprise*, 200.

2 Forbes and O'Beirne, *Technical Development*, 95, 102-8.

3 Ibid., 110-14, 118; Beaton, *Enterprise*, 200-2, 204-6.

4 Forbes and O'Beirne, *Technical Development*, 120-5; Beaton, *Enterprise*, 202-5, 552; MS, 'Fifty Years, Shell Petroleum Corporation Geophysical Research Laboratory 1936–Shell Development Company Bellaire Research Center 1986', 6-7. Marland Oil probably adopted seismology at the instigation of its Vice-President E&P, Van der Gracht.

5 Forbes and O'Beirne, *Technical Development*, 131-40.

6 Ibid., 165-72.

7 Ibid., *Technical Development*, 179-93; Beaton, *Enterprise* 103-4; SHA 49/180-1, Kessler to Guest, 30 September 1925 (McDuffie appointment, temporary sojourn in The Hague, and future assignments to other operating areas).

8 Beaton, *Enterprise*, 244.

9 Forbes and O'Beirne, *Technical Development*, 332; Beaton, *Enterprise*, 240; SHA board minutes Bataafsche, 14 February 1918

10 Beaton, *Enterprise*, 244; the Dubbs was both more simple to build and maintain, and it gave a better yield of gasoline.

11 Beaton, *Enterprise*, 244-7; Forbes and O'Beirne, *Technical Development*, 335-42; SHA board minutes Bataafsche, 13 July 1922 (budget for cracking trials Balik Papan); 24 April 1925 (20 per cent gasoline from cracking in 1923); SLA SC46/1, Vincent, MS 'One hundred years of Shell refining 1891-1991', 3 (25 Dubbs units 1927).

12 SHA 190C/23A, paper Forbes,

'Asfaltbitumen', 8-9 (contacting foreign asphalt experts); Forbes and O'Beirne, *Technical Development*, 426-7 (Group representatives on international white oil committees).

13 Homburg, Small and Vincken, 'Van carbo- naar petrochemie', 337-40.

14 SRTCA, Abbott Room, manuscript Duinmaijer and Groenveld about the history of the laboratory, budgets laboratory 1914-18. the 1918 budget was almost 100,000 guilders, just over £8,300, up from 30,000 guilders (£2,500) in 1914, a substantial increase even considering wartime inflation.

15 SRTCA, Abbott Room, manuscript Duinmaijer and Groenveld about the history of the laboratory; Schweppe, *Research aan het IJ*, 28-32.

16 SHA 49/737, memo Brocades Zaalberg 13 January 1934.

17 Homburg, Small and Vincken, 'Van carbo- naar petrochemie', 344-53; SHA 190C/40, Bataafsche sold its share for 600,000 guilders, having invested a total of 3 million in the company, so the board clearly wanted to get rid of it.

18 Cf. SHA 190C/23A, paper Forbes, 'Asfaltbitumen'; 190C/34A, paper Forbes, 'Benzine'; 190C/466, paper Forbes 'Witte oliën'; 190C/251, paper Forbes 'Smeeroliebedrijf'.

19 SHA 190C/34A, Forbes, 'Benzine', 15, 16, 18, which gives more details than Forbes and O'Beirne, *Technical Development*, 394-6.

20 According to Deterding, Rolls Royces ran best on Shell No. 1 Spirit: SHA 49/317, Deterding to Colijn, 3 December 1917. Some straight-run gasoline had octane numbers as high as 75; Charles Lindbergh allegedly crossed the Atlantic on aviation fuel with octane number 73, MS, 'Shell Oil Company's Research and War Production 1943', 39.

21 SHA 190C/34A, Forbes, 'Benzine', 8-9.

22 Ibid., 8-10, 12; Reynolds, *Ricardo*, 143-6; Beaton, *Enterprise*, 549-50; Gabriëls, *Koninklijke Olie*, 94; Delft initially concentrated on drafting specifications for Diesel fuel.

23 From this point of departure, octane numbers developed into performance indicators, so there could be more than 100 octane gasoline, or 100 octane gasoline with only 45 per cent iso-octane. Cf. Beaton, *Enterprise*, 581.

24 Forbes and O'Beirne, *Technical Development*, 369-70; SHA 190C/250, Forbes, 'Extractiemethoden' is more detailed than the text in the book, cf. for instance on the extracts 6-7, 14.

25 SHA 190C/34A, Forbes, 'Benzine', 11, 13, 16, 17.

26 Beaton, *Enterprise*, 342.

27 SHA 49/180-1, Kessler to De Kok, 25 August 1927.

28 SHA 190C/34A, Forbes, 'Benzine', 13, 14.

29 Larson/Knowlton/Popple, *New Horizons*, 161.

30 SHA 190C/34A, Forbes, 'Benzine', 13, 17; Beaton, *Enterprise*, 412-15.

31 SHA 190C/34A, Forbes, 'Benzine', 21-4; Schweppe, *Research aan het IJ*, 60.

32 Beaton, *Enterprise*, 412-15; SHA 10/234, Airey to Deterding, 21 November 1930 (gasoline without ethyl now uncompetitive); 49/74-2, memo Aviation Department, London 30 March 1933 (need to add lead to aircraft gasoline means building ethyl mixing plants at strategic points), Dooijewaard to Caland, 1 April 1933 (lead the only way to give the market what it wants).

33 Reynolds, *Ricardo*, 158-65; SLA 119/11/17 and 119/11/23, Anglo-Saxon board memos, 15 November 1927 and 18 December 1933; SHA 49/161 and 49/162, contracts with Ricardo concerning Diesel engines, 1927-1935; 49/164-169 and /174-175, negotiations and agreements with engine manufacturers in various countries. The importance of Ricardo's innovation becomes fully clear by the assertion that, in the 1980s, some 90 per cent of the world's diesel cars and commercial vehicles had Comet-derived combustion chambers: Reynolds, *Ricardo*, 164.

34 Ibid., *Ricardo*, 146-7.

35 SHA 190B/138, memoirs Späth, 13-4, 19; 8/1170, memo take-over Stern-Sonneborn 19 July 1924; Flieger, *Gelben Muschel*, 107-15; SHA 190C/250, memo Forbes, 'Extractiemethoden', 15-17; 190C/251, memo Forbes 'Smeeroliebedrijf', 6-12, 16-21; Forbes and O'Beirne, *Technical Development*, 408-20.

36 190C/251, memo Forbes 'Smeeroliebedrijf', 15, 26-7; Beaton, *Enterprise*, 409-11; SLA 119/11/24, board memos Anglo-Saxon, memo 17 May 1935 (extraction plants Shell Haven), 119/11/27, memo 19 April 1939 (building Stanlow).

37 Detailed discussion in SHA 190C/23A, paper Forbes, 'Asfaltbitumen' and 190C/466, paper Forbes 'Witte oliën'.

38 At least in gasoline manufacturing, it took some time before blending had become a faultless procedure: SHA 49/74-3, Asiatic Singapore to Bataafsche, 21 September 1935, 49/74-4, Bataafsche to Balik Papan, 14 October 1935, 49/74-5, BIM to the Amsterdam laboratory, 26 July 1937, the Amsterdam laboratory to BIM, 4 August 1937.

39 Beaton, *Enterprise*, 157-9.

40 SHA 49/180-1, memo extension research work Group, 5 April 1927; Forbes and O'Beirne, *Technical Development*, 456-74; Beaton, *Enterprise*, 502-7; Homburg, Small and Vincken, 'Van carbo- naar petrochemie', 344-53. In 1927, Pyzel estimated that the Group's US companies wasted the gas equivalent of 10,000 barrels of fuel daily: SHA 49/180-1, memo extension Group research, 5 April 1927.

41 Hayes, *Industry and Ideology*, 36-8; Knowlton/Larson/Popple, *New horizons*, 154-7.

42 Royal Dutch Annual Report 1926, 16-8; Spitz, *Petrochemicals*, 37.

43 De Vries, *Hoogovens*, 320-1; Homburg, Small and Vincken, 'Van carbo- naar petrochemie', 352; memoirs Kessler in *Olie*, June 1957. Teagle had travelled to London from the talks with Farben in Heidelberg on 9 August; the meeting took place on the 18th: Wall and Gibb, *Teagle*, 301.

44 Gibb and Knowlton, *Resurgent Years*, 537; Beaton, *Enterprise*, 513.

45 Forbes and O'Beirne, *Technical Development*, 456; Homburg, Small and Vincken, 'Van carbo- naar petrochemie', 354.

46 Reader, *ICI*, II, 163-164.

47 Spitz, *Petrochemicals*, 33.

48 Homburg, Small and Vincken, 'Van carbo- naar petrochemie', 348-9; Spitz, *Petrochemicals*, 35, 37.

49 SHA 49/180-2, Kessler to De Kok, 16 June 1930.

50 SHA 49/180, Kessler to De Kok, 25 August 1927. Quoted in Forbes and O'Beirne, *Technical Development*, 456-7.

51 Underlining his conviction that research and chemicals needed to be exempted from short-term considerations, Kessler wrote to De Kok in August 1930 emphasizing that the spending cuts then introduced by Bataafsche should not harm these functions unduly, whereupon

De Kok assured him that he would keep this in mind: SHA 8/960, Kessler to De Kok, 21 August 1930; De Kok to Kessler, 23 August 1930.

52 Quoted in Forbes and O'Beirne, *Technical Development*, 458.

53 Ibid., 460, 464, 472; Homburg, Small and Vincken, 'Van carbo- naar petrochemie', 351-355; Schweppe, *Research aan het IJ*, 32-37. SHA 190Y/1038, brochure Intellectual Property Services 1917-97; Beaton, *Enterprise*, 507. In its first twenty years of existence, Shell Development spent only 1.4 per cent of its total expenses of $64.6 million on acquiring patents: SHA 2/80, Highlights Shell Development Company, 1928-48. Kessler was keen to keep the Amsterdam laboratory a Bataafsche department, and not turn it into a separate company, fearing that this would become too isolated from business considerations: SHA 49/180-1, Kessler to S. A. Guest, 30 September 1925. Guest, employed by Astra, had advocated intensifying Group research efforts in three long letters to Kessler, emphasizing the need to have the research organization separate from the managerial and executive organization.

54 SHA 8/1869, organization patent department. By 1934 the department employed a staff of 26. See also 190Y/1038, brochure Intellectual Property Services 1917-97. On the initial considerations about patent management 49/178, general patent policy; it typically took two years before the overlaps with the patent activities of a London department were sorted out.

55 Forbes and O'Beirne, *Technical Development*, 460, 464, 472; Homburg, Small and Vincken, 'Van carbo- naar petrochemie', 351-355; Schweppe, *Research aan het IJ*, 33-37. From 1928, research budgets were also scrutinized more closely by directors. In January 1928, neither De Kok nor Kessler knew exactly how much Bataafsche spent on the Amsterdam laboratory, but they soon made sure to know, and also how much the research had yielded in commercial benefits: SHA 49/180-2, Kessler to De Kok, 20 January 1928. SHA 190Y/1038, brochure Intellectual Property Services 1917-97; Beaton, *Enterprise*, 507. On pesticides SRTCA manuscript Duinmaijer and Groenveld chapter 1921-29, 3; SHA 49/79, Braybrook to the Amsterdam laboratory, 3 January 1940, 11/22, Asiatic to BIM, 18 April 1940 (Dutch version film on fruit protection). Shell sold insecticides in the US since the late 1920s: Beaton, *Enterprise*, 408.

56 SHA 49/180-1, Kessler to S. A. Guest, 30 September 1925.

57 SHA 49/151, 15/215 (Simplex and Research Agreements); 49/23-1, Van Eck to De Booy, 20 July 1938 for the gist of the agreement and the date of the first agreement. The agreement was signed in 1932 and deemed to have run from January 1st, 1929: Beaton, *Enterprise*, 547.

58 SHA 49/180-1, memo extension Group research work 5 April 1927, defining the separation of work between the US and Amsterdam as California to do the more technical research, and Amsterdam the semi-technical and scientific work. Shell Development was moulded on the examples of the Standard Oil Development Company, which Pyzel visited with the specific intention to get information about its organization. Pyzel probably also visited the General Motors research organization with the same intention. SHA Agenda's board BPM, 1939, separate table with research spending 1938 and budget 1939; Larson, Knowlton and Popple, *New Horizons*, 174; SRTCA, manuscript Duinmaijer and Groenveld, 1929-40, 1 (staff in 1940); Beaton, *Enterprise*, 620 (staff Emeryville 1940); SHA 8/1869, patent department organization; SHA 49/180-1, correspondence on the siting and motives behind the setting up of Shell Development.

59 Beaton, *Enterprise*, 517-18, 531-3; Spitz, *Petrochemicals*, 85-6; SHA 2/80, Highlights Shell Development Company, 1928-48.

60 Beaton, *Enterprise*, 408.

61 SHA 15/152, De Kok to Kessler, 15 October 1931; 111/545, report February 1932; 111/655, report July 1932; board minutes BPM, memo 28 July 1954; Pradier, *Shell France*, 22-3, 90-8. Butagaz's signature blue bottle was later also successfully applied to Camping Gaz, a company selling canisters of gas and matching small stoves and lights to generations of Europeans holidaying in tents and caravans. URG had a 70 per cent stake in Camping Gaz.

62 The crucial importance of nitrogen fixation may be deduced from the recent estimate that almost half of the nitrogen atoms in the proteins of human bodies came at one time or another from an ammonia factory: *The Economist*, 24 December 2005, 27.

63 'I have joined the company to become a director, and that Guus would really have the same desire speaks for itself' ('Ik ben bij de Mij. om directeur te worden en dat Guus au fond hetzelfde verlangen zou koesteren spreekt vanzelf'), Dolph Kessler to Margot Kessler-De Lange, 30 January 1911, *Tussen moeder en zoon*, 219.

64 The Group did have a few other conspicuous cases in which members of the same family came to occupy prominent managerial positions, as with the Deterding brothers, the Engles in the London office and of course father and son Loudon. In order to avoid nepotism managers instituted a special vetting procedure for relatives of employees. In 1939 Godber issued a formal directive detailing this, but the ban on employing close relatives probably operated earlier than that. SHA 11/22, Memos Van Eck, folder no. 1, Godber to Engle and others, 6 July 1939.

65 De Vries, 'J. B. Kessler', 295.

66 SHA 190C/350, Survey of Group Russian interests by Kessler, 1916, p. 77; De Vries, 'J. B. Kessler', 296; reminiscenses Kessler in *Olie*, June 1957, retirement in *Olie* July 1961, obituary in *De ingenieur* 84 (1972) A 1067. Dolph may have thought it better for his own career not to have his brother join the company, too: Dolph Kessler to Margot Kessler-De Lange, 15 January 1910, published in *Tussen moeder en zoon*, 175, also referring to Deterding's categorical statement that he did not want Guus in the company, repeated in a letter of 30 January 1911, ibid. 218.

67 Cf. R. A. Alkmaar, De Lange papers, no inventory numbers available at the time of writing, Dolph's diary entry for 12 December 1914. Guus Kessler had married Anna Françoise Stoop, daughter of Adriaan Stoop's eldest brother François, the senior partner in the family banking firm. SHA 5/335, Colijn to Deterding, 24 April 1915, about Kessler's intended position with the Residu Gas Maatschappij. Lane considered Dolph himself to blame for his failing to achieve his aims: NA CAMT 132 AQ 162, correspondence 1915, Lane to Weill,

8 April 1915.

68 SHA 190D/637/2, Deterding to Colijn, 10 February 1916 (employing Kessler if possible, 'if only out of respect to his father'); Colijn to Deterding, 16 February 1916.

69 SHA 190C/52, memo 21 September 1927, outlining the basic agreement between BPM and Hoogovens.

70 In 1932 BPM technical staff carefully considered switching to the Haber–Bosch process, only to reject it. Forbes and O'Beirne, *Technical Development*, 503-7.

71 De Vries, *Hoogovens*, 322-3. De Vries' reference to the Mekog proposal being discussed in the board of the Royal Dutch-Shell Group is erroneous, since of course no such organ existed. The proposal will have been discussed in the Bataafsche board, but the minutes for the later 1920s are missing, so we have not been able to ascertain why some board members opposed the proposals. However, even De Kok was initially sceptical about the venture: SHA 49/180, De Kok to Kessler, 24 August 1927.The Hoogovens board held out to get a higher profit share, which rendered their gas supply very profitable indeed. Mekog paid Hoogovens the market price at which the gas was sold to municipal gas companies, but used only 40 per cent of its calorific value. After extracting the hydrogen Mekog returned the gas to Hoogovens, which could then sell again the remaining 60 per cent calorific value still present. This proved to be so profitable that at one point managers wanted to expand the coking plant and iron foundry simply because the offgases yielded such good revenues. Forbes and O'Beirne, *Technical Development*, 504; Dankers and Verheul, *Hoogovens*, 48.

72 SHA 8/1682; Homburg, Small and Vincken, 'Van carbo- naar petrochemie', 352; De Vries, *Hoogovens*, 323-5.

73 Forbes and O'Beirne, *Technical Development*, 504-5; De Vries, *Hoogovens*, 323-5.

74 To his dismay, Pyzel did not become president of Shell Chemical, presumably because he was thought to lack the administrative qualities required: SHA 15/59, correspondence Pyzel; SHA 15/215, Godber to De Kok, 4 December 1930; SHA 49/180-2, memo discussions Kessler, Gallagher, and Pyzel about the nitrogen fixation plant, 2 November 1928; id. about the setting up of Shell Chemical, 13 November 1928.

75 Beaton, *Enterprise*, 522.

76 Homburg, Small and Vincken, 'Van carbo- naar petrochemie', 352-3; Beaton, *Enterprise*, 522; Forbes and O'Beirne, *Technical Development*, 515-6. By 1934, the Pyzels' family affair had become irksome to Godber, who no longer wanted them to work together: SHA 15/236, Godber to De Kok, 11 May 1934.

77 Larson, Knowlton and Popple, *New Horizons*, 156-7.

78 SHA 49/150, a comprehensive file on the hydrogenation negotiations; SHA 15/59, Deterding to De Kok, 19 October 1929 (IGF as wolf); Homburg, Small and Vincken, 'Van carbo- naar petrochemie', 353.

79 SHA minutes BPM 2 May 1930, giving the outlines of the deal; on the negotiations and detailed cost calculations about the deal SHA 49/150; Homburg, Small and Vincken, 'Van carbo- naar petrochemie', 353. By 1931, ICI had spent about 1.25 million pounds, or 3.6 million dollars, on hydrogenation research, yet the Billingham plant was nowhere being finished: Reader, *ICI*, II, 175.

80 Bamberg, *BP*, II, 180; Homburg, Small and Vincken, 'Van carbo- naar petrochemie', 353.

81 Cf. Hayes, *Industry and Ideology*, 38.

82 SHA 111/1904, memo BPM commercial patent policy, 14 February 1936.

83 SHA BPM minutes, 9 January, 29 August 1930 (first details of the UOP deal). The Group appears to have hesitated about the UOP deal in April 1930, thinking that Jersey had a better process which would make Dubbs obsolete, cf SLA 119/3/7, minutes Anglo-Saxon, 16 April 1930. Beaton, *Enterprise*, 256-8, 570-3; Hengstebeck, *Petroleum processing*, 148-9. The weakness of patents was also demonstrated when the Soviet government approached UOP for a Dubbs licence, presumably during the 1920s. As a UOP shareholder, the Group opposed this as a matter of principle, so the Russians simply built and used Dubbs installations without paying royalties. The question resurfaced in 1937, when UOP negotiated with the Soviet government over a licence to an unspecified cracking process. By then the Group agreed that UOP might as well make some money since the Russians would use the process anyway. SHA 15/4, correspondence Godber; SHA 15/302 on the negotiations with Soviet Russia about taking part in IHP. For the competition between UOP and Kellogg, SHA 49/46. On the cat cracking pool SHA 11/21-2, Larson/Knowlton/Popple, *New Horizons*, 166-169, Bamberg, *BP* II, 194-195.

84 SHA 15/302, memo IHP, 10 August 1935; Country Volumes Italy, vol. 2, ANIC to IHP, 28 March 1936, memo on patent position IHP, 20 June 1936.

85 The most modern process for coal hydrogenation is claimed to be economically viable if crude prices are 25 dollars a barrel or more: 'Steenkoololie', *NRC-Handelsblad*, 11 September 2005. During most of the 1920s and 1930s, crude prices ranged between one and two dollars a barrel, with lows of 65 cents.

86 Reader, *ICI*, II, 170-181. The Group presented a memo to the British Government arguing against subsidizing coal hydrogenation on the grounds of exorbitant costs and subsequently refused to take a part in it: SHA 15/302, Corbett to De Kok, 5 December 1932, Van Eck to Wilkinson (New York), telegram 18 February 1937.

87 SHA 11/24, Van Embden to De Booy, 28 December 1939. Japan had probably acquired the Fischer-Tropsch patents in 1937: SHA 11/21, Department AGT/AMN to Van Eck, 9 June 1937.

88 Karlsch and Stokes, *Faktor Öl*, 139, 152-3.

89 190C/251, memo Forbes 'Smeerolie-bedrijf', 23-5; Forbes and O'Beirne, *Technical Development*, 417.

90 SHA 190C/34A, Forbes, 'Benzine', 18, 20-2; SLA 119/11/26 board memos Anglo-Saxon, memo 29 June 1938; SHA 49/861-13 (Ploesti); SHA 10/617, memo May 1938 on the services provided by the BPM. By 1945, Shell Oil had a reforming capacity of 9 million barrels a year on a total gasoline output of 51 million barrels: Annual Report 1945, 10.

91 Beaton, *Enterprise*, 534-5, 564-8. Bataafsche wrote to the London Aviation Department in August 1934 asking whether it would be a good idea to make a new aircraft fuel with it for demanding customers such as KLM, in which case Shell Chemical could increase its production of iso-octane: SHA 49/74-3, Bataafsche to London, 6 August 1934. SHA 49/69 (comprehensive file on polymerization). The Stanlow plant was

partly built because the British Air Ministry did not want to be dependent on supplies from the Continent in case of war: SLA SC7/92/10/2 Vol. 2, Godber to De Kok, 4 January 1937, memo Hill to Godber, 31 December 1936. SHA 10/607, memo 10 July 1935 (proposal to set up the Pernis iso-octane plant), 11/24, memo CI department July 1939 (Pernis investment and profits). The original proposal had estimated the pay-back time at four years.

92 SHA 15/180, Brylinski to Kessler, 2 October 1936; 11/21-1, minutes meeting 17 March 1938; 11/21-2, Riedemann to Godber, 16 June 1938, Godber to Riedemann, 17 June 1938, Kessler to De Booy, 6 March 1939.

93 SHA 49/74, Aviation Department to Bataafsche, 12 May 1936.

94 Schweppe, *Research aan het IJ*, 65; Bamberg, *BP*, II, 204. Bataafsche's application was dated 16 July; Anglo-Iranian's 29 July.

95 Beaton, *Enterprise*, 592 (butadiene); SHA 49/24 (contract with IGF); 49/79 (complete file on ester salts and the negotiations with IGF); 10/617, memo May 1938 on the organization of The Hague central office (products to be made from slack wax); SLA 119/11/25, board memos Anglo-Saxon, memo 21 December 1937 on the building of the Stanlow Teepol plant; 119/11/26, board memos Anglo-Saxon, memo 3 August 1938 (Stanlow Teepol plant). On the development of Teepol: Van der Most et al., 'Synthetische producten', 367-86.

96 Gerretson, *Geschiedenis*, III, 253-4. In 1902, Shell used the Rising Sun brand for kerosene in Asia, except for Japan, where the company sold it under the Horse and Anchor brand: AN CAMT 132 AQ 154, Deterding to Bnito, 24 December 1902. These brands may, of course, have been used alongside others, such as the Fish brand current in Singapore in 1922: Moey, *Shell Endeavour*, 33.

97 SLA 120/35/5, minutes executive committee Asiatic, 5 April 1907, 'It was decided to ask the "Shell" Co. to register the "Shell" brand in all countries of Europe and the East, and to confer upon the Asiatic the right to use the same'.

98 SLA 120/35/5, minutes executive committee Asiatic, 20 February 1907, registration in France and in Italy; 29 July 1907 (Belgium); 30 January 1908 (UK); 15 September 1908 (Germany); 1 October 1908 (Hong Kong and East Africa); 15 November 1908 (Denmark); 12 January 1912 (Jamaica). The pecten was first registered in the United States in 1909 for gasoline and extended in 1914 to cover a range of oil products: Shell Oil Houston Documents Room, legal memorandum by James J. Mullen, 'Shell Oil Company/Scallop Corporation Trademark Matter; Use of the Pecten Symbol', 3 September 1980.

99 SLA 119/11/3, 4/5, 78/2, board memos Anglo Saxon, 14 September 1918.

100 AN CAMT 132 AQ 198 folder correspondence 1908, Marcus Samuel to Lane, 27 November 1908.

101 Henriques, *Samuel*, 479.

102 Hidy and Hidy, *Pioneering*, 577.

103 Montgomery, *Down Many a Road* 18-19; the Irish branch of the General Petroleum Company for example published its first ads in the *Irish Motor Magazine* for April 1906.

104 Montgomery, *Down Many a Road*, 22.

105 Howarth, *Century* 83, 87; *Shell News*, May 1952, 9; Murray, *Go Well*, 33.

106 Howarth, *Century*, 103.

107 Montgomery, *Down Many a Road*, 26 mentions three gasoline grades and their prices in 1914: Shell, Shell II (slightly cheaper) and Crown (the cheapest); Gabriëls, *Koninklijke Olie*, endflaps, for advertising of candles, vaseline, and turpene.

108 However, as late as 1922 Asiatic in Singapore still sold kerosene under the Fish brand: Moey, *Shell Endeavour*, 33.

109 Cf. the plate in Gabriëls, *Koninklijke Olie* on the endflaps.

110 BIM also moved its head office from Rotterdam to The Hague, in the vicinity of the central office there. De Goey, 'Deterding', 71-73; SHA, *De Bron* 1931, 90, six years ago: 'den naam "Autoline" vervangen door "Shell" (...) 'In een mum van tijd was Shell erin', 'gele pompen', 'gele Shell reclame borden', 'Shell zag men overal'. Apparently the corporate style was then not yet a uniform template covering all aspects of the organization. When in 1926 Anglo-Saxon voted a budget to buy cars for the regional marketing inspectors in the Netherlands, the board decided that the cars 'should all be painted the same distinctive colour, preferably red, similar to the inspectors' cars of Shell-Mex Ltd', SLA 119/3/6, board minutes Anglo-Saxon, 3 November 1926. The network overhaul and the introduction of the Shell brand in the Netherlands probably coincided with a similar operation in Germany, where until then Rhenania had sold gasoline under the Stellin brand: Flieger, *Gelben Muschel*, 98-9, 120-3.

111 See Chapter 1 for the efforts to mask Royal Dutch involvement with Rhenania. The Californian company was initially called American Gasoline Company and would have had a Shell name but for the fact that a Shell Petroleum Company already existed there, run by two Shell brothers: Beaton, *Enterprise*, 64-5. However, Deterding expressly wanted the Midwest operations not to have associations with the rest of the Group, giving Mark Abrahams free rein to choose a name: SHA 195/33-3, Gulbenkian to Deterding, 23 January 1917.

112 See the charts in Gerretson, *History*, III., facing 288, and IV, facing 174.

113 Beaton, *Enterprise*, 348.

114 SHA 8/204-2, Deterding to Kessler, 28 April 1925.

115 SHA 8/204-2, De Kok to Kessler, 6 August 1928.

116 The list of operating companies in Bank of England Archives G1/482. We have added Rhenania, omitted from the list as an enemy company, to the count. Cf. Howarth, *Century*, 248.

117 Gibb/Knowlton, *Resurgent years*, 494-5. At the time of writing Jersey Standard's successor company Exxon Mobil still did not have a single brand for gasoline, service stations in the US selling Exxon, whereas those in Europe continue to sell Esso.

118 SHA 49/74-2, Dooijewaard to Caland, 19 February 1932.

119 Beaton, *Enterprise* 77-8.

120 SHA 190C/370, Memo 'Information received from Mr. D. Pyzel about the Shell colours and trade mark' 12 November 1956. The Californian organization was also credited for developing the yellow Shell flag with a red pecten, probably first used on service stations in 1915.

121 Beaton, *Enterprise*, 792.

122 Ibid., 273-4.

123 Ibid., 275.

124 Temporary National Economic Committee (TNEC), *Control of the Petroleum Industry by Major Oil Companies* (Washington: Government Printing Office, 1941), 57

125 TNEC, *Control*, 50

126 Beaton, *Enterprise* 283.

127 SLA 119/11/17, board memos Anglo-Saxon, memo 20 March 1928. The same was true for Germany, where after the currency stabilization of November 1923 Jersey began expanding its service station network, followed by the Group: Flieger, *Gelben Muschel*, 98-9.

128 SLA 119/11/18, Anglo-Saxon board memos, memo 24 October 1928.

129 SLA 119/11/17, Anglo-Saxon board memos, memo 28 March 1928.

130 *De Bron* 1931, 90.

131 SHA 10/523-1, De Jongh to Rhenania-Ossag, 5 and 14 November 1932. Bataafsche managers in the Dutch East Indies received instructions to fly the Shell flag side by side with the Royal Dutch flag: Wouters, *Shell Tankers*, 85

132 See *Shell Magazine*, July 1934, 313-4, explaining the philosophy behind the posters in detail.

133 Montgomery, *Down many a road*, 224-8.

134 On marketing campaigns in Germany see Gries, 'Geistige Landnahme'.

135 In 1983 the Barbican Art gallery organized a retrospective featuring some of the best specimens, a tribute to both the company which commissioned the posters and the artists who designed them.

136 Beaton, *Enterprise*, opposite 305.

137 Hillier, *New Fame*, 66-73.

138 Gabriëls, *Koninklijke Olie*, 81 (first film commissioned 1924); *Olie* 1965 No. 6, 169-241 (documentary films); Howarth, *Century*, 168.

139 SHA 11/22-1, Darch to Van Eck (programme Film Unit).

140 Klemperer, *Zeugnis*, 25 April, p. 345; Klemperer refers as well to the maps and guide books which formed part of the same campaign. Cf. Gries, 'Geistige Landnahme' for a discussion of the campaign and the Shell marketing strategy in Germany. The German Aral company also used film to promote its gasoline; a few months earlier Klemperer had watched one of their films: *Zeugnis*, 24 November 1936, p. 321. As early as 1932, Asiatic contemplated making films on general themes to which local material would be added to increase their appeal in the countries where the films would be used: SHA 10/523-1, Asiatic to Rhenania-Ossag, 18 June 1932.

141 SHA 11/22-1, Godber to A. E. Moore, 29 December 1938.

142 SHA 11/22-1, Asiatic to BIM, 18 April 1940.

143 SLA 119/3/8, board minutes Anglo-Saxon, 18 September 1929 (aviation departments in Japan, South Africa, and India, each with a light aircraft); Beaton, *Enterprise*, 404-5.

144 SHA board minutes Bataafsche, 15 September 1920, 9 February 1922; like Royal Dutch, the airline had also obtained the royal warrant from its inception, though this time without a court intrigue.

145 Dierikx, *Blauw*, 41-5; Dierikx, *Begrensde*, 80-100.

146 SLA SC7/92/10/2 Vol. 2, memo Aviation Department 18 February 1938.

147 SHA 49/74-1, Plesman to De Kok, 25 September 1928.

148 SHA 49/74-1, Hill to De Kok, 29 February and 3 May 1928, note Hill for Deterding, May 1929; 49/74-2, Hill to Bataafsche Surabaya, 3 April 1933, KLM to BIM, 19 October 1933, Gould to Godber, 6 June 1934.

149 SHA 49/74-1, Hill to De Kok, 2 and 10 April, 3 May 1928; Asiatic to KLM, 30 January 1929; 49/74-2, memo Hill for Deterding, May 1929; memo Delft for BPM, 19 and 21 October 1931; Bataafsche to Aviation Department, 20 March 1933; Dooijewaard to Caland, 1 April 1933; Hill to Bataafsche Surabaya, 3 April 1933; 49/74-3, Aviation Department to Caland, 8 March 1934; Bataafsche to laboratory Amsterdam, 13 April 1934; Delft to Bataafsche, 26 April 1935; 10/503-6 and -7, Sandkuyl to Bataafsche The Hague, 27 May and 7 August 1937 (discount of 25 per cent on KLM tickets).

150 London marketing was not entirely convinced of the need for publication, commenting in March 1928 that 'It is necessary for sales people to be able to exercise a certain amount of bluff and they can do so best if they have not too much exact information as to possible effects of ingredients in the asphalt', quoted in SHA 190C/23A, paper Forbes, 'Asfaltbitumen', 6, 8.

151 SHA 190C/23A, paper Forbes, 'Asfaltbitumen', 16-17; *Petroleum handbook*, 366; Flieger, *Gelben Muschel*, 133-4.

152 SHA 190C/23A, paper Forbes, 'Asfaltbitumen', 32-3; Beaton, *Enterprise*, 474-5.

153 We are indebted to Jan Verloop for spelling these maxims out for us.

Chapter 7

1 Quoted in Royal Dutch Annual Report 1929, 8-10; the full text in *Special Annual Meeting* Bulletin of the API, a copy in SHA 195/2. The 1928 Annual Report had already rehearsed most of these themes. In 1932 Cadman, the CEO of Anglo-Persian, touched on them as well in his API address: Yergin, *Prize*, 266.

2 Beaton, *Enterprise*, 315-6; Yergin, *Prize*, 267.

3 SHA BPM minutes, 7 March 1930 (price cuts); file 8/960 folder 1, Jacobson to managers, 3 and 17 July 1929.

4 SLA GHC/USA/D9/1/1, Corbett (London) to Fraser, (St Louis), 4 May 1934.

5 Yergin, *Prize*, 244-7, 265.

6 Royal Dutch Annual Report 1929, 16-7, and an article signed by Deterding in the *Daily Telegraph* 15 December 1930, quoted in Wouters, *Shell Tankers*, 76.

7 SHA 15/132, Deterding to De Kok, 3 May 1932, plus other correspondence and press cuttings on the matter.

8 De Vries, *Hoogovens*, 328; Beaton, *Enterprise*, 524.

9 Beaton, *Enterprise*, 524-525; SHA 15/45, Godber to Kessler, 1 February 1933. By 1939, Shell Chemical still had only a modest turnover of $4 million, $2.3 million in fertilizers and $1.6 million in solvents: Bundesarchiv Berlin, Reichswirtschaftsministerium (BB RWM) R87-5951, Anhang 3, page 27.

10 Royal Dutch Annual Report for 1931, giving the exchange rate at which the loss was calculated as 8.485 guilders to the pound, whereas before the par rate had been 12.09. The amount of 288 million guilders represented twenty per cent of the company's total assets ultimo 1930. To limit its loss, Bataafsche split its London assets of £24 million into a loan account of £14 million at the devalued exchange rate and a £10 million sterling loan to Anglo-Saxon, which was valued at 12 guilders to the pound on the spurious grounds that this money had been invested in fixed assets which supposedly retained their original value: SHA 15/112, auditors' report on Bataafsche for 1931; 15/133-1, Van Wijk to Deterding, 30 July 1930, 15/133-2, memo Sterling loans Anglo-Saxon 4 December 1935; 15/134, correspondence on inter-Group financial relations.

11 SHA 15/227, minutes Bataafsche 14 May 1930; Agnew to De Kok, 19 May 1930.

12 SHA 190C/266, staff numbers at The Hague central office.

13 SLA 119/11/23, board memos Anglo-Saxon, memo redevelopment Great St Helen's, 14 January 1931.

14 Wouters, *Shell Tankers*, 77. The initiative came from London; Deterding informed De Kok only after having informed London managers of the directors' decision, SHA 15/239, Deterding to De Kok, 13 August 1931.

15 Beaton, *Enterprise*, 362-3.

16 SHA 15/220, travel reports Godber.

17 Annual Reports Shell Union, 1929-31.

18 SHA monthly reports, 1930-31; BPM Annual Reports, 1931-32; Royal Dutch Annual Reports, 1931-32. The practice of selling equipment and hiring it back can be deduced from SHA 190A/116-1, cost comparisons Indonesia. The item does not appear not in 1929, but it does in the figures for 1937. Expenses in Indonesia dropped from 94 million guilders in 1929 to 30 million in 1936: SHA 190A/116/1. Comparative costs 1911-35 in SHA 10/445.

19 SHA 8/960-2, correspondence on the reactions to the redundancies.

20 SHA minutes BPM 2 May 1930; file 8/960, BPM The Hague to general manager Batavia, telegram 6 September 1930.

21 Royal Dutch Annual Reports 1929-33.

22 SHA 10/36, 3rd and 4th reports on efficiency measures at Curaçao; 15/3, Van Wijk to Agnew and Zulver, 7 June 1932; minutes BPM, 10 April 1930; Van Soest, *Olie*, 315, 323; Wouters, *Shell Tankers*, 77.

23 Wouters, *Shell Tankers*, 77.

24 SHA 190Y/863, Vijftig jaar Shell op Curaçao, 15.

25 SHA 12/177, memo Public Relations and Group staff, summer 1944.

26 Royal Dutch Annual Reports, 1931-38.

27 SHA 190A/123, to calculate their respective shares, parties had agreed on a sliding scale. With dividends of 0-25 per cent, the colonial government would get 60 per cent, the Group 40. Between 25 and 35 per cent, the split was 65:35; for 35 and 45 per cent, a 70:30 split; 45-55 per cent, 75:25; 55 per cent and above, 80:20. From 1928 to 1934 dividends ranged between 2 and 6 per cent overall. For 1935, it was 15.6 per cent (for the government) to 9.9 (for the Group); 1936: 40.8 to 24.0; 1937: 130.7 to 50.3; 1938: 132.3 to 50.9; 1939: 103 to 39.8; 1940: 125.8 to 48.4.

28 Wouters, *Shell Tankers*, 74.

29 Beaton, *Enterprise*, 364-9, 784-5.

30 SLA GHC/USA/D9/1/1-4, correspondence 1933-34.

31 Beaton, *Enterprise*, 426-36.

32 An accountant inspecting Pladju in 1922 asked for cost price figures and received the answer that these were not compiled at all by the Indonesian administrations: SHA 8/999-12, report of a visit to Pladju 9-17 October 1922.

33 Shell Oil Houston, interview Spaght, 17.

34 By way of speaking, Shell Union managers would say that the company marketed in 50 per cent of the geographical area, but reached 90 per cent of the people: Shell Oil Houston, interview Spaght, 17.

35 Beaton, *Enterprise*, 426-36, and the

number of retail outlets on 792. After an inspection trip in 1939, Van Eck still considered the Shell Union companies as underperforming, notably in marketing: SHA 15/219, notes Van Eck 21 July 1939, and memorandum 4 July 1939. However, gasoline prices in California were much higher, pushing up profits there: BB RWM R87-5951, Anhang 3, 18.

36 The data in SHA 190A/116/1. Only Romania failed to supply data to the required template, but Astra did send in regular refinery cost price reports (cf. SHA 49/795), which shows the extent to which the organization failed to use the available data.

37 BB RWM R87-5950, p. 19; R87-5951, Anhang 2, 10. Exploration drilling in Indonesia was already reaching 3,000-3,400 metres. By contrast, the wells in Egypt were at 800-900 metres and those in Iraq only 500 metres.

38 Cf. SHA 10/503, correspondence general manager-The Hague, 1936-38.

39 For all coordination efforts, there was still no regular exchange of research results between the main installations in Indonesia as late as 1935: SHA 49/16, BPM The Hague to research managers Indonesia and Singapore, 2 November 1935. Despite repeated efforts, complaints about a lack of research co-ordination continued to surface for another two years.

40 SHA 8/1868, organization geological services, 1928-33, memo 27 January 1930; 12/177, memo Public Relations and Group staff, summer 1944. The scheme set up just prior to the Japanese invasion envisaged a two-year course, the top students then going to the US for further university and practical training.

41 SHA 190A/116-1.

42 *Petroleum Times*, 10 October 1931, 475-8; 12 March 1932, 279-83.

43 Quoted in Yergin, *Prize*, 265.

44 SHA 15/4, Godber to Airey, 23 November 1931.

45 SHA 15/4, Godber to Van der Woude, 18 March 1932.

46 Jersey had planned to have its foreign markets to be entirely supplied from foreign production, and no longer from US exports, by 1935: SHA 49/119, Airey (New York) to Agnew, 17 October 1934.

47 SHA 15/133-1, Engle to Van Wijk, 4 July 1927, for the Group's pattern of purchases in 1926.

48 SHA 15/4, Godber to Van der Woude, 28 April 1932.

49 SLA SC7/A22/22, Airey to Agnew, 23 February 1929.

50 SLA SC7/A22/22, Airey to De Kok, 22 November 1929.

51 SLA SC7/A22/22, De Kok to Airey, 25 November 1929.

52 SHA 10/388, Agnew to De Kok, 14 October 1932.

53 A copy of the draft agreement in SHA Country Volumes Spain Volume 1, with letter Godber to De Kok, 1 May 1934, and also in Italy Vol. 2. Cf. Bamberg, *BP* II, 115. Following this general agreement, the three companies appointed a special travelling team of auditors to carry out periodical checks on delivery figures against the As-Is agreement covering the market concerned: SHA 10/549, De Jongh to Rhenania-Ossag, 24 August 1934.

54 SHA 10/42 (As-Is Curaçao), 10/639 (Netherlands), 15/212 (Argentinia), Country Volume Spain, Godber to de Kok, 1 May 1934 (As-Is agreement Spain), 10/454-2 (Dutch East Indies). On the fuel oil bunker discussions Kessler to Wilkinson, 27 October 1933, in SHA 10/454-2. Discussions with Socony Vacuum on an As-Is agreement in SHA 10/454-3, Wilkinson to Godber, 11 October 1934. SHA 15/200, minutes two-party discussions 14 August 1935 (procedure differences and Central Committee), one such three-party conference, about Yugoslavia, included Socony Vacuum rather than Anglo-Iranian: SHA 10/584, three party conference 25 April 1938; Bamberg, *BP* II, 114-15; Larson, Knowlton and Popple, *New horizons*, 310-11. A special As-Is department at St Helen's Court monitored these agreements but, as with other departments, no records survive. The instruction from London to keep As-Is details secret meant that at least one area manager in the Dutch East Indies did not put the agreement there into practice, since he was not allowed to brief his local managers as to the purpose and intention of the agreement. Consequently, competition there continued unabated until, urged by complaints from the Jersey representative, London revised its instructions: SHA 10/454-2, memo 28 December 1932.

55 SHA 15/109 (formation of SMBP); Bamberg, *BP* II, 119, 129-30.

56 SHA 190C/34A, memo Forbes, 'Benzine', 17; Forbes does not mention a specific date for this agreement, but the context indicates that it must have been around 1932.

57 SLA 119/11/24, board memos Anglo-Saxon, memo 27 March 1934.

58 SHA 15/202, two-and three-party discussions; specifically on refineries in consumption countries, SHA 49/664, Fenwick to Sluyterman van Loo, 13 July 1935 (joint resistance with AIOC to Italian Government's wish for a refinery); 15/181; 10/581, Kessler to New York, 1 May 1935 (joint action successful in thwarting refinery plans in Japan, Denmark and Ireland) and 49/23, Kessler to De Kok, 12 September 1938.; cf. Bamberg, *BP* II, 115.

59 SHA 49/19, a comprehensive file about product swaps with Jersey Standard and the NKPM; 49/119, a thin file about specific exchanges with Jersey; 15/110, a proposed collaboration between the refineries on Curaçao and Aruba.

60 SHA 10/523-1, Rudeloff to De Kok, 13 September 1932, commenting that the new cartel agreement might just give the participating companies some peace until outsiders would come in once again.

61 SHA 15/114, De Kok to Kessler, 1 December 1933, with annex estimating Asiatic returns for 1933 in comparison to 1932 for asphalt, kerosene and gasoline.

62 Retail prices from Centraal Kantoor voor de Statistiek, Mededeeling No. 148, Prijzen, indexcijfers en wisselkoersen op Java 1913-37, Batavia 1938.

63 SHA 15/114, annex to De Kok to Kessler, 1 December 1933. In July 1931, the general manager in the Dutch East Indies Sandkuyl asked Bataafsche in The Hague for comparative data on the Group's gasoline prices across Asia, so he could defend himself against the accusations that prices in Indonesia were higher than elsewhere. After conferring with Agnew, De Kok replied that he could not give him these figures and did not see the point of making comparisons anyway, since pricing policy involved so many factors which outsiders would misinterpret that it would be counterproductive to do so: SHA 10/454-1, Sandkuyl to De Kok, 21 July 1931, De Kok to Sandkuyl, 24 August 1931. Clearly De Kok preferred to leave his general manager in the dark as well about the extent to which the Indonesian

market was exploited.

64 SHA49/19-1, memo TL Department, 16 December 1935.

65 BB RWM R87-5950, 44-5.

66 Cf. Bamberg, *BP* II, 116-17.

67 Larson, Knowlton and Popple, *New Horizons*, 313-14; Yergin, *Prize*, 268.

68 SHA 49/19-1, memo Godber 1 December 1937.

69 SLA 119/3/9, board minutes Anglo-Saxon, 2 January 1933, 3 January 1934.

70 Middlemiss, *Tankers*, 44-7. The new ships offered considerable fuel savings; fuel consumption of the Group's fleet averaged 14.1 tons of fuel a day in 1933, down from 19.47 tons in 1928, underlining the effect of the economy drive during the crisis: SLA 119/11/24, board memos Anglo-Saxon, memo fuel consumption fleet, 14 February 1934.

71 SLA 119/11/23, board memos Anglo-Saxon, memo 14 January 1931, redevelopment Great St Helen's (quote); 119/11/24, memo 8 May 1936 (resumption redevelopment).

72 SLA 119/11/26, memos 1 and 7 June, 21 July, 10 and 30 August, 15 September, 24 and 31 October 1938, with nice coloured plans indicating the Group's premises and the properties bought.

73 BB RWM R87-5951, Anhang 1, 3, 17, 29, Anhang 2, 30, 60.

74 BB RWM R87-5951, Anhang 2, 19-22 for a good overview of the Pernis installations at the end of the 1930s.

75 *Van Rotterdam Charlois naar Rotterdam Pernis*, 26-33; Bank of England Archives (BEA) G1/482, memo 28 December 1942, Annex C (refining capacity). At 1.5 million tons, Pernis equalled Group refining capacity in Romania.

76 Van Soest, *Olie*, 358-361; BB RWM R87-5950, 28-30, R87-5951, Anhang 4, 56-7, 60.

77 Source for Table 7.1: SHA 15/133-1.

78 BB RWM R87-5951, Anhang 2, 25.

79 BB RWM R87-5950, 38-9.

80 BEA G1/482, memo 3 December 1942 on the Group's overall position, Annex D. Having been drafted during the war, the document omitted to mention Group sales in Germany, Austria, and Czechoslovakia, but Italy did appear in it, since the operations there had been sequestered in July 1940. For unknown reasons, sales in the US do not appear either. We have added sales in Germany, Austria, and Czechoslovakia for 1938 found in Ministry of Finance, The Hague (MFH) GS 86, code 1.822.145.3, Bataafsche to Ministry of Finance, June 1945, annex, and taken Shell Union's production and crude purchases in 1939 from Beaton, *Enterprise* 784-5, as a rough gauge for US product sales, which results in a probable understatement. The marketing data figured in the 1942-43 discussions Bearsted, Godber and Legh-Jones had with Bank of England officials about efforts to change the ownership balance in the Group to 50:50, about which see Volume 2, Chapter 1.

81 SHA 15/151, memo London Area Management F, 17 August 1937, with documentation. Average profits for Jupiter over 1926-36 were 3.75 per cent, those for Rhenania-Ossag 1.65 per cent, this latter figure depressed by exceptional depreciations during 1935 and 1936 totalling RM 19 million.

82 SHA 49/664, Fenwick to Sluyterman van Loo, 13 July 1935 (joint resistance with APOC to Italian Government's wish for a refinery); 15/181; 10/581, Kessler to New York, 1 May 1935 (joint action successful in thwarting refinery plans in Japan, Denmark and Ireland) and 49/23, Kessler to De Kok, 12 September 1938.

83 BEA G1/482, memo 28 December 1942, annexes C (refining capacity) and D (volume sales), using crude supplies to Germany as proxy for refining capacity there.

84 SHA 11/22, Godber to Van Eck, 11 October 1939.

85 A typical example of the scepticism towards the state companies was a remark on the newly established Italian oil company AGIP, stating that 'the A.G.I.P., everyone admits, cannot last possibly more than a year', SLA 119/11/15, board memos Anglo-Saxon, memo Fenwick 29 September 1926.

86 SHA Country Volumes Spain, vol. 2, s.v. Monopoly.

87 Ibid., s.v. Monopoly, cable from H. E. Bedford (Jersey Standard), 30 August 1927.

88 Ibid., s.v. Monopoly.

89 Correljé and Holman, 'Spaanse oliemonopolie'.

90 SHA, Country Volumes Spain, vol. 2, s.v. Monopoly, memo 13 June 1931 mentions that the total sum paid for the expropriation of the oil companies, 4.2 million pounds, 'was sent out of the country. This heavy drain was undoubtedly responsible for the starting of the slump of the peseta'.

91 SLA SC7/92/9/4, memo Fenwick to Godber, 30 November 1936. The Group continued to watch the Spanish market closely in order to follow where imports came from: SHA 10/726. Jersey and the Group applied the same policy of supply refusal when Japan imposed an oil monopoly in its puppet state of Manchukuo on the Chinese mainland in 1934: SHA 15/2, Dutch Colonial ministry to De Kok, 29 May 1935, 10/454-3, correspondence about the embargo, NA Kew FO371/2775/F9027, memo F. E. W. Barnett 6 September 1941.

92 SLA SC7/92/9/4, memo Fenwick for Godber, 30 November 1936.

93 SHA 8/1585, board minutes Bataafsche, 12 December 1929; SLA Boxes HR, Blair report 1959, 12-3; thus technically El Aguila was not a British company as Yergin, *Prize*, 275 has it. La Corona remained in existence as a shipping company.

94 BB RWM R87/5951, Anhang 5, 25; SLA GHC/Mex/C1-2, London to Van Hasselt, 3 September 1937.

95 SLA GHC/Mex/B2-1, memo 27 June 1938 on Mexican Eagle's financial performance.

96 SLA GHC/Mex/D35-2, memo 9 October 1934 for a catalogue of the grievances.

97 Van Vuurde, *Países Bajos*, 92.

98 SLA GHC/Mex/D35-1, memo J. D. Bowles to Godber, 6 September 1934; GHC/Mex/D35-2, memo Davidson, 25 September 1935; GHC/Mex/D36, an unsigned memo from December 1935 judging a suspension of exports jointly with Jersey Standard inadvisable as likely to incite violence.

99 SLA Boxes HR, Blair report 1959, 5, 12-3, mentioning Astra Romana as another company, apart from the Western European marketing companies, in which most of the senior posts were held by locals. As may be seen, El Aguila did not find itself in a crossfire between local management and London as Yergin, *Prize*, 274, describes on the basis of diplomatic gossip. Cf. Van Vuurde, *Países Bajos*, 91-2.

100 Van Vuurde, *Países Bajos*, 92-3, quoting an amount of 14 million pesos or $3.9 million.

101 Larson, Knowlton and Popple, *New Horizons*, 128-31; Yergin, *Prize*, 274-5.

102 SLA GHC/Mex/B2-1, memo 27 June 1938 on Mexican Eagle's financial performance.

103 SLA SC7/92/9/2 vol. 2, memo Godber

about a telephone conversation with Van Hasselt, 17 August 1937.

104 SLA GHC/Mex/C1-2, Davidson and Van Hasselt to Godber, 22 December 1937.

105 SLA GHC/Mex/C1-2, Godber to De Booy, 22 December 1937.

106 SLA SC7/92/9/4, Godber to Van Hasselt, 24 May 1937 (provisional agreement 1936), Van Vuurde, *Países Bajos*, 95 (agreement 1937).

107 Deterding's outburst against the El Aguila general manager quoted in Yergin, *Prize*, 274, is conspicuous for being an exception. Following the nationalization, the Cowdray Estate and the Group commissioned the English novelist Evelyn Waugh to make a trip to Mexico with the object of writing a travel book telling 'the story of Mexico, politically, economically and, particularly, from the oil point of view'. Waugh travelled via New York to be briefed at Asiatic Corporation, all in the deepest secrecy to guard his cover. In 1939, he published his account under the title *Robbery under Law, the Mexican Object Lesson*, retitled for the US market to the less inflammatory *Mexico: An Object Lesson*. SLA SC7/92/10/4-4, Godber to Wilkinson, 20 July 1938; Brennan, 'Greene, Waugh, Mexico'.

108 SHA 10/581, Asiatic Ltd., London, to Asiatic Corporation New York, 12 September 1934.

109 SHA 12/525, Dubbs royalties (capacity La Spezia 1929); 190C/386, Group processing capacity in Europe, 1938; 10/581, management Italy, memo for the Italian government, 29 October 1934 (doubling cracking capacity).

110 SHA 10/51, De Graan (Group general manager Italy) to Fenwick (London), 9 October 1934.

111 SHA Country Volumes Italy, vol. 2, s.v. Anglo-Persian Agreement; Bamberg, *BP* II, 115-6; Larson, Knowlton and Popple, *New Horizons*, 335-6.

112 SHA 10/581, Asiatic Ltd. to Asiatic Corp. New York, 12 September 1934 (Japan), Asiatic Ltd. to Asiatic Corp., 15 September 1934 (Argentina), Asiatic Corp. to Asiatic Ltd., 19 March 1935 (Japan), Kessler to Asiatic Corp., 1 May 1935 (Denmark, Ireland, Japan).

113 SHA 10/581, Asiatic Corp. to Asiatic Ltd. , 13 September 1934.

114 SHA 10/581, Asiatic Ltd. to Asiatic Corp., 15 September 1934. The Group was not always so principled itself, however. During the summer of 1933 the Rhenania-Ossag manager Kruspig and Kessler had talks with German officials about the government's plans for expanding the refinery capacity in that country, much to the indignation of Jersey and APOC, who considered this a breach of the agreement to act jointly in this and other matters in Germany. Kessler defended his action by saying that he considered building refineries a lesser evil compared to the building of new coal hydrogenation plants: Bamberg, *BP* II, 131-2, Karlsch and Stokes, *Faktor Öl*, 165-7.

115 SHA Country Volumes Italy, Vol. 2, s.v. SIO and UIL., for the acquisition of Gulf's fuel oil and lube oil businesses, in addition to SHA 10/581.

116 SHA 10/581, Asiatic Ltd. to Asiatic Corp., 15 March 1935.

117 SHA 10/581, Kessler to Asiatic Corp., 1 May 1935, Asiatic Corp. to Asiatic Ltd., 20 May 1935.

118 SHA 49/664, Fenwick to Sluyterman van Loo, 13 July 1935.

119 SHA 10/581, Asiatic Ltd. to Asiatic Corp., 15 August 1935, Asiatic Corp. to Asiatic Ltd., 20 August 1935.

120 This decision was probably reached around the middle of October, when Agnew and Bearsted were in New York for talks with Jersey: SHA 10/581, Asiatic Ltd. to Nafta Italiana, 17 October 1935, Godber to Agnew, 19 October 1935, Asiatic Corp. to Asiatic Ltd., 9 November 1935, Airey to Godber, 15 October, and 18 November 1935, Asiatic Ltd. to Asiatic Corp., 22 and 27 November 1935. Cf. Larson, Knowlton and Popple, *New Horizons*, 336.

121 BEA G1/482, memo 3 December 1942 on the Group's overall position, Annex D. Jersey's volume rose by 47% between 1927 and 1938: Larson, Knowlton and Popple, *New Horizons*, 324.

122 SLA SC7/92/9/4, memo Fenwick to Godber, 30 November 1936.

123 SLA SC7/92/9/4, Godber to R.W. Sellers, 4 December 1936, Godber to Riedemann, 7 December 1936; SC7/92/9/3, Godber to H.B. Heath Eves (APOC), 16 December 1936. In August 1937, however, the companies refused to supply 50,000 tons of aviation gasoline for reasons unknown: SLA SC7/92/9/2, H. E. Bedford (Jersey Standard) to Godber, 19 August 1937, Godber to Bedford, 20 August 1937.

124 BEA, G1/482, memo 28 December 1942, Annex D.

125 Larson, Knowlton and Popple, *New Horizons*, 339-40.

126 Karlsch and Stokes, *Faktor Öl*, 165-6, referring to a row in the summer of 1933 about talks between Kessler and Kruspig and government officials about plans to expand German refinery capacity, which Jersey and APOC took as going behind their backs.

127 Karlsch and Stokes, *Faktor Öl*, 191. The gasoline market share was actually slightly higher, if one takes into account the 6.3% share of Deutsche Gasolin, in which the Group had a 25 per cent stake, DAPG (25%) and IG Farben (50%) holding the rest. This company sold IG Farben's synthetic product.

128 SHA 15/151, memo Asiatic 30 July 1934.

129 SHA 15/151, financial data Rhenania-Ossag 1926-36. RM 190,000 equalled 112,700 guilders or £9,300; RM 2.4 million about 1.4 million guilders or £115,600; RM 40 million 23.7 million guilders or £2 million. Flieger, *Gelben Muschel*, 147, 149 (head office Hamburg).

130 Karlsch/Stokes, *Faktor Öl*, 148.

131 SHA 15/151, financial data Rhenania-Ossag 1926-36.

132 SHA 15/151, De Kok to Kessler, 7 September 1931 (Minister of Finance Colijn agrees to 60:40 clearing split), memos De Jongh, 29 August and 4 September 1934 (German imports 2/3rd via Dutch-German clearing and 1/3rd Romanian-German clearing), Knoops to De Jongh, 5 September 1934 (shifting exports to Germany from Bataafsche to Asiatic) De Kok to Colijn, 4 February 1935 (doors for exporting to Germany). The relative ease with which the Group surmounted the currency problems contrasted sharply with the difficulties encountered by Anglo-Persian and its German subsidiary Olex, see Forbes, *Doing Business*, 150-2, 154-5.

133 SHA 15/151, Van Eck to Kessler, 29 July 1937.

134 SHA 49/47, Bataafsche to Proefstation Delft, 12 April 1937; SLA 119/11/27, board memos Anglo-Saxon, memo 19 April 1939 (58,000 tons lube oil imported from Rhenania); Karlsch and Stokes, *Faltor Öl*, 179 (tariff 1935).

135 SHA 10/540, board minutes Bataafsche, 29 January 1937 (RM 5 million for syntholube plant Harburg), 8 September 1938 (RM 27 million for expanding lube oil

production Grasbrook, Reisholz, Harburg); SHA 10/549, memo 4 January 1939 (technical advice from Amsterdam laboratory results in 50 per cent higher production at the Freital Voltol lube oil factory).

136 Middlemiss, *Tankers*, 132-8; BB RWM R 3101-15235, Abteilung Handels- und Devisenfragen, file Rhenania-Ossag, Tankers.

137 SHA 11/20, Van Eck to Deterding, 29 July 1937, Van Eck to Kruspig, 20 October 1938.

138 Barkai, *Boycott*, 13-32.

139 SHA 190D/803, (Rudeloff?) to Kessler, 1 March 1933 (bomb in Königberg), Rudeloff to Reichskanzlei, 30 March 1933 (Westphalian service stations), Rudeloff to Kessler, 3 April 1933 (works council demands), Rudeloff to Kessler, 4 April 1933 (immediate leave Franken), Franken to Rudeloff, 5 April 1933 (emigration). Rhenania-Ossag's Hamburg works council had been recently elected, all nine members now being Nazis.

140 SHA 190D/803, (Rudeloff?) to Kessler, 1 March 1933 (denial violence), Rudeloff to De Kok and to Deterding, 30 March 1933 (denials in foreign press), De Kok to Rudeloff, 3 April 1933 (denial published in *De Telegraaf*), Rudeloff to Deterding, 3 April 1933 (grateful to government for disciplined boycott). Deterding was against publishing the denials in the Group's house magazines, because these were not intended to serve political ends: Rudeloff to De Kok, 6 April 1933.

141 SHA 190D/803, Rudeloff to Kessler, 4 April 1933.

142 Karlsch and Stokes, *Faktor Öl*, 161. The transfers would appear to have concerned a limited number of Jewish employees, for following the Nuremberg Laws in 1935 the Rhenania-Ossag board appointed a commission to deal with 'special staff problems', presumably referring to a perceived need to dismiss or transfer other Jewish staff.

143 SHA 10/525, Rudeloff to Kessler, 30 June 1933 (overhaul necessary for the future, third member not immediately necessary), Rudeloff to Hogrewe, 30 June 1933 (resignations Stern, Sonneborn, Hogrewe, De Jongh, Knoops), Rudeloff to Knoops, 7 July 1933 (resignation); Karlsch and Stokes, *Faktor Öl*, 161. In 1937, there appears to have been pressure on the Dutch employees at Rhenania-Ossag to leave, Kruspig assuring Van Eck that he would protect their interests: SHA 10/552, Van Eck to De Booy, 22 March 1937.

144 James, *Nazi Dictatorship*, 40-7; Wubs, *Unilever*, 51-3. Cf. Turner, *General Motors*, 16-7, for General Motors in 1934 reforming the supervisory board of Opel to minimize the apparent foreign influence. On general questions of big business in Nazi Germany, see Nicosia and Hener, *Business and History*, and Kobrak, *European Business*.

145 Such a fundamental matter would have been debated at the Bataafsche board, which concerned itself with the smallest details of Rhenania-Ossag's business, but the minutes for 1933 have not survived. SHA 10/525, Rudeloff to Kessler, 30 June 1933, suggests that Kessler demurred at the thoroughness of the overhaul, Rudeloff writing to him that this was necessary because of the expected *Gleichschaltung*, i.e. legislation barring Jews from holding directorships and imposing mandatory party members as directors on companies.

146 During the 1933 row about talks with German government over refinery capacity, a DAPG representative told his APOC counterpart that he considered the Group's position influenced by fanatical Nazis at Rhenania-Ossag: Karlsch and Stokes, *Faktor Öl*, 166. Given the circumstances and the board changes then underway at Rhenania-Ossag, this uttering looks more like an angry outburst than a balanced assessment of the situation.

147 Cf. Overy, 'Transportation'.

148 SHA Country Files Germany vol. 3, s.v. shareholders.

149 SLA 119/11/24, board memos Anglo-Saxon, report Rhenania-Ossag sales January–October 1933.

150 Karlsch/Stokes, *Faktor Öl*, 167-8, 180; SHA 15/151, memo Asiatic 30 July 1934, a copy of which also in NA Kew FO C5279. According to this memo the Group was initially in favour of this deal; we have been unable to ascertain why the other companies were against. See also Forbes, *Doing Business* 149-53; contrary to his suggestion, the demand for increased stocks came from the German government, and did not issue from a meeting between Deterding and a high official at the *Reichswirtschaftsamt*.

151 Karlsch and Stokes, *Faktor Öl*, 168-9, 180-1.

152 SHA 10/541-2, Kessler to Van Wijk, 19, 23, and 25 September 1935, Van Wijk to Kessler, 21 and 24 September 1935.

153 SHA 15/302, memo Kruspig sent 29 October 1936. Karlsch and Stokes, *Faktor Öl* 193, describe the participation from the Group and Jersey Standard as motivated by a wish to find a suitable investment for surplus marks held by their German subsidiaries, but the wish to regain some control over the synthetic gasoline market would appear to have been a more powerful motive. Their suggestion that IG Farben dissimulated the true intent of the project by emphasizing that the plant would use residue and not coal is contradicted by Kruspig's memo, which mentions that it would use both.

154 SHA 15/302, memo Kruspig 29 October 1936. Van Eck's later protestations that the Group had only joined the project at the invitation of IG Farben and Jersey Standard would appear to be incorrect; Kruspig participated in the discussions almost from the start.

155 SHA 15/302, Kessler to De Kok, 2 November 1936.

156 SHA 11/21-1 and -2, memo 26 April 1937 (Italy), letter to Van Eck, 9 June 1937 (Japan), minutes meeting 17 March 1938, Riedemann to Godber, 16 and 20 June 1938, Godber to Riedemann, 17 June 1938, Van der Woude to London, 28 May 1938 (France), Homburg, Small and Vincken, 'Carbochemie', 354 (Netherlands). The latter ascribe the Group's reluctance to participate in hydrogenation plants to the disappointments with Bergius during the early 1920s, but the documents on such projects make it clear that the high cost was the prime motive.

157 SHA 15/302, Van Eck to Wilkinson, 18 February 1937, outlining the project. Throughout later negotiations Van Eck never missed an opportunity to emphasize that the Group had joined at the behest of others, which suggest that managers had finally agreed to it with the greatest reluctance; cf. correspondence in SHA 11/20.

158 SHA 15/151, Van Eck to Kessler, 29 July 1937, referring to the possibility that under the Four-Year Plan Rhenania-Ossag's sales would likely become restricted to lube oil and perhaps asphalt. BB RWM R87-5951, p. 41, R87-5953, Anhang 13 (gasoline sales Germany 1937-38).

159 SHA 15/302, Van Eck to Wilkinson, 18

February 1937; Karlsch and Stokes, *Faktor Öl*, 194. Delbrück Schickler & Co. represented the Group, Deutsche Länderbank AG Jersey Standard, and not the other way around, as Karlsch and Stokes write.

160 SHA 11/20, memo Van Eck, 2 February 1939 (precarious position), 13 February 1939 (true position); 11/21-2, Van Eck to Kruspig, 1 February 1939; BB RWM R3101-18238, memo 23 August 1938 (costs now estimated at RM 300 million); BB Deutsche Bank R 8119F-P170, P171, P 1759, copious documentation about the difficulties concerning the bonds.

161 SHA 11/20, Van Eck to De Kok, 18 April 1939, noting that since several months the credit of £330,000 was being paid back at a rate of £70,000 a month.

162 SHA 11/20, Van Eck to De Kok, 18 April 1939, 11/22, Van Eck to Godber, 9 December 1938; SLA GHC/Mex/B2-2, memo 3 November 1938. The Group and Jersey had earlier attempted, unsuccessfully, to prevent the imports from Mexico. DAPG was forced to process Mexican crude in the autumn of 1938.

163 BB RWM R 3101-18238, Von Heemskerck to Römer, 28 January 1939. Jersey Standard's manager Riedemann had warned Van Eck about the implications of the German joint-stock company law, but Van Eck had not wanted to believe him: ibidem, Riedemann to Van Eck, 25 January 1939, Van Eck to Riedemann, 27 January 1939. Officials had to use similar pressure on the majority of companies involved in the expansion of synthetic gasoline production: Karlsch and Stokes, *Faktor Öl*, 199.

164 SHA 11/21-2, Van Eck to Von Heemskerck, 17 February 1939. In replying to Van Eck's capitulation, Von Heemskerck gave him a heavy hint that the Rhenania-Ossag board would have to be reinforced with German businessmen if the company were to avoid further trouble. Kruspig had meanwhile warned Van Eck that the Dutchmen working in senior management positions would have to be replaced by Germans as well for military reasons: SHA 10/525, Von Heemskerck to Van Eck, 21 March 1939, Van Eck to De Kok, 24 March 1939. The main target was probably I. J. F. Reydon, who as Rhenania-Ossag's technical manager must have been regarded by the Nazis as a security risk. Security reasons also seem to have led to a thorough reconsideration of the regular exchange of technical data, notably on lube oil: SHA 10/549, memo 12 December 1938. To avoid the need for consulting the Bataafsche board about the new commitments to Pölitz, Van Eck and De Booy contrived to have the money approved by the Anglo-Saxon board, as Group Treasurer, in the form of a loan by Rhenania-Ossag to Delbrück Schickler: SHA 11/21-2, De Booy to Van Eck, 21 October 1938, Van Eck to De Booy, 29 October 1938.

165 Karlsch and Stokes, *Faktor Öl*, 196; *http://www.police.pl/historiager.html* (forced labour camps).

166 Cf. for instance SHA 15/133, for Deterding's correspondence about Group finance dealings to counteract exhange rate fluctuations.

167 RA Alkmaar De Lange collection, Deterding to Kessler, 7 December 1930 (even setting the date of 1 July 1931 and prompting the Royal Dutch commissarissen to start preparations for his succession), Deterding to Kessler, 19 January 1933, Guépin to Kessler, 18 April 1935.

168 Yergin, *Prize*, 274 (outburst). Deterding's correspondence in SHA for the 1930s is nearly all of a private nature, i.e. concerning currency policy or general economic policy, and business letters from him on major issues are rare indeed.

169 SHA 195/101.

170 Griffiths, *Netherlands*, 35-6; Royal Dutch Annual Reports, 1932, 1933; Hendrix, *Deterding*, 262.

171 SHA 15/263, Deterding to Kessler, 26 December 1933; RA Alkmaar De Lange papers (no inventory numbers at time of writing), Deterding to Kessler, 27 June 1936; SHA 195/97-2, correspondence about currency matters during the summer of 1933; Langeveld, *Colijn*, II, 90, 111.

172 SHA 195/22-5, Deterding to Rudeloff, 23 April 1931; 195/22-6, idem 25 April 1932; 195/97-1, Deterding to Rudeloff, 14 March 1932.

173 The efforts of Hendrix in his Deterding biography to whitewash his evident sympathy for Nazism simply fail to convince. As a specimen of Deterding's convictions, see Naylor, *Oil man*, 114, where he confesses to a desire to shoot idlers on sight. Deterding used the term Fascist in a letter to W. M. Westerman, 10 April 1933, SHA 195/22-6. In 1942 an SS official in the Netherlands wrote a report on Deterding based on a conversation with W. Dijt, who had known Deterding well. According to the report, Deterding claimed to be a Fascist, but 'he never really succeeded in mastering the Nazi set of ideas and convictions (Gedankengut)', NIOD 77/801, letter to Rauter, 7 July 1942.

174 SHA 195/22-7, Deterding to Rudeloff, 4 February 1933, worth quoting for the remark '*Hitler & politics*. It looks as if after all the whole movement will do a lot of good, especially as it is directed against Communism'; id. 15 March 1933, with the radio lectures idea, commenting that these 'would certainly mean an enormous indirect support to Hitler's marvellous stand against Communism who is after all the first courageous public man to announce openly that he means to root out the Communists at any cost, being a menace to all civilisation as we understand it'. Rebuffed, Deterding then donated money to the Berlin Pergamon museum, to which he had made donations before: SHA 195/22-7, Deterding to Rudeloff, 27 March 1933. NIOD No. 207/FOSD 1584 382385/ 382457, also in BB Reichskanzlei R 43, No. II/1461, Reichskanzlei to Rudeloff, 23 March 1933, with memo 18 March 1933; Deterding to Craven, 15 November 1933, SHA 195/101-4 and quoted in Langeveld, *Colijn*, II, 126. In this letter, Deterding refers to Hitler and Mussolini as men who have 'done really a great deal to bring about the dawn of world understanding which I think has started today'. He does not refer to the visit in a letter to Craven dated 10 November, so the meeting presumably occurred between the 10th and the 15th.

175 SHA 195/97-2, Deterding to Leon, 2 May 1933, describing the meeting with Schacht at the German embassy in Paris; quoted in Hendrix, *Deterding*, 270.

176 Claims such as Wennekes, *Aartsvaders*, 366-7, that Deterding knew Rosenberg since 1921 and entertained him at his Norfolk home would appear to be untrue: cf SHA 195/97-5, Deterding to Rosenberg, 27 April 1934, referring to his gratitude in having just made Rosenberg's acquaintance. He did know Rosenberg's reputation, though; cf 195/97-1, Deterding to Rudeloff, 14 March 1932, commenting

that 'the name Rosenberg is about the worst introduction for anything'.

177 NA Kew FO 371/18868, C6788. The London FO considered this man, Horst Obermüller, to be a 'well-known apologist of the Nazi regime and a promoter of Anglo-German friendship'.

178 Seraphim, *Tagebuch Rosenberg*, 38, 46, 139-40; Hendrix, *Deterding*, 282-3, 286.

179 Deterding to Groeninx, 14 July 1934, quoted in Langeveld, *Colijn* II, 126.

180 NIOD No. 207/FOSD 1584 382385/ 382457, also in BB Reichskanzlei R 43, No. II/1461, Reichskanzlei to Rudeloff, 23 March 1933, with memo 18 March 1933.

181 Hendrix, *Deterding* 282-3; NIOD, 77/801, letter to Rauter, 7 July 1942.

182 Quoted in Turner, *Big Business*, 271.

183 SHA 15/263, De Kok to Rothschilds, 4 and 14 September 1933; Rothschilds to De Kok, 11 September 1933; Deterding claimed that Gulbenkian, whom he accused of having led the bear raid on Royal Dutch two years before, was behind the loan proposal, SHA 195/97-3, Deterding to Serruys, 16 November 1933.

184 Cf. SHA 15/193, 190C/196.

185 SHA 15/263, Rothschilds to Deterding, 19 September 1933; Deterding to Rothschilds, 21 September 1933; to De Kok, Deterding contradicted his earlier letter, writing that at the time the Bnito had been a good deal for both parties: Deterding to De Kok, 21 September 1933.

186 SHA 15/263, Kessler to Robert de Rothschild, 24 October 1933; Robert de Rothschild to Kessler, 31 October 1933; Kessler to Deterding, 22 December 1933; Deterding to Kessler, 26 December 1933. By that time Deterding realized that he had better raise the matter at the Bataafsche board in January, of which no minutes survive. At the second interview, Baron Edmond had plainly told Kessler that he questioned Deterding's mental health; moreover, the Rothschilds could not condone his insults to France and his support for Hitler, however much they valued his achievements for the Group. CAMT Roubaix, 132 AQ 168, minutes meeting Edmond de Rothschild with Kessler, 19 December 1933.

187 Cf. RA Alkmaar, De Lange collection, Deterding to Kessler, 4 February 1925, 7 December 1930. The sharply anti-Semitic remarks attributed to Deterding in a Dutch Nazi periodical in November 1940 (quoted in De Jong, *Koninkrijk*, I, 394) appear to have been a fabrication: E. de Jong, 'Was Deterding antisemiet?', in: *Haagse Post*, 7 March 1969. SHA 195/22-7, Deterding to Rudeloff, 27 March 1933, shows that Deterding did nurse a suspicion against Jews as being allegedly in such large numbers supporters of Communism. Deterding does not appear to have been interested in eugenics and racial theories. In December 1933, he turned down a request for support from a Dutch eugenics association: SHA 195/16.

188 SHA 195/22-7, Deterding to Rudeloff, 15 and 27 March 1933.

189 Cf. E. de Jong, 'Was Deterding antisemiet?', in *Haagse Post*, 7 March 1969.

190 NIOD No. 207/FOSD 1584 382385/ 382457, also in BB Reichskanzlei R 43, No. II/1461, Reichskanzlei to Rudeloff, 23 March 1933, with memo 18 March 1933.

191 Earlier that year, De Kok had acted for Deterding in a similar way, sending cuttings from British newspapers downplaying the anti-Semitism in Hitler Germany to the editor of a Dutch right-wing newspaper with a view to getting them published: SHA 15/277, correspondence Deterding-De Kok, April–June 1933.

192 SHA 190B/21, Deterding to Kessler, 29 December 1938; SHA 15/132, Deterding to De Kok, 3 May 1932; Langeveld, *Colijn*, II, 176.

193 Langeveld, *Colijn*, II, 124-5.

194 However, Deterding did have a hand in the smear campaign against the naturalization of Mannheimer, and he probably also helped to engineer the intrige to bring down Prime Minister Colijn, whose currency policy he hated, with allegations of an affair with a German woman: Hendrix, *Deterding*, 272-6; Langeveld, *Colijn*, II, 173, 181, 208.

195 Pool and Pool, *Who Financed Hitler?*, 322-3, for instance, give an uncritical survey of the rumours about loans, coming to the remarkable conclusion that 'With so many sources agreeing on the matter, there can be little doubt that Deterding financed Hitler. All that remains uncertain is the exact sum of money.' A similar line of reasoning in Wennekes, *Aartsvaders*, 366-7. Roberts, *Most Powerful Man* is a good compendium of Deterding rumours and stories. Richardi and Schumann, *Geheimakte*, 76, link Deterding more specifically to Röhm's SA. For a more judicious review of the evidence see Turner, *Big Business*, 270-1.

196 CF. SHA 15/268, correspondence from 1931 explicitly rejecting the practical possibility of such a monopoly.

197 Deterding's rejection of them quoted in De Jong, *Koninkrijk*, I, 395; one donation of 1,000 guilders spotted as an exception by Langeveld, *Colijn*, II, 126.

198 De Jong, *Koninkrijk*, I, 273, 396-7, showing that De Kok also donated a small sum to the paper, but gave a far larger sum to a charity helping Protestant-Jewish refugees from Germany; SLA 119/11/23, memo Russian department 14 June 1932, about the support given to Dr. Edouard Luboff and his *Anglo-Russian News*. Deterding's sponsoring of the periodical *De Waag* may have been tied to his funding of a fledgling political movement in The Hague called the Rijksunie, Van der Boom, *The Hague*, 77-8. Deterding's annoyance with the Dutch press led him in 1935 to offer substantial financial support for setting up a new paper: E. de Jong, 'Was Deterding antisemiet?', in *Haagse Post*, 7 March 1969.

199 NIOD No. 207/FOSD 1584 382385/ 382457, also in BB Reichskanzlei R 43, No. II/1461, Reichskanzlei to Rudeloff, 23 March 1933, with memo 18 March 1933.

200 Hirschfeld, *Herinneringen*, 92-5; a full discussion in Krips-Van der Laan, 'Plan Deterding'.

201 Turner, *Big Business*, 270-1.

202 Wennekes, *Aartsvaders*, 367, misquoting an entry in the Goebbels diaries and misdating this to 12 and 13 January 1936. The complete edition of the Goebbels diaries makes it patently clear that the entries are for 1937, not 1936, and deal with Deterding's swap with Dutch food generating RM 40 million for the Winterhilfswerk (written down by Goebbels as W.H.W., which Wennekes wrongly transcribes as Deterding's initials H.W.A.). Goebbels then mentions that RM 30 million of the RM 100 million total donated by the public to the Winterhilfswerk would be syphoned off for the Volkswagen factory: Fröhlich, *Tagebücher Goebbels*, 3/2: 325, 327. Thus technically some of Deterding's money went towards the factory, but as a result of misappropriation by the Nazi government, not because of Deterding's intention.

203 RA Alkmaar, De Lange collection, note

Guépin 8 April 1936, quoting Deterding's optimism about Germany; SHA 10/541-2, Kessler to Van Wijk, 19 September 1935, Van Wijk to Kessler, 21 September 1935; SHA SHA 15/151, exports into Germany and France.

204 NA Kew FO 371/17769, C3591, Phipps to FO, 7 June 1934; C5177, minutes 31 July 1934.

205 The Romanian Chargé d'Affaires in London had enquired at the Foreign Office for confirmation of the story on 24 May: FO371/17769 C5177, minutes 17 July 1934. The French Embassy in Berlin had heard the story as well: FO371/17769 C4030, Phipps to Foreign Office, 20 June 1934. In both cases the quid pro quo was again rumoured to be a monopoly for the Group in Germany.

206 Seraphim, *Tagebuch Rosenberg*, 38, 46, 139-40.

207 SHA 15/151, exports into Germany and France.

208 There had been a discussion between Kessler and Deterding about limiting Rhenania's credit outstanding to £600,000 and, if necessary, to cut back supplies: RA Alkmaar De Lange collection, memo Guépin, 8 April 1936. Cf. SHA 190C/8, Rhenania-Ossag's outstanding debt to Asiatic of RM 1.6 million was expected to be cleared by September 1934, memo De Jongh discussions 29 August 1934. The limit was later lowered to £330,000 and in 1939 the Group had even decided to abolish the entire credit: SHA 11/20, Van Eck to De Kok, 18 April 1939.

209 NA Kew FO371/17769, C 5280, minute R. F. Wigram, 8 August 1934.

210 NA The Hague 2.21.095 De Jonge papers No. 25, De Kok to De Jonge, 23 December 1931.

211 NA Kew FO 371/18868, C6788, Faulkner to Vansittard, 30 September 1935.

212 RA Alkmaar, De Lange papers (no inventory numbers at the time of writing), Guépin to Kessler, Kessler to Agnew, Deterding, and Rudeloff, Deterding to Rudeloff, all 18 April 1935.

213 J. B. A. Kessler III, the son of J.B.A. Kessler Jr, to Stephen Howarth, 26 October 1998; RA Alkmaar De Lange papers (no inventory numbers at the time of writing), Guépin to Kessler and memo Guépin, 8 April 1936 (Rhenania correspondence).

214 RA Alkmaar, De Lange papers (no inventory numbers at the time of writing), Deterding to Kessler, 27 June 1936. Rumours that Deterding had intentions to settle in the Netherlands had excited one notary public sufficiently to enquire whether Deterding would not be interested in buying Soestdijk palace in Baarn, the late Queen Mother's residence which Queen Wilhelmina rarely used: SHA 195/16. The year before Deterding had bought his second Mecklenburg estate, Dobbin, from the Prince Consort's estate.

215 Sampson, *Seven Sisters*, 96, and Wennekes, *Aartsvaders*, 371-2, state that Deterding was forced to go, but they cite no evidence. Oral tradition from Loudon's son supports this: Dr. J. B. A. K. Kessler III to Stephen Howarth, 26 October 1998. However, Kessler puts the crucial meeting in September or October 1936, whereas Deterding had effectively withdrawn already in April.

216 SHA 195/2, Sandkuyl to Oppenheim, 27 October 1936; Adrian Corbett, a manager at the London office with a long career in the Group, was equally surprised: RA Alkmaar De Lange papers (no inventory numbers at the time of writing), Corbett to Kessler, 31 October 1936, a copy of which in SHA 190D/730.

217 NA The Hague 2.21.095, No. 55, De Jonge to Kessler, 9 January 1931, Kessler to De Jonge, 12 January 1931, De Jonge to Deterding, 14 February 1931, Loudon to De Jonge, 14 March 1931.

218 Corbett felt 'in his bones' that Deterding had played a trick: RA Alkmaar De Lange papers (no inventory numbers at the time of writing), Corbett to Kessler, 31 October 1936, a copy of which in SHA 190D/730.

219 In 1937, Godber and De Booy set up a scheme for a regular exchange between British trainees in The Hague and Dutch trainees in London with a view to having the latter then sent to the US, thus helping to break down this odd barrier: SLA SC7/92/9/2 Vol 2, 189/L, Godber to De Booy, 26 January 1937; SHA 49/23, Kessler to De Booy, 26 February 1937, Van Eck to De Booy, 26 February 1937. Schweppe, *Research aan het IJ*, 51.

220 RA Alkmaar De Lange collection, De Jonge, probably to Kessler, 9 January 1931; Kessler to De Jonge, 12 January 1931; De Jonge to Deterding, 14 February 1931.

221 SHA Royal Dutch board minutes, 8 October 1936.

222 For a rare exception see 119/3/9, minutes Anglo-Saxon 17 October 1934, spelling out a procedure for dealing with differences of opinion within the board.

223 SLA 119/3/7, Minutes Anglo-Saxon, 28 October 1936. The Bataafsche board already operated with proxies during the 1920s.

224 SHA 190B/21, Kessler to Collot d'Escury (NHM), 23 December 1938; Kessler to Deterding, 26 December 1938.

225 SHA 49/23, De Booy to Godber, 28 September 1938.

226 A full report in *De Bron*, 28 February 1939. The suggestion by Gilbert, *Churchill*, v, Companion Part 3, 1286, footnote 2, that Deterding committed suicide, is not supported by any evidence.

227 Wennekes, *Aartsvaders*, 372; De Jong, *Koninkrijk*, I, 273.

228 Hendrix, *Deterding*, 296 (newspaper article).

229 The bust in The Hague was apparently removed in 1945, and replaced by a plaque commemorating De Kok: information from Mr Pieter Folmer, and E. de Jong, 'Was Deterding antisemiet?', in *Haagse Post*, 7 March 1969.

230 Deterding's death immediately sparked speculations about his estate and the possibility that Nazi Germany might get hold of his Royal Dutch shares and use them to take hold of the company. These rumours died down only when Colijn assured the British envoy in The Hague, Sir Neville Bland, about the preference shares construction which protected Royal Dutch against any takeover attempt: NA Kew FO371/23087. The Royal Dutch directors held these shares for the duration of their tenure, and consequently Deterding had lost his shares upon his resignation.

Appendix: sources of figures and tables

Royal Dutch/Shell, BP and Standard oil production

All the information dating from 1890 to 1939 relates to gross production of crude oil (which includes royalties); oil production figures for Standard are exclusive of royalties, but have been corrected to include royalties (estimated at one-eighth of production).* European companies used to measure oil production in metric tons a year; their American counterparts in barrels a day. According to Royal Dutch/Shell one barrel a day is, depending on the specific gravity of the crude oil, equivalent to approximately 50 to 55 metric tons a year. All the production figures quoted have been converted using an average of these two numbers: one barrel a day is calculated at 52.5 metric tons a year.

Royal Dutch/Shell's production figures are based on the Royal Dutch Annual Reports, and for Shell Transport for the period before 1907, on the Annual Reports of that company. The Annual Reports also contain detailed breakdowns of production for each country and region. The comparable figures for BP are from Ferrier, *The History*, 271, 370, 601, and Bamberg, *The History*, 69 and 242; and for Standard, Hidy and Hidy, *Pioneering*, 374-5; Gibb and Knowlton, *The Resurgent Years*, 676-7 and Larson, Knowlton and Popple, *New Horizons*. World production is derived from Etemad et al, *World Energy Production.*

Share prices, market capitalization and other financial data

The main source for financial data of Shell Transport and Royal Dutch were the Annual Reports of both companies, which contained information on gross and net profits, the structure and size of balance sheets etc. Additional data was collected on share prices (and exchange rates between the guilder, pound sterling and dollar) of both Royal Dutch and Shell Transport from *De Telegraaf*, 1903-1916 and *Nieuwe Rotterdamsche Courant*, 1903-1916, and from *Van Os Effectenboek*, 1911-1939. Financial data for BP and Standard is from the same company histories mentioned under oil production: Ferrier, *The History*, Bamberg, *The History* (BP); and Hidy and Hidy, *Pioneering*; Gibb and Knowlton, *The Resurgent Years*, and Larson, Knowlton and Popple, *New Horizons* (Standard). In addition, financial data has been obtained from the *Oil and Petroleum Manual* (1921-1927) and the *Oil and Petroleum Yearbook* (1928-1971/2).

Figures 7.2 and 7.3 are based on the Annual Reports of Royal Dutch and Shell Transport, taking the revenues of the two holding companies – which mainly consisted of dividends from the operating companies – as an approximation for net profits. Figure 7.3 is based on the Annual Reports of the Bataafsche and Shell Union for operating profits and revenues respectively, and on the crude supply figures from Royal Dutch Annual Reports.

See volume IV for the actual data underlying these tables.

*Hidy and Hidy 1955: 375.

Abbreviations

A

AG Aktien Gesellschaft (joint-stock company)
AGIP Azienda Generali Italiana di Petroli
AGM annual general meeting of shareholders
AGNS Allied General Nuclear Services
AIOC Anglo-Iranian Oil Company
AN CAMT Archives Nationaux, Centre des Archives du Monde du Travail (Roubaix, France)
API American Petroleum Institute
APOC Anglo-Persian Oil Company
Avgas aviation fuel

B

B/d, bpd barrels per day
BASF Badische Anilin- und Soda-Fabriken
BB RWM Bundesarchiv Berlin, Reichswirtschaftsministerium
BB Bundesarchiv Berlin
BEA Bank of England Archives, London (UK)
BEF British Expeditionary Force
BHP Broken Hill Proprietary Company
BIM Bataafsche Import Maatschappij
BNOC British National Oil Corporation
BP British Petroleum
BPC Basrah Petroleum Company
BPM Bataafsche Petroleum Maatschappij

C

CBE Commander of the Order of the British Empire
CEI Compagnie d'Esthétique Industrielle
CEO chief executive officer
CEP Current Estimated Potential
CERA Cambridge Energy Research Associates
CFCs chlorofluorocarbons
CFO Chief Financial Offices
CFP Compagnie Française des Pétroles
CIF cost, insurance, freight
CMD Committee of Managing Directors
CNOOC China National Offshore Oil Corporation
CONCAWE Conservation of Clean Air and Water, Western Europe
COT Curaçao Oil Terminal
CPIM Curaçaosche Petroleum Industrie Maatschappij
CPMR Pipeline Mozambique Rhodesia Company
CSM Curaçaosche Scheepvaart Maatschappij
CSV Compañia Shell de Venezuela

D

DAPG Deutsch-Amerikänische Petroleum Gesellschaft
DEA Deutsche Erdöl Aktiengesellschaft
DNB Nederlandsche Bank, Amsterdam (Netherlands)
DSM Dutch State Mines
DWT Deadweight Tonnes

E

E&P exploration and production
EC European Community
EEC European Economic Community
EGM extraordinary general meeting
Elf Essence et Lubrifiants français
ENI Ente Nazionali Indrocarburi
EP, E&P Exploration and Production
EPU Europäische Petroleum Union
ERAP Entreprise de Recherches et d'Activités Pétrolières
Expro Exploration and Production

F

FCE Fletcher Challenge Energy
FIH free in harbour
FOB free on board
FSA Financial Services Authority
FTC Federal Trade Commission

G

GRT Gross Registered Tonnes
GTL Gas to Liquids

H

HDNP Historisch Documentatiecentrum voor het Nederlands Protestantisme (Vrije Universiteit, Amsterdam, the Netherlands)
HMG His/Her Majesty's Government
HR Human Resources
HTGR High Temperature Gas-cooled Reactor

I

ICC International Chamber of Commerce
IHECC International Hydrogenation Engineering and Chemical Company
IHP International Hydrogenation Patents Company
IMCO Intergovernmental Maritime Consultative Organisation
INOC Iraq National Oil Company
IPC Iraq Petroleum Company

K

KBE Knigt Commander of the Order of the British Empire
KNPM Koninklijke Nederlandse Petroleum Maatschappij
KOC Kuwait Oil Company
KPM Koninklijke Paketvaart Maatschappij
KSLA Koninklijke Shell Laboratorium Amsterdam

L

LEAP Leadership and Performance
LNG Liquid Natural Gas
LPG Liquified Petroleum Gas
LWR Light Water Reactor

M

Mekog Maatschappij tot Exploitatie van Kooks-Oven Gassen
MFH Ministry of Finance, The Hague (the Netherlands)
Mogas automobile fuel
Mori Market & Opinion Research International
Mosop Movement for the Survival of the Ogoni People

N

NA Kew National Archives, Kew (UK)
NA The Hague National Archives, The Hague (the Netherlands)
NAM Nederlandse Aardolie Maatschappij
NGO Non-governmental organization
NHM Nederlandsche Handel-Maatschappij
NIIHM Nederlandsch-Indische Industrie- en Handel-Maatschappij
NIOC National Iranian Oil Company
NIOD Nederlands Instituut voor Oorlogsdocumentatie (Amsterdam, the Netherlands)
NIT Nederlandsch-Indische Tankstoomboot Maatschappij
NKPM Nederlandsche Koloniale Petroleum Maatschappij
NNPC Nigerian National Petroleum Corporation
NSB Nationaal Socialistische Beweging
NT New Technology Ventures Division
NTB Non-Traditional Business
NV Naamloze Vennootschap (joint-stock company)
NVD New Venture Divisions

O

OAPEC Organization of Arab Petroleum Exporting Countries
OBE Officer of the Order of the British Empire
OECD Organization for Economic Cooperation and Development
OPC Oil Price Collapse
OPEC Organizaton of Petroleum Exporting Countries
OVA Overhead Value Analysis

P

PA Public Affairs
PDO Petroleum Development Oman
PDVSA Petróleos de Venezuela S.A.
Pemex Petróleos Mexicanos
PET Polyethylene Terephthalate
PPAG Petroleum Produkte Aktien Gesellschaft
PVC Polyvinyl Chloride

R

R&D Research and Development
RA Alkmaar Regionaal Archief Alkmaar (the Netherlands)
RAF Royal Air Force
RD Royal Dutch
RD/S Royal Dutch/Shell
RDS Royal Dutch Shell plc
RIS Republik Indonesia Serikat
RM Reichsmark
ROACE Return on Average Capital Employed
RTZ Rio Tinto-Zinc Corporation
RVI Retail Visual Identity
RWM Reichswirtschaftsministerium

S

SAPREF Shell and BP South African Petroleum Refineries
SASOL South African Synthetic Oil Ltd
SCORE Service Companies Operations Review Exercise
SEC Securities and Exchange Commission
SHA Shell Archives, The Hague (the Netherlands)
SHAC Shell High Activity Catalyst
SHOP Shell Higher Olefins Process
SIEP Shell International E&P
Sietco Shell International Eastern Trading Company
SIMEX Singapore International Monetary Exchange
Sinopec China Petroleum & Chemical Corporation
SIPC Shell International Petroleum Company Ltd.
SIPM Shell Internationale Petroleum Maatschappij NV
SIS Shell International Shipping
Sitco Shell International Trading Company
SLA Shell Archives, London (UK)
SM/PO Styrene Monomer/Propylene Oxide
SMBP Shell-Mex & BP Ltd.
SMDS Shell Middle Distillate Synthesis
SOC Standard Oil Company
SOCAR State Oil Company of Azerbaijan Republic
Socony Standard Oil Company of New York
Stanvac Standard-Vacuum Oil Company

T

TBA Tyres, Batteries, and Accessories
TCP Tri-chresyl Phospate
TEL Tetra-Ethyl Lead
TINA There Is No Alternative
TNT Tri-Nitro Toluene
TPC Turkish Petroleum Company
TVP True Vapour Phase

U

UMWA United Mine Workers of America
UNC United Nuclear Corporation
UOP Universal Oil Products

V

VLCC Very Large Crude Carrier
VOC Venezuelan Oil Concessions Ltd.
VU HDNP Historisch Documentatiecentrum voor het Nederlands Protestantisme (Vrije Universiteit, Amsterdam, the Netherlands)

W

WOCANA World outside the Communist area and North America
WTI West Texas Intermediate

Y

YPF Yacimientos Petroliferos Fiscales

Bibliography

Annual Reports of the 'Shell' Transport and Trading Company, 1898-1940.

Annual Reports of Royal Dutch..., see N. V. Koninklijke...

Bamberg, J. H., *The History of the British Petroleum Company*, II: *The Anglo-Iranian years, 1928-1954* (Cambridge: Cambridge University Press, 1994).

Beaton, K., *Enterprise in oil: a history of Shell in the United States* (New York: Appleton-Century-Crofts, 1957).

Becker, J. J., *La première guerre mondiale* (Paris: MA Editions, 1985).

Bérenger, H., *Le pétrole et la France* (Paris: Flammarion 1920).

Boele, C., and Laar, P. T. van de *Geschiedenis Koninklijke Van Ommeren NV, 1839-1999* (Rotterdam: Koninklijke Vopak, 2001).

Bosboom, N., *In moeilijke omstandigheden, augustus 1914-mei 1917* (Gorinchem: Noorduyn, 1933).

Breman, J., *Koelies, planters en koloniale politiek: het arbeidsregime op de grootlandbouwondernemingen aan Sumatra's oostkust in het begin van de twintigste eeuw* (Dordrecht: Foris, 1987).

Brennan, M., 'Graham Greene, Evelyn Waugh and Mexico', *Renascence, Essays on Values in Literature* 55 (2002), 7-23.

Brown, J. C., *Oil and Revolution in Mexico* (Berkeley: University of California Press, 1993).

Chapman, S. D., *Merchant Enterprise in Britain: From the Industrial Revolution to World War 1* (Cambridge and New York: Cambridge University Press, 1992).

Chernow, R., *Titan: The Life of John D. Rockefeller, Sr.* (New York: Random House, 1998).

CKS, *Prijzen, Indexcijfers en Wisselkoersen op Java 1913-1937*, Mededeeling no. 148 (Batavia, Centraal Kantoor voor de Statistiek, 1938).

Clémentel, E., *La France et la politique économique interalliée* (Paris: Les Presses Universitaires de France, 1931).

Corum, J. S, *The Luftwaffe, Creating the Operational Air War, 1918-1940* (Lawrence: Kansas University Press, 1997).

Corley, T. A. B., *A History of the Burmah Oil Compony, 1886-1924* (London: William Heinemann Ltd., 1983).

Correljé, A. F., and Holman, O., 'Opkomst en ondergang van het Spaanse oliemonopolie: de Nederlandse betrokkenheid', *Negotiation Magazine*, 3 (1989), 121-9.

Creveld, M. L. van, *Supplying War: Logistics from Wallenstein to Patton* (Cambridge: Cambridge University Press, 1978).

DeNovo, J. A., 'Petroleum and the United States navy before World War 1', *The Mississippi Valley Historical Review* (1954), 641-56.

Deterding, H. W. A., *An International Oilman: As Told to Stanley Naylor* (London and New York: Harper & Brothers, 1934).

Dierikx, M. L. J., *Begrensde horizonten, de internationale luchtvaartpolitiek van Nederland in het Interbellum* (Zwolle: Tjeenk Willink, 1988).

— *Blauw in de lucht, Koninklijke Luchtvaart Maatschappij, 1919-1999* (The Hague: SDU, 1999).

DiNardo, R. L., and Bay, A., 'Horse-drawn Transport in the German Army', *Journal of Contemporary History*, 23 (1988), 129-42.

Dixon, C., 'Sea Shell', *The Mariner's Mirror* 79 (1993), 119.

Driel, H. van, *Vier eeuwen veembedrijf: de voorgeschiedenis van Pakhoed 1616-1967* (Rotterdam: Koninklijke Pakhoed, 1992).

Duinmaijer, J., and Groeneveld, C., *Geschiedenis van het Koninklijke/Shell-laboratorium Amsterdam* (Amsterdam, 1957).

Embden, P. H., *Money Powers of Europe in the Nineteenth and Twentieth Centuries* (London: Low, Marston, 1937).

Etemad, B., Bairoch, P., Luciani, J., and Totain, J.-C., *World energy production 1800-1985* (Geneva: Libraire Droz, 1991).

Fenton, C. A., 'Ambulance Drivers in France and Italy: 1914-1918', *American Quarterly*, 4 (1951), 326.

Ferrier, R. W., *The History of the British Petroleum Company*, vol. 1: *The Developing Years 1901-1932* (Cambridge: Cambridge University Press, 1982).

Flieger, H., *Unter der gelben Muschel, die Geschichte der Deutschen Sgell* (Düsseldorf: Verlag Deutsche Wirtschaftsbiographien, 1961).

Foley, P., 'Petroleum Problems of the World War', *Proceedings US Naval Institute*, 50 (1924), 1802-32.

Forbes, N., *Doing Business with the Nazis: Britain's Financial and Economic Relations with Germany, 1931-1939* (London: Frank Cass, 2000).

Forbes, R. J., and O'Beirne, D. R., *The Technical Development of the Royal Dutch/Shell, 1890-1940* (Leiden: Brill, 1957).

Friedensburg, F., *Das Erdöl im Weltkrieg* (Stuttgart: Enke, 1939).

Fröhlich, E. (ed.), *Die Tagebücher von Joseph Goebbels* Vol 3/2, March 1936–February 1937 (Munich: K. G. Saur, 2001).

Gabriëls, H., *Koninklijke Olie: de eerste honderd jaar 1890-1990* (The Hague: Shell Internationale Petroleum Maatschappij, 1990).

Geljon, P. A., *De algemene banken en het effectenbedrijf 1860-1914* (Amsterdam: NIBE, 2005).

Gerretson, F. C., *Geschiedenis der 'Koninklijke'*, 5 vols. (Baarn: Bosch & Keuning, 1932-71).

— *History of the Royal Dutch*, 4 vols., 2nd edn. (Leiden: E. J. Brill, 1958).

Gibb, G. S., and E. H. Knowlton, E. H., *The Resurgent Years: History of Standard Oil Company (New Jersey). 1911-1927* (New York: Harper & Brothers, 1956).

Gilbert, M., *Winston S. Churchill*, vol. v: *Companion Part 3, Documents: The Coming of War 1936-1939* (London: Heinemann, 1982).

Goey, F. de, 'Henri Deterding, Royal Dutch/Shell and the Dutch Market for Petrol, 1902-1946', *Business History*, 44 (2002), 55-84.

Graaff, B. G. J. de, '*Kalm temidden van woedende golven': het ministerie van kolonien en zijn taakomgeving, 1912-1940* (The Hague: SDU, 1997).

Haber, S., Maurer, N. and Razo, Armando, 'When the law does not matter: the Rise and Decline of the Merxican Oil Industry', *Journal of Economic History*, 63, 1 (2003) 1-33.

Hansen, P. H. and Kobrak, C., *European Business, Dictatorship, and Political Risk* (New York: Berghahn, 2004).

Hayes, P., *Industry and ideology, IG Farben in the Nazi-era* (Cambridge: Cambridge University Press, 1987).

Hendrix, P., *Henri Deterding, de Koninklijke, de Shell en de Rothschilds* (The Hague: SDU, 1996).

Hengstebeck, R. J., *Petroleum processing, principles and applications* (New York: McGraw-Hill, 1959).

Henniker, A. M., *Transportation on the Western Front: 1914-1918* (London: HMSO, 1937).

Henriques, R., *Marcus Samuel, First Viscount Bearsted and Founder of The "Shell" Transport and Trading Company, 1853-1927* (London: Barrie and Rockliff, 1960).

— *Sir Robert Waley Cohen, 1877-1952: A Biography* (London: Secker & Warburg, 1966).

Henssen, E. W. A., *Gerretson en Indië* (Groningen: Wolters-Noordhof, 1983).

— *Uit de geschiedenis der Nederlandsche geologische wetenschappen* (Groningen: STYX Publications, 1995).

Heuzé, P., *Les camions de la victoire* (Paris: La renaissance du Livre, 1920).

Hidy, R. W., and Hidy, M. E., *Pioneering in Big Business, History of the Standard Oil Company (New Jersey). 1882-1911* (New York: Harper & Brothers, 1955).

Homburg, E., Rip, A., and Small, J. B., 'Chemici, hun kennis en de industrie', in H. W. Lintsen a.o. (ed.), *Techniek in Nederland in de twintigste eeuw* (Zutphen: Walburg Pers 2000), II. 299-316.

— Small, J. B., and Vincken, P. F. G., 'Van carbo- naar petrochemie, 1910-1940' in H. W. Lintsen a.o. (ed.), *Techniek in Nederland in de twintigste eeuw* (Zutphen: Walburg Pers, 2000), II. 333-58.

Howarth, S., *Sea Shell: The Story of Shell's British Tanker Fleets 1892-1992* (London: Thomas Reed, 1992).

— *A Century in Oil: the "Shell" Transport and Trading Company 1897-1997* (London: Weidenfeld & Nicolson, 1997).

James, H., *The Nazi Dictatorship and the Deutsche Bank* (Cambridge: Cambridge University Press, 2004).

Jensen, W. G.,'The Importance of Energy in the First and Second World Wars', *The Historical Journal*, 11 (1968), 538-54.

Jeremy, D. J. and Shaw, C. (eds.), 'Frederick Lane', in *Dictionary of British Business Biography* (London: Butterworths, 1984), III. 652-4.

Jones, G. G., 'The British Government and the Oil Companies, 1912-1924, the search for an oil policy', *Historical Journal*, 20 (1977), 647-72.

— *The State and the Emergence of the British Oil Industry* (London: Macmillan, 1981).

— 'Lane', in D. J. Jeremy (ed.), *Dictionary of British Business Biography*, III (London: Butterworths, 1985), 652-4.

— *Merchants to Multinationals: British Trading Companies in the Nineteenth and Twentieth Centuries* (Oxford: Oxford University Press, 2000).

Jones, H. O., and Wootton, H. A., 'The Chemical Composition of Petroleum from Borneo', *Transactions Journal of the Chemical Society*, 91 (1907), 1146.

Kamphues, A., 'Na Rhemrev. Arbeidsomstandigheden op de Westerse ondernemingen in de buitengewesten van Nederlandse-Indie', *Economisch- en sociaalhistorisch jaarboek* 51 (1988), 299-337.

Karlsch, R., and Stokes, R. G., *Faktor Öl: die Mineralölwirtschaft in Deutschland 1859-1974* (Munich: Beck, 2003).

Kent, M., *Oil and Empire: British Policy and Mesopotamian Oil 1900-1920* (London: Macmillan, 1976).

Klein, P. W., 'Colijn en de "Koninklijke"', in J. de Bruijn and H. J. Langeveld (eds.), *Colijn, bouwstoffen voor een biografie* (Kampen: Kok, 1994).

Klemperer, V., *Ich will Zeugnis ablegen bis zum letzten* (Berlin: Aufbau Verlag, 1995).

N.V. Koninklijke Nederlandsche Maatschappij tot Exploitatie van Petroleumbronnen in Nederlandsch-Indië, *Verslag over...* (The Hague: Sijthoff, 1891-1940).

Korthals Altes, W. L., *Prices (non-rice), 1814-1940* (Amsterdam: Royal Tropical Institute, 1994).

Krips-Van der Laan, H. M. F., 'Het plan Deterding en de landbouwpolitiek', *Bijdragen en mededelingen betreffende de geschiedenis der Nederlanden*, 107 (1992) 459-85.

Langeveld, H., *Hendrikus Colijn 1869-1944, 1, 1869-1933: Dit leven van krachtig handelen* (Amsterdam: Balans, 1998).

— *Hendrikus Colijn 1869-1944, II, 1933-1944: Schipper naast God* (Amsterdam: Balans, 2004).

Larson, H. M., Knowlton, E. H., and Popple, C. S., *New Horizons: History of Standard Oil Company (New Jersey). 1927-1950* (New York: Harper & Row, 1971).

Laux, J. M., 'Trucks in the West during the First World War', *Journal of Transport History*, 6 (1985) 64-70.

Leeuw, A. S. de, *Nederland in de wereldpolitiek* (Zeist: De Torentrans, 1936).

Lévy, A., *Le Commerce et l'industrie du pétrole en France* (Paris, 1923).

McBeth, B. S., *Juan Vincente Gómez and the Oil Companies in Venezuela, 1908-1935* (Cambridge: Cambridge University Press, 1983).

Meertens, P. J. (ed.), *Biografisch Woordenboek van het socialisme en de arbeidersbeweging in Nederland* (Amsterdam: Stichting Beheer IISG, 1953-2003).

Michielsen, R., *100 jaar Van Ommeren*, 2 vols. (Rotterdam: Nijgh & Van Ditmar, 1939).

Middlemiss, N. L., *The Anglo-Saxon/Shell Tankers* (Newcastle upon Tyne: Shield, 1990).

Mitchell, B. R., *European Historical Statistics 1750-1970* (London: Macmillan, 1978).

— and Jones, H.G., *Second Abstract of British Historical Statistics* (Cambridge: Cambridge University Press, 1971).

Moey, N., *The Shell Endeavour, the first 100 years in Singapore* (Singapore: Shell Companies in Singapore, 1991).

Montgomery, B., *Down Many a Road: The Story of Shell in Ireland* (Dublin: Dreoilín, 2002).

Morrow, J. H., *Building German Air Power 1909-1914* (Knoxville: University of Tennessee Press, 1976).

— *The Great War: An Imperial History* (London and New York: Routledge, 2004).

Most, F. van der, Homburg, E., Hooghoff, P., and van der Selm, A., 'Nieuwe synthetische producten: plastics en wasmiddelen na de Tweede Wereldoorlog', in H. W. Lintsen (ed.), *Techniek in Nederland in de twintigste eeuw*, ii: *Delfstoffen, energie, chemie* (Zutphen: Walburg pers, 2000), 359-75.

Murray, R., *Go Well: One Hundred Years of Shell in Australia* (Melbourne: Hargreen Publishing Co., 2001).

Nicosia, F. R. and Huener, J., *Business and Industry in Nazi Germany* (New York: Berghahn, 2004).

Nouschi, A., *La France et le pétrole de 1924 à nos jours* (Paris: Picard, 2001).

Oil and Petroleum Manual (London: *Financial Times*, 1921-1927).

Oil and Petroleum Yearbook (London: *Financial Times*, 1928-1971-72).

Overy, R., 'Transportation and Rearmament in the Third Reich', *The Historical Journal*, 16 (1973) 389-409.

Pearton, M., *Oil and the Romanian State* (Oxford: Clarendon, 1971).

Pradier, H., *Shell France, un double anniversaire 1919-1989, 1948-1988* (Shell Information No. 289, Jan.–Feb.–March 1989).

Putten, F. P. van der, *Corporate Behaviour and Political Risk: Dutch Companies in China, 1903-1941* (Leiden: CNWS, 2001).

Rady, V., 'Godber', in D. J. Jeremy (ed.), *Dictionary of British Business Biography*, II (London: Butterworths, 1984), 583-5.

Reader, W. J., *Imperial Chemical Industries, a History*, ii: *The First Quarter Century, 1927-1952* (Oxford: Oxford University Press, 1975).

Reynolds, J., *Engines and Enterprise: The Life and Work of Sir Harry Ricardo* (Stroud: Sutton 1999).

Richardi, H. G., and Schumann, K., *Geheimakte Gehrlich/Bell, Röhms Pläne für ein Reich ohne Hitler* (Munich: W. Ludwig, 1993).

Carl Coke Rister *Oil!, titan of the Southwest*, 2nd print (Norman, OK: Oklahoma Press, 1949)

Samuel, M., *Liquid Fuel: Its Production, Application and Uses* (London: 'Shell' T&T Co., 1899).

Schenk, H., *Mergers, Efficient Choice, and International Competitiveness. Bandwagon Behaviour and Industrial Policy Implications* (Cheltenham: Edward Elgar, 2002).

Schouteeten, V., *Les chevaux pendant la grande guerre* (s.n.s.l. 1994).

Schweppe, H. J., *Research aan het IJ. LBPMA 1914 -KSLA 1989* (Amsterdam: Shell Research, 1989).

Seidenzahl, F., *100 Jahre Deutsche Bank 1870-1970* (Frankfurt am Main: Deutsche Bank Aktiengesellschaft, 1970).

Seraphim, H. G., *Das politische Tagebuch Alfred Rosenbergs* (Göttingen: Musterschmidt Verlag, 1956).

Singleton, J., 'Britain's Military Use of Horses 1914-1918', *Past and Present*, 139, (1993), 178-203.

Smith, P. G. A., *The 'Shell' that Hit Germany Hardest* (London: Shell Marketing Company, n.a.).

Smits, J. P., and Gales, B. P. A. 'Olie en gas', in H. W. Lintsen a.o. (ed.), *Techniek in Nederland in de twintigste eeuw* (Zutphen: Walburg Pers 2000), II. 67-90.

Soest, J. van, *Olie als water. De Curacaosche economie in de eerste helft van de twintigste eeuw* (Zutphen: De Walburg Pers, 1977).

Sternheim, A., *De kroniek van dr. A. Sternheim: halfmaandelijksch tijdschrift voor economie, financiën, statistiek, bedrijfshuishoudkunde* (1925-46), 3/81.

Sumida, J. T., 'British Naval Operational Logistics 1914-1918', *Journal of Military History*, 57 (1993), 447-480.

Suzuki, T., *Japanese Government Loan Issues on the London Capital Market, 1870-1913* (London: Athlone Press, 1994).

Taselaar, A. P., *De Nederlandse koloniale lobby: ondernemers en de Indische politiek, 1914-1940* (Leiden: CNWS, 1998).

Temporary National Economic Committee, *Control of the Petroleum Industry by Major Oil Companies* (Washington: Government Printing Office, 1941).

Turner, H. A., *General Motors and the Nazis, the Struggle for Control of Opel, Europe's Biggest Car Maker* (New Haven: Yale University Press, 2005).

Veen, F. R. van, *Willem van Waterschoot van der Gracht 1873-1943: een biografie* (Delft: Delftse Universitaire Pers, 1996).

Vincent, J. W., 'The History of Shell's United Kingdom Oil Refineries 1914-1959' (SLA, SC46/1 June 1959).

Voorst Vader-Duyckinck Sander, H. van, *Leven en laten leven: een biografie van ir. Adriaan Stoop 1856-1935* (Haarlem: Schuyt & Co., 1994).

Vries, Joh. de, *Hoogovens IJmuiden 1918-1968: ontstaan en groei van een basisindustrie* (IJmuiden: Koninklijke Hoogovens, 1968).

—— 'J. B. A. Kessler Sr.', in J. Charité (ed.), *Biografisch woordenboek van Nederland* I (The Hague: Nijhof, 1979), 296-7.

Vuurde, R. van, Los Paises Bajos, el pétroleo y la Revolución Mexicana, 1910-1950 (Amsterdam: Thela 1997).

—— *Nederland, olie, en de Mexicaanse revolutie* (Utrecht, 1994).

Wall, Bennet H. and Gibb, G. S., *Teagle of Jersey Standard* (New Orleans: Tulane University, 1974).

Wilkins, M., *The History of Foreign Investment in the United States, 1914-1945* (Cambridge, Mass.: Harvard University Press, 2004).

Wouters, W., *Shell Tankers: van koninklijke afkomst* (Rotterdam: Shell Tankers BV, 1984).

Wubs, B., *Unilever between Reich and Empire* (Ph.D. thesis, Rotterdam, 2006; London: Routledge, forthcoming).

Yergin, D., *The Prize: The Epic Quest for Oil, Money and Power* (London: Simon & Schuster, 1991).

Zwet, J. R. van, 'Hoe de "Koninklijke" Koninklijk werd', *De Nederlandsche Bank, Onderzoeksrapport WO&E* No. 670, December 2001.

—— *President in Indië en Nederland: Mr N.P. van den Berg als centraal bankier* (Leiden, 2004).

Illustration credits

The publisher has made every effort to contact all those with ownership rights pertaining to the illustrations. Nonetheless, should you believe that your rights have not been respected, please contact Boom Publishers, Amsterdam.

IISG: International Institute of Social History, Amsterdam
IWM: Imperial War Museum, London
KIT: Royal Tropical Institute, Amsterdam
KITLV: Royal Netherlands Institute of Southeast Asian and Caribbean Studies, Leiden
SHA: Shell Archive, The Hague
SLA: Shell Archive, London
SRTCA: Shell Research and Technology Centre, Amsterdam

Introduction

1 Beaulieu National Motor Museum/Shell Advertising Art Collection/F.C. Harrison, 1928
2 Beaulieu National Motor Museum/Shell Advertising Art Collection/ artist unknown, 1920's

Chapter 1

1 SHA, 190F/KON 237
2 SHA, 190F/ 39
3 SHA, 190F/ 39
4 SHA, 190F/ 39
5 Shell Photographic Services/Shell Int. Ltd./Shell Int. Ltd.
6 Shell Photographic Services/Shell Int. Ltd.
7 KIT
8 SHA, from: *Century of exploration*
9 Shell Photographic Services/Shell Int. Ltd.
10 Shell Photographic Services/Shell Int. Ltd.
11 KIT
12 SHA, 190F/ 39
13 SHA, 190F/ 39
14 SHA, 190F/ 39
15 SHA, 190F/ 39
16 Deli Courant, 1902
17 Sumatra Post, 1902
18 Collection Jan Breman
19 SHA, 190C/141
20 Collection Collection Rijksmuseum, Amsterdam
21 Shell Photographic Services/Shell Int. Ltd.
22 Collection Collection Rijksmuseum, Amsterdam
23 SHA, 190F/ 39
24 SHA, 190F/ 39
25 SHA, 190F/ 39
26 Collection Collection Rijksmuseum, Amsterdam
27 Collection Collection Rijksmuseum, Amsterdam
28 Collection Collection Rijksmuseum, Amsterdam
29 Collection Collection Rijksmuseum, Amsterdam
30 Shell Photographic Services/Shell Int. Ltd.
31 Spaarnestad Photo
32 Shell Photographic Services/Shell Int. Ltd.
33 Arjen van Susteren
34 SHA, 190F/200
35 Shell Photographic Services/Shell Int. Ltd.
36 Shell Photographic Services/Shell Int. Ltd.
37 From: Henriques, *Bearsted*
38 SHA, 190A/107-1
39 Shell Photographic Services/Shell Int. Ltd.
40 From: Howarth, *Century*, and Shell Photographic Services/Shell Int. Ltd. (flags)
41 Arjen van Susteren
42 Arjen van Susteren
43 Shell Photographic Services/Shell Int. Ltd.
44 SHA, Shell Tankerarchief 1473
45 SHA, Shell Tankerarchief 1473
46 From: Gerretson, *History* II, 90
47 Paul Maas/Eric van Rootselaar, based on: Gerretson, *History* II, 91
48 From: Forbes, *Technical development*
49 SHA, 190F/ 35
50 Shell Photographic Services/Shell Int. Ltd.
51 Shell Photographic Services/Shell Int. Ltd.
52 SHA, 190F/ 34
53 SHA, 190F/KON
54 SHA, 190F/ 39
55 SHA, 190F/ 39
56 SHA, Portraits
57 Collection Collection Rijksmuseum, Amsterdam
58 Collection Collection Rijksmuseum, Amsterdam
59 SHA, 3/186
60 Shell Photographic Services/Shell Int. Ltd.
61 Arjen van Susteren, based on: Gerretson, *History* II, 156
62 SHA, 190A/107-1
63 Shell Photographic Services/Shell Int. Ltd.

64 SHA, Portraits
65 SHA, Portraits
66 Shell Photographic Services/Shell Int. Ltd.
67 Getty Images
68 Getty Images
69 From: Rister, *Oil! Titan of the South-West*
70 Shell Photographic Services/Shell Int. Ltd.
71 Shell Photographic Services/Shell Int. Ltd.
72 Arjen van Susteren
73 SHA, 190F/Doos IV, United States
74 From: Forbes, *Technical development*
75 Arjen van Susteren, based on:: Gerretson, *History* II, 173
76 SHA, 190F/Doos II-E
77 SHA, 190F/Doos II-E
78 SHA, 190F/Doos II-E
79 SHA, 190F/Doos II-E
80 SHA, Correspondence Deterding 195/110
81 Shell Photographic Services/Shell Int. Ltd.
82 Shell Photographic Services/Shell Int. Ltd.
83 SLA, 141/11/1
84 SHA, 999/16
85 Collection Collection Rijksmuseum, Amsterdam

Chapter 2

1 SHA, 190F/ 125
2 Joost Jonker
3 SLA, News clippings 1909
4 SHA, Portraits
5 SHA, Portraits
6 Shell Photographic Services/Shell Int. Ltd.
7 Shell Photographic Services/Shell Int. Ltd.
8 Shell Photographic Services/Shell Int. Ltd.
9 Shell Photographic Services/Shell Int. Ltd.
10 Shell Photographic Services/Shell Int. Ltd.
11 Shell Photographic Services/Shell Int. Ltd.
12 Fundação Calouste Gulbenkian, Lisbon
13 Collection Collection Rijksmuseum, Amsterdam
14 Collection Collection Rijksmuseum, Amsterdam
15 Collection Collection Rijksmuseum, Amsterdam
16 Collection Collection Rijksmuseum, Amsterdam
17 Collection Collection Rijksmuseum, Amsterdam
18 SLA, 190F/ Deterding
19 Shell Photographic Services/Shell Int. Ltd.
20 Shell Photographic Services/Shell Int. Ltd.
21 From: BPM, *Leven en werken in de bedrijven der N.V. Bataafsche Petroleum Maatschappij*
22 SHA, 190F/ 39
23 SHA, Correspondence Deterding 195/112
24 SHA, 190F/ 39
25 SHA, 190F/ 39
26 Arjen van Susteren, based on: Gerretson, *History* II, 156
27 Paul Maas/Eric van Rootselaar, based on: Beaton, *Enterprise*, 86
28 KITLV
29 KITLV
30 Collection Ton Kuyer/Photo: Hans van den Bogaard
31 Collection Ton Kuyer/Photo: Hans van den Bogaard
32 Shell Australia
33 Shell Photographic Services/Shell Int. Ltd.
34 SHA, 190F/KON 400
35 SHA, 190F/KON 511
36 Shell Photographic Services/Shell Int. Ltd.
37 Shell Photographic Services/Shell Int. Ltd.
38 Shell Photographic Services/Shell Int. Ltd.
39 Shell Photographic Services/Shell Int. Ltd.
40 SHA,190F/KON 44
41 SHA,190F/201
42 KITLV
43 KITLV
44 Arjen van Susteren
45 Collection Collection Rijksmuseum, Amsterdam
46 SHA, 190F/ 127
47 SHA, 190F/ 127
48 SHA, 190F/ 127
49 SLA, News clippings 1910
50 SHA, Correspondence Deterding 195/111
51 Arjen van Susteren
52 Collection Ton Kuyer/Photo: Hans van den Bogaard
53 Shell Photographic Services/Shell Int. Ltd.
54 SLA, News clippings 1911
55 SHA, 190F/ 125
56 SHA, 190F/ 124
57 SHA, 190F/ 124
58 SHA, 190F/ 125
59 SHA, 190F/ 126
60 SHA, 190F/ 126
61 SLA, News clippings 1912
62 SHA, Shell Tankerarchief 708
63 SHA, Shell Tankerarchief 708
64 SHA, 190F/KON
65 SHA, 190F/KON
66 SHA, 190F/KON 45, 46
67 From: Schweppe, *Research*
68 SHA, 190F/KON
69 Studio Marise Knegtmans, based on: Forbes, *Technical development*, 368
70 SHA, 190F/ 38
71 SHA, 190F/202
72 SHA, 195/34
73 SHA, 195/37
74 Beaulieu National Motor Museum/Shell Advertising Art Collection/ artist unknown
75 Beaulieu National Motor Museum/Shell Advertising Art Collection/ artist unknown
76 Beaulieu National Motor Museum/Shell Advertising Art Collection/ artist unknown
76 Beaulieu National Motor Museum/Shell Advertising Art Collection/ artist unknown
77 Beaulieu National Motor Museum/Shell Advertising Art Collection/ artist unknown
78 Shell Photographic Services/Shell Int. Ltd.
79 Collection Ton Kuyer/Photo: Hans van den Bogaard

Chapter 3

1 Spaarnestad Photo
2 SLA, 120/3/1
3 Legermuseum, Delft
4 SHA, 190F/39
5 KITLV
6 SHA, 190F/KON 550
7 Getty Images
8 IWM, E-AUS757
9 IWM, E-AUS921
10 Studio Marise Knegtmans, based on: Gerretson, *History* IV, 174
11 Studio Marise Knegtmans, based on: Gerretson, *History* III, 288
12 Mary Evans Picture Library, UK
13 Arjen van Susteren
14 Arjen van Susteren
15 Shell Photographic Services/Shell Int. Ltd.
16 Getty Images
17 Getty Images
18 Spaarnestad Photo
19 Getty Images
20 IWM, Q48217
21 IWM, Q43463
22 IWM, HU60867
23 Spaarnestad Photo
24 Spaarnestad Photo
25 IWM, HU70333
26 Spaarnestad Photo
27 Arbeiderspers
28 Spaarnestad Photo
29 Shell Photographic Services/Shell Int. Ltd.
30 Spaarnestad Photo
31 Spaarnestad Photo
32 Spaarnestad Photo
33 IWM, Q20616
34 Library of Congress, Washington
35 Getty Images
36 Spaarnestad Photo
37 Spaarnestad Photo
38 Spaarnestad Photo
39 Spaarnestad Photo
40 IWM, HU52582

41 Getty Images
42 Getty Images
43 Shell Houston, AVRP0002181
44 Paul Maas/Eric van Rootselaar, based on: Beaton, *Enterprise*, 93
45 Shell Photographic Services/Shell Int. Ltd.
46 Shell Photographic Services/Shell Int. Ltd.
47 Getty Images
48 Arjen van Susteren
49 Getty Images
50 Spaarnestad Photo
51 IWM, Q20638
52 IWM, Q20637
53 IWM, Q20639
54 SLA, *St. Helen's Court Bulletin,* vol.5
55 Getty Images
56 Getty Images
57 SHA, 190F/200
58 SHA, 190G/110
59 SHA, 190F/110
60 SHA, 8/1863-1
61 SLA, *St. Helen's Court Bulletin,* vol. 3
62 SLA, *St. Helen's Court Bulletin,* vol. 3
63 SLA, *St. Helen's Court Bulletin,* vol. 4
64 SHA, 190C/452
65 SHA, 195/33-3
66 Getty Images
67 Getty Images

Chapter 4

1 Shell Photographic Services/Shell Int. Ltd.
2 Shell Photographic Services/Shell Int. Ltd.
3 Arjen van Susteren
4 Shell Houston, AVRP0002696
5 From: Beaton, *Enterprise*
6 Shell Houston, AVRP0008538
7 Shell Houston, AVRP0002696
8 SHA, 190F/KON 441
9 SHA, 190F/KON 440
10 Shell Houston, AVRP0002183
11 Collection Guus Kessler jr., Amsterdam
12 Shell Photographic Services/Shell Int. Ltd.
13 SHA, 190F/Doos IV United States
14 Getty Images/Russell Lee
15 Getty Images/Carl Mydans
16 SHA, 190F/Doos IV Venezuela
17 Arjen van Susteren
18 Shell Photographic Services/Shell Int. Ltd.
19 SLA, Albums
20 SLA, Albums
21 SLA, Albums
22 SHA, 195/172
23 SHA, 195/172
24 SHA, 190F/Doos IV Venezuela
25 SHA, 190F/Doos IV Venezuela
26 SHA, 190F/Doos IV Venezuela
27 SHA, 190F/Doos IV Venezuela
28 SHA, 190F/Doos IV Venezuela
29 Arjen van Susteren, based on: Forbes, *Technical development*, 565
30 Shell Photographic Services/Shell Int. Ltd.
31 Arjen van Susteren, based on: Gibb, *Resurgent years*
32 Getty Images/Henry Guttmann
33 Getty Images
34 Getty Images
35 SHA, 190F/98
36 SHA, 190F/46-2
37 SHA, 190F/47-A
38 SHA, 190F/98
39 SHA, 190F/47-A
40 KITLV
41 SHA, KON 203
42 SHA, 195/168
43 SHA, KON 204
44 SHA, 195/168
45 SHA, 195/168
46 Getty Images
47 Getty Images
48 Getty Images
49 David King Collection
50 Getty Images
51 Getty Images
52 Getty Images
53 Spaarnestad Photo
54 IISG
55 IISG
56 IISG
57 From: Gibb, *Resurgent years*
58 Shell Photographic Services/Shell Int. Ltd.
59 From: Howarth, *Century*
60 Stephen Howarth
61 Stephen Howarth
62 SLA, Albums

Chapter 5

1 SHA, 190F/18-2
2 Bildarchiv Preussischer Kulturbesitz, Berlin
3 KITLV
4 SHA, 195/168
5 SHA, 195/168
6 Fundação Calouste Gulbenkian, Lisbon
7 Fundação Calouste Gulbenkian, Lisbon
8 SHA, 195/33-4
9 SHA, *Minjak,* vol. 16, 1936
10 SHA, KON 304
11 SHA, *De Bron*, Special: *In memoriam*, 1939
12 SLA, *Pipeline,* vol.1 (1920)
13 SLA, *Pipeline,* vol. 1(1926)
14 SLA, *Pipeline,* vol. 2 (1921)
15 SLA, *Pipeline,* vol. 3 (1922)
16 SHA, *Maandblad van het personeel der verbonden Petroleum-maatschappijen*, 7/8, 1921
17 SHA, *De Bron*, Special: *Deterding Cup*, 1934
18 SHA, *De Bron*, Special: *In memoriam*, 1939
19 SHA, *Minjak*, vol. 1 (1920/'21)
20 SHA, *Minjak*, vol. 1 (1920/'21)
21 SHA, 190F/Doos 2-H
22 SHA, 190F/203
23 SHA, Minjak, vol.2 (1922/'23)
24 Shell Houston, AVRP0002713

Chapter 6

1 SLA, Albums
2 From: Forbes, *Technical development*
3 From: Forbes, *Technical development*
4 Paul Maas/Eric van Rootselaar, based on: Forbes, *Technical development*, 173-175
5 Shell Photographic Services/Shell Int. Ltd.
6 Shell Photographic Services/Shell Int. Ltd.
7 Shell Photographic Services/Shell Int. Ltd
8 SHA, 190F/35-B
9 SHA, 190F/35-B
10 From: Forbes, *Technical development*, 114
11 Studio Marise Knegtmans, based on: Forbes, *Technical development*, 111
12 SHA, 190F/35-B
13 Studio Marise Knegtmans, based on: Forbes, *Technical development*, 117
14 From: Beaton, *Enterprise*, 204-205
15 SLA, Albums
16 SLA, Albums
17 SHA, 190F/96
18 SLA, Album
19 SLA, Albums
20 SLA, Albums
21 SLA, Albums
22 SLA, Albums
23 IWM, Q33850
24 Shell Photographic Services/Shell Int. Ltd.
25 Paul Maas/Eric van Rootselaar, based on: Beaton, *Enterprise*, 245
26 SRTCA
27 SRTCA
28 Shoreham Technical Centre, UK
29 From: Forbes, *Technical development*, 397
30 From: Gabriëls, *Koninklijke Olie*
31 IISG
32 Getty Images
33 Collection Ton Kuyer/Photo: Hans van den Bogaard
34 SLA, Albums
35 Collection Ton Kuyer/Photo: Hans van den Bogaard
36 SHA, 190F/96
37 SHA, 190F/96
38 SHA, 190F/96
39 SHA, 190F/96
40 Shell Houston, AVRP0001256
41 Shell Nederland Raffinaderij, Hoogvliet Rotterdam

42 Shell Nederland Raffinaderij, Hoogvliet Rotterdam
43 Shell Nederland Raffinaderij, Hoogvliet Rotterdam
44 Collection Guus Kessler jr., Amsterdam
45 Collection Collection Rijksmuseum, Amsterdam
46 Shell Photographic Services/Shell Int. Ltd.
47 Studio Marise Knegtmans, based on: Forbes, *Technical development*, 492
48 SHA, 190F/17-B
49 SRTCA
50 Shell Photographic Services/Shell Int. Ltd.
51 SLA, News clippings, 1911
52 Shell Australia
53 SLA, Albums
54 SLA, Albums
55 SHA, 190F/102
56 SLA, Albums
57 SHA, 190F/102
58 Beaulieu National Motor Museum/Shell Advertising Art Collection/artist unknown, 1920
59 Beaulieu National Motor Museum/Shell Advertising Art Collection/Barnett Freedman, 1932
60 SLA, Albums
61 SHA, KON 278
62 IISG
63 Collection Ton Kuyer/Photo: Hans van den Bogaard
64 Beaulieu National Motor Museum/Shell Advertising Art Collection/Tom Purvis, 1925
65 Shell Photographic Services/Shell Int. Ltd.
67 Beaulieu National Motor Museum/Shell Advertising Art Collection/ artist unknown, 1923
68 SLA, Albums
69 SLA, Albums
70 Shell Photographic Services/Shell Int. Ltd.
71 SHA, KON 519
72 Shell Houston, AVRP0002898
73 SLA, Albums
74 Shell Houston, AVRP0002898
75 SLA, Albums
76 Shell Houston, AVRP0002898
77 KITLV
78 Beaulieu National Motor Museum/Shell Advertising Art Collection/ artist unknown, 1925
79 Beaulieu National Motor Museum/Shell Advertising Art Collection/Barker, 1925
80 Beaulieu National Motor Museum/Shell Advertising Art Collection/Tristan Hillier, 1936
81 Beaulieu National Motor Museum/Shell Advertising Art Collection/artist unknown, 1934
82 Beaulieu National Motor Museum/Shell Advertising Art Collection/Drake Brookshaw, 1933
83 Beaulieu National Motor Museum/Shell Advertising Art Collection/Eckersley-Lombers
84 Beaulieu National Motor Museum/Shell Advertising Art Collection/Richard Guyatt, 1939
85 Beaulieu National Motor Museum/Shell Advertising Art Collection/Charles Mozley, 1938
86 Shell Photographic Services/Shell Int. Ltd.
87 Shell Photographic Services/Shell Int. Ltd.
88 Shell Photographic Services/Shell Int. Ltd.
89 Shell Photographic Services/Shell Int. Ltd.
90 Bildarchiv Preussischer Kulturbesitz, Berlin
91 Shell Photographic Services/Shell Int. Ltd.
92 From: Beaton, *Enterprise*
93 Beaulieu National Motor Museum/Shell Advertising Art Collection/Dacres Adams, 1929
94 From: Beaton, *Enterprise*
95 SHA, 190F/121
96 SHA, 190F/102
97 Reclamearsenaal
98 Beaulieu National Motor Museum/Shell Advertising Art Collection/artist unknown, 1920s
99 Stadsarchief Amsterdam
100 SHA, KON 403
101 SHA, 190F/105
Tins: Collection Ton Kuyer/Photo: Hans van den Bogaard

Chapter 7

1 Corbis/C. Delius
2 SHA, 15/132
3 SHA, 8/960
4 SHA, 8/960
5 Getty Images
6 Corbis
7 Mary Evans Picture Library, UK
8 Getty Images
9 Getty Images
10 Getty Images
11 SHA, 190F/203
12 Spaarnestad Photo
13 Corbis
14 SHA, 8/160
15 Getty Images
16 SHA, 190F/66-1
17 SHA, 190F/66-4
18 SHA, KON 202
19 SHA, 190F/66-4
20 Arjen van Susteren, based on: SHA, 190F/66-4
21 SHA, 8/1868-1
22 SLA, Albums
23 SLA, Albums
24 SLA, Albums
25 SLA, Albums
26 Corbis
27 Getty Images
28 Shell Australia
29 SHA, 15/239
30 SHA, Shell Tankerarchief 1430
31 SHA, Shell Tankerarchief 1430
32 SHA, Shell Tankerarchief 1430
33 SLA, 119/11/23
34 SHA, 190F/17-B
35 Getty Images
36 Corbis
37 Corbis
38 Corbis
39 Corbis
40 SHA, 10/581
41 Corbis
42 Corbis
43 Hollandse Hoogte/Magnum/Robert Capa
44 Hollandse Hoogte/Magnum/Robert Capa
45 Hollandse Hoogte/Magnum/David Seymour
46 Hollandse Hoogte/Magnum/Robert Capa
47 Hollandse Hoogte/Magnum/Robert Capa
48 Hollandse Hoogte/Magnum/Robert Capa
49 Getty Images
50 Rue des Archives, Paris
51 Rue des Archives, Paris
52 SHA, 10/523-2
53 SHA, 10/523-2
54 SHA, 10/523-2
55 SHA, 10/523-2
56 Corbis
57 Getty Images
58 SHA, 195/13
59 SHA, 195/109
60 Spaarnestad Photo
61 IISG
62 IISG
63 SHA, 15/263
64 SHA, 15/277
65 Spaarnestad Photo
66 Hitler Historical Museum
67 Getty Images/Heinrich Hoffmann
68 From: Henriques, *Bearstead*
69 SHA, 195/13
70 Spaarnestad Photo
71 SHA, *De Bron*, Special: *In memoriam*, 1939
72 SHA, *De Bron*, Special: *In memoriam*, 1939
73 SHA, 195/13
74 Spaarnestad Photo

Index

Page numbers in grey indicate illustrations or maps.
Page numbers followed by a slash and a number refer to notes.
Names of ships, newspapers and periodicals are in *italics*.